Lecture Notes in Artificial Intelligence   3245

Edited by J. G. Carbonell and J. Siekmann

Subseries of Lecture Notes in Computer Science

T0223562

Einoshin Suzuki   Setsuo Arikawa (Eds.)

Discovery Science

7th International Conference, DS 2004
Padova, Italy, October 2-5, 2004
Proceedings

 Springer

Series Editors

Jaime G. Carbonell, Carnegie Mellon University, Pittsburgh, PA, USA
Jörg Siekmann, University of Saarland, Saarbrücken, Germany

Volume Editors

Einoshin Suzuki
Yokohama National University
Electrical and Computer Engineering
79-5 Tokiwadai, Hodogaya, Yokohama 240-8501, Japan
E-mail: suzuki@ynu.ac.jp

Setsuo Arikawa
Kyushu University, Department of Informatics
Hakozaki 6-10-1, Higashi-ku, Fukuoka 812-8581, Japan
E-mail: arikawa@i.kyushu-u.ac.jp

Library of Congress Control Number: 2004112902

CR Subject Classification (1998): I.2, H.2.8, H.3, J.1, J.2

ISSN 0302-9743
ISBN 3-540-23357-1 Springer Berlin Heidelberg New York

Springer is a part of Springer Science+Business Media

springeronline.com

© Springer-Verlag Berlin Heidelberg 2004
Printed in Germany

Typesetting: Camera-ready by author, data conversion by Olgun Computergrafik
Printed on acid-free paper   SPIN: 11324614   06/3142   5 4 3 2 1 0

Preface

This volume contains the papers presented at the 7th International Conference on Discovery Science (DS 2004) held at the University of Padova, Padova, Italy, during 2–5 October 2004.

The main objective of the discovery science (DS) conference series is to provide an open forum for intensive discussions and the exchange of new information among researchers working in the area of discovery science. It has become a good custom over the years that the DS conference is held in parallel with the International Conference on Algorithmic Learning Theory (ALT). This co-location has been valuable for the DS conference in order to enjoy synergy between conferences devoted to the same objective of computational discovery but from different aspects. Continuing the good tradition, DS 2004 was co-located with the 15th ALT conference (ALT 2004) and was followed by the 11th Symposium on String Processing and Information Retrieval (SPIRE 2004). The agglomeration of the three international conferences together with the satellite meetings was called Dialogues 2004, in which we enjoyed fruitful interaction among researchers and practitioners working in various fields of computational discovery. The proceedings of ALT 2004 and SPIRE 2004 were published as volume 3244 of the LNAI series and volume 3246 of the LNCS series, respectively.

The DS conference series has been supervised by the international steering committee chaired by Hiroshi Motoda (Osaka University, Japan). The other members are Alberto Apostolico (University of Padova, Italy and Purdue University, USA), Setsuo Arikawa (Kyushu University, Japan), Achim Hoffmann (UNSW, Australia), Klaus P. Jantke (DFKI, Germany), Masahiko Sato (Kyoto University, Japan), Ayumi Shinohara (Kyushu University, Japan), Carl H. Smith (University of Maryland, College Park, USA), and Thomas Zeugmann (Hokkaido University, Japan).

In response to the call for papers 80 submissions were received. The program committee selected 20 submissions as long papers and 20 submissions as regular papers, of which 19 were submitted for publication. This selection was based on clarity, significance, and originality, as well as relevance to the field of discovery science. This volume consists of two parts. The first part contains the accepted long papers, and the second part contains the accepted regular papers.

We appreciate all individuals and institutions who contributed to the success of the conference: the authors of submitted papers, the invited speakers, the tutorial speakers, the steering committee members, the sponsors, and Springer. We are particularly grateful to the members of the program committee for spending their valuable time reviewing and evaluating the submissions and for participating in online discussions, ensuring that the presentations at the conference were of high technical quality. We are also grateful to the external additional referees for their considerable contribution to this process.

Last but not least, we express our deep gratitude to Alberto Apostolico, Massimo Melucci, Angela Visco, and the Department of Information Engineering of the University of Padova for their local arrangement of Dialogues 2004.

October 2004   Setsuo Arikawa
                                                                    Einoshin Suzuki

Organization

Conference Chair

Setsuo Arikawa   Kyushu University, Japan

Program Committee

Einoshin Suzuki (Chair)	Yokohama National University, Japan
Elisa Bertino	University of Milan, Italy
Wray Buntine	Helsinki Institute of Information Technology, Finland
Vincent Corruble	University of Pierre et Marie Curie, Paris, France
Manoranjan Dash	Nanyang Technological University, Singapore
Luc De Raedt	Albert Ludwigs University of Freiburg, Germany
Andreas Dress	Max Planck Institute for Mathematics in the Sciences, Leipzig, Germany
Sašo Džeroski	Jožef Stefan Institute, Slovenia
Tapio Elomaa	Tampere University of Technology, Finland
Johannes Fürnkranz	Technical University of Darmstadt, Germany
Gunter Grieser	Technical University of Darmstadt, Germany
Fabrice Guillet	École Polytechnique of the University of Nantes, France
Mohand-Said Hacid	University of Claude Bernard, Lyon, France
Achim Hoffmann	University of New South Wales, Australia
Eamonn Keogh	University of California, Riverside, USA
Ramamohanarao Kotagiri	University of Melbourne, Australia
Aleksandar Lazarević	University of Minnesota, USA
Michael May	Fraunhofer Institute for Autonomous Intelligent Systems, Germany
Hiroshi Motoda	Osaka University, Japan
Jan Rauch	University of Economics, Prague, Czech Republic
Domenico Saccá	ICAR-CNR and University of Calabria, Italy
Tobias Scheffer	Humboldt University of Berlin, Germany
Rudy Setiono	National University of Singapore, Singapore
Masayuki Takeda	Kyushu University, Japan
Kai Ming Ting	Monash University, Australia
Ljupčo Todorovski	Jožef Stefan Institute, Slovenia
Hannu Toivonen	University of Helsinki, Finland
Akihiro Yamamoto	Kyoto University, Japan
Djamel A. Zighed	Lumière University, Lyon, France

Local Arrangements

Melucci Massimo   University of Padova

Subreferees

Helena Ahonen-Myka
Fabrizio Angiulli
Hideo Bannai
Maurice Bernadet
Sourav S. Bhowmick
Steffen Bickel
Ulf Brefeld
Christoph Bscher
Lourdes Peña Castillo
Narendra S. Chaudhari
Damjan Demšar
Isabel Drost
Alfredo Garro
Vivekanand Gopalkrishnan
Mounira Harzallah
Daisuke Ikeda
Akira Ishino
Matti Kääriäinen
Branko Kavšek
Kristian Kersting
Heidi Koivistoinen

Jussi Kujala
Kari Laasonen
Sau Dan Lee
Remi Lehn
Jussi T. Lindgren
Elio Masciari
Taneli Mielikäinen
Phu Chien Nguyen
Riccardo Ortale
Martijn van Otterlo
Luigi Palopoli
Clara Pizzuti
Juho Rousu
Alexandr Savinov
Francesco Scarcello
Alexandre Termier
Charles-David Wajnberg
Tetsuya Yoshida
Bernard Ženko

Sponsors

Department of Information Engineering of the University of Padova
Yokohama National University
Research Institute on High Performance Computing and Networking,
    Italian National Research Council (ICAR-CNR)

Table of Contents

Long Papers

Pattern Mining

Classification

Outlier Detection

Clustering

Feature Construction and Generation

Knowledge Acquisition

Discovery Science in Reality

Regular Papers

Pattern Mining

Machine Learning Algorithms

Web Mining

Applications of Predictive Methods

Interdisciplinary Approaches

Author Index

Predictive Graph Mining

Andreas Karwath and Luc De Raedt

Albert-Ludwigs-Universität Freiburg, Institut für Informatik,
Georges-Köhler-Allee 079, D-79110 Freiburg, Germany
{karwath,deraedt}@informatik.uni-freiburg.de

Abstract. Graph mining approaches are extremely popular and effective in molecular databases. The vast majority of these approaches first
derive interesting, i.e. frequent, patterns and then use these as features
to build predictive models. Rather than building these models in a two
step indirect way, the SMIREP system introduced in this paper, derives
predictive rule models from molecular data directly. SMIREP combines
the SMILES and SMARTS representation languages that are popular
in computational chemistry with the IREP rule-learning algorithm by
Fürnkranz. Even though SMIREP is focused on SMILES, its principles
are also applicable to graph mining problems in other domains. SMIREP
is experimentally evaluated on two benchmark databases.

1 Introduction

In recent years, the problem of graph mining in general, and its application to
chemical and biological problems, has become an active research area in the
field of data-mining. The vast majority of graph mining approaches first derives interesting, i.e. frequent, patterns and then uses these as features to build
predictive models. Several approaches have been suggested [1–8] for the task
of identifying fragments which can be used to build such models. The earliest
approaches to compute such fragments are based on techniques from inductive
logic programming [1]. Whereas inductive logic programming techniques are theoretically appealing because of the use of expressive representation languages,
they exhibit significant efficiency problems, which in turn implies that their application has been restricted to finding relatively small fragments in relatively
small databases. Recently proposed approaches to mining frequent fragments in
graphs such as gSpan [5], CloseGraph [9], FSG [2], and AGM [7] are able to mine
complex subgraphs more efficiently. However, the key difficulty with the application of these techniques is – as for other frequent pattern mining approaches –
the number of patterns that are generated. Indeed, [6] report on 10^6 of patterns
being discovered. Furthermore, frequent fragments are not necessarily of interest
to a molecular scientist. Therefore, [3] and [6] propose approaches that take into
account the classes of the molecules. Kramer *et al.* compute all simple patterns
that are frequent in the positives (or actives) and infrequent in the negatives (or
inactives) [3]. Inokuchi *et al.* compute correlated patterns [6] . While Inokuchi *et
al.* claim that the discovered patterns can be used for predictive purposes, they
do not report on any predictive accuracies.

E. Suzuki and S. Arikawa (Eds.): DS 2004, LNAI 3245, pp. 1–15, 2004.
© Springer-Verlag Berlin Heidelberg 2004

The approach taken here is different: SMIREP produces predictive models (in the form of rule-sets) directly. SMIREP combines SMILES, a chemical representation language that allows the representation of complex graph structures as strings, with IREP [10, 11], a well-known rule-learner from the field of machine learning. SMIREP has been applied to two benchmark data sets (the mutagenicity dataset [12] and the AIDS Antiviral Screening Database [3]) and the experiments show that SMIREP produces *small* rule sets containing possibly complex fragments, that SMIREP is competitive in terms of predictive accuracy and that SMIREP is quite efficient as compared to other data mining systems.

Although, the SMIREP system is tailored towards the chemical domain, the approach can be employed as a general approach to build predictive models for graph mining domains.

The paper is organized as follows: in section 2 we give an overview of SMILES and SMARTS as language for chemical compounds and fragments as well as their applicability to other structured data, like non-directed graphs; section 3 gives an overview of the SMIREP system; in section 4, we report on experiments and findings conducted with SMIREP;finally, in section 5, we touch upon related work and conclude.

2   SMILES and SMARTS

SMILES (Simplified Molecular Input Line Entry System) [13] is a linear string representation language for chemical molecules. The SMILES language is commonly used in computational chemistry and is supported by most major software tools in the field, like the commercial Daylight toolkit or the Open-Source Open-Babel library.

The SMILES notations of chemical compounds are comprised of atoms, bonds, parathesis, and numbers:

- *Atoms:* Atoms are represented using their atomic symbols. E.g. C for carbon, N for nitrogen, or S for sulfur. For aromatic atoms, lower case letters are used, and upper case letters otherwise. Atoms with two letter symbols, like chlorine (Cl) or bromine (Br), are always written with the first letter in upper case and the second letter can be written upper or lower case. With a rare few exceptions, hydrogen atoms are not included in the string representation of a molecule.
- *Bonds:* Four basic bond types are used in the SMILES language: single, double, triple, and aromatic bonds, represented by the symbols: '-', '=', '#', and ':' respectively. Single and aromatic bonds are usually omitted from SMILES strings. Not belonging to the four basic bonds are ionic bonds, or *disconnections*, represented by a '.'.
- *Branches:* Branches are specified by enclosing brackets, "(" and ")", and indicate side-structures. A branch can, and often does, contain other branches.
- *Cycles:* Cyclic structures are represented by breaking one bond in each ring. The atoms adjacent to the bond obtain the same number. E.g. 'cccccc' denotes a (linear) sequence of six aromatic carbons and 'c1ccccc1' denotes

a *ring* of six carbons. Here, we refer to the numbers indicating the cyclic structures of a compound as *cyclic link numbers*. These cyclic link numbers are not unique within the a SMILES representation of a molecule. E.g. 'c1ccccc1Oc2ccccc2' or 'c1ccccc1Oc1ccccc1' are valid notations for the same molecule.

To give an example, figure 1 shows the drug Guanfacine,which is used to treat high blood pressure (http://www.guanfacine.com/). It consist of one aromatic ring and several branches. This is reflected in its SMILES notation 'c1(Cl)cccc(Cl) c1CC(=O)NC(=N)N', where the first substring reaching from the first 'c1' to the second 'c1' code for the ring structure.

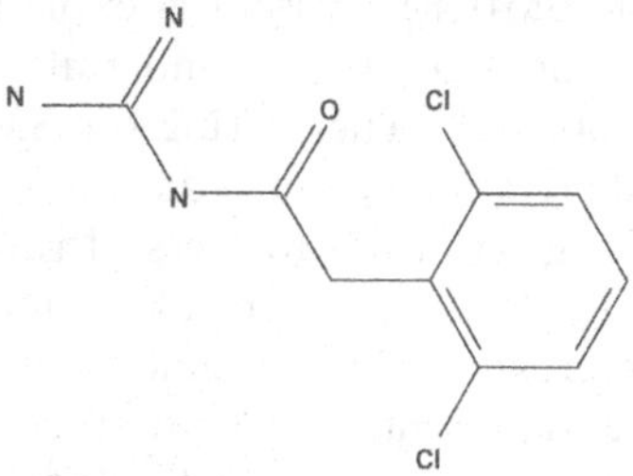

Fig. 1. A 2D graphical representation of Guanfacine. Its SMILES string representation is 'c1(Cl)cccc(Cl)c1CC(=O)NC(=N)N'.

The aim of the SMIREP system is to find predictive rules, or fragments, to differentiate active from non-active compounds. To represent such rules, we use the SMARTS language (http://www.daylight.com). SMARTS is a matching language for SMILES strings. While SMILES is a language representing molecules, SMARTS is a language representing fragments. In principle, SMARTS is derived from SMILES to represent (sub-)fragments. Almost all SMILES strings are valid SMARTS search strings. Furthermore, SMARTS allows the use of wildcards and more complicated constraints. Here, we use a subset of the SMARTS pattern language, that is, we use the SMILES notation for fragments and allow the atom wildcard '*' at the beginning of some matches. E.g. instead of a pattern '=O', the pattern '*=O' is used.

Although, SMILES and SMARTS have been developed as representation languages for chemical compounds and fragments, the SMILES syntax is applicable to other undirected graphs. Vertices can be seen as atoms, while edges can be modeled by bonds. This allows the construction of arbitrary non-cyclic graphs like trees. Furthermore, using the notation of cyclic link numbers, more complex data can be modeled. An example is given in figure 2. The figure is modeled using only one type of vertices ('C') and one type of bond ('-'), but makes use of the cyclic link numbers. In the SMILES notation this would be expressed as 'C-12345C-6C-1C-2C-3C-4C-56' or 'C12345C6C1C2C3C4C56' for short.

This mapping between vertices and edges to atoms and bonds should, in principle, allow SMILES to be used for a wide range of other structured data, e.g. they could be used in general link mining.

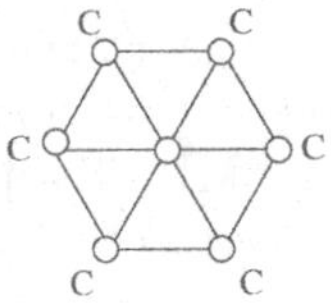

Fig. 2. The graphical representation of the graph 'C-12345C-6C-1C-2C-3C-4C-56', using a 'C' for the vertices and and implicitly a '-' for edges.

3   Algorithms

In the following sections, the SMIREP system is explained in detail. SMIREP adopts ideas of the IREP algorithm [10, 11], and tailors the original IREP to a predictive graph mining approach. The setting for SMIREP is similar as for IREP, the system learns predictive rules given a database of positive and negative examples. In detail, SMIREP grows rules for sets of randomly selected subsets of positive and negative examples, e.g. chemical compounds considered to be active or inactive. Rules are grown, using a randomly selected set of seeds and their corresponding SMILES representation as a starting point. The SMILES strings are decomposed into sub-fragments. All sub-fragments are stored in a labeled tree, the *fragment tree*. The leaf nodes of this tree, so called *ground fragments* (see below for details), are used to grow rules by either adding new ground fragments to a rule or refining sub-fragments in the rule according to the fragment tree. For a given seed, the *best* rule is selected according to the propositional version of FOIL's information gain measure [14] and reported back to the main SMIREP algorithm. The rules grown for all seeds are immediately pruned using a randomly selected pruning set, and the best rule of all pruned rules is selected and added to the rule set. The individual steps and terms are discussed in detail in the following sections.

3.1   Fragment Decomposition

The starting point for growing a rule are the ground fragments of a compounds SMILES string representation and its corresponding fragment tree.

As undirected graphs in general, and therefore SMILES strings in particular, have no unique representation, some canonical form is usually necessary to compare chemical compounds in graph notation. In SMIREP, the SMILES representation returned by the OpenBabel toolkit (http://openbabel.sourceforge.net) as the canonical form. This is only done once for each compound.

3.2   Fragments

For the decomposition of a SMILES representation, two different fragment types are of interest: linear and cyclic fragments. Linear fragments (*LFs*) are fragments defined by branches (or round brackets in the string), and cyclic fragments (*CFs*) are defined by their numbered links within the SMILES representation of a molecule. Linear ground fragments (*LGFs*), are the LFs of a molecule representation containing no branches. Cyclic ground fragments (*CGFs*) are SMILES substring with a full cyclic structure. CGFs contain all atoms, bonds, redundant and non-redundant branching brackets), and all cyclic structure elements between the cyclic link numbers, including the corresponding atoms with their cyclic link numbers.

To give an example for both LGFs and CGFs, consider the SMILES representation string S = 'c1c2c (ccc1) nc1c (cccc1c2Nc1ccc (cc1)S (=O))'. The set of LGFs for this SMILES representation is (in order of appearance from left to right): LGFs(S) = {'c1c2c', 'ccc1', 'nc1c', 'cccc1c2Nc1ccc', 'cc1', 'S', '=O'} and the corresponding set of CGFs is: CGFs(S) = {'c1c2c(ccc1)','c2c(ccc1)nc1c (cccc1c2', 'c1c(cccc1', 'c1ccc(cc1'}. The corresponding 2D representation of this compound is given in 3.

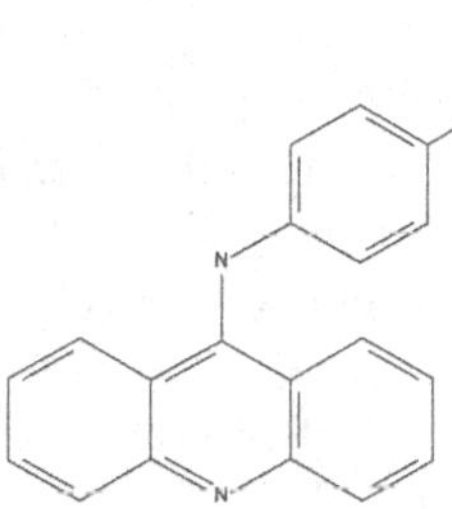

Fig. 3. Example 1: The 2D representation of the molecule having the SMILES code: 'c1c2c(ccc1)nc1c(cccc1c2Nc1ccc(cc1)S(=O))'.

3.3   Decomposition

For the decomposition of a SMILES string we have used an algorithm *DECOMPOSE* (not shown here). DECOMPOSE constructs a tree of (sub-)fragments of a SMILES representation of a molecule. It does so by repeatedly splitting substrings into linear and cyclic fragments, until no further splits are possible.

As input, the algorithm takes a SMILES string converted to the so-called internal SMILES representation. The internal representation assigns unique identifiers to all cyclic link numbers. This includes marking a cyclic link as either *start* (denoted by an '{')or *end* (denoted by a '}'), and mapping a cyclic link number to a *unique cyclic link number*. This is necessary, as after decomposition, the fragments are re-combined by the system. Without assigning unique identifiers to the link numbers, the constructed SMARTS string for searching the database would produce wrong hits and therefore wrong results.

To simplify the extraction of CFs, the internal representation modifies the SMILES string regarding to the order of atoms and cyclic link numbers: it reverses all substrings containing an atom followed by cyclic link number at the beginning of a cyclic structure, e.g. *cc1cccc1* to *c1ccccc1*. Given the original SMILES string: S = 'c1c2c(ccc1) nc1c(cccc1c2Nc1ccc(cc1)S(=O))', the internal

representation would be: $S_{int} = $ '$\{_0c\{_1cc(ccc)_0\}n\{_2cc(cccc)_2c\}_1N\{_3cccc(cc)_3\}S$
$(=O))$', where S_{int} denotes the string in internal representation. DECOMPOSE
uses this internal representation and repeatedly splits substrings into linear and
cyclic fragments, while building up the fragment tree. A node in the tree contains
a SMILES substring, a link to its parent node, zero or more child nodes, and a
label, indicating from which part of a linear or cyclic split the node originated.

Initially, all cyclic fragments of a SMILES string are found. This is done by
simply parsing S_{int} to find all strings s of the form '$\{_N T\}_N$' where '{' and '}'
are matching opening and closing brackets according to their unique identifier
N, and T is a string of symbols. For every matching substring s, a node n_s is
constructed containing a unique identifier as label, the substring s as data, and no
children. For this a method $findCycles$ is used. All nodes are added to the root
node as children. Similarly, the same is done to find linear fragments. The current
SMILES string S_{int} is matched against the pattern $A(B)C$ where A, B and C are
substrings and '(' and ')' are corresponding brackets. E.g., consider the SMILES
string '$CC(=OC(OC)C)CC(=O)C$'. The above method would split this into A
$= $ 'CC', $B = $ '$=OC(OC)C$', and $C = $ '$CC(=O)C$'. Similarly to $findCycles$,
the method to find linear fragments is called $findBranches$. Both methods are
called recursively on the found fragments, adding the new fragments as children
to the previous level fragments, stopping if no more more CFs or LFs can be
found.

To illustrate the basic decomposition performed by DECOMPOSE, consider
the SMILES fragment representation $S = $ '$c1$ $c2c(ccc1)ncc(cc2NccccS$ $(=O))$' or
$S_{int} = $ '$\{_0c$ $\{_1cc(ccc$ $\}_0)n\{_2cc(cccc$ $\}_2c$ $\}_1NccccS$ $(=O))$'. The whole fragment tree
constructed by DECOMPOSE is depicted in figure 4, and the 2D representation
in figure 5.

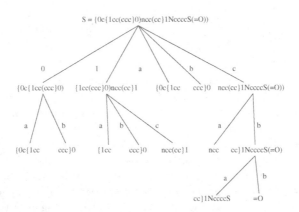

Fig. 4. The resulting fragment tree given the SMILES string $S = $ '$c1c2c(ccc1)ncc$
$(cc2NccccS(=O))$'.

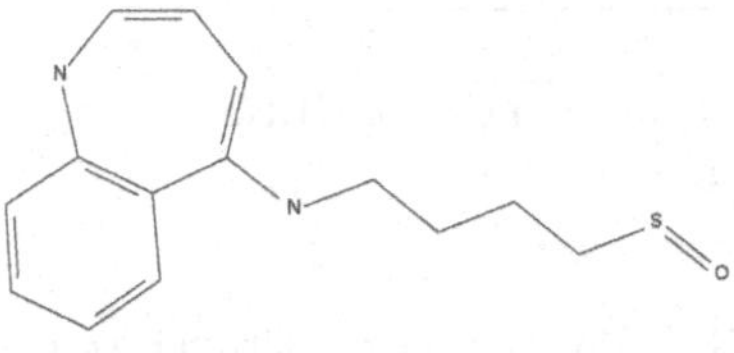

Fig. 5. Example 2. The 2D representation of the molecule having the SMILES code: 'c1c2c(ccc1) nc1c(cccc1c2Nc1ccc(cc1)S(=O))'.

3.4   SMIREP

Like the original IREP algorithm [10], SMIREP employs a growing algorithm, GROW (see below), to grow rules according to a randomly selected set of examples (GrowPos and GrowNeg), and prunes the found rule immediately on a corresponding pruning example set (PrunePos and PrunNeg). After the pruning phase, SMIREP subsequently adds the pruned rule to a rule set. In the SMIREP system, the *growing* algorithm (GROW) specializes fragment strings in SMILES format using refinement operators. The refinement operators are described in detail later in this section. While in IREP, rules consist only of conjunctions of attribute-value pairs, in SMIREP rules consist of conjunction of SMARTS pattern strings. An example rule would be 'NC' ∧ 'cc1C=Ccc1cc(cc)', stating that pattern 'NC' and pattern 'cc1C=Ccc1cc(cc)' have to occur in the compound simultaneously. As mentioned earlier, the SMIREP system uses a modified version of the original IREP algorithm, namely: rules are grown for a set of seeds only. k seeds are selected randomly from the list of *GrowPos*. The default number of seeds used here was arbitrarily set to $k = 10$.

The output of the SMIREP system, is a set of rules, predicting active compounds. A single rule in the rule set, consists of conjunctions of SMARTS pattern strings, their corresponding subfragments in the tree and their refinement history. The SMIREP algorithm is depicted as algorithm 1.

3.5   The Refinement Operator

The growing of rules in SMIREP is done using refinement operators, making the individual rule more specific. The refinement operator r used here, refines a rule by using two different elementary refinements, *Lengthening* and *Ascending*. Both elementary refinement operators refine an existing rule by specializing it. The elementary refinements are:

– *Lengthening*: A ground fragment is added to the existing conjunctions of fragments.
– *Ascending*: The last fragment is refined with respect to its label and parent. Given the labels, the following patterns are constructed (according to their labels):
  - a: Construct patterns $A(B)$ and AC, with label ab and ac respectively.
  - b: Construct new pattern $A(B)$ with label ba.

Algorithm 1 SMIREP.

1: /* INPUT: Databases *Pos* and *Neg* in SMILES */
2: /* OUTPUT: A set of rules for *Pos* */
3: RuleSet := {}
4: **while** Pos ≠ {} **do**
5:     split (Pos,Neg) into (GrowPos, GrowNeg, PrunePos, PruneNeg)
6:     select randomly k seeds ∈ GrowPos
7:     PrunedRules := {}
8:     **for all** seed in seeds **do**
9:        GrownRules := GROW(seed, GrowPos, GrowNeg)
10:       PrunedRule := PRUNE(GrownRules, PrunePos, PruneNeg)
11:       PrunedRules = PrunedRules ∪ PrunedRule
12:    **end for**
13:    select BestRule in PrunedRules by score
14:    **if** error rate of BestRule on (PrunPos, PruneNeg) < 50% **then**
15:       RuleSet := RuleSet ∪ {Rule}
16:       remove examples covered by Rule from (Pos, Neg)
17:    **else**
18:       return RuleSet
19:    **end if**
20: **end while**

- c: Construct new pattern AC with label ac.
- i, where i is an integer: Construct new pattern, where the pattern is the parent's pattern.
- ab, ac, ba, or ca: Construct pattern $A(B)C$, where $A(B)C$ is the parent of A, B, or C.
- r, no construction of a new pattern, the fragment was a root node.

The *Ascending* elementary refinement only succeeds, if the refinement operator is able to construct a new pattern. This is not the case for patterns, or better fragments, representing the root node (labeled r). In these cases, the refinement operator fails and returns no new refinement. The refinement operator specializes a sub-fragment by either adding a new conjunction or refining an existing one, according to its label and parent. Please note, that in some cases the produced pattern can start with a bond. An additional atom wildcard is added to the pattern to transform it to a valid SMARTS search string. This refinement operator is used in the algorithm GROW.

Other graph mining approaches, like *gSpan* [5] or *AprioriAGM* [6] apply refinement operators resulting in minimal refinement steps. In contrast, the refinement operator used here, does not result in minimal refinements in graphs, it is rather comparable to *macros* within the ILP framework.

3.6 Growing Rules

The initial set of rules for growing, contains only the most general rule possible. For a SMARTS search, as well as the SMIREP system, this is the atom wildcard pattern ('*'). Rules are made more specific either by the *Lengthening* or by the *Ascending* operators. However, for the most general rule there is no possible *Ascending* refinement. Therefore, in the first step, the refinement operation returns all rules returned by the *Lengthening* operation, e.g. all rules containing a ground fragment. These rules become the current rule set. Each rule is scored according to a propositional version of the *FOIL gain* [14].

Algorithm 2 GROW.

```
1: /* INPUT Seeds, GrowPos, GrowNeg */
2: /* OUTPUT */
3: BestRules := {}
4: for all Seed in Seeds do
5:    Tree := DECOMPOSE(Seed)
6:    Fragments := groundFragments(Tree)
7:    CurrentRules := {}
8:    repeat
9:       CurrentRules := refine(CurrentRules, Tree, Fragments)
10:      score CurrentRules
11:      CurrentRules := kBest(CurrentRules, k)
12:      BestRule := bestRule(CurrentRules)
13:   until Cover(BestRule(GrowNeg)) = 0 or BestRule does not improve
14:   return BestRule
15: end for
```

As refining all possible fragments of a molecule is computationally too expensive, only the k-best rules of the current rule set, are selected for further refinement, according to their score. Arbitrarily, k has been set to 10. The original growing algorithm in [10, 11] stops refining a rule, when no more negative examples are covered by the rule. In the work presented here, we have loosened this restrictive stopping condition for two reasons: First, it is theoretically possible, that a compound considered active is a subfragment of a compound considered to be inactive, resulting in an infinite loop in GROW. Second, as only the k-best scoring rules are selected for further refinement, the search for "better" patterns can get stuck in a local minimum, never reaching the "best" pattern, the initial seed fragment. In our implementation, GROW stops on refining a rule when either no negative examples are covered or the coverage of the current best rule concerning the negative examples does not decrease compared to the best rule from the previous refinement round. Please note, this is a conservative measure. The system does not consider potentially better rules, which might have occurred in later rounds.

3.7 Pruning

After growing a rule, the rule is immediately pruned. The pruning of a rule is performed by deleting any final sequence of refinements from a rule, and choosing the rule maximizing the following scoring function: $v^*(Rule, PrunePos, PruneNeg) \equiv \frac{p-n}{p+n}$, where p are the positive examples covered by the rule in the pruning set, and n are the negative examples covered by the rule in the pruning set, as suggested by Cohen [11].

The deletion of final refinements is done in reverse to the original refinement: if the rule resulted from a *lengthening* operation, the final condition is deleted; if the rule resulted from a *ascending* operation, the last refinement is undone stepping further down the tree. The process is repeated until no more refinements can be be undone or one single fragment is left.

3.8   Extension to SMIREP

In many cases, additional information like numerical attributes is available about chemical compounds. We have modified our original SMIREP system to incorporate such extra information. The modification is done, by first discretising numerical attributes using equal frequency binning, using a default bin size of $s_{bin}=10$, and then use the attribute-value information during the growing stage. For each seed, all bins are evaluated, resulting in attributes like '$X \leq num$' or '$X > num$', where X is a numerical attribute and num a bin border. This information is used like a fragment during the growing stage. The new rule types produced by SMIREP are than a combination of SMILES fragments and intervals for numerical attributes.

4   Implementation and Experiments

The main SMIREP system has been developed in the programming language Python (version 2.2). Python allows rapid prototype development, due to a wide range of available libraries. The SMARTS matching has been done using the open-source chemical library OpenBabel (version 1.100, http://openbabel. sourceforge.net). All experiments were run on a PC running Suse Linux 8.2 with an Intel Pentium IV - 2.8 GHz CPU and 2 GB of main memory.

4.1   Mutagenesis Data

The first experiment evaluates the SMIREP approach using the mutagenicity benchmark database [12, 15]. The database was originally used to compare the ILP system PROGOL with attribute-value learners like linear regression based approaches. Here, we compare the results from SMIREP to the ones reported in the original paper.

The database contains structural information as well as chemical properties for a number of chemical compounds. These properties are ϵ_{LUMO} (Lowest unoccupied Molecule Orbital) and $logP$ (Octanol-water partition coefficient), both are often used in QSAR studies. The database is divided into two sets: a regression friendly set with 188 entries and a regression unfriendly set with 42 entries. In the first experiment (SMIREP), only the structural information in form of SMILES strings was employed. The comparison of SMIREP to PROGOL can be seen in table 1. Although, PROGOL outperforms SMIREP on both test databases, we believe that SMIREP is still a valid approach, as no extensive background knowledge, like ILP's language bias nor chemical properties

Table 1. The table shows the predictive accuracies of SMIREP compared to PROGOL (taken from [12]) on the two different datasets. The results on the regression friendly (188) dataset where calculated on a 10-fold cross validation. The accuracies reported on the regression unfriendly dataset (42) are based on leave one out validation.

Method	188	42
PROGOL+NS+S1	0.82	0.83
PROGOL+NS+S2	0.88	0.83
SMIREP	0.78	0.81
SMIREP+NS	0.85	0.82

were employed, still producing comparable results. In the second experiment (SMIREP + NS), the SMILES strings and numerical attributes (ϵ_{LUMO} and $logP$) where used together with the extended version of SMIREP. On the 188 dataset the accuracy of SMIREP improved, still not reaching the predictive accuracy of PROGOL + NS + S2, while nearly matching the predictive accuracy of both PROGOL experiments in the 42 test database.

4.2 HIV Data

For the second main experiment, we have used the DTP AIDS Antiviral Screening Database, originating from the NCI's development therapeutics program NCI/NIH (http://dtp.nci.nih.gov/). The database consists of syntactically correct SMILES representations of 41,768 chemical compounds [3]. Each data entry is classified as either active (CA), moderately active (CM), or inactive (CI). A total of 417 compounds are classified as active, 1,069 as moderately active, and 40,282 as inactive.

For the first experiment active-moderate, or AM for short, we have applied SMIREP to build a predictive models for the databases CA and CM in a ten-fold cross-validation. On average, it took 648 seconds to induce rules for each fold. The precision, recall, number of rules and predictive accuracy is given in table 2. The default accuracy using a majority vote is 72.1%. The produced rules vary for each fold, but often involve a rule containing the SMILES pattern 'N=[N]=N'. This is not surprising, as this is an acidic Subfragment of Azidothymidine (AZT).

Table 2. The results of the active-moderate and active-inactive test for SMIREP. *Training* corresponds to the 90% training set and *Testing* to the 10% test set. The numbers in brackets indicate the standard deviation. The accuracy for the AI test is not reported (see text).

	Active-Moderate(AM)		Active-Inactive(AI)	
	Training	Testing	Training	Testing
Precision	0.823 (0.03)	0.736 (0.06)	0.846 (0.03)	0.725 (0.07)
Recall	0.483 (0.08)	0.481 (0.09)	0.276 (0.07)	0.249 (0.10)
Num Rules	19.4 (5.7)	19.4 (5.7)	11.5 (5.6)	11.5 (5.6)
Accuracy		0.80 (0.04)		N/A

In previous studies derivatives of AZT have already been identified as potential drug targets. Another interesting rule occurred in a number of folds, with pattern 'Cc1ccoc1C'. In the training set, 11 positive and no negative example are covered and in the test set five positive and no negative examples are covered. The pattern's 2D representation is depicted in figure 6, pattern 1.

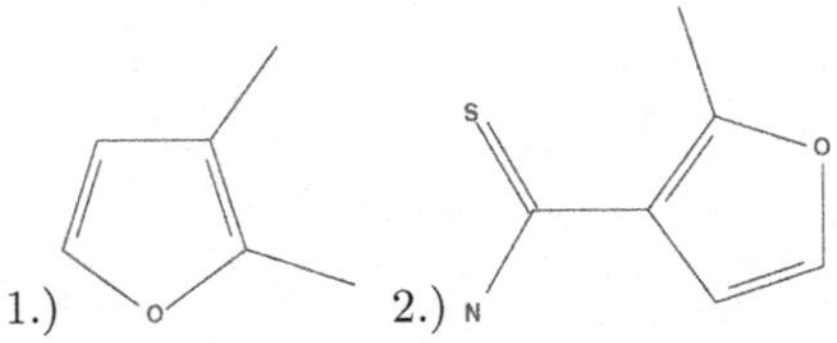

Fig. 6. A 2D graphical representation of an example rules found by SMIREP. Pattern 1 was found in the AM experiment. The rule contained only one single pattern ('Cc1ccoc1C'). Pattern 2 was found in AI experiment. The rule contained only one single pattern ('NC(=S)c1ccoc1C').

The second experiment involving the HIV database was performed on the set of active against inactive chemical compounds using a 10-fold cross validation (AI experiment). On average, SMIREP took 7279 seconds, for each fold. In this experiment the average precision and recall on the test is lower to that obtained in the AM experiment. Here, we have not recorded the average prediction accuracy on the test set, as the default accuracy is 99.0% and a comparison would not result in any gain of information. One of the best rules found (pattern 2), was is a superstructure of the pattern 1 in figure 6 active-moderate experiment.

Interestingly, the recall decreases dramatically when compared to the AM experiment. We believe that this is due to the stopping criterion of a 50% error rate on the last best rule during the learning. In Fürnkranz work, two different stopping criteria where suggested: either a fixed cut of value of 50% error rate or the rule's accuracy is below the default accuracy. Here we have chosen the former, as using the default distribution would allow rules to be learned, having an error rate of up to 99%.

Although, the reported average recall and precision of SMIREP in the AI experiment are relatively low, we believe that valuable patterns have been found. Further investigation is needed, to verify the effectiveness of SMIREP and the produced rules.

5   Discussion and Conclusions

Several approaches to mine graphs for predictive rules in the chemical domain have been proposed [12, 16, 17, 2]. The SubdueCL system [16, 17] is able to induce predictive rules by graph-based relational learning (GBCL). SubdueCL is able to find Substructures in data, by extending candidate subgraphs by one edge searching for Substructures that best compress the graph given, using the MDL principle. The SubdueCL system has been applied to the PTE challenge data [17]. However, it has not been used to learn concepts for large databases like the HIV Database. Another approach, FSG [2], constructs features based on sub-graph discovery and uses support vector machines (SVM) for the classification task. While in this work, a 5-fold cross validation was performed, the results seem to be comparable, slightly favoring the FSG approach. Still, we believe that SMIREP is a valid approach, as a very low number of rules are learned, enabling the examination of these rules by human experts. In [18, 19], the authors report on a system, DT-GBI, combining graph based induction of patterns with decision trees. The method constructs a decision tree, by pairwise chunking of GBI using the constructed patterns as attributes in the tree. The method employs a beam search to keep the computational complexity at a tolerant level. To some intent, SMIREP is similar to DT-GBI, by using a heuristic search as well as not inducing all possible frequent patterns. DT-GBI can be applied to a variety of graph structured data, not only to chemical compounds. DT-GBI was evaluated on a DNA dataset, therefore, no direct comparison to SMIREP are possible here. In future, this comparison has to be made.

As mentioned in section 2, the SMILES language can be used for other types of structured data. Hence we are convinced, that SMIREP can also be applied to other types of predictive graph mining, e.g. in the mining of messenger RNAs [20]. In a first step, a simple mapping between vertices and edges directly to atoms and bonds might allow the SMIREP system, using OpenBabel as matching library, to be used without alterations. However, this can only be done for problems having no more than four edge label types, or more vertice types than chemical atoms. For richer data structures, exceeding this limitations, a generic graph matching library would have to be employed.

The SMIREP system could be improved by using a different, more sophisticated, stopping condition during the learning stage. A further line of improvement lies in the upgrade from IREP to RIPPER [11] as underlying learning algorithm, repeatedly growing and pruning the found rules. Initial experiments on the mutagenesis database showed that the inclusion of numerical attributes, like ϵ_{LUMO} and $logP$ can further improve the SMIREP system. Further experiments on other datasets are needed to the usefulness of the approach.

A novel approach to predictive graph mining was presented. SMIREP combines principles of the chemical representation language SMILES with the inductive rule learner IREP. As compared to contemporary approaches to graph mining, SMIREP directly builds a *small* set of predictive rules, whereas approaches such as gSpan, closeGraph [9], FSG [2], and AGM [7] typically find a *large* set of patterns satisfying a minimum frequency threshold, which are not necessarily predictive. Frequency based approaches attempt also to *exhaustively* search the space for *all* patterns that satisfy the frequency constraint. SMIREP – in contrast – is a *heuristic* approach to find approximations of the most *predictive* rules. In this regard, SMIREP is more akin to the inductive logic programming approaches that induce rules from positive and negative examples, cf. [21]. On the one hand, these approaches are – in theory – more powerful because they employ Prolog, a programming language. On the other hand, they are – in practice – much more limited because expressive power comes at a computational cost. Indeed, molecular applications of inductive logic programming have so far not been able to induce rules containing large fragments, e.g. Warmr [1] only discovered rules of up to 7 literals in the PTE challenge [22] in a small data set. So, SMIREP tries to combine principles of inductive logic programming with those of graph mining.

Even though SMIREP was introduced in the context of molecular applications, it is clearly also applicable to general predictive graph mining problems. The authors hope that SMIREP will inspire further work on predictive graph mining.

Acknowledgments

This work has been partially supported by the EU IST project cInQ (Consortium on Inductive Querying). The authors would further like to thank Ross D. King for supplying the original mutagenesis dataset, and Christoph Helma and David P. Enot for helpful discussions and suggestions.

References

1. Dehaspe, L.: Frequent Pattern Discovery in First-Order Logic. K. U. Leuven (1998)
2. Deshpande, M., Kuramochi, M., Karypis, G.: Frequent sub-structure-based approaches for classifying chemical compounds. In: Proc. ICDM-03. (2003) 35–42
3. Kramer, S., De Raedt, L., Helma, C.: Molecular feature mining in HIV data. In Provost, F., Srikant, R., eds.: Proc. KDD-01, New York, ACM Press (2001) 136–143
4. Zaki, M.: Efficiently mining frequent trees in a forest. In Hand, D., Keim, D., Ng, R., eds.: Proc. KDD-02, New York, ACM Press (2002) 71–80
5. Yan, X., Han, J.: gspan: Graph-based substructure pattern mining. In: Proc. ICDM-02. (2002)
6. Inokuchi, A., Kashima, H.: Mining significant pairs of patterns from graph structures with class labels. In: Proc. ICDM-03. (2003) 83–90
7. Inokuchi, A., Washio, T., Motoda, H.: Complete mining of frequent patterns from graphs: Mining graph data. Machine Learning 50 (2003) 321–354
8. Kuramochi, M., Karypis, G.: Frequent subgraph discovery. In: Proc. ICDM-01. (2001) 179–186
9. Yan, X., Han, J.: Closegraph: Mining closed frequent graph patterns. In: Proc. KDD-03. (2003)
10. Fürnkranz, J., Widmer, G.: Incremental reduced error pruning. In Cohen, W.W., Hirsh, H., eds.: Proc. ICML 1994, Morgan Kaufmann (1994) 70–77
11. Cohen, W.W.: Fast effective rule induction. In: Proc. ICML 1995, Morgan Kaufmann (1995) 115–123
12. King, R.D., Muggleton, S., Srinivasan, A., Sternberg, M.J.E.: Structure-activity relationships derived by machine learning: The use of atoms and their bond connectivities to predict mutagenicity by inductive logic programming. Proc. of the National Academy of Sciences 93 (1996) 438–442

13. Weininger, D.: SMILES, a chemical language and information system 1. Introduction and encoding rules. J. Chem. Inf. Comput. Sci. **28** (1988) 31–36
14. Quinlan, J.R.: Learning logical definitions from relations. Machine Learning **5** (1990) 239–266
15. Srinivasan, A., Muggleton, S., Sternberg, M.E., King, R.D.: Theories for mutagenicity: a study of first-order and feature based induction. A.I. Journal **85** (1996) 277–299
16. Cook, Holder: Graph-based data mining. ISTA: Intelligent Systems & their applications **15** (2000)
17. Gonzalez, J.A., Holder, L.B., Cook, D.J.: Experimental comparison of graph-based relational concept learning with inductive logic programming systems. In: Proc. ILP 2003. Volume 2583 of Lecture Notes in Artificial Intelligence., Springer-Verlag (2003) 84–100
18. Warodom, G., Matsuda, T., Yoshida, T., Motoda, H., Washio, T.: Classifier construction by graph-based induction for graph-structured data. In: Advances in Knowledge Discovery and Data Mining. Volume 2637 of LNAI., Springer Verlag (2003) 52–62
19. Geamsakul, W., Matsuda, T., Yoshida, T., Motoda, H., Washio, T.: Constructing a decision tree for graph structured data. In: Proc. MGTS 2003, http://www.ar.sanken.osaka-u.ac.jp/MGTS-2003CFP.html (2003) 1–10
20. Horvath, T., Wrobel, S., Bohnebeck, U.: Relational instance-based learning with lists and terms. Machine Learning **43** (2001) 53–80
21. Muggleton, S.: Inverting entailment and Progol. In: Machine Intelligence. Volume 14. Oxford University Press (1995) 133–188
22. Srinivasan, A., King, R.D., Bristol, D.W.: An assessment of ILP-assisted models for toxicology and the PTE-3 experiment. In Džeroski, S., Flach, P., eds.: Proc. ILP-99. Volume 1634 of LNAI., Springer-Verlag (1999) 291–302

An Efficient Algorithm for Enumerating Closed Patterns in Transaction Databases

Takeaki Uno[1], Tatsuya Asai[2,*], Yuzo Uchida[2], and Hiroki Arimura[2]

[1] National Institute of Informatics, 2-1-2, Hitotsubashi, Chiyoda-ku, Tokyo, Japan
uno@nii.jp
[2] Department of Informatics, Kyushu University, 6-10-1, Hakozaki, Fukuoka, Japan
{t-asai,y-uchida,arim}@i.kyushu-u.ac.jp

Abstract. The class of closed patterns is a well known condensed representations of frequent patterns, and have recently attracted considerable interest. In this paper, we propose an efficient algorithm LCM (Linear time Closed pattern Miner) for mining frequent closed patterns from large transaction databases. The main theoretical contribution is our proposed *prefix-preserving closure extension* of closed patterns, which enables us to search all frequent closed patterns in a depth-first manner, in linear time for the number of frequent closed patterns. Our algorithm do not need any storage space for the previously obtained patterns, while the existing algorithms needs it. Performance comparisons of LCM with straightforward algorithms demonstrate the advantages of our prefix-preserving closure extension.

1 Introduction

Frequent pattern mining is one of the fundamental problems in data mining and has many applications such as association rule mining [1, 5, 7] and condensed representation of inductive queries [12]. To handle frequent patterns efficiently, equivalence classes induced by the occurrences had been considered. Closed patterns are maximal patterns of an equivalence class.

This paper addresses the problems of enumerating all frequent closed patterns. For solving this problem, there have been proposed many algorithms [14, 13, 15, 20, 21]. These algorithms are basically based on the enumeration of frequent patterns, that is, enumerate frequent patterns, and output only those being closed patterns. The enumeration of frequent patterns has been studied well, and can be done in efficiently short time [1, 5]. Many computational experiments supports that the algorithms in practical take very short time per pattern on average.

However, as we will show in the later section, the number of frequent patterns can be exponentially larger than the number of closed patterns, hence the computation time can be exponential in the size of datasets for each closed pattern on average. Hence, the existing algorithms use heuristic pruning to cut off non-closed frequent patterns. However, the pruning are not complete, hence they still have possibilities to take exponential time for each closed pattern.

---

* Presently working for Fujitsu Laboratory Ltd., e-mail: asai.tatsuya@jp.fujitsu.com

E. Suzuki and S. Arikawa (Eds.): DS 2004, LNAI 3245, pp. 16–31, 2004.

Moreover, these algorithms have to store previously obtained frequent patterns in memory for avoiding duplications. Some of them further use the stored patterns for checking the "closedness" of patterns. This consumes much memory, sometimes exponential in both the size of both the database and the number of closed patterns. In summary, the existing algorithms possibly take exponential time and memory for both the database size and the number of frequent closed patterns. This is not only a theoretical observation but is supported by results of computational experiments in FIMI'03[7]. In the case that the number of frequent patterns is much larger than the number of frequent closed patterns, such as BMS-WebView1 with small supports, the computation time of the existing algorithms are very large for the number of frequent closed patterns.

In this paper, we propose a new algorithm LCM (Linear time Closed pattern Miner) for enumerating frequent closed patterns. Our algorithm uses a new technique called *prefix preserving extension* (ppc-extension), which is an extension of a closed pattern to another closed pattern. Since this extension generates a new frequent closed pattern from previously obtained closed pattern, it enables us to completely prune the unnecessary non-closed frequent patterns. Our algorithm always finds a new frequent closed pattern in linear time of the size of database, but never take exponential time. In the other words, our algorithm always terminates in linear time in the number of closed patterns. Since any closed pattern is generated by the extension from exactly one of the other closed patterns, we can enumerate frequent closed patterns in a depth-first search manner, hence we need no memory for previously obtained patterns. Thereby, the memory usage of our algorithm depends only on the size of input database. This is not a theoretical result but our computational experiments support the practical efficiency of our algorithm. The techniques used in our algorithm are orthogonal to the existing techniques, such as FP-trees, look-ahead, and heuristic preprocessings. Moreover, we can add the existing techniques for saving the memory space so that our algorithm can handle huge databases much larger than the memory size.

For practical computation, we propose *occurrence deliver, anytime database reduction*, and *fast ppc-test*, which accelerate the speed of the enumeration significantly. We examined the performance of our algorithm for real world and synthesis datasets taken from FIMI'03 repository, including the datasets used in KDD-cup [11], and compare with straightforward implementations. The results showed that the performance of the combination of these techniques and the prefix preserving extension is good for many kinds of datasets.

In summary, our algorithm has the following advantages.

· Linear time enumeration of closed patterns
· No storage space for previously obtained closed patterns
· Generating any closed pattern from another unique closed pattern
· Depth-first generation of closed patterns
· Small practical computational cost of generation of a new pattern

The organization of the paper is as follows. Section 2 prepares notions and definitions. In the Section 3, we give an example to show the number of frequent patterns can be up to exponential to the number of closed patterns. Section 4 explains the existing schemes for closed pattern mining, then present the prefix-preserving closure extension and our algorithm. Section 5 describes several improvements for practical use. Section 6 presents experimental results of our algorithm and improvements on synthesis and re-alworld datasets. We conclude the paper in Section 7.

2   Preliminaries

We give basic definitions and results on closed pattern mining according to [1, 13, 6].

Let $\mathcal{I} = \{1, \ldots, n\}$ be the set of *items*. A *transaction database* on $\mathcal{I}$ is a set $\mathcal{T} = \{t_1, \ldots, t_m\}$ such that each t_i is included in $\mathcal{I}$. Each t_i is called a *transaction*. We denote the total size of $\mathcal{T}$ by $||\mathcal{T}||$, i.e., $||\mathcal{T}|| = \sum_{t \in \mathcal{T}} |t|$. A subset P of $\mathcal{I}$ is called a *pattern* (or *itemset*). For pattern P, a transaction including P is called an *occurrence* of P. The *denotation* of P, denoted by $\mathcal{T}(P)$ is the set of the occurrences of P. $|\mathcal{T}(P)|$ is called the *frequency* of P, and denoted by $frq(P)$. For given constant $\theta \in \mathsf{N}$, called a *minimum support*, pattern P is *frequent* if $frq(P) \geq \theta$. For any patterns P and Q, $\mathcal{T}(P \cup Q) = \mathcal{T}(P) \cap \mathcal{T}(Q)$ holds, and if $P \subseteq Q$ then $\mathcal{T}(Q) \supseteq \mathcal{T}(P)$.

Let $\mathcal{T}$ be a database and P be a pattern on $\mathcal{I}$. For a pair of patterns P and Q, we say P and Q are *equivalent* to each other if $\mathcal{T}(P) = \mathcal{T}(Q)$. The relationship induces equivalence classes on patterns. A maximal pattern and a minimal pattern of an equiva-lence class, w.r.t. set inclusion, are called a *closed pattern* and *key pattern*, respectively. We denote by $\mathcal{F}$ and $\mathcal{C}$ the sets of all frequent patterns and the set of frequent closed patterns in $\mathcal{T}$, respectively.

Given set $\mathcal{S} \subseteq \mathcal{T}$ of transactions, let $\mathcal{I}(\mathcal{S}) = \bigcap_{T \in \mathcal{S}} T$ be the set of items common to all transactions in $\mathcal{S}$. Then, we define the *closure* of pattern P in $\mathcal{T}$, denoted by $Clo(P)$, by $\bigcap_{T \in \mathcal{T}(P)} T$. For every pair of patterns P and Q, the following properties hold (**Pasquier et al.[13]**).

(1) If $P \subseteq Q$, then $Clo(P) \subseteq Clo(Q)$.
(2) If $\mathcal{T}(P) = \mathcal{T}(Q)$, then $Clo(P) = Clo(Q)$.
(3) $Clo(Clo(P)) = Clo(P)$.
(4) $Clo(P)$ is the unique smallest closed pattern including P.
(5) A pattern P is a closed pattern if and only if $Clo(P) = P$.

Note that a key pattern is not the unique minimal element of an equivalence class, while the closed pattern is unique. Here we denote the set of frequent closed patterns by $\mathcal{C}$, the set of frequent patterns by $\mathcal{F}$, the set of items by $\mathcal{I}$, and the size of database by $||\mathcal{T}||$.

For pattern P and item $i \in P$, let $P(i) = P \cap \{1, \ldots, i\}$ be the subset of P consisting only of elements no greater than i, called the *i-prefix* of P. Pattern Q is a *closure extension* of pattern P if $Q = Clo(P \cup \{i\})$ for some $i \notin P$. If Q is a closure extension of P, then $Q \supset P$, and $frq(Q) \leq frq(P)$.

3   Difference Between Numbers of Frequent Patterns and Frequent Closed Patterns

This section shows that the the number of frequent patterns can be quite larger than the number of frequent closed patterns. To see it, we prove the following theorem. Here an irredundant transaction database is a transaction database such that

· no two transactions are the same,
· no item itself is an infrequent pattern, and
· no item is included in all transactions.

Intuitively, if these conditions are satisfied, then we can neither contract nor reduce the database.

Theorem 1. *There are infinite series of irredundant transaction databases $\mathcal{T}$ such that the number $|\mathcal{F}|$ of frequent patterns is exponential in m and n while the number $|\mathcal{C}|$ of frequent closed patterns is $O(m^2)$, where m is the number of transactions in $\mathcal{T}$ and n is the size of the itemset on which the transactions are defined. In particular, the size of $\mathcal{T}$ is $\Theta(nm)$.*

Proof. Let n be any number larger than 4 and m be any even number satisfying that $n - (\lceil \lg m \rceil + 2)$ is larger than both n^ϵ and m^ϵ for a constant ϵ. Let

$X = \{1, ..., n - 2(\lceil \lg m \rceil + 2)\}$
$Y_1 = \{n - 2(\lceil \lg m \rceil + 2) + 1, ..., n - (\lceil \lg m \rceil + 2)\}$
$Y_2 = \{n - (\lceil \lg m \rceil + 2) + 1, ..., n\}$
$\mathcal{J}_1 :=$ a subset of $2^{Y_1} \setminus \{\emptyset, Y_1\}$ of size $m/2 - 1$ without duplications
$\mathcal{J}_2 :=$ a subset of $2^{Y_2} \setminus \{\emptyset, Y_2\}$ of size $m/2 - 1$ without duplications.

Since $|Y_1| = |Y_2| = \lceil \lg m \rceil + 2$, such $\mathcal{J}_1$ and $\mathcal{J}_2$ always exist. Then, we construct a database as follows.

$$\mathcal{T} := \begin{array}{l} \{X \cup Y_1 \cup S | S \in \mathcal{J}_2\} \\ \cup \{X \cup Y_2 \cup S | S \in \mathcal{J}_1\} \\ \cup \{Y_1 \cup Y_2\} \\ \cup \{X\} \end{array} .$$

The database has m transactions defined on an itemset of size n. The size of database is $\Theta(nm)$.

We set the minimum support to $m/2$. We can see that no transactions are the same, no item is included in all transactions, and any item is included in at least $m/2$ transactions. Let $\mathcal{F}$ and $\mathcal{C}$ be the set of frequent patterns and the set of frequent closed patterns, respectively.

Since any transaction in $\mathcal{T}$ except for $Y_1 \cup Y_2$ includes X, any subset of X is a frequent pattern. Since $|X| = n - (\lceil \lg m \rceil + 2)$ is larger than both n^ϵ and m^ϵ, we see that $|\mathcal{F}|$ is exponential in both n and m.

On the other hand, any frequent closed pattern includes X. Hence, any frequent closed pattern is equal to $X \cup S$ for some $S \subseteq Y_1 \cup Y_2$. Since $|Y_1 \cup Y_2| = 2(\lceil \lg m \rceil + 2)$, we have

$$|\mathcal{C}| \leq 2^{2(\lceil \lg m \rceil + 2)}$$
$$= 2^{2 \lg m + 6}$$
$$= 64m^2.$$

Therefore, $|\mathcal{C}| = O(m^2)$.   □

From the theorem, we can see that frequent pattern mining based algorithms can take exponentially longer time for the number of closed patterns. Note that such patterns may appear in real world data in part, because some transactions may share a common large pattern.

4   Algorithm for Enumerating Closed Patterns

We will start with the existing schemes for closed pattern enumeration.

4.1   Previous Approaches for Closed Pattern Mining

A simple method of enumerating frequent closed patterns is to enumerate all frequent patterns, classify them into equivalence classes, and find the maximal pattern for each equivalent class. This method needs $O(|\mathcal{F}|^2)$ time and $O(|\mathcal{F}|)$ memory, since pairwise comparisons of their denotations are needed to classify frequent patterns. Although the number of comparisons can be decreased by using some alphabetical sort of denotations such as radix sort, it still needs $O(|\mathcal{F}|)$ computation. As we saw, $|\mathcal{F}|$ is possibly quite larger than $|\mathcal{C}|$, hence this method consumes a great deal of time and memory.

Some state-of-the-art algorithms for closed pattern mining, such as CHARM [21] and CLOSET [15], use heuristic pruning methods to avoid generating unnecessary non-closed patterns. Although these pruning methods efficiently cut off non-closed pattens, the number of generated patterns is not bounded by a polynomial of $|\mathcal{C}|$.

Pasquier et al.[13] proposed the use of closure operation to enumerate closed patterns. Their idea is to generate frequent patterns, and check whether the patterns are closed patterns or not by closure operation. Although this reduces the storage space for non-closed patterns, the algorithm still requires $|\mathcal{C}|$ space. They actually generate frequent key patterns instead of frequent patterns, to reduce the computational costs. Thus, the computation time is linear in the number of frequent key patterns, which is less than $|\mathcal{F}|$ but can be up to exponential in $|\mathcal{C}|$.

4.2   Closed Patterns Enumeration in Linear Time with Small Space

Our algorithm runs in linear time in $|\mathcal{C}|$ with storage space significantly smaller than that of Pasquier et al.'s algorithm [13], since it operates no non-closed patterns using depth-first search The following lemmas provide a way of efficiently generating any closed pattern from another closed pattern. In this way, we construct our basic algorithm. Then, the basic algorithm is improved to save the memory use by using ppc extension.

Lemma 1. *Let P and Q, $P \subseteq Q$ be patterns having the same denotation, i.e., $\mathcal{T}(P) = \mathcal{T}(Q)$. Then, for any item $i \notin P$, $\mathcal{T}(P \cup \{i\}) = \mathcal{T}(Q \cup \{i\})$.*

Proof. $\mathcal{T}(P \cup \{i\}) = \mathcal{T}(P) \cap \mathcal{T}(\{i\}) = \mathcal{T}(Q) \cap \mathcal{T}(\{i\}) = \mathcal{T}(Q \cup \{i\}).$ □

Lemma 2. *Any closed pattern $P \neq \perp$ is a closure extension of other closed patterns.*

> **Algorithm Closure_version**
> 1. $\mathcal{D} := \{\perp\}$
> 2. $\mathcal{D}' := \{\, Clo(P \cup \{i\}) \mid P \in \mathcal{D}, i \in \mathcal{I} \backslash P \,\}$
> 3. **if** $\mathcal{D}' = \emptyset$ **then output** $\mathcal{D}$; **halt**
> 4. $\mathcal{D} := \mathcal{D} \cup \mathcal{D}'$; **go to** 2

Fig. 1. Basic algorithm for enumerating frequent closed patterns.

Proof. Let Q be a pattern obtained by repeatedly removing items from P until its denotation changes, and i be the item removed last. Then, $Clo(Q \cup \{i\}) = P$. Such a pattern must exist since $P \neq \perp$. Since $\mathcal{T}(Q) \neq \mathcal{T}(Q \cup \{i\})$, $i \notin Clo(Q)$. From Property 1, $Clo(Q \cup \{i\}) = Clo(Clo(Q) \cup \{i\})$. Thus, P is a closure extension of $Q \cup \{i\}$.   □

Through these lemmas, we can see that all closed patterns can be generated by closure extensions to closed patterns. It follows the basic version of our algorithm, which uses levelwise (breadth-first) search similar to Apriori type algorithms [1] using closed expansion instead of tail expansion. We describe the basic algorithm in Figure 1. Since the algorithm deals with no non-closed pattern, the computational cost depends on $|\mathcal{C}|$ but not on $|\mathcal{F}|$. However, we still need much storage space to keep $\mathcal{D}$ in memory.

A possible improvement is to use depth-first search instead of Apriori-style levelwise search. For enumerating frequent patterns, Bayardo [5] proposed an algorithm based on *tail extension*, which is an extension of a pattern P by an item larger than the maximum item of P. Since any frequent pattern is a tail extension of another unique frequent pattern, the algorithm enumerates all frequent patterns without duplications in a depth-first manner, with no storage space for previously obtained frequent patterns. This technique is efficient, but cannot directly be applied to closed pattern enumeration, since a closed pattern is not always a tail-extension of another closed pattern.

We here propose *prefix-preserving closure extension* satisfying that any closed pattern is an extension of another unique closed pattern unifying ordinary closure-expansion and tail-expansion. This enables depth-first generation with no storage space.

4.3 Prefix-Preserving Closure Extension

We start with definitions. Let P be a closed pattern. The *core index* of P, denoted by $core\_i(P)$, is the minimum index i such that $\mathcal{T}(P(i)) = \mathcal{T}(P)$. We let $core\_i(\perp) = 0$.

Here we give the definition of ppc-extension. Pattern Q is called a *prefix-preserving closure extension (ppc-extension)* of P if

(i) $Q = Clo(P \cup \{i\})$ for some $i \in P$, that is, Q is obtained by first adding i to P and then taking its closure,
(ii) item i satisfies $i \notin P$ and $i > core\_i(P)$, and
(iii) $P(i-1) = Q(i-1)$, that is, the $(i-1)$-prefix of P is preserved.

Actually, ppc-extension satisfies the following theorem. We give an example in Fig. 2.

transaction database

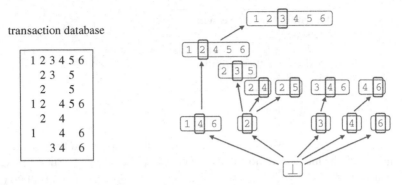

Fig. 2. Example of all closed patterns and their ppc extensions. Core indices are circled.

Theorem 2. *Let $Q \neq \perp$ be a closed pattern. Then, there is just one closed pattern P such that Q is a ppc-extension of P.*

To prove the theorem, we state several lemmas.

Lemma 3. *Let P be a closed pattern and $Q = Clo(P \cup \{i\})$ be a ppc-extension of P. Then, i is the core index of Q.*

Proof. Since $i > core\_i(P)$, we have $\mathcal{T}(P) = \mathcal{T}(P(i))$. From Lemma 2, $Clo(P(i) \cup \{i\}) = Clo(P \cup \{i\}) = Clo(Q)$, thus $core\_i(Q) \leq i$. Since the extension preserves the i-prefix of P, we have $P(i-1) = Q(i-1)$. Thus, $Clo(Q(i-1)) = Clo(P(i-1)) = P \neq Q$. It follows that $core\_i(Q) > i - 1$, and we conclude $core\_i(Q) = i$. □

Let Q be a closed pattern and $\mathcal{P}(Q)$ be the set of closed patterns such that Q is their closure extension. We show that Q is a ppc-extension of a unique closed pattern of $\mathcal{P}(Q)$.

Lemma 4. *Let $Q \neq \perp$ be a closed pattern, and $P = Clo(Q(core\_i(Q) - 1))$. Then, Q is a ppc-extension of P.*

Proof. Since $\mathcal{T}(P) = \mathcal{T}(Q(core\_i(Q)-1))$, we have $\mathcal{T}(P \cup \{i\}) = \mathcal{T}(Q(core\_i(Q)-1) \cup \{i\}) = \mathcal{T}(Q(core\_i(Q)))$. This implies $Q = Clo(P \cup \{i\})$, thus Q satisfies condition (i) of ppc-extension. Since $P = Clo(Q(core\_i(Q) - 1))$, $core\_i(P) \leq i-1$. Thus, Q satisfies condition (ii) of ppc-extension. Since $P \subset Q$ and $Q(i-1) \subseteq P$, we have $P(i-1) = Q(i-1)$. Thus, Q satisfies condition (iii) of ppc-extension. □

Proof of Theorem 2: From Lemma 4, there is at least one closed pattern P in $\mathcal{P}(Q)$ such that Q is a ppc-extension of P. Let $P = Clo(Q(core\_i(Q) - 1))$. Suppose that there is a closed pattern $P' \neq P$ such that Q is a ppc-extension of P'. From lemma 3, $Q = Clo(P' \cup \{i\})$. Thus, from condition (iii) of ppc-extension, $P'(i-1) = Q(i-1) = P(i-1)$. This together with $\mathcal{T}(P) = \mathcal{T}(P(i-1))$ implies that $\mathcal{T}(P) \supset \mathcal{T}(P')$. Thus, we can see $\mathcal{T}(P'(i-1)) \neq \mathcal{T}(P')$, and $core\_i(P') \geq i$. This violates condition (ii) of ppc-extension, and is a contradiction. □

Algorithm LCM($\mathcal{T}$:transaction database, θ:support)
  1. **call** ENUM_CLOSEDPATTERNS($\bot$) ;
Procedure ENUM_CLOSEDPATTERNS(P: frequent closed pattern)
  2. **if** P is not frequent **then Return**;
  2. **output** P;
  3. **for** $i = core\_i(P) + 1$ **to** $|\mathcal{I}|$
  4. $Q = Clo(P \cup \{i\})$;
  5.   **If** $P(i-1) = Q(i-1)$ **then**   // Q is a ppc-extension of P
  6.   **Call** ENUM_CLOSEDPATTERNS(Q);
  7. **End for**

Fig. 3. Description of Algorithm LCM.

From this theorem, we obtain our algorithm LCM, described in Figure 3, which generate ppc-extensions for each frequent closed pattern.

Since the algorithm takes $O(||\mathcal{T}(P)||)$ time to derive the closure of each $P \cup \{i\}$, we obtain the following theorem.

Theorem 3. *Given a database $\mathcal{T}$, the algorithm LCM enumerates all frequent closed patterns in $O(||\mathcal{T}(P)|| \times |\mathcal{I}|)$ time for each pattern P with $O(||\mathcal{T}||)$ memory space.*

The time and space complexities of the existing algorithms [21, 15, 13] are $O(||\mathcal{T}|| \times |\mathcal{F}|)$ and $O(||\mathcal{T}|| + |\mathcal{C}| \times |\mathcal{I}|)$, respectively. As we saw in the example in Section 3, the difference between $|\mathcal{C}|$ and $|\mathcal{F}|$, and the difference between $|\mathcal{C}| \times |\mathcal{I}|$ and $||\mathcal{T}||$ can be up to exponential. As compared with our basic algorithm, the ppc extension based algorithm exponentially reduces the memory complexity when $O(|\mathcal{C}|)$ is exponentially larger than $||\mathcal{T}||$. In practice, such exponential differences often occur (see results in Section 6). Thus, the performance of our algorithm possibly exponentially better than the existing algorithms in some instances.

5   Reducing Practical Computation Time

The computation time of LCM described in the previous section is linear in $|\mathcal{C}|$, with a factor depending on $||\mathcal{T}|| \times |\mathcal{I}|$ for each closed pattern $P \in \mathcal{C}$. However, this still takes a long time if it is implemented in a straightforward way. In this section, we propose some techniques for speeding up frequency counting and closure operation. These techniques will increase practical performance of the algorithms and incorporated into the implementations used in the experiments in Section 6 of the paper, although independent of our main contribution. In Figure 7, we describe the details of LCM with these practical techniques.

5.1   Occurrence Deliver

Occurrence deliver reduces the construction time for $\mathcal{T}(P \cup \{i\})$, which is used for frequency counting and closure operation. This technique is particularly efficient for

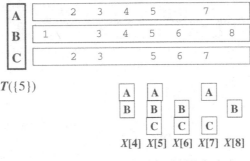

$T(\{5\})$

$X[4]$ $X[5]$ $X[6]$ $X[7]$ $X[8]$

Occurrences of $\{5,i\}$ for i=4,...,8

Fig. 4. Occurrence deliver: build up denotations by inserting each transaction to each its member.

sparse datasets, such that $|T(P \cup \{i\})|$ is much smaller than $|T(P)|$ on average. In a usual way, $T(P\cup\{i\})$ is obtained by $T(P)\cap T(P\cup\{i\})$ in $O(|T(P)|+|T(P\cup\{i\})|)$ time (this is known as *down-project* []). Thus, generating all ppc-extensions needs $|\mathcal{I}|$ scans and takes $O(||T(P)||)$ time.

Instead of this, we build for all $i = core\_i(P),\ldots,|\mathcal{I}|$ denotations $T(P \cup \{i\})$, simultaneously, by scanning the transactions in $T(P)$. We initialize $X[i] := \emptyset$ for all i. For each $t \in T(P)$ and for each $i \in t; i > core\_i(P)$, we insert t to $X[i]$. Then, each $X[i]$ is equal to $T(P \cup \{i\})$. See Fig. 4 for the explanation. This correctly computes $T(P\cup\{i\})$ for all i in $O(||T(P)||)$. Table 1 shows results of computational experiments where the number of item-accesses were counted. The numbers are deeply related to the performance of frequency counting heavily depending on the number of item-accesses.

Table 1. The accumulated number of item-accesses over all iterations.

Dataset and support	connect,65%	pumsb,80%	BMS-webview2,0.2%	T40I10D100K,5%
Straightforward	914074131	624940309	870850439	830074845
Occurrence deliver	617847782	201860874	17217493	73406900
Reduction factor	1.47	3.09	50.5	11.3

5.2   Anytime Database Reduction

In conventional practical computations of frequent pattern mining, the size of the input database is reduced by removing infrequent items i such that $T(\{i\}) < \theta$, and then merging the same transactions into one. Suppose that we are given the database on the left hand side of Fig. 5, and minimum support 3.

In the database, items 4, 6, 7, and 8 are included in at most two transactions, hence patterns including these items can not be frequent. Items 1 and 2 are included in all transactions, hence any frequent patterns can contain any subset of $\{1,2\}$. It follows that items 1 and 2 are redundant. Hence, we remove items 1, 2, 4, 6, 7, and 8 from the database. After the removal, we merge the same transactions into one. We then record the multiplicity of transactions. Consequently, we have a smaller database, right hand side of Fig. 5, with fewer items and fewer transactions, which includes the same set

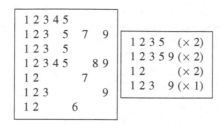

Fig. 5. Example of anytime database reduction.

of frequent patterns. We call this operation *database reduction* (or *database filtering*). Many practical algorithms utilize this operation at initialization. Database reduction is known to be efficient in practice, especially for large support [7].

Here we propose *anytime database reduction*, which is to apply database reduction in any iteration. For an iteration (invokation) of Enum_ClosedPatterns inputting pattern P, the following items and transactions are unnecessary for any of its descendant iterations:

(1) transactions not including P,
(2) transactions including no item greater than $core\_i(P)$,
(3) items included in P,
(4) items less than or equal to $core\_i(P)$, and
(5) items i satisfies that $frq(P \cup \{i\}) < \theta$ (no frequent pattern includes i).

In each iteration, we restrict the database to $\mathcal{T}(P)$, apply database reduction to remove such items and transactions, merge the same transactions, and pass it to child iterations.

Anytime database reduction is efficient especially when support is large. In the experiments, in almost all iterations, the size of the reduced database is bounded by a constant even if the patterns have many occurrences. Table 2 lists simple computational experiments comparing conventional database reduction (applying reduction only at initialization) and anytime database reduction. Each cell shows the accumulated number of transactions in the reduced databases in all iterations.

Table 2. Accumulated number of transactions in database in all iterations.

Dataset and support	connect, 50%	pumsb, 60%	BMS-WebView2, 0.1%	T40I10D100K, 0.03%
Database reduction	188319235	2125460007	2280260	1704927639
Anytime database reduction	538931	7777187	521576	77371534
Reduction factor	349.4	273.2	4.3	22.0

We also use anytime database reduction for closure operation. Suppose that we have closed pattern P with core index i. Let transactions $t_1, ..., t_k \in \mathcal{T}(P)$ have the same i-suffix, i.e., $t_1 \cap \{i, ..., |\mathcal{I}|\} = t_2 \cap \{i, ..., |\mathcal{I}|\} =, ..., = t_k \cap \{i, ..., |\mathcal{I}|\}$.

Lemma 5. *Let $j; j < i$ be an item, and j' be items such that $j' > i$ and included in all $t_1, ..., t_k$. Then, if j is not included in at least one transaction of $t_1, ..., t_k$, $j \notin Clo(Q)$ holds for any pattern Q including $P \cup \{i\}$.*

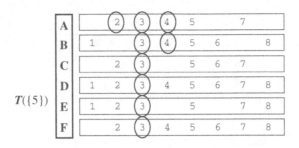

$T(\{5\})$

Fig. 6. Transaction A has the minimum size. Fast ppc test accesses only circled items while closure operation accesses all items.

Proof. Since $\mathcal{T}(Q)$ includes all $t_1, ..., t_k$, and j is not included in one of $t_1, ..., t_k$. Hence, j is not included in $Clo(Q)$. □

According to this lemma, we can remove j from $t_1, ..., t_k$ from the database if j is not included in at least one of $t_1, ..., t_k$. By removing all such items, $t_1, ..., t_k$ all become $\bigcap_{h=1,...,k} t_h$. Thus, we can merge them into one transaction similar to the above reduction. This reduces the number of transactions as much as the reduced database for frequency counting. Thus, the computation time of closure operation is shorten drastically. We describe the details on anytime database reduction for closure operation.

1. Remove transactions not including P
2. Remove items i such that $frq(P \cup \{i\}) < \theta$
3. Remove items of P
4. Replace the transactions $T_1, ..., T_k$ having the same i-suffix by the intersection, i.e., replace them by $\bigcap\{T_1, ..., T_k\}$.

5.3 Fast Prefix-Preserving Test

Fast prefix-preserving test (fast ppc-test) efficiently checks condition (iii) of ppc-extension for $P \cup \{i\}$. A straightforward way for this task is to compute the closure of $P \cup \{i\}$. This usually takes much time since it accesses all items in the occurrences of P. Instead of this, we check for each j whether $j; j < i, j \notin P(i-1)$ is included in $Clo(P \cup \{i\})$ or not. Item j is included in $Clo(P \cup \{i\})$ if and only if j is included in every transaction of $\mathcal{T}(P \cup \{i\})$. If we find a transaction not including j, then we can immediately conclude that j is not included in every transaction of $\mathcal{T}(P)$. Thus, we do not need to access all items in occurrences. In particular, we have to check items j in occurrence t^* of the minimum size, since other items can not be included in every transaction of $\mathcal{T}(P)$. Fig. 6 shows an example of accessed items by fast ppc test.

This results $O(\sum_{j \in t^* \setminus P} |\mathcal{T}(P \cup \{i\} \cup \{j\})|)$ time algorithm, which is much faster than the straightforward algorithm with $O(\sum_{j < i} |\mathcal{T}(P \cup \{i\} \cup \{j\})|)$ time. Table 3 shows the results of computational experiments comparing the number of accesses between closure operation and fast ppc test.

Table 3. Accumulated number of accessed items in all iterations.

Dataset and support	connect,60%	pumsb,75%,	BMS-WebVeiw2,0.1%	T40I10D100K,0.1%
Closure operation	1807886455	741205890	50313395	2093327534
Fast ppc test	2333551	2703318	127722	1701748
Reduction factor	774.7	274.1	393.9	1230

Algorithm LCM ($\mathcal{T}$:transaction database, θ:support)
1. **call** ENUM_CLOSEDPATTERNS($\mathcal{T}, \perp, \mathcal{T}$) ;

Procedure ENUM_CLOSEDPATTERNS ($\mathcal{T}$: transaction database,
                P:frequent closed pattern, Occ: transactions including P)
2. **Output** P;
3. Reduce $\mathcal{T}$ by *Anytime database reduction*;
4. Compute the frequency of each pattern $P \cup \{i\}, i > core\_i(P)$
                by *Occurrence deliver* with P and Occ ;
5. **for** $i := core\_i(P) + 1$ **to** $|\mathcal{I}|$
6. $Q := Clo(P \cup \{i\})$;
7.   **If** $P(i-1) = Q(i-1)$ and Q is frequent **then**   // Q is a ppc-extension of P
                Call ENUM_CLOSEDPATTERNS(Q);
8. **End for**

The version of fast-ppc test is obtained by replacing the lines 6 and 7 by
6.   **if** *fast-ppc test* is true for $P \cup \{i\}$ and $P \cup \{i\}$ is frequent **then**
                Call ENUM_CLOSEDPATTERNS(Q);

Fig. 7. Algorithm LCM with practical speeding up.

In fact, the test requires an adjacency matrix (sometimes called a *bitmap*) representing the inclusion relation between items and transactions. The adjacency matrix requires $O(|\mathcal{T}| \times |\mathcal{I}|)$ memory, which is quite hard to store for large instances. Hence, we keep columns of the adjacency matrix for only transactions larger than $|\mathcal{I}|/\delta$, where δ is a constant number. In this way, we can check whether $j \in t$ or not in constant time if j is large, and also in short time by checking items of t if t is small. The algorithm uses $O(\delta \times ||\mathcal{T}||)$ memory, which is linear in the input size.

6 Computational Experiments

This section shows the results of computational experiments for evaluating the practical performance of our algorithms on real world and synthetic datasets. Fig. 8 lists the datasets, which are from the FIMI'03 site (http://fimi.cs.helsinki.fi/): retail, accidents; IBM Almaden Quest research group website (T10I4D100K); UCI ML repository (connect, pumsb); (at http://www.ics.uci.edu/~mlearn/MLRepository.html) Click-stream Data by Ferenc Bodon (kosarak); KDD-CUP 2000 [11] (BMS-WebView-1, BMS-POS) (at http://www.ecn.purdue.edu/KDDCUP/).

Dataset	#items	#Trans	AvTrSz	$|\mathcal{F}|$	$|\mathcal{C}|$	support (%)
BMS-POS	1,657	517,255	6.5	122K–33,400K	122K–21,885K	0.64–0.01
BMS-Web-View1	497	59,602	2.51	3.9K–NA	3.9K–1,241K	0.1–0.01
T10I4D100K	1,000	100,000	10.0	15K–335K	14K–229K	0.15–0.025
kosarak	41,270	990,000	8.10	0.38K–56,006K	0.38K–17,576K	1–0.08
retail	16,470	88,162	10.3	10K–4,106K	10K–732K	0.32–0.005
accidents	469	340,183	33.8	530–10,692K	530–9,959K	70–10
pumsb	7,117	49,046	74.0	2.6K–NA	1.4K–44,453K	95–60
connect	130	67,577	43.0	27K–NA	3.4K–8,037K	95–40

Fig. 8. Datasets: AvTrSz means average transaction size.

To evaluate the efficiency of ppc extension and practical improvements, we implemented several algorithms as follows.

- freqset: algorithm using frequent pattern enumeration
- straight: straightforward implementation of LCM (frequency counting by tuples)
- occ: LCM with occurrence deliver
- occ+dbr: LCM with occurrence deliver and anytime database reduction for both frequency counting and closure
- occ+fchk: LCM with occurrence deliver, anytime database reduction for frequency counting, and fast ppc test

The figure also displays the number of frequent patterns and frequent closed patterns, which are written as #freqset and #freq closed. The algorithms were implemented in C and compiled with gcc 3.3.1. The experiments were run on a notebook PC with mobile Pentium III 750MHz, 256MB memory. Fig. 9 plots the running time with varying minimum supports for the algorithms on the eight datasets.

From Fig. 9, we can observe that LCM with practical optimization (occ, occ+dbr, occ+fchk) outperforms the frequent pattern enumeration-based algorithm (freqset). The speed up ratio of the ppc extension algorithm (straight) against algorithm freqset totally depends on the ratio of #freqset and #freq. closed. The ratio is quite large for several real world and synthetic datasets with small supports, such as BMS-WebView, retails, pumsb, and connect. On such problems, sometimes the frequent pattern based algorithm takes quite long time while LCM terminates in short time.

Occurrence deliver performs very well on any dataset, especially on sparse datasets, such as BMS-WebView, retail, and IBM datasets. In such sparse datasets, since $\mathcal{T}(\{i\})$ is usually larger than $\mathcal{T}(P \cup \{i\})$, occurrence deliver is efficient.

Anytime database reduction decreases the computation time well, especially in dense datasets or those with large supports. Only when support is very small, closed to zero, the computation time were not shorten, since a few items are eliminated and few transactions become the same. In such cases, fast ppc test performs well. However, fast ppc test does not accelerate speed so much in dense datasets or large supports.

For a detailed study about the performance of our algorithm compared with other algorithms, consult the companion paper [18] and the competition report of FIMI'03 [7]. Note that the algorithms we submitted to [18, 7] were old versions, which does not include anytime database reduction, thus they are slower than the algorithm in this paper.

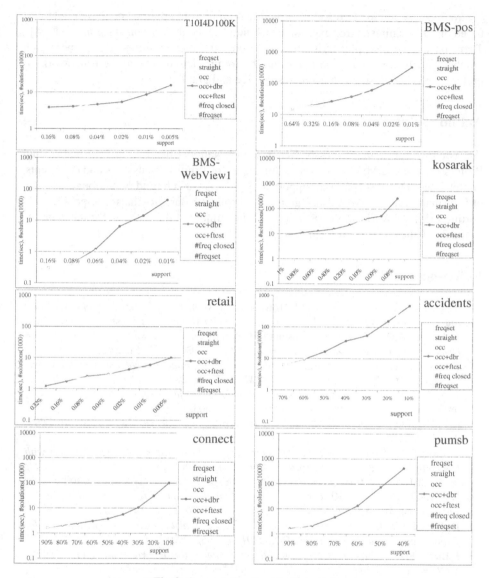

Fig. 9. Computation time for datasets.

7   Conclusion

We addressed the problem of enumerating all frequent closed patterns in a given transaction database, and proposed an efficient algorithm LCM to solve this, which uses memory linear in the input size, i.e., the algorithm does not store the previously obtained patterns in memory. The main contribution of this paper is that we proposed prefix-preserving closure extension, which combines tail-extension of [5] and closure operation of [13] to realize direct enumeration of closed patterns.

We recently studied frequent substructure mining from ordered and unordered trees based on a deterministic tree expansion technique called the *rightmostexpansion* [2–4]. There have been also pioneering works on closed pattern mining in sequences and graphs [17, 19]. It would be an interesting future problem to extend the framework of prefix-preserving closure extension to such tree and graph mining.

Acknowledgment

We would like to thank Professor Ken Satoh of National Institute of Informatics and Professor Kazuhisa Makino of Osaka University for fruitful discussions and comments on this issue. This research was supported by group research fund of National Institute of Informatics, Japan. We are also grateful to Professor Bart Goethals, and people supporting FIMI'03 Workshop/Repository, and the authors of the datasets for the datasets available by the courtesy of them.

References

1. R. Agrawal,H. Mannila,R. Srikant,H. Toivonen,A. I. Verkamo, *Fast Discovery of Association Rules*, In *Advances in Knowledge Discovery and Data Mining*, MIT Press, 307–328, 1996.
2. T. Asai, K. Abe, S. Kawasoe, H. Arimura, H. Sakamoto, S. Arikawa, *Efficient Substructure Discovery from Large Semi-structured Data*, In *Proc. SDM'02*, SIAM, 2002.
3. T. Asai, H. Arimura, K. Abe, S. Kawasoe, S. Arikawa, *Online Algorithms for Mining Semi-structured Data Stream*, In *Proc. IEEE ICDM'02*, 27–34, 2002.
4. T. Asai, H. Arimura, T. Uno, S. Nakano, *Discovering Frequent Substructures in Large Unordered Trees*, In *Proc. DS'03*, 47–61, LNAI 2843, 2003.
5. R. J. Bayardo Jr., *Efficiently Mining Long Patterns from Databases*, In Proc. SIGMOD'98, 85–93, 1998.
6. Y. Bastide, R. Taouil, N. Pasquier, G. Stumme, L. Lakhal, *Mining Frequent Patterns with Counting Inference*, *SIGKDD Explr.*, 2(2), 66–75, Dec. 2000.
7. B. Goethals, *the FIMI'03 Homepage*, http://fimi.cs.helsinki.fi/, 2003.
8. E. Boros, V. Gurvich, L. Khachiyan, K. Makino, *On the Complexity of Generating Maximal Frequent and Minimal Infrequent Sets*, In *Proc. STACS 2002*, 133-141, 2002.
9. D. Burdick, M. Calimlim, J. Gehrke, *MAFIA: A Maximal Frequent Itemset Algorithm for Transactional Databases*, In *Proc. ICDE 2001*, 443-452, 2001.
10. J. Han, J. Pei, Y. Yin, *Mining Frequent Patterns without Candidate Generation*, In *Proc. SIGMOD'00*, 1-12, 2000
11. R. Kohavi, C. E. Brodley, B. Frasca, L. Mason, Z. Zheng, *KDD-Cup 2000 Organizers' Report: Peeling the Onion*, *SIGKDD Explr.*, 2(2), 86-98, 2000.
12. H. Mannila, H. Toivonen, *Multiple Uses of Frequent Sets and Condensed Representations*, In *Proc. KDD'96*, 189–194, 1996.
13. N. Pasquier, Y. Bastide, R. Taouil, L. Lakhal, *Efficient Mining of Association Rules Using Closed Itemset Lattices*, *Inform. Syst.*, 24(1), 25–46, 1999.
14. N. Pasquier, Y. Bastide, R. Taouil, L. Lakhal, *Discovering Frequent Closed Itemsets for Association Rules*, In *Proc. ICDT'99*, 398-416, 1999.
15. J. Pei, J. Han, R. Mao, *CLOSET: An Efficient Algorithm for Mining Frequent Closed Itemsets*, In *Proc. DMKD'00*, 21-30, 2000.
16. R. Rymon, *Search Through Systematic Set Enumeration*, In *Proc. KR-92*, 268–275, 1992.

17. P. Tzvetkov, X. Yan, and J. Han, *TSP: Mining Top-K Closed Sequential Patterns*, In *Proc. ICDM'03*, 2003.
18. T. Uno, T. Asai, Y. Uchida, H. Arimura, *LCM: An Efficient Algorithm for Enumerating Frequent Closed Item Sets*, In *Proc. IEEE ICDM'03 Workshop FIMI'03*, 2003. (Available as CEUR Workshop Proc. series, Vol. 90, http://ceur-ws.org/vol-90)
19. X. Yan and J. Han, *CloseGraph: Mining Closed Frequent Graph Patterns*, In *Proc. KDD'03*, ACM, 2003.
20. M. J. Zaki, *Scalable Algorithms for Association Mining*, *Knowledge and Data Engineering*, 12(2), 372–390, 2000.
21. M. J. Zaki, C. Hsiao, CHARM: An Efficient Algorithm for Closed Itemset Mining, In *Proc. SDM'02*, SIAM, 457-473, 2002.

Finding Optimal Pairs of Cooperative and Competing Patterns with Bounded Distance

Shunsuke Inenaga[1], Hideo Bannai[2], Heikki Hyyrö[3], Ayumi Shinohara[3,5],
Masayuki Takeda[4,5], Kenta Nakai[2], and Satoru Miyano[2]

[1] Department of Computer Science, P.O. Box 26 (Teollisuuskatu 23),
FIN-00014 University of Helsinki, Finland
inenaga@cs.helsinki.fi
[2] Human Genome Center, Institute of Medical Science,
The University of Tokyo, Tokyo 108-8639, Japan
{bannai,knakai,miyano}@ims.u-tokyo.ac.jp
[3] PRESTO, Japan Science and Technology Agency (JST)
helmu@cs.uta.fi
[4] SORST, Japan Science and Technology Agency (JST)
[5] Department of Informatics, Kyushu University 33, Fukuoka 812-8581, Japan
{ayumi,takeda}@i.kyushu-u.ac.jp

Abstract. We consider the problem of discovering the optimal pair of substring patterns with bounded distance α, from a given set S of strings. We study two kinds of pattern classes, one is in form $p \wedge_\alpha q$ that are interpreted as *cooperative* patterns within α distance, and the other is in form $p \wedge_\alpha \neg q$ representing *competing* patterns, with respect to S. We show an efficient algorithm to find the optimal pair of patterns in $O(N^2)$ time using $O(N)$ space. We also present an $O(m^2 N^2)$ time and $O(m^2 N)$ space solution to a more difficult version of the optimal pattern pair discovery problem, where m denotes the number of strings in S.

1 Introduction

Pattern discovery is an intensively studied sub-area of Discovery Science. A large amount of effort was paid to devising efficient algorithms to extract interesting, useful, and even surprising substring patterns from massive string datasets such as biological sequences [1, 2]. Then this research has been extended to more complicated but very expressive pattern classes such as subsequence patterns [3, 4], episode patterns [5, 6], VLDC patterns [7] and their variations [8].

Another interesting and challenging direction of this research is discovery of *optimal pattern pairs*, whereas the above-mentioned algorithms are only for finding optimal *single* patterns. Very recently, in [9] we developed an efficient $O(N^2)$ time algorithm for finding optimal pairs of substring patterns combined with any Boolean functions such as $\wedge$ (AND), $\vee$ (OR) and $\neg$ (NOT), where N denotes the total string length in the input dataset. For instance, the algorithm allows to find pattern pairs in form $p \wedge q$, which can be interpreted as two sequences with *cooperative* functions. Some developments have been made for

E. Suzuki and S. Arikawa (Eds.): DS 2004, LNAI 3245, pp. 32–46, 2004.
© Springer-Verlag Berlin Heidelberg 2004

finding cooperative pattern pairs with a certain distance α, denoted, for now, by $p \wedge_\alpha q$, in terms of *structured motifs* [10, 11] and *proximity patterns* [12, 13].

This paper extends special cases in the work of [9] by producing a generic algorithm to find the optimal pairs of *cooperative* patterns in form $p \wedge_\alpha q$, and *competing* patterns in form $p \wedge_\alpha \neg q$. We develop a very efficient algorithm that runs in $O(N^2)$ time using only $O(N)$ space for *any* given threshold parameter $\alpha \geq 0$, hence not increasing the asymptotic complexity of [9] which does not deal with any bounded distance between two patterns. The pattern pairs discovered by our algorithm are optimal in that they are guaranteed to be the highest scoring pair of patterns with respect to a given scoring function. Our algorithm can be adjusted to handle several common problem formulations of pattern discovery, for example, pattern discovery from positive and negative sequence sets [1, 14, 8, 15], as well as the discovery of patterns that *correlate* with a given numeric attribute (e.g. gene expression level) assigned to the sequences [16–18, 2, 19].

The efficiency of our algorithm comes from the uses of helpful data structures, *suffix trees* [20] and *sparse suffix trees* [21, 22]. We also present an efficient implementation that uses *suffix arrays* and *lcp arrays* [23] to simulate bottom-up traversals on suffix trees and sparse suffix trees.

Comparison to Related Works. Marsan and Sagot [10] gave an algorithm which is limited to finding pattern pairs in form $p \wedge_\alpha q$. Although the time complexity is claimed to be $O(Nm)$, where m denotes the number of strings in the dataset, a significant difference is that the length of each pattern is fixed and is regarded as a constant, whereas our algorithm does not impose such a restriction. Arimura et al. [12, 13] presented an algorithm to find the optimal (α, d)-proximity pattern that can be interpreted into a sequence $p_1 \wedge_\alpha \cdots \wedge_\alpha p_d$ of d patterns with bounded distance α. Although the expected running time of their algorithm is $O(\alpha^{d-1} N (\log N)^d)$, the worst case running time is still $O(\alpha^d N^{d+1} \log N)$. Since in our case $d = 2$, it turns out to be $O(\alpha^2 N^3 \log N)$ which is rather bad compared to our $O(N^2)$ time complexity. In addition, it is quite easy to extend our algorithm to finding (α, d)-proximity pattern in $O(N^d)$ time using only $O(N)$ space. Since the algorithm of [13] consumes $O(dN)$ space, our algorithm improves both time and space for extracting proximity patterns. On the other hand, an algorithm to discover pattern pairs in form $p \wedge_\alpha \neg q$ was given recently in [24], in terms of *missing patterns*. As the algorithm performs in $O(n^2)$ time using $O(n)$ space for a single input string of length n, our algorithm generalizes it for a set of strings with the same asymptotic complexity.

None of the above previous algorithms considers the fact that if one pattern of pair $p \wedge_\alpha q$ is very long, its ending position may be *outside* of the region specified by α, or for pattern pair in form $p \wedge_\alpha \neg q$, it may *invade* the α-region with respect to some other occurrence of p. Since such pattern pairs seem to be of less interest for some applications, we design a version of our algorithm which permits us to find pattern pair $p \wedge_\alpha q$ such that both p and q occur *strictly within* the α-region of each occurrence of p, and pattern pair $p \wedge_\alpha \neg q$ such that q is *strictly outside* the α-region of any occurrences of p. This version of our algorithm runs in $O(m^2 N^2)$ time using $O(m^2 N)$ space.

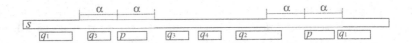

Fig. 1. From pattern p in string s, q_1 begins within α-distance, q_2 begins outside α-distance, q_3 occurs within α-distance, and q_4 occurs outside α-distance.

2   Preliminaries

2.1   Notation

Let Σ be a finite alphabet. An element of Σ^* is called a *string*. Strings x, y, and z are said to be a *prefix*, *substring*, and *suffix* of string $w = xyz$, respectively. The length of a string w is denoted by $|w|$. The empty string is denoted by ε, that is, $|\varepsilon| = 0$. The i-th character of a string w is denoted by $w[i]$ for $1 \leq i \leq n$, and the substring of a string w that begins at position i and ends at position j is denoted by $w[i:j]$ for $1 \leq i \leq j \leq n$. For any set S, let $|S|$ denote the cardinality of S. Let $[i, j]$ denote the set of consecutive integers $i, i+1, \ldots, j$.

For strings p and s, let $Beg(p, s)$ denote the set of all beginning positions of p in s, e.g.,

$$Beg(p, s) = \{i \mid s[i:i + |p| - 1] = p\}.$$

Also, let $Cov(p, s)$ denote the set of all positions in s covered by p, e.g.,

$$Cov(p, s) = \cup_{j \in Beg(p,s)} [j, j + |p| - 1].$$

For strings p, s and threshold value $\alpha \geq 0$, let us define set $D(p, s, \alpha)$ and of integers by

$$D(p, s, \alpha) = (\cup_{j \in Beg(p,s)} [j - \alpha, j + \alpha]) \cap [1, |s|].$$

That is, $D(p, s, \alpha)$ consists of the regions inside s that are closer than or equal to α positions from the beginning position of some occurrence of p. The shaded regions illustrated in Fig. 1 are the regions in $D(p, s, \alpha)$.

If $|Beg(q, s) \cap D(p, s, \alpha)| \geq 1$, that is, if one or more occurrences of q in s are within α positions from at least one occurrence of p in s, then q is said to *begin within α-distance* from p, and otherwise *begin outside α-distance*. Similarly, if $|Cov(q, s) \cap D(p, s, \alpha)| \geq 1$, that is, if there is at least one overlap between an occurrence of p and a region in $D(p, s, \alpha)$, then q is said to *occur within α-distance* from p, and otherwise *occur outside α-distance*. See also Fig. 1.

Let $\psi(p, s)$ be a Boolean matching function that has the value $\texttt{true}$ if p is a substring of s, and $\texttt{false}$ otherwise. We define $\langle p, F, q \rangle$ as a *Boolean pattern pair* (or simply *pattern pair*) which consists of two patterns p, q and a Boolean function $F : \{\texttt{true}, \texttt{false}\} \times \{\texttt{true}, \texttt{false}\} \to \{\texttt{true}, \texttt{false}\}$. We say that a single pattern or pattern pair π *matches* string s if and only if $\psi(\pi, s) = \texttt{true}$.

For α-distance, we define four Boolean matching functions F_1, F_2, F_3, and F_4 as follows: F_1 is such that $\psi(\langle p, F_1, q \rangle, s) = \texttt{true}$ if $\psi(p, s) = \texttt{true}$ and q begins within α-distance from p, and $\psi(\langle p, F_1, q \rangle, s) = \texttt{false}$ otherwise. F_2 is such that

$\psi(\langle p, F_2, q \rangle, s) = \text{true}$ if $\psi(p, s) = \text{true}$ and q begins outside α-distance from p, and $\psi(\langle p, F_2, q \rangle, s) = \text{false}$ otherwise. F_3 is such that $\psi(\langle p, F_3, q \rangle, s) = \text{true}$ if $\psi(p, s) = \text{true}$ and q occurs within α-distance from p, and $\psi(\langle p, F_3, q \rangle, s) = \text{false}$ otherwise. F_4 is such that $\psi(\langle p, F_4, q \rangle, s) = \text{true}$ if $\psi(p, s) = \text{true}$ and q occurs outside α-distance from p, and $\psi(\langle p, F_4, q \rangle, s) = \text{false}$ otherwise. In the sequel, let us write as:

$$\langle p, F_1, q \rangle = p \wedge_{b(\alpha)} q, \quad \langle p, F_2, q \rangle = p \wedge_{b(\alpha)} \neg q,$$
$$\langle p, F_3, q \rangle = p \wedge_{c(\alpha)} q, \quad \langle p, F_4, q \rangle = p \wedge_{c(\alpha)} \neg q.$$

Note that, by setting $\alpha \geq |s|$, we always have $D(p, s, \alpha) = [1, |s|]$. Therefore, the above pattern pairs are generalizations of pattern pairs such as $(p \wedge q)$ and $(p \wedge \neg q)$ that are considered in our previous work [9].

Given a set $S = \{s_1, \ldots, s_m\}$ of strings, let $M(\pi, S)$ denote the subset of strings in S that π matches, that is,

$$M(\pi, S) = \{s_i \in S \mid \psi(\pi, s_i) = \text{true}\}.$$

We suppose that each $s_i \in S$ is associated with a numeric attribute value r_i. For a single pattern or pattern pair π, let $\sum_{M(\pi, S)} r_i$ denote the sum of r_i over all s_i in S such that $\psi(\pi, s_i) = \text{true}$, that is,

$$\sum_{M(\pi, S)} r_i = \sum_{s_i \in S} (r_i \mid \psi(\pi, s_i) = \text{true}).$$

As S is usually fixed in our case, we shall omit S where possible and let $M(\pi)$ and $\sum_{M(\pi)} r_i$ be a shorthand for $M(\pi, S)$ and $\sum_{M(\pi, S)} r_i$, respectively.

2.2   Problem Definition

In general, the problem of finding a good pattern from a given set S of strings refers to finding a pattern π that maximizes some suitable scoring function *score* with respect to the strings in S. We concentrate on *score* that takes parameters of type $|M(\pi)|$ and $\sum_{M(\pi)} r_i$, and assume that the score value computation itself can be done in constant time if the required parameter values are known. We formulate the pattern pair discovery problem as follows:

Problem 1. Given a set $S = \{s_1, \ldots, s_m\}$ of m strings, where each string s_i is assigned a numeric attribute value r_i, a threshold parameter $\alpha \geq 0$, a scoring function *score* : $\mathbf{R} \times \mathbf{R} \to \mathbf{R}$, and a Boolean Function $F \in \{F_1, \ldots, F_4\}$, find the Boolean pattern pair $\pi \in \{\langle p, F, q \rangle \mid p, q \in \Sigma^*\}$ that maximizes $score(|M(\pi)|, \sum_{M(\pi)} r_i)$.

The specific choice of the scoring function depends highly on the particular application, for example, as follows: The positive/negative sequence set discrimination problem [1, 14, 8, 15] is such that, given two disjoint sets of sequences S_1 and S_2, where sequences in S_1 (the positive set) are known to have some biological function, while the sequences in S_2 (the negative set) are known not to, find pattern pairs which match more sequences in S_1, and less in S_2. Common scoring functions that are used in this situation include the entropy information gain, the Gini index, and the chi-square statistic, which all are essentially functions of $|M(\pi, S_1)|$, $|M(\pi, S_2)|$, $|S_1|$ and $|S_2|$. The correlated patterns problem [16–18, 2, 19] is such that we are given a set S of sequences, with a numeric attribute value r_i associated with each sequence $s_i \in S$, and the task is to find pattern pairs whose occurrences in the sequences correlate with their numeric attributes. In this framework, scoring functions such as mean squared error, which is a function of $|M(\pi)|$ and $\sum_{M(\pi)} r_i$, can be used.

2.3   Data Structures

We intensively use the following data structures in our algorithms. The efficiency of our algorithms, in both time and space consumptions, comes from the use of these data structures.

The *suffix tree* [20] for a string s is a rooted tree whose edges are labeled with substrings of s, satisfying the following characteristics. For any node v in the suffix tree, let $l(v)$ denote the string spelled out by concatenating the edge labels on the path from the root to v. For each leaf v, $l(v)$ is a distinct suffix of s, and for each suffix in s, there exists such a leaf v. Each leaf v is labeled with $pos(v) = i$ such that $l(v) = s[i : |s|]$. Also, each internal node has at least two children, and the first character of the labels on the edges to its children are distinct. It is well known that suffix trees can be represented in linear space and constructed in linear time with respect to the length of the string [20].

The *generalized suffix tree* (*GST*) for a set $S = \{s_1, \ldots, s_m\}$ of strings is basically the suffix tree for the string $s_1\$_1 \cdots s_m\$_m$, where each $\$_i$ ($1 \le i \le m$) is a distinct character which does not appear in any of the strings in the set. However, all paths are ended at the first appearance of any $\$_i$, and each leaf is labeled with id_i that specifies the string in S the leaf is associated with. For set $\{\texttt{abab}, \texttt{aabb}\}$, the corresponding GST is shown in Fig. 2. GSTs are also linear time constructible and can be stored in linear space with respect to the total length of the strings in the input set [20].

The *sparse suffix tree* (*SST*) [21, 22] for a string s is a rooted tree which represents a subset of the suffixes of s, i.e., a suffix tree of s from which the *dead leaves* (and corresponding internal nodes) are removed, where the dead leaves stand for the leaves that correspond to the suffixes *not* in the subset. On the other hand, the leaves existing in the SST are called the *living leaves*. An SST can be constructed in linear time by once building the full suffix tree and pruning dead leaves from it. (Direct construction of SSTs was also considered in [21, 22] but the pruning approach is sufficient for our purposes.)

The *suffix array* [23] A_s for a string s of length n, is a permutation of the integers $1, \ldots, n$ representing the lexicographic ordering of the suffixes of s. The value $A_s[i] = j$ in the array indicates that $s[j : n]$ is the ith suffix in the lexicographic ordering. The *lcp array* for a string s, denoted lcp_s, is an array of integers representing the longest common prefix lengths of adjacent suffixes in the suffix array, that is,

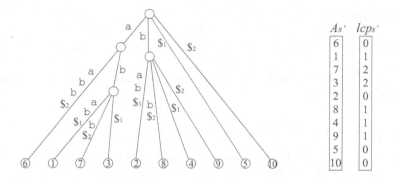

Fig. 2. The tree is the GST for set {abab, aabb}. Each leaf y is labeled by $pos(y)$ with respect to the concatenated string $s' = $ abab$\$_1$aabb$\$_2$. The left and right arrays are the suffix and lcp arrays of string s', respectively.

$$lcp_s[i] = \max\{k \mid s[A_s[i-1]:A_s[i-1]+k-1] = s[A_s[i]:A_s[i]+k-1]\}.$$

Note that each position i in A_s corresponds to a leaf in the suffix tree of s, and $lcp_s[i]$ denotes the length of the path from the root to leaves of positions $i-1$ and i for which the labels are equivalent. Fig. 2 shows the suffix array $A_{s'}$ and the lcp array $lcp_{s'}$ of string $s' = $ abab$\$_1$aabb$\$_2$.

Recently, three methods for constructing the suffix array directly from a string in linear time have been developed [25–27]. The *lcp* array can be constructed from the suffix array also in linear time [28]. It has been shown that several (and potentially many more) algorithms which utilize the suffix tree can be implemented very efficiently using the suffix array together with its *lcp* array [28, 29]. This paper presents yet another good example for efficient implementation of an algorithm based conceptually on full and sparse suffix trees, but uses the suffix and *lcp* arrays.

The *lowest common ancestor lca(x, y)* of any two nodes x and y in a tree is the deepest node which is common to the paths from the root to both of the nodes. The tree can be pre-processed in linear time to answer the lowest common ancestor (*lca-query*) for any given pair of nodes in constant time [30]. In terms of the suffix array, the *lca*-query is almost equivalent to a *range minimum query* (*rm-query*) on the *lcp* array, which, given a pair of positions i and j, $rmq(i, j)$ returns the position of the minimum element in the sub-array $lcp[i:j]$. The *lcp* array can also be pre-processed in linear time to answer the *rm*-query in constant time [30, 31].

In the sequel, we will show a clever implementation of our algorithms which simulates bottom-up traversals on SGSTs using suffix and *lcp* arrays. Our algorithms need no explicit prunings of leaves from the SGSTs which means that the suffix and *lcp* arrays do not have to be explicitly modified or reconstructed. Therefore our algorithms are both time and space efficient in practice.

3   Algorithms

In this section we present our efficient algorithms to solve Problem 1 for Boolean functions $F_1, \ldots, F_4$. Let $N = |s_1\$_1 \cdots s_m\$_m|$, where $S = \{s_1, \ldots, s_m\}$. The algorithms calculate scores for all possible pattern pairs, and output a pattern pair of the highest score.

3.1   Algorithm for Single Patterns

For a single pattern, we need to consider only patterns of form $l(v)$, where v is a node in the GST for S: If a pattern has a corresponding path that ends in the middle of an edge of the GST, it will match the same set of strings as the pattern corresponding to the end node of that same edge, and hence the score would be the same. Below, we recall the $O(N)$ time algorithm of [9] that computes $\sum_{M(l(v))} r_i$ for all single pattern candidates of form $l(v)$. The algorithm is derived from the technique for solving the *color set size problem* [32]. Note that we do not need to consider separately how to compute $|M(l(v))|$: If we give each attribute r_i the value 1, then $\sum_{M(l(v))} r_i = |M(l(v))|$.

For any node v in the GST of S, let $LF(v)$ be the set of all leaves in the subtree rooted by v, $c_i(v)$ be the number of leaves in $LF(v)$ with the label id_i, and

$$\sum_{LF(v)} r_i = \sum (c_i(v)r_i \mid \psi(l(v), s_i) = \text{true}).$$

For any such v, $\psi(l(v), s_i) = \text{true}$ for at least one string s_i. Thus, we have

$$\sum_{M(l(v))} r_i = \sum (r_i \mid \psi(l(v), s_i) = \text{true})$$

$$= \sum_{LF(v)} r_i - \sum ((c_i(v) - 1)r_i \mid \psi(l(v), s_i) = \text{true}).$$

We define the above subtracted sum to be a *correction factor*, denoted by $corr(l(v), S) = \sum ((c_i(v) - 1)r_i \mid \psi(l(v), s_i) = \text{true})$. The following recurrence

$$\sum_{LF(v)} r_i = \sum (\sum_{LF(v')} r_i \mid v' \text{ is a child of } v)$$

allows us to compute the values $\sum_{LF(v)} r_i$ for all v during a linear time bottom-up traversal of the GST. Now we need to remove the redundancies, represented by $corr(l(v), S)$, from $\sum_{LF(v)} r_i$.

Let $I(id_i)$ be the list of all leaves with the label id_i in the order they appear in a depth-first traversal of the tree. Clearly the lists $I(id_i)$ can be constructed in linear time for all labels id_i. We note four properties in the following proposition:

Proposition 1. *For the GST of S, the following properties hold:*

(1) The leaves in $LF(v)$ with the label id_i form a continuous interval of length $c_i(v)$ in the list $I(id_i)$.

(2) If $c_i(v) > 0$, a length-$c_i(v)$ interval in $I(id_i)$ contains $c_i(v) - 1$ adjacent (overlapping) leaf pairs.
(3) If $x, y \in LF(v)$, the node $lca(x, y)$ belongs to the subtree rooted by v.
(4) For any $s_i \in S$, $\psi(l(v), s_i) = $ true if and only if there is a leaf $x \in LF(v)$ with the label id_i.

We initialize each node v to have a correction value 0. Then, for each adjacent leaf pair x, y in the list $I(id_i)$, we add the value r_i into the correction value of the node $lca(x, y)$. It follows from properties (1) - (3) of Proposition 1, that now the sum of the correction values in the nodes of the subtree rooted by v equals $(c_i(v) - 1)r_i$. After repeating the process for each of the lists $I(id_i)$, property (4) tells that the preceding total sum of the correction values in the subtree rooted by v becomes

$$\sum((c_i(v) - 1)r_i \mid \psi(l(v), s_i) = \text{true}) = corr(l(v), S).$$

Now a single linear time bottom-up traversal of the tree enables us to cumulate the correction values $corr(l(v), S)$ from the subtrees into each node v, and at the same time we may record the final values $\sum_{M(l(v))} r_i$.

The preceding process involves a constant number of linear time traversals of the tree, as well as a linear number of constant time lca-queries (after a linear time preprocessing). Hence the values $\sum_{M(l(v))} r_i$ are computed in linear time.

3.2   Algorithms for Pattern Pairs with α-Distance

The algorithm described in Section 3.1 permits us to compute $\sum_{M(l(v))} r_i$ in $O(N)$ time. In this section, we concentrate on how to compute the values $\sum_{M(\pi)} r_i$ for various types of pattern pair π with bounded distance α. In so doing, we go over the $O(N)$ choices for the first pattern, and for each fixed pattern $l(v_1)$, we compute $\sum_{M(\pi)} r_i$ where π consists of $l(v_1)$ and second pattern candidate $l(v)$. The result of this section is summarized in the following theorem:

Theorem 1. *Problem 1 is solvable in $O(N^2)$ time using $O(N)$ space for $F \in \{F_1, F_2\}$, and $O(m^2 N^2)$ time and $O(m^2 N)$ space for $F \in \{F_3, F_4\}$.*

Below, we will give the details of our algorithm for each type of pattern pair.

For Pattern Pairs in Form $p \wedge_{b(\alpha)} q$. Let $D(l(v_1), S, \alpha) = \cup_{s_i \in S} D(l(v_1), s_i, \alpha)$. It is not difficult to see that $D(l(v_1), S, \alpha)$ is $O(N)$ time constructible (see Theorem 9 of [24]). We then mark as 'living' every leaf of the GST whose position belongs to $D(l(v_1), S, \alpha)$. This aims at conceptually constructing the *sparse generalized suffix tree (SGST)* for the subset of suffixes corresponding to $D(l(v_1), S, \alpha)$, and this can be done in $O(N)$ time. Then, we additionally label each string $s_i \in S$, and the corresponding leaves in the SGST, with the Boolean value $\psi(l(v_1), s_i)$. This can be done in $O(N)$ time. Now the trick is to cumulate the sums and correction factors *only* for the nodes existing in the SGST, and *separately* for different values of the additional label. The end result is that we will obtain the values $\sum_{M(\pi)} r_i = \sum(r_i \mid \psi(\pi, s_i) = \texttt{true})$ in linear time, where π denotes the pattern pair $l(v_1) \wedge_{b(\alpha)} l(v)$ for all nodes v in the conceptual SGST. Since there are only $O(N)$ candidates for $l(v_1)$, the total time complexity is $O(N^2)$. The space requirement is $O(N)$, since we repeatedly use the same GST for S and the additional information storage is also linear in N at each phase of the algorithm.

For Pattern Pairs in Form $p \wedge_{b(\alpha)} \neg q$. Let us denote any pattern pair in form $p \wedge_{b(\alpha)} \neg q$ by $\overline{\pi}$. For the pattern pairs of this form, it stands that

$$\sum_{M(\overline{\pi})} r_i = \sum(r_i \mid \psi(\overline{\pi}, s_i) = \texttt{true})$$

$$= \sum(r_i \mid \psi(p, s) = \texttt{true}) - \sum_{M(\pi)} r_i,$$

where π denotes $p \wedge_{b(\alpha)} q$. For each fixed first pattern $l(v_1)$, $\sum_{M(\pi)} r_i$ and $\sum(r_i \mid \psi(l(v_1), s) = \texttt{true})$ can be computed in $O(N)$ time, and hence $\sum_{M(\overline{\pi})} r_i$ can be computed in $O(N)$ time as well. Since we have $O(N)$ choices for the first pattern, the total time complexity is $O(N^2)$, and the space complexity is $O(N)$.

For Pattern Pairs in Form $p \wedge_{c(\alpha)} q$. There are two following types of patterns that have to be considered as candidates for the second pattern q:

(1) pattern q which has an occurrence beginning at some position in $D(l(v_1), S, \alpha)$.
(2) pattern q which has an occurrence beginning at some position in $[1, N] - D(l(v_1), S, \alpha)$, but overlapping with some positions in $D(l(v_1), S, \alpha)$.

Now, we separately treat these two kinds of patterns. In so doing, we consider two SGSTs, one consists only of the leaves x such that $pos(x) \in D(l(v_1), S, \alpha)$, and the other consists only of the leaves y such that $pos(y) \in [1, N] - D(l(v_1), S, \alpha)$. Denote these by $SGST_1$ and $SGST_2$, respectively.

For integer i with $1 \leq i \leq m$, let $N_i = |s_1 \$_1 \cdots s_i \$_i|$. For any leaf y of $SGST_2$ with id_i, we define $len(y)$ as follows:

$$len(y) = \begin{cases} N_{i-1} + h - \alpha - pos(y) & \text{if } pos(y) < N_{i-1} + k - \alpha, \\ N_i - pos(y) & \text{otherwise,} \end{cases}$$

where h denotes the minimum element of $Beg(l(v_1), s_i)$ satisfying $N_{i-1} + h > pos(y)$, and $k = \max(Beg(l(v_1), s_i))$. Fig. 3 illustrates $len(y)$ with respect to s_i and $l(v_1)$. Computing $len(y)$ can be done in constant time for each leaf y, after a linear time preprocessing of scanning $D(l(v_1), S, \alpha)$ from right to left. For any node v of $SGST_2$, let $SLF_i(v)$ denote the list of all leaves that are in the subtree of $SGST_2$ rooted at v and are labeled with id_i. Then, we compute $len_i(v) = \min_{y \in SLF_i(v)}(len(y))$ for every internal node v in $SGST_2$. For all nodes

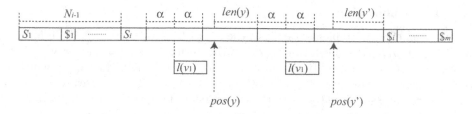

Fig. 3. For two leaves y, y' of $SGST_2$, $len(y)$ and $len(y')$ with respect to string $s_i \in S$ and the first pattern $l(v_1)$.

v and $i = 1, \ldots, m$, $len_i(v)$ can be computed in $O(mN)$ time using $O(mN)$ space. Notice that only nodes v such that for some i, $len_i(v) < |l(v)|$ in $SGST_2$, can give candidates for the second pattern, since the occurrences of the substrings represented by the other nodes are completely outside $D(l(v_1), S, \alpha)$.

Consider an internal node v' and its parent node v in $SGST_2$. Assume the length of the edge between v and v' is more than one. Suppose the subtree of $SGST_2$ rooted at v' contains a leaf with id_i. Now consider an 'implicit' node t that is on the edge from v to v'. Then, even if $len_i(v') < |l(v')|$, it might stand that $len_i(t) \geq |l(t)|$ since $len_i(t) = len_i(v')$ and $|l(t)| < |l(v')|$ always hold. Note that in this case we have to take into account such implicit node t for the second pattern candidate, independently from v'. On the other hand, if we locate t so that $len_i(t) = |l(t)|$, we will have $len_i(t') > |l(t')|$ for any implicit node t' upper than t in the edge, and will have $len_i(t'') < |l(t'')|$ for any implicit node t'' lower than t in the edge. Thus this single choice of t suffices. Since the subtree can contain at most m distinct id's, each edge can have at most m such implicit nodes on it, and because $SGST_2$ has $O(N)$ edges, the total number of the implicit nodes we have to treat becomes $O(mN)$. What still remains is how to locate these implicit nodes in $O(mN)$ time. This is actually feasible by using the so-called *suffix links* and the *canonization* technique introduced in [33]. For each of these implicit nodes we have to maintain m number of information as mentioned in the above paragraph, and hence in total we are here required $O(m^2N)$ time and space.

Cumulating the sums and correction factors is done in two rounds: For simplicity, let us denote pattern pair $l(v_1) \wedge_{c(\alpha)} l(v)$ by π. First we traverse $SGST_1$ and for each encountered node v, calculate the value $\sum_{M(\pi),1} r_i$ and store it in v. Then, we traverse $SGST_2$ and for each encountered node w calculate the value $\sum_{M(\pi),2} r_i$ if necessary, while checking whether $len_i(w) < |l(w)|$ or not, and store it in w but *separately* from $\sum_{M(\pi),1} r_i$. Then, the final value $\sum_{M(\pi)} r_i = \sum_{M(\pi),1} r_i + \sum_{M(\pi),2} r_i$. Since we go over $O(N)$ choices for the first pattern, the resulting algorithm requires $O(m^2N^2)$ time and $O(m^2N)$ space.

For Pattern Pairs in Form $p \wedge_{c(\alpha)} \neg q$. As similar arguments to pattern pairs in form $p \wedge_{b(\alpha)} \neg q$ hold, it can also be done $O(m^2N^2)$ time and $O(m^2N)$ space.

3.3   Implementation

We show an efficient implementation of the algorithms for pattern pairs in form $p \wedge_{b(\alpha)} q$ and $p \wedge_{b(\alpha)} \neg q$, both run in $O(N^2)$ time using $O(N)$ space as previously explained. We further modify the implementation of the algorithm in [9], which is an extension of the Substring Statistics algorithm in [28], that uses the suffix and lcp arrays to simulate the bottom up traversals of the generalized suffix tree. We note that the simulation does not increase the asymptotic complexity of the algorithm, but rather it is expected to increase the efficiency of the traversal, as confirmed in previous works [34, 28, 29].

Fig. 4 shows a pseudo-code for solving the Color Set Size problem with pruned edges using a suffix array, which is used in the algorithm of the previous subsection. There are two differences from [9]: (1) the correction factors are set to $lca(i, j)$ of consecutive 'living' leaves i, j ($i < j$) corresponding to the same string. Since the simulation via suffix arrays does not explicitly construct all internal nodes of the suffix tree, we use another array CF to hold the correction factors. The correction factor for these leaves i, j is summed into $CF[rmq(i + 1, j)]$. (2) 'Dead' leaves are ignored in the bottom-up traversal. This can be done in several ways, the simplest which assigns 0 to the weights of each 'dead' leaf. However, although the worst-case time complexity remains the same, such an algorithm would report *redundant* internal nodes that only have at most a single 'living' leaf in its subtree. A more complete pruning can be achieved, as shown in the pseudo-code, in the following way: for any consecutive 'living' leaves i, j ($i < j$) in the suffix array, we add up all $CF[k]$ for $i < k \le j$ which is then added to the node corresponding to $lca(i, j)$. The correctness of this algorithm can be confirmed by the following argument: For a given correction factor $CF[k]$ ($i < k \le j$), since the query result came from consecutive 'living' leaves of the same string, they must have been from leaves i', j', such that $i' \le i$ and $j \le j'$. This means that $lcp[k] = lcp[rmq(i' + 1, j')] \le lcp[rmq(i + 1, j)] \le lcp[k]$. Therefore $lcp[rmq(i' + 1, j')] = lcp[rmq(i + 1, j)] = lcp[k]$, and $lca(i', j') = lca(i, j)$, which shows that $CF[k]$ should be summed into the node with depth $lcp[rmq(i+1, j)]$.

4   Computational Experiments

We tested the effectiveness of our algorithm using the two sets of predicted 3'UTR processing site sequences provided in [35], which are constructed based on the microarray experiments in [36] that measure the degradation rate of yeast mRNA. One set S_f consists of 393 sequences which have a fast degradation rate ($t_{1/2} < 10$ minutes), while the other set S_s consists of 379 predicted 3'UTR processing site sequences which have a slow degradation rate ($t_{1/2} > 50$ minutes). Each sequence is 100 nt long, and the total length of the sequences is $77,200$ nt. For the scoring function, we used the chi-squared statistic, calculated by $(|S_f| + |S_s|)(\mathtt{tp} * \mathtt{tn} - \mathtt{fp} * \mathtt{fn})^2 / (\mathtt{tp} + \mathtt{fn})(\mathtt{tp} + \mathtt{fp})(\mathtt{tn} + \mathtt{fp})(\mathtt{tn} + \mathtt{fn})$ where $\mathtt{tp} = |M(\pi, S_f)|$, $\mathtt{fp} = |S_f| - \mathtt{tp}$, $\mathtt{tn} = |S_s| - \mathtt{fn}$, and $\mathtt{fn} = |M(\pi, S_s)|$.

For $b(\alpha)$ and $1 \le \alpha \le 30$, the best score was 48.3 ($p < 10^{-11}$) for the pair AUA $\wedge_{b(10)} \neg$UGUA, which matched 159/393 and 248/379 fast and slowly degrading

```
1   Let Stack = {(0, -1, 0)} be the stack;
2   cc := 0; iplus1 = 0;
3   foreach j = 1,..., n + 1 do:
4       cc := cc + CF[j];
5       (L, H, C) = top(Stack);
6       if j is a living leaf
7           (Lj, Hj, Cj) := (j, lcp[rmq(iplus1, j)], 0);
8           while (H > Hj) do:
9               (L, H, C) := pop(Stack);
10              report (L, H, C); /* s[A[L] : A[L] + H - 1] has count C */
11              Cj := C + Cj;
12              (L, H, C) := top(Stack);
13          if (H < Hj) then
14              push((Lj, Hj, Cj + cc), Stack);
15          else /* H = Hj */
16              (L, H) := pop(Stack);
17              push((L, H, C + Cj + cc), Stack);
18          push((j, N - A[j] + 1, 1), Stack);
19      cc := 0; iplus1 = j + 1;
```

Fig. 4. Core of the algorithm for solving the Color Set Size problem with pruned edges using a suffix array. We assume each leaf (position in array) has been labeled 'dead' or 'living', and the correction factors for 'living' leaves are stored in the array CF. A node in the suffix tree together with $|M(l(v))|$ is represented by a three-tuple (L,H,C).

sequences respectively, meaning that this pattern pair is more frequent in the slowly degrading sequences. This result is similar to the best pattern of the form $p \wedge \neg q$ obtained in [9], which was UGUA $\wedge \neg$AUCC with score 33.9 (240/393 and 152/379 ($p < 10^{-8}$)), appearing more frequently in the fast degrading sequences. In fact, UGUA is known as a binding site of the PUF protein family which plays important roles in mRNA regulation [37]. Of the 268/393 and 190/379 sequences that contain both AUA and UGUA, their distances were farther than $\alpha = 10$ in only 37/268 of fast degrading sequences, while the number was 67/190 for slowly degrading sequences. This could mean that AUA is another sequence element whose distance from UGUA influences how efficiently UGUA functions.

We did not observe any notable difference in computational time compared to the algorithm in [9], that the marking of each leaf as 'dead' or 'living' could cause. For a given value of α, computation took around 2140 seconds for the above dataset on a PC with Xeon 3 GHz Processor running Linux.

5   Concluding Remarks

In this paper we studied the problem of finding the optimal pair of substring patterns p, q with bounded distance α, from a given set S of strings, which is an extension of the problem studied in [9]. We developed an efficient algorithm that finds the best pair of *cooperative* patterns in form $p \wedge_{b(\alpha)} q$, and *competing*

Fig. 5. From pattern p in string s, q_1 begins within α-gap, q_2 begins outside α-gap, q_3 occurs within α-gap, and q_4 occurs outside α-gap.

patterns in form $p \wedge_{b(\alpha)} \neg q$, which performs in $O(N^2)$ time using $O(N)$ space, where N is the total length of the strings in S. For the more difficult versions of the problem, referred to finding $p \wedge_{c(\alpha)} q$ and $p \wedge_{c(\alpha)} \neg q$, we gave an algorithm running in $O(m^2 N^2)$ time and $O(m^2 N)$ space, where $m = |S|$. An interesting open problem is if the m-factors can be removed from the complexities.

Any pairs output from our algorithms are guaranteed to be optimal in the pattern classes, in the sense that they give the highest score due to the scoring function. Our algorithms are adapted to various applications such as the *positive/negative sequence set discrimination problem* for which the entropy information gain, the Gini index, and the chi-square statistic are commonly used as the scoring function, and the *correlated patterns problem* for which the mean squared error can be used.

It is possible to use our algorithms for bounded *gaps* α (see Fig. 5). Since the ending positions of the first patterns have to be considered in this case, we will naturally have $O(N^2)$ choices for the first pattern, and this fact suggests the use of suffix tries rather than suffix trees for the first patterns. Still, there are only $O(N)$ choices for the second pattern, and thus for the pattern pairs with respect to $b(\alpha)$, we can modify our algorithm to solve the problems in $O(N^3)$ time using $O(N^2)$ space. We remark that for $c(\alpha)$, a modified algorithm finds the optimal pattern pair in $O(m^2 N^3)$ time using $O(N^2 + m^2 N)$ space.

References

1. Shimozono, S., Shinohara, A., Shinohara, T., Miyano, S., Kuhara, S., Arikawa, S.: Knowledge acquisition from amino acid sequences by machine learning system BONSAI. Transactions of Information Processing Society of Japan **35** (1994) 2009–2018
2. Bannai, H., Inenaga, S., Shinohara, A., Takeda, M., Miyano, S.: Efficiently finding regulatory elements using correlation with gene expression. Journal of Bioinformatics and Computational Biology **2** (2004) 273–288
3. Baeza-Yates, R.A.: Searching subsequences (note). Theoretical Computer Science **78** (1991) 363–376
4. Hirao, M., Hoshino, H., Shinohara, A., Takeda, M., Arikawa, S.: A practical algorithm to find the best subsequence patterns. In: Proc. 3rd International Conference on Discovery Science (DS'00). Volume 1967 of LNAI., Springer-Verlag (2000) 141–154
5. Mannila, H., Toivonen, H., Verkamo, A.I.: Discovering frequent episode in sequences. In: Proc. 1st International Conference on Knowledge Discovery and Data Mining, AAAI Press (1995) 210–215

6. Hirao, M., Inenaga, S., Shinohara, A., Takeda, M., Arikawa, S.: A practical algorithm to find the best episode patterns. In: Proc. 4th International Conference on Discovery Science (DS'01). Volume 2226 of LNAI., Springer-Verlag (2001) 435–440

7. Inenaga, S., Bannai, H., Shinohara, A., Takeda, M., Arikawa, S.: Discovering best variable-length-don't-care patterns. In: Proc. 5th International Conference on Discovery Science (DS'02). Volume 2534 of LNCS., Springer-Verlag (2002) 86–97

8. Takeda, M., Inenaga, S., Bannai, H., Shinohara, A., Arikawa, S.: Discovering most classificatory patterns for very expressive pattern classes. In: Proc. 6th International Conference on Discovery Science (DS'03). Volume 2843 of LNCS., Springer-Verlag (2003) 486–493

9. Bannai, H., Hyyrö, H., Shinohara, A., Takeda, M., Nakai, K., Miyano, S.: Finding optimal pairs of patterns. In: Proc. 4th Workshop on Algorithms in Bioinformatics (WABI'04). (2004) to appear.

10. Marsan, L., Sagot, M.F.: Algorithms for extracting structured motifs using a suffix tree with an application to promoter and regulatory site consensus identification. J. Comput. Biol. **7** (2000) 345–360

11. Palopoli, L., Terracina, G.: Discovering frequent structured patterns from string databases: an application to biological sequences. In: 5th International Conference on Discovery Science (DS'02). Volume 2534 of LNCS., Springer-Verlag (2002) 34–46

12. Arimura, H., Arikawa, S., Shimozono, S.: Efficient discovery of optimal word-association patterns in large text databases. New Generation Computing **18** (2000) 49–60

13. Arimura, H., Asaka, H., Sakamoto, H., Arikawa, S.: Efficient discovery of proximity patterns with suffix arrays (extended abstract). In: Proc. the 12th Annual Symposium on Combinatorial Pattern Matching (CPM'01). Volume 2089 of LNCS., Springer-Verlag (2001) 152–156

14. Shinohara, A., Takeda, M., Arikawa, S., Hirao, M., Hoshino, H., Inenaga, S.: Finding best patterns practically. In: Progress in Discovery Science. Volume 2281 of LNAI., Springer-Verlag (2002) 307–317

15. Shinozaki, D., Akutsu, T., Maruyama, O.: Finding optimal degenerate patterns in DNA sequences. Bioinformatics **19** (2003) 206ii–214ii

16. Bussemaker, H.J., Li, H., Siggia, E.D.: Regulatory element detection using correlation with expression. Nature Genetics **27** (2001) 167–171

17. Bannai, H., Inenaga, S., Shinohara, A., Takeda, M., Miyano, S.: A string pattern regression algorithm and its application to pattern discovery in long introns. Genome Informatics **13** (2002) 3–11

18. Conlon, E.M., Liu, X.S., Lieb, J.D., Liu, J.S.: Integrating regulatory motif discovery and genome-wide expression analysis. Proc. Natl. Acad. Sci. **100** (2003) 3339–3344

19. Zilberstein, C.B.Z., Eskin, E., Yakhini, Z.: Using expression data to discover RNA and DNA regulatory sequence motifs. In: The First Annual RECOMB Satellite Workshop on Regulatory Genomics. (2004)

20. Gusfield, D.: Algorithms on Strings, Trees, and Sequences. Cambridge University Press (1997)

21. Andersson, A., Larsson, N.J., Swanson, K.: Suffix trees on words. Algorithmica **23** (1999) 246–260

22. Kärkkänen, J., Ukkonen, E.: Sparse suffix trees. In: Proceedings of the Second International Computing and Combinatorics Conference (COCOON'96). Volume 1090 of LNCS., Springer-Verlag (1996) 219–230

23. Manber, U., Myers, G.: Suffix arrays: a new method for on-line string searches. SIAM Journal on Computing **22** (1993) 935–948
24. Inenaga, S., Kivioja, T., Mäkinen, V.: Finding missing patterns. In: Proc. 4th Workshop on Algorithms in Bioinformatics (WABI'04). (2004) to appear.
25. Kim, D.K., Sim, J.S., Park, H., Park, K.: Linear-time construction of suffix arrays. In: 14th Annual Symposium on Combinatorial Pattern Matching (CPM'03). Volume 2676 of LNCS., Springer-Verlag (2003) 186–199
26. Ko, P., Aluru, S.: Space efficient linear time construction of suffix arrays. In: 14th Annual Symposium on Combinatorial Pattern Matching (CPM'03). Volume 2676 of LNCS., Springer-Verlag (2003) 200–210
27. Kärkkäinen, J., Sanders, P.: Simple linear work suffix array construction. In: 30th International Colloquium on Automata, Languages and Programming (ICALP'03). Volume 2719 of LNCS., Springer-Verlag (2003) 943–955
28. Kasai, T., Arimura, H., Arikawa, S.: Efficient substring traversal with suffix arrays. Technical Report 185, Department of Informatics, Kyushu University (2001)
29. Abouelhoda, M.I., Kurtz, S., Ohlebusch, E.: The enhanced suffix array and its applications to genome analysis. In: Second International Workshop on Algorithms in Bioinformatics (WABI'02). Volume 2452 of LNCS., Springer-Verlag (2002) 449–463
30. Bender, M.A., Farach-Colton, M.: The LCA problem revisited. In: Proc. 4th Latin American Symp. Theoretical Informatics (LATIN'00). Volume 1776 of LNCS., Springer-Verlag (2000) 88–94
31. Alstrup, S., Gavoille, C., Kaplan, H., Rauhe, T.: Nearest common ancestors: a survey and a new distributed algorithm. In: 14th annual ACM symposium on Parallel algorithms and architectures (SPAA'02). (2002) 258–264
32. Hui, L.: Color set size problem with applications to string matching. In: Proc. 3rd Annual Symposium on Combinatorial Pattern Matching (CPM'92). Volume 644 of LNCS., Springer-Verlag (1992) 230–243
33. Ukkonen, E.: On-line construction of suffix trees. Algorithmica **14** (1995) 249–260
34. Kasai, T., Lee, G., Arimura, H., Arikawa, S., Park, K.: Linear-time longest-common-prefix computation in suffix arrays and its applications. In: 12th Annual Symposium on Combinatorial Pattern Matching (CPM'01). Volume 2089 of LNCS., Springer-Verlag (2001) 181–192
35. Graber, J.: Variations in yeast 3'-processing cis-elements correlate with transcript stability. Trends Genet. **19** (2003) 473–476 http://harlequin.jax.org/yeast/turnover/.
36. Wang, Y., Liu, C., Storey, J., Tibshirani, R., Herschlag, D., Brown, P.: Precision and functional specificity in mRNA decay. Proc. Natl. Acad. Sci. **99** (2002) 5860–5865
37. Wickens, M., Bernstein, D.S., Kimble, J., Parker, R.: A PUF family portrait: 3'UTR regulation as a way of life. Trends Genet. **18** (2002) 150–157

Mining Noisy Data Streams via a Discriminative Model

Fang Chu, Yizhou Wang, and Carlo Zaniolo

University of California, Los Angeles CA 90095, USA

Abstract. The two main challenges typically associated with mining data streams are concept drift and data contamination. To address these challenges, we seek learning techniques and models that are robust to noise and can adapt to changes in timely fashion. In this paper, we approach the stream-mining problem using a statistical estimation framework, and propose a discriminative model for fast mining of noisy data streams. We build an ensemble of classifiers to achieve adaptation by weighting classifiers in a way that maximizes the likelihood of the data. We further employ robust statistical techniques to alleviate the problem of noise sensitivity. Experimental results on both synthetic and real-life data sets demonstrate the effectiveness of this new discriminative model.

Keywords: Data mining, discriminative modelling, robust statistics, logistic regression

1 Introduction

Recently there is a substantial research interest on learning data streams: the type of data that arrives continuously in high volume and speed. Applications involving stream data abound, such as network traffic monitoring, credit card fraud detection and stock market trend analysis. The work presented here focuses on the primary challenges of data stream learning: concept drift and data contamination.

Data in traditional learning tasks is stationary. That is, the underlying concept that maps the attributes to the class labels is unchanging. With data streams, however, the concept is not stable but drift with time due to changes in the environment. For example, customer purchase preferences change with season, availability of alternatives or services. Often the changes make the model learned from old data inconsistent with the new data, and updating of the model is necessary. This is generally known as *concept drift* [18].

Data contamination is a practical problem, introduced either due to unreliable data sources or during data transmission. Noise can greatly impair data learning. It is an even more severe problem in the context of data streams because it interferes with change adaptation. If an algorithm is too greedy to adapt to concept changes, it may overfit noise by mistakenly interpreting it as data from a new concept. If one is too robust to noise, it may overlook the changes and adjust slowly.

In this paper, we propose a novel discriminative model for adaptive learning of noisy data streams. The goal is to combine adaptability to concept drift and robustness to noise. The model produced by the learning process is in the form of an ensemble. Our work contributes in two aspects: (1) we formulate the classifier weighting problem as a regression problem, targeting at likelihood maximization of data from current concept; and, (2) we integrate outlier detection into the overall model learning.

E. Suzuki and S. Arikawa (Eds.): DS 2004, LNAI 3245, pp. 47–59, 2004.

2   Related Work

The problem of concept drift has been addressed in both machine learning and data mining communities. The first system capable of handling concept drift was STAG-GER [14], which maintains a set of concept descriptions. Later on, there were the instance-based algorithm IB3 [2], which stores a set of instances and eliminate those when they become irrelevant or out-dated. The FLORA family [18] keeps a rule system in step with the hidden concept through a window of recent examples, and alters the window size according to changes in prediction accuracy and concept complexity. Although these algorithms introduced an interesting problem and proposed valuable insights, they were developed and tested only on very small datasets. It is not very clear how suitable they are for data streams which are on a significantly larger scale than previously studied.

Several scalable learning algorithms designed for data streams were proposed recently. They either maintain a single model incrementally, or an ensemble of base learners. Work belonging to the first category includes Hoeffding tree [9], which grows a decision tree node by splitting an attribute only when that attribute is statistically predictive. The statistics is recorded and periodically checked to discover possible concept changes. However, many examples are needed to decide a split, hence the algorithm requires a large number of training samples to reach a fair performance. Domeniconi and Gunopulos [7] designed an incremental support vector machine algorithm for continuous learning. Experimental results suggest that SVM is not the ideal choice for fast learning on stream data.

Now we focus on ensemble-based approaches. Kolter and Maloof [10] proposed to track concept drift by an ensemble of experts. The poor experts are weighted low or discarded, and the good experts are updated using recent examples. One problem with this approach is the incremental base learner. Incremental algorithms are not readily available, and their updating is not easy without restricting assumptions that are impractical in many situations. For example, an assumption is made in [10] that the values of each attribute follow certain Gaussian distribution.

Other effective ensemble-based approaches simply partition the data stream into sequential blocks of fixed size and learn an ensemble from these blocks. Once a base model is constructed, it will never be updated with new examples. Two types of voting schemes are adopted. Street et al. [16] let their ensemble vote uniformly, while Wang et al. [17] prefer a weighted voting. These two approaches, [16] and [17], are closely related to what we will present in this paper, as our method also build an ensemble from sequential blocks. In our comparative study, we will refer to these two methods as *Bagging* and *Weighted Bagging*, respectively[1].

What is missing from the above mentioned work is a mechanism dealing with noise. Although there have been a number of off-line algorithms [1, 13, 5, 11, 6] for noise identification, or often indistinguishably called *outlier* detection, the concept drift problem creates a substantial gap between stream data and the existing techniques. In addition, we address the problem of finding a general distance metric. Popular distance func-

---

[1] These methods conform to the traditional bagging [4] and online bagging ensembles [12], where training samples are not weighted.

tions include the Euclidean distance or other distribution-sensitive functions, which are unable to treat categorical values.

Our work differs in two aspects from previous approaches to outlier detection. First, we tightly integrate outlier detection into the model learning process, since outliers also drift as concept drifts away. Secondly, the distance metric is derived from the classification performance of the ensemble, instead of a function defined in the data space. The adaptive model learning and outlier detection therefore mutually reinforce each other. An accurate model helps to identify the outliers. On the other hand, by correctly identifying and eliminating the outliers, a more accurate model can be computed.

In section 3 and section 4, we will describe the discriminative model with regard to adaptation and robustness, respectively. Section 5 gives the model formulation and computation. Extensive experimental results will be shown in section 6.

3   Adaptation to Concept Drift

Ensemble weighting is the key to fast adaptation. In this section we show that this problem can be formulated as a statistical optimization problem solvable by logistic regression.

We first look at how an ensemble is constructed and maintained. The data stream is simply broken into small blocks, then a classifier can be learned from each block. The ensemble is comprised of the most recent K classifiers. Old classifiers retire sequentially by age. Besides training examples, evaluation examples (also within known class labels) are needed for weighting classifier. If training data is sufficient, part of it can be reserved for evaluation; otherwise, random samples of the most recent data blocks can serve the purpose. We only need to make the two data sets as synchronized as possible. When sufficient data is collected for training and evaluation, we do the following operations: (1) learn a new classifier from the training block; (2) replace the oldest classifier in the ensemble with this newly learned; and (3) use the evaluation data to weigh the ensemble.

The rest of this section will give a formal description of ensemble weighting. For simplicity, a two-class classification setting is considered, but the treatment can be extended to multi-class tasks.

The evaluation data is represented as

$$(\mathcal{X}, \mathcal{Y}) = \{(\mathbf{x}_i, y_i); i = 1, \cdots, N\}$$

with $\mathbf{x}_i$ a vector valued sample attribute and $y_i \in \{0, 1\}$ the sample class label. We assume an ensemble of classifiers, denoted in a vector form as

$$\mathbf{f} = (f_1(\mathbf{x}), \cdots, f_K(\mathbf{x}))^T$$

where each $f_k(\mathbf{x})$ is a classifier function producing a value for the belief on a class. The individual classifiers in the ensemble may be weak or out-of-date. It is the goal of our discriminative model $\mathcal{M}$ to make the ensemble strong by weighted voting. Classifier weights are model parameters, denoted as

$$\mathbf{w} = (w_1, \cdots, w_K)^T$$

where w_k is the weight associated with classifier f_k. The model $\mathcal{M}$ also specifies for decision making a weighted voting scheme, that is,

$$\mathbf{w}^T \cdot \mathbf{f}$$

Because the ensemble prediction $\mathbf{w}^T \cdot \mathbf{f}$ is a continuous value, yet the class label y_i to be decided is discrete, a standard approach is to assume that y_i conditionally follows a Bernoulli distribution parameterized by a latent score η_i:

$$y_i|\mathbf{x}_i; \mathbf{f}, \mathbf{w} \sim \text{Ber}(q(\eta_i))$$
$$\eta_i = \mathbf{w}^T \cdot \mathbf{f} \tag{1}$$

where $q(\eta_i)$ is the logit transformation of η_i:

$$q(\eta_i) \triangleq \text{logit}(\eta_i) = \frac{e^{\eta_i}}{1 + e^{\eta_i}}$$

Eq.(1) states that y_i follows a Bernoulli distribution with parameter q, thus the posterior probability is

$$p(y_i|\mathbf{x}_i; \mathbf{f}, \mathbf{w}) = q^{y_i}(1 - q)^{1 - y_i} \tag{2}$$

The above description leads to optimizing classifier weights using logistic regression. Given a data set $(\mathcal{X}, \mathcal{Y})$ and an ensemble $\mathbf{f}$, the logistic regression technique optimizes the classifier weights by maximizing the likelihood of the data. The optimization problem has a closed-form solution which can be quickly solved. We postpone the detailed model computation till section 5.

Logistic regression is a well-established regression method, widely used in traditional areas when the regressors are continuous and the responses are discrete [8]. In our work, we formulate the classifier weighting problem as an optimization problem and solve it using logistic regression. In section 6 we shows that such a formulation and solution provide much better adaptability for stream data mining, as compared to other weighting schemes such as the intuitive weighting based on accuracies. (Refer to Fig.1-2, section 6 for a quick reference.)

4   Robustness to Outliers

The reason regression is adaptive to changes is that it always tries to fit the data from the current concept. But, noise overfitting is a potential problem. Therefore, we propose the following outlier detection method as an integral part of the model learning.

We interpret outliers as samples with very small likelihoods under a given data model. The goal of learning is to compute a model that best fits the bulk of data, that is, the non-outliers. Outliers are hidden information in this problem. This suggest us to solve the problem under the EM framework, using a robust statistics formulation.

In Section 3, we have described a data set as $\{(\mathbf{x}_i, y_i), i = 1, \cdots, N\}$, or $(\mathcal{X}, \mathcal{Y})$. This is an *incomplete* data set, as the outlier information is missing. A *complete* data set is a triplet

$$(\mathcal{X}, \mathcal{Y}, \mathcal{Z})$$

where

$$\mathcal{Z} = \{z_1, \cdots, z_N\}$$

is a hidden variable that distinguishes the outliers from the clean ones. $z_i = 1$ if $(\mathbf{x}_i, y_i)$ is an outlier, $z_i = 0$ otherwise. This $\mathcal{Z}$ is not observable and needs to be inferred. After the values of $\mathcal{Z}$ are inferred, $(\mathcal{X}, \mathcal{Y})$ can be partitioned into a clean sample set

$$(\mathcal{X}_0, \mathcal{Y}_0) = \{(\mathbf{x}_i, y_i, z_i), \mathbf{x}_i \in \mathcal{X}, y_i \in \mathcal{Y}, z_i = 0\}$$

and an outlier set

$$(\mathcal{X}_\phi, \mathcal{Y}_\phi) = \{(\mathbf{x}_i, y_i, z_i), \mathbf{x}_i \in \mathcal{X}, y_i \in \mathcal{Y}, z_i = 1\}$$

It is the samples in $(\mathcal{X}_0, \mathcal{Y}_0)$ that all come from one underlying distribution, and are used to fit the model parameters.

To infer the outlier indicator $\mathcal{Z}$, we introduce a new model parameter λ. It is a threshold value of sample likelihood. A sample is marked as an outlier if its likelihood falls below λ. This λ, together with $\mathbf{f}$ (classifier functions) and $\mathbf{w}$ (classifier weights) discussed earlier, constitutes the complete set of parameters of our discriminative model $\mathcal{M}$, denoted as $\mathcal{M}(\mathbf{x}; \mathbf{f}, \mathbf{w}, \lambda)$.

5   Our Discriminative Model

In this section, we give the model formulation followed by model computation. The symbols used are summarized in table 1.

Table 1. Summary of symbols used.

$(\mathbf{x}_i, y_i)$	a sample, with $\mathbf{x}_i$ the sample attribute, y_i the sample class label,
$(\mathcal{X}, \mathcal{Y})$	an incomplete data set with outlier information missing,
$\mathcal{Z}$	a hidden variable,
$(\mathcal{X}, \mathcal{Y}, \mathcal{Z})$	a complete data set with outlier information,
$(\mathcal{X}_0, \mathcal{Y}_0)$	a clean data set,
$(\mathcal{X}_\phi, \mathcal{Y}_\phi)$	an outlier set,
$\mathcal{M}$	the discriminative model,
$\mathbf{f}$	a vector of classifier function, a model parameter,
$\mathbf{w}$	a vector of classifier weights, a model parameter,
λ	a threshold of likelihood, a model parameter.

5.1   Model Formulation

Our model is a four-tuple representation $\mathcal{M}(\mathbf{x}; \mathbf{f}, \mathbf{w}, \lambda)$. Given an evaluation data set $(\mathcal{X}, \mathcal{Y})$, an ensemble of classifiers $\mathbf{f} = (f_1(\mathbf{x}), \cdots, f_K(\mathbf{x}))^T$, we want to achieve two objectives.

1. To infer about the hidden variable $\mathcal{Z}$ that distinguishes non-outliers $(\mathcal{X}_0, \mathcal{Y}_0)$ from outliers $(\mathcal{X}_\phi, \mathcal{Y}_\phi)$.
2. To compute the optimal fit for model parameters $\mathbf{w}$ and λ in the discriminative model $M(\mathbf{x}; \mathbf{f}, \mathbf{w}, \lambda)$.

An assumption is made that each non-outlier sample $(\mathbf{x}_i, y_i) \in (\mathcal{X}_0, \mathcal{Y}_0)$ is drawn from an independent identical distribution belonging to a probability family characterized by parameters $\mathbf{w}$, denoted by a density function $p((\mathbf{x}, y); \mathbf{f}, \mathbf{w})$. The problem is to find the values of $\mathbf{w}$ that maximizes the likelihood of $(\mathcal{X}_0, \mathcal{Y}_0)$ in the probability family. As customary, we use log-likelihood to simplify the computation:

$$\log\, p((\mathcal{X}_0, \mathcal{Y}_0)|\mathbf{f}, \mathbf{w})$$

A parametric model for outlier distribution is not available, due to the highly irregularity of the outlier data. Therefore, we use instead a non-parametric statistics based on the number of outliers $(\|(\mathcal{X}_\phi, \mathcal{Y}_\phi)\|)$. Then, the problem becomes an optimization problem. The score function to be maximized involves two parts: (i) the log-likelihood term for clean data $(\mathcal{X}_0, \mathcal{Y}_0)$, and (ii) a penalty term for outliers $(\mathcal{X}_\phi, \mathcal{Y}_\phi)$. That is:

$$(\mathbf{w}, \lambda)^* = \arg\max_{(\mathbf{w}, \lambda)} \left(\log\, p((\mathcal{X}_0, \mathcal{Y}_0)|\mathbf{f}, \mathbf{w}) - \zeta((\mathcal{X}_\phi, \mathcal{Y}_\phi); \mathbf{w}, \lambda) \right) \qquad (3)$$

where the penalty term, which penalizes having too many outliers, is defined as

$$\zeta((\mathcal{X}_\phi, \mathcal{Y}_\phi); \mathbf{w}, \lambda) = e \cdot \|(\mathcal{X}_\phi, \mathcal{Y}_\phi)\| \qquad (4)$$

$\mathbf{w}$ and λ affect ζ implicitly. The value of e is empirical. If e is too large, outliers will tend to be misclassified as inliers to avoid a large penalty. If e is too small, many inliers will be treated as outliers and excluded from the clean data model fitting, leading to a seemingly good model that actually under-fits the true data. In our experiments we set $e \in (0.2, 0.3)$.

After expanding the log-likelihood term, we have:

$$\log\, p((\mathcal{X}_0, \mathcal{Y}_0)|\mathbf{f}, \mathbf{w})$$

$$= \sum_{\mathbf{x}_i \in \mathcal{X}_0} \log\, p((\mathbf{x}_i, y_i)|\mathbf{f}, \mathbf{w})$$

$$= \sum_{\mathbf{x}_i \in \mathcal{X}_0} \log\, p(y_i|\mathbf{x}_i; \mathbf{f}, \mathbf{w}) + \sum_{\mathbf{x}_i \in \mathcal{X}_0} \log\, p(\mathbf{x}_i)$$

After we absorb $\sum_{\mathbf{x}_i \in \mathcal{X}_0} \log\, p(\mathbf{x}_i)$ into the penalty term $\zeta((\mathcal{X}_\phi, \mathcal{Y}_\phi); \mathbf{w}, \lambda)$, and replace the likelihood in Eq.(3) with the logistic form (Eq.(2)), then the optimization goal becomes finding the best fit $(\mathbf{w}, \lambda)^*$.

$$(\mathbf{w}, \lambda)^* = \arg \max_{(\mathbf{w}, \lambda)} \Big(\sum_{\mathbf{x}_i \in \mathcal{X}_0} (y_i \ q + (1 - y_i)(1 - q))$$

$$+ \zeta((\mathcal{X}_\phi, \mathcal{Y}_\phi); \mathbf{w}, \lambda) \Big) \tag{5}$$

The score function to be maximized is not differentiable because of the non-parametric penalty term. We have to resort to a more elaborate technique based on the Expectation-Maximization (EM) [3] algorithm to solve the problem.

5.2   Model Inference and Learning

The main goal of model computation is to infer the missing variables and compute the optimal model parameters, under the EM framework. The EM in general is a method for maximizing data likelihood in problems where data is incomplete. The algorithm iteratively performs an Expectation-Step (*E-Step*) followed by an Maximization-Step (*M-Step*) until convergence. In our case,

1. E-Step: to impute / infer the outlier indicator $\mathcal{Z}$ based on the current model parameters $(\mathbf{w}, \lambda)$.
2. M-Step: to compute new values for $(\mathbf{w}, \lambda)$ that maximize the score function in Eq.(3) with current $\mathcal{Z}$.

Next we will discuss how to impute outliers in E-Step, and how to solve the maximization problem in M-Step. The M-Step is actually a Maximum Likelihood Estimation (MLE) problem.

E-Step: Impute Outliers

With the current model parameters $\mathbf{w}$ (classifier weights), the model for clean data is established as in Eq.(1), that is, the class label (y_i) of a sample $\mathbf{x}_i$ follows a Bernoulli distribution parameterized with the ensemble prediction for this sample $(\mathbf{w}^{\mathbf{T}} \cdot \mathbf{f}(\mathbf{x}_i))$. Thus, y_i's log-likelihood $\log p(y_i|\mathbf{x}_i; \mathbf{f}, \mathbf{w})$ can be computed from Eq.(2).

Note that the line between outliers and clean samples is drawn by λ, which is computed in the previous M-Step. So, the formulation of imputing outliers is straightforward:

$$z_i = \text{sign}\big(\log \ p(y_i|\mathbf{x}_i; \mathbf{f}, \mathbf{w}) - \lambda\big) \tag{6}$$

where

$$\text{sign}(x) = \begin{cases} 1 \text{ if } x < 0 \\ 0 \text{ otherwise} \end{cases}$$

M-Step: MLE

The score function (in Eq.(5)) to be maximized is not differentiable because of the penalty term. We consider a simple approach for an approximate solution. In this approach, the computation of λ and $\mathbf{w}$ is separated.

1. λ is computed using the standard K-means clustering algorithm on log-likelihood $\log p(y_i|\mathbf{x}_i; \mathbf{f}, \mathbf{w})$. The cluster boundaries are candidates of likelihood threshold λ^*, which separates outliers from clean data. There is a tradeoff between efficiency and accuracy when choosing the value of K. In our experiments, we set $K = 3$. A larger K value slows down the computation, but does not necessarily gains much on accuracy.
2. By fixing each of the candidate λ^*, $\mathbf{w}^*$ can be computed using the standard MLE procedure. Running an MLE procedure for each candidate λ^*, and the maximum score given by Eq.(5) will identify the best fit of $(\mathbf{w}, \lambda)^*$.

The standard MLE procedure for computing $\mathbf{w}$ is described as follows. Taking the derivative of the non-outlier likelihood with respect to $\mathbf{w}$ and set it to zero, we have

$$\frac{\partial}{\partial \mathbf{w}} \sum_{y_i \in \mathcal{Y}_o} \left(y_i \frac{e^{\eta_i}}{1 + e^{\eta_i}} + (1 - y_i) \frac{1}{1 + e^{\eta_i}} \right) = 0$$

To solve this equation, we use the Newton-Raphson procedure, which requires the first and second derivatives. For clarity of notation, we use $h(\mathbf{w})$ to denote the first derivative of clean data likelihood function with regard to $\mathbf{w}$. Starting from $\mathbf{w}_t$, a single Newton-Raphson update is

$$\mathbf{w}_{t+1} = \mathbf{w}_t - \left(\frac{\partial^2 h(\mathbf{w}_t)}{\partial \mathbf{w} \partial \mathbf{w}^T} \right)^{-1} \frac{\partial h(\mathbf{w}_t)}{\partial \mathbf{w}}$$

Here we have

$$\frac{\partial h(\mathbf{w})}{\partial \mathbf{w}} = \sum_{y_i \in \mathcal{Y}_o} (y_i - q) \mathbf{f}(\mathbf{x}_i)$$

and,

$$\frac{\partial^2 h(\mathbf{w})}{\partial \mathbf{w} \partial \mathbf{w}^T} = - \sum_{y_i \in \mathcal{Y}_o} q(1 - q) \mathbf{f}(\mathbf{x}_i) \mathbf{f}^T(\mathbf{x}_i)$$

The initial values of $\mathbf{w}$ are important for computation convergence. Since there is no prior knowledge, we can set $\mathbf{w}$ to be uniform initially.

6   Experiments

We use both synthetic data and a real-life application to evaluate the model's adaptability to concept shifts and robustness to noise. Our model is compared with two previous approaches: *Bagging* and *Weighted Bagging*. We show that although the empirical weighting in *Weighted Bagging* [17] performs better than unweighted voting, the robust regression weighting method is more superior, in terms of both adaptability and robustness.

C4.5 decision trees are used in our experiments, but in principle our method does not require any specific base learning algorithm.

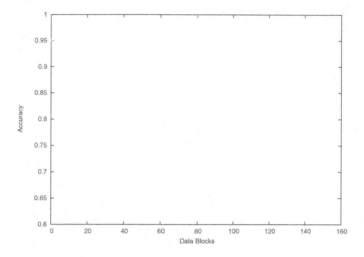

Fig. 1. Adaptability comparison of the three ensemble methods on data with abrupt shifts.

6.1   Data Sets

Synthetic Data

In the synthetic data set for controlled study, a sample $(\mathbf{x}, y)$ has three independent features $\mathbf{x} =< x_1, x_2, x_3 >$, $x_i \in [0, 1]$, $i = 0, 1, 2$. Geometrically, samples are points in a 3-dimension unit cube. The real class boundary is a sphere defined as

$$B(\mathbf{x}) = \sum_{i=0}^{2}(x_i - c_i)^2 - r^2 = 0$$

where $\mathbf{c} =< c_1, c_2, c_3 >$ is the center of the sphere, r the radius. $y = 1$ if $B(\mathbf{x}) \leq 0$, $y = 0$ otherwise. This learning task is not easy, because the feature space is continuous and the class boundary is non-linear.

To simulate a data stream with concept drift between adjacent blocks, we move the center $\mathbf{c}$ of the sphere that defines the class boundary. The movement is along each dimension with a step of $\pm\delta$. The value of δ controls the level of shifts from small, moderate to large, and the sign of δ is randomly assigned independently along each dimension. For example, if a block has $\mathbf{c} = (0.40, 0.60, 0.50)$, $\delta = 0.05$, the sign along each direction is $(+1, -1, -1)$, then the next block would have $\mathbf{c} = (0.45, 0.55, 0.45)$. The values of δ ought to be in a reasonable range, to keep the portion of samples that change class labels reasonable. In our setting, we consider a concept shift small if δ is around 0.02, and relatively large if δ around 0.1.

To study the model robustness, we insert noise into the training data sets by randomly flipping the class labels with a probability of p. Clean testing data sets are used in all the experiments for accuracy evaluation.

The experiments shown below are obtained when the block size equals 2k, but similar results are obtained for other block sizes(1k, 4k, 8k, etc.).

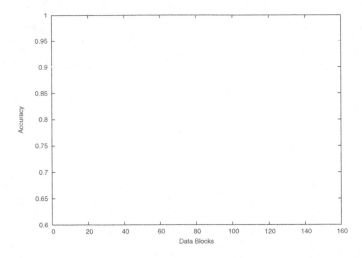

Fig. 2. Adaptability comparison of the three ensemble methods with mixed changes.

Credit Card Data

The real-life application is to build a weighted ensemble for detection of fraudulent transactions in credit card transactions, data contributed by a major credit card company. A transaction has 20 features, including the transaction amount, the time of the transaction, and so on. Detailed data description is given in [15, 17]. Same as in [17], concept drift in our work is simulated by sorting transactions by the transaction amount.

6.2   Evaluation of Adaptation

In this subsection we compare our robust regression ensemble method with *Bagging* and *Weighted Bagging*. Concept shifts are simulated by the movement of the class boundary center, which occur between adjacent data blocks. The moving distance δ along each dimension controls the magnitude of concept drift. We have two sets of experiments with varying δ values, both have abrupt changes occurring every 40 blocks, which means that the abrupt concept changing points are block 40, 80 and 120, where δ is around 0.1. In one experiment, data remains stationary between these changing points. In the other experiment, small shifts are mixed between abrupt ones, with $\delta \in (0.005, 0.03)$. The percentage of positive samples fluctuates between $(41\%, 55\%)$.

As shown in Fig.1 and Fig.2, our robust regression model always gives the highest performance. The unweighted bagging ensembles have the worst predictive accuracy. Both bagging methods are seriously impaired at concept changing points. The robust regression, on the other hand, is able to catch up with the new concept quickly.

In the above experiments, noises are around 10%. Experiments with varying noise levels are discussed in the next subsection. In terms of learning speed, robust regression is comparable to, although a little slower than, the bagging ensembles. For example, the total learning time on 40 blocks, each containing 2000 samples, is 76 seconds for bagging, 90 seconds for weighted bagging, and 110 seconds for robust logistic regression.

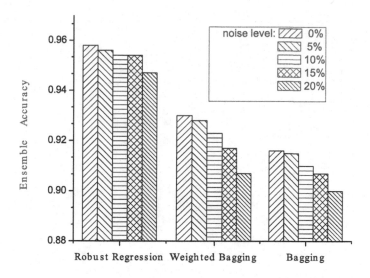

Fig. 3. Robustness comparison of the three ensemble methods under varying noise levels.

6.3 Robustness in the Presence of Outliers

Noise is the major source of outliers. Fig. 3 shows the ensemble performance for the different noise levels: 0%, 5%, 10%, 15% and 20%. The accuracy is averaged over 100 runs spanning 160 blocks, with small gradual shifts between blocks. We can make two major observations here:

1. The robust regression ensembles are the most accurate for all the different noise levels, as clearly shown in Fig. 3.
2. Robust regression also gives the least performance drops when noise increases. This conclusion is confirmed using paired t-test at 0.05 level. In each case when noise level increases by 10%, 15% or 20%, the decrease in accuracy produced by robust regression is the smallest, and the differences are statistically significant.

The robustness results from the fact that the noisy samples/outliers are treated as hidden information in the model learning. Once inferred, the outliers are excluded from the clean data model fitting. The experiments on synthetic data sets indeed prove that a majority of noisy samples are captured in the learning procedure. In one of the experiments, for instance, we inserted 10% noise into a 2K data block. Our model identified 217 outliers, among which 189 were true noisy samples. The false positive rate was low. This confirms experimentally that our robust regression technique is good at separating data caused by concept drift from noisy data.

6.4 Experiments on Real Life Data

In the credit card application, we build a classification model to detection of fraudulent transactions. A transaction has 20 features including the transaction amount, the time of

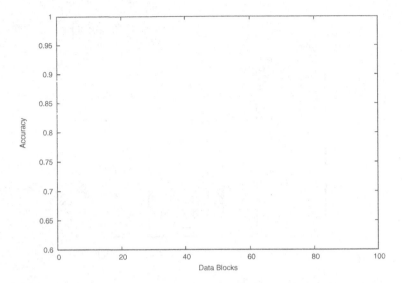

Fig. 4. Performance comparison of the three ensembles on credit card data. Concept drift is simulated by sorting the transactions by transaction amount.

the transaction, etc. We study the ensemble performance using varying block size (1k, 2k, 3k and 4k). We show one experiment in Fig.4 with a block size of 1k. The curve shows fewer and smaller drops in accuracy with robust regression than with the other methods. These drops actually occur when the transaction amount jumps.

7   Summary

In this paper, we propose a discriminative model that is highly adaptive to concept changes and is robust to noise. The model produces a weighted ensemble. The weights of classifiers are computed by logistic regression technique, which ensures good adaptation. Furthermore, outlier detection is integrated into the model, so that classifier weight training involves only the clean data, which leads to the robustness of the resulting ensemble. For outlier detection, we assume that a clean sample's belonging to certain class follows a Bernoulli distribution, which is parameterized by the ensemble prediction. Outliers can thus be identified as samples each with a very small likelihood. The classifier weights are thus estimated to maximize the data likelihood of the clean samples.

Compared with recent works [16, 17], the experimental results show that our discriminative model achieves higher accuracy, adapts to underlying concept shifts more promptly, and is less sensitive to noise. These benefits are attributed to the classifier weighting scheme using logistic regression, and the integral outlier detection technique.

References

1. C. Aggarwal and P. Yu. Outlier detection for high dimensional data. In *Int'l Conf. Management of Data (SIGMOD)*, 2001.
2. D. Aha, D. Kibler, and M. Albert. Instance-based learning algorithms. In *Machine Learning 6(1), 37-66*, 1991.
3. J. Bilmes. A gentle tutorial on the em algorithm and its application to parameter estimation for gaussian mixture and hidden markov models. In *Technical Report ICSI-TR-97-021*, 1998.
4. L. Breiman. Bagging predictors. In *Machine Learning 24(2), 123-140*, 1996.
5. M. Breunig, H. Kriegel, R. Ng, and J. Sander. LOF: identifying density-based local outliers. In *Int'l Conf. Management of Data (SIGMOD)*, 2000.
6. C. Brodley and M. Friedl. Identifying and eliminating mislabeled training instances. In *Artificial Intelligence, 799-805*, 1996.
7. C. Domeniconi and D. Gunopulos. Incremental support vector machine construction. In *Int'l Conf. Data Mining (ICDM)*, 2001.
8. T. Hastie, R. Tibshirani, and J. Friedman. *The Elements of Statistical Learning, Data Mining,Inference and Prediction*. Springer, 2000.
9. G. Hulten, L. Spencer, and P. Domingos. Mining time-changing data streams. In *Int'l Conf. on Knowledge Discovery and Data Mining (SIGKDD)*, 2001.
10. J. Kolter and M. Maloof. Dynamic weighted majority: A new ensemble method for tracking concept drift. In *Int'l Conf. Data Mining (ICDM)*, 2001.
11. J. Kubica and A. Moore. Probabilistic noise identification and data cleaning. In *Int'l Conf. Data Mining (ICDM)*, 2003.
12. N.C. Oza and S. Russell. Online bagging and boosting. In *Artificial Intelligence and Statistics, 105-112*, 2001.
13. S. Ramaswamy, R. Rastogi, and K. Shim. Efficient algorithms for mining outliers from large data sets. In *Int'l Conf. Management of Data (SIGMOD)*, 2000.
14. J. Schlimmer and F. Granger. Beyond incremental processing: Tracking concept drift. In *Proc. of Int'l Conf. on Artificial Intelligence 502-507*, 1986.
15. S. Stolfo, W. Fan, W. Lee, A. Prodromidis, and P. Chan. Credit card fraud detection using meta-learning: Issues and initial results. In *AAAI-97 Workshop on Fraud Detection and Risk Management*, 1997.
16. W. Street and Y. Kim. A streaming ensemble algorithm (sea) for large-scale classification. In *Int'l Conf. on Knowledge Discovery and Data Mining (SIGKDD)*, 2001.
17. H. Wang, W. Fan, P. Yu, and J. Han. Mining concept-drifting data streams using ensemble classifiers. In *Int'l Conf. on Knowledge Discovery and Data Mining (SIGKDD)*, 2003.
18. G. Widmer and M. Kubat. Learning in the presence of concept drift and hidden contexts. In *Machine Learning 23, 69-101*, 1996.

CorClass: Correlated Association Rule Mining for Classification

Albrecht Zimmermann and Luc De Raedt

Institute of Computer Science, Machine Learning Lab, Albert-Ludwigs-University
Freiburg, Georges-Köhler-Allee 79, 79110 Freiburg, Germany
{azimmerm,deraedt}@informatik.uni-freiburg.de

Abstract. A novel algorithm, CorClass, that integrates association rule mining with classification, is presented. It first discovers all correlated association rules (adapting a technique by Morishita and Sese) and then applies the discovered rule sets to classify unseen data. The key advantage of CorClass, as compared to other techniques for *associative classification*, is that CorClass directly finds the associations rules for classification by employing a branch-and-bound algorithm. Previous techniques (such as CBA [1] and CMAR [2]) first discover all association rules satisfying a minimum support and confidence threshold and then post-process them to retain the best rules.
CorClass is experimentally evaluated and compared to existing associative classification algorithms such as CBA [1], CMAR [2] and rule induction algorithms such as Ripper [3], PART [4] and C4.5 [5].

1 Introduction

Rule discovery is popular in the field of machine learning and data mining. Rule learning techniques have largely focussed on finding compact sets of rules that can be used for classification. On the other hand, approaches within data mining have concentrated on finding all rules that satisfy a set of constraints (typically based on support and confidence). Recently, these two approaches have been integrated in that a number of different research groups have developed tools for classification based on association rule discovery (associative classification). These tools such as CBA [1] and CMAR [2] in a first phase typically discover all association rules and then post-process the resulting rule sets in order to retain only a small number of suitable rules. Several studies [6, 1, 2] have shown that the obtained rule sets often perform better or comparable to those obtained using more traditional rule learning algorithms such as PART[4], Ripper[3] or CN2 [7].

We present a novel approach to associative classification, called CorClass (correlated classification). CorClass directly aims at finding the k best correlated association rules for classification, adapting a technique due to Morishita and Sese[8]. The advantage is that this is not only more efficient (no post-processing is necessary) but also more elegant in that it is a direct approach. To the best of the authors' knowledge, it is the first time that Morishita and Sese's framework is being applied to discovery of classification rules.

E. Suzuki and S. Arikawa (Eds.): DS 2004, LNAI 3245, pp. 60–72, 2004.
© Springer-Verlag Berlin Heidelberg 2004

When using association rules for classification, one needs to employ a strategy for combining the predictions of different association rules. In [6] several approaches have been evaluated and compared to existing techniques.

The remainder of the paper is structured as follows. In the next section the basic terminology used throughout the paper is introduced. In section 3 we explain how an extension of Morishita's approach to a single multi-valued target attribute can be derived and explain the algorithm used for mining the rules. Different strategies for combining derived rules following [6] are introduced in section 4. In section 5, the experimental setup and results will be presented and we conclude in section 6.

2 Preliminaries

2.1 Association Rule Mining

Let $\mathcal{A} = \{A_1, ..., A_d\}$ a set of attributes, $\mathcal{V}[A] = \{v_1, ..., v_j\}$ the domain of attribute A, $\mathcal{C} = \{c_1, ..., c_t\}$ a set of possible class values for class attribute C. An instance e is then a tuple $\langle v_1, ..., v_d \rangle$ with $v_i \in \mathcal{V}[A_i]$. The multiset of instances $\mathcal{E} = \{e_1, ..., e_n\}$ is called a data set.

Definition 1 (Literal) *A **literal** is an attribute-value-pair of the form $A = v, v \in \mathcal{V}[A]$. An instance $\langle v_1, ..., v_d \rangle$ satisfies a literal $A_i = v$ iff $v_i = v$.*

Definition 2 (Association rule) *An **association rule** r is of the form $b_r \Rightarrow h_r$ with b_r of the form $l_1 \wedge ... \wedge l_i$ called the **rule body** and h_r of the form $l'_1 \wedge ... \wedge l'_j$, $b_r \cap h_r = \emptyset$ the **rule head**. An instance e satisfies b_r iff it satisfies all literals in b_r and it satisfies r iff it satisfies h_r as well.*

Definition 3 (Support and Confidence) *For a given rule r of the form $b_r \Rightarrow h_r$,*

$$sup(r) = |\{e \mid e \in \mathcal{E}, e \text{ satisfies } r\}|$$

*is called the **support** of r on D,*

$$conf(r) = \frac{sup(r)}{|\{e \mid e \in \mathcal{E}, e \text{ satisfies } b_r\}|}$$

*the **confidence** of r.*

2.2 Association Rules for Classification

When learning in a classification setting, one starts from a databases of instances and a target class attribute. The goal is then to induce a set of rules with the class attribute in the rule head. In this context, only association rules relating the rule body to a certain class value are of interest. In the literature on associative classification the term *class association rule* has been introduced to distinguish such rules from regular association rules whose head may consist of an arbitrary conjunction of literals.

Definition 4 *(Class Association Rule)* *A* ***class association rule*** *r is of the form $b_r \to h_r$ with b_r of the form $l_1 \wedge \dots \wedge l_r$ called the **rule body** and h_r of the form $C = c_i$ the **rule head**, where C is the class attribute.*

Satisfaction of a rule is treated as above. For an unclassified example satisfying b_r the rule predicts class c_i. Support and confidence are defined as above. For convenience we will write $b_r \Rightarrow c$ for $b_r \Rightarrow C = c$ and $sup(c)$ for $sup(C = c)$ from now on. Note that *confidence* is called *accuracy* in the context of rule induction.

Rule sets can then be applied to classify examples. To this aim, one combines the predictions of all rules whose body satisfies the example. If there is only one rule, then the corresponding value in the head is predicted; if there is no rule body satisfying the example, then a default class is predicted; and if there are multiple rule bodies satisfying the example, then their predictions must be combined. Various strategies for realizing this are discussed in section 4.

Many different rule learning algorithms are presented in the literature, cf. [7, 5, 4, 3]. A relatively recent approach to rule learning is *associative classification*, which combines well-developed techniques for association rule mining such as [9, 10] with post-processing techniques. Several such techniques are discussed in the next section.

3   Methodology

3.1   Current Approaches to Associative Classification

In association rule mining the usual framework calls for the user to specify two parameters, sup_{min} and $conf_{min}$. The parameter sup_{min} defines the minimum number of instances that have to satisfy a rule r for r to be considered interesting. Similarly $conf_{min}$ sets a minimum confidence that a rule has to achieve.

In [1], the first approach to associative classification, rules are mined with $sup_{min} = 1\%$ and $conf_{min} = 50\%$ using *apriori* [9]. Once all class association rules satisfying these constraints are found, the rules are ranked based on confidence, support and generality and an obligatory pruning step based on database coverage is performed. Rules are considered sequentially in descending order and training instances satisfying a rule are marked. If a rule classifies at least one instance correctly, the instances satisfying it are removed. An optional pruning step based on a pessimistic error estimate [5], can also be performed. The resulting set of classification rules is used as an ordered decision list.

In [2], the mining process uses similar support and confidence thresholds and employs *FP-growth* [10]. Pruning is again performed after mining all rules and involves a database coverage scheme, and also removes all rules that do not correlate positively with the class attribute. Classification of an example is decided by a weighted combination of the values of rules it satisfies. The weight is derived by calculating the χ^2-value of the rules and normalized by the maximum χ^2-value a rule *could* have to penalize rules favoring minority classes. For a more thorough discussion see [2].

In [6], both *apriori* (with similar parameters to the CBA setting) and *predictive apriori* [11] are employed. In *predictive apriori*, the confidence of rules is corrected by support and rules with small support are penalized, since those will probably not generalize well. For a more in-depth explanation we refer the reader to the work of Scheffer [11]. The resulting rule sets are employed both pruned (in the manner used by CBA) and unpruned. Classification is decided using different schemes that are described in section 4.

There are a few problems with the support-confidence framework. If the minimal support is too high, highly predictive rules will probably be missed. Is the support threshold set too low, the search space increases vastly, prolonging the mining phase, and the resulting rule set will very likely contain many useless and redundant rules. Deciding on a good value of sup_{min} is therefore not easy. Also, since the quality of rules is assessed by looking at their confidence, highly specialized rules will be preferred. Previous works [11, 2] attempt to correct this by penalizing minority rules during the mining and prediction step respectively. Finally, for the support based techniques it is necessary to mine **all** rules satisfying the support- and confidence-thresholds and extract the set of prediction rules in a post-processing step.

Correlation measures have, as far as we know, so far not been used as search criteria. The reason lies in the fact that measures such as *entropy gain*, χ^2 etc. are neither anti-monotonic nor monotonic, succinct or convertible and therefore do not lend themselves as pruning criteria very well. Morishita et al in [8] introduced a method for calculating upper bounds on the values attainable by specializations of the rule currently considered. This effectively makes convex correlation measures anti-monotonic and therefore pruning based on their values possible.

Correlation measures quantify the difference between the conditional probability of the occurence of a class in a subpopulation and the unconditional occurence in the entire data set. They typically normalize this value with the size of the population considered. This means that predictive rules will be found that are not overly specific. Additionally setting a threshold for pruning rules can be based on significance assessments, which may be more intuitive than support. The upper bound finally allows dynamic raising of the pruning threshold, differing from the fixed minimal support used in existing techniques. This will result in earlier termination of the mining process. Since the quality criterion for rules is used directly for pruning, no post-processing of the discovered rule set is necessary.

In the next section we will briefly sketch Morishita's approach and outline how to extend it to a multi-value class variable. This will then be used to find the rules sets for classification.

3.2   Correlation Measures and Convexity

Let $n = |\mathcal{E}|$, $m = sup(c)$ for a given class value c and $x = sup(b_r)$, $y = sup(r)$ for given rule r of the form $b_r \Rightarrow c$. A contingency table reflecting these values is shown in table 1. Since virtually all correlation measures quantify the difference

Table 1. Contingency table for $b_r \Rightarrow c$.

	c	$\neg c$	
b_r	$sup(r) = y$	$sup(b_r \Rightarrow \neg c)$	$sup(b_r) = x$
$\neg b_r$	$sup(\neg b_r \Rightarrow c)$	$sup(\neg b_r \Rightarrow \neg c)$	$sup(\neg b_r)$
	$sup(c) = m$	$sup(\neg c)$	n

between expected and observed distributions, such measures σ can, for fixed $\mathcal{E}$, be viewed as functions $f(x, y) : \mathbb{N}^2 \to \mathbb{R}$.

The tuple $\langle x, y \rangle$ characterizing a rule's behavior w.r.t. a given data set is called a *stamp point*.

The set of *actual* future stamp points S_{act} of refinements of r is unknown until these refinements are created and evaluated on the data set. But the current stamp point bounds the set of *possible* future stamp points S_{poss}. For the 2-dimensional case they fall inside the parallelogram defined by the vertices $\langle 0, 0 \rangle$, $\langle y, y \rangle$, $\langle x - y, 0 \rangle$, $\langle x, y \rangle$, cf. [8].

Quite a few correlation functions are convex (χ^2, *information gain, gini index, interclass variance, category utility, weighted relative accuracy* etc.). Convex functions take their extreme values at the points on the convex hull of their domain. So by evaluating f at the convex hull of S_{poss} (the vertices of the parallelogram mentioned above), it is possible to obtain the extreme values bounding the values of σ attainable by refinements of the r. Furthermore, since $\langle 0, 0 \rangle$ characterizes a rule that is not satisfied by a single instance and $\langle x, y \rangle$ a rule of higher complexity that conveys no additional information, the upper bound of $\sigma(r')$ for any specialization r' of r is calculated as $max\{f(x - y, 0), f(y, y)\}$. The first of the two terms evaluates a rule that covers no example of the class c and in a two-class problem therefore perfectly classifies $\neg c$, while the second term evaluates a rule that has 100% accuracy with regard to c.

3.3   Multi-valued Target Attribute

A similar contingency table to the one shown in table 1 can be constructed for more than two classes (which cannot be treated as c and $\neg c$ anymore). An example for such a table is shown in table 2.

The body defined by the vertices $\langle 0, 0, 0 \rangle$, $\langle x - y_1, 0, y_2 \rangle$, $\langle x - y_2, y_1, 0 \rangle$, $\langle x - (y_1 + y_2), 0, 0 \rangle$, $\langle y_1 + y_2, y_1, y_2 \rangle$, $\langle y_2, 0, y_2 \rangle$, $\langle y_1, y_1, 0 \rangle$, $\langle x, y_1, y_2 \rangle$ encloses all stamp points that can be induced by specializations of the current rule. This can of course be extended to a higher number of classes. It has to be kept in mind though that 2^t number of points will be induced for t classes.

Similar to the 2-dimensional approach, an upper bound on the future value of class association rules derived from the current rule can be calculated. For the remainder of this paper we will refer to this upper bound for a given rule as $ub_\sigma(r)$. Following the argument for the 2-dimensional case, the tuples $\langle 0, 0, 0 \rangle$ and $\langle x, y_1, y_2 \rangle$ can be ignored.

Table 2. Contingency table for three class values.

	c_1	c_2	c_3	
b_r	$sup(b_r \Rightarrow c_1) = y_1$	$sup(b_r \Rightarrow c_2) = y_2$	$sup(b_r \Rightarrow c_3) = x - (y_1 + y_2)$	$sup(b_r) = x$
$\neg b_r$	$sup(\neg b_r \Rightarrow c_1)$	$sup(\neg b_r \Rightarrow c_2)$	$sup(\neg b_r \Rightarrow c_3)$	$sup(\neg b_r) = n - x$
	$sup(c_1) = m_1$	$sup(c_2) = m_2$	$sup(c_3) = n - (m_1 + m_2)$	n

3.4   The CorClass Algorithm

The upper bound allows for two types of pruning w.r.t. the actual mining process. First, the user can specify a threshold which might be based on e.g. significance for χ^2. It should be easier to find a meaningful threshold than it is when looking for a minimal support that does not lead to the creation of too many rules but at the same time captures many interesting and useful ones. Second, the goal can be to mine for a user specified maximum number k of rules in which case the threshold is raised dynamically. We will only describe the k-best algorithm here since deriving the threshold-based algorithm should be straightforward. The algorithm is listed in figure 1.

The algorithm starts from the most general rule body (denoted by $\top$). We use an optimal refinement operator (denoted by ρ in the listing). This refinement operator is defined as follows:

Definition 5 *(Optimal Refinement Operator) Let $\mathcal{L}$ a set of literals, $\prec$ a total order on $\mathcal{L}$, $\tau \in \mathbb{R}$.*

$$\rho(r) = \{r \wedge l_i \mid l_i \in \mathcal{L}, ub_\sigma(l_i) \geq \tau, \forall l \in r : l \prec l_i\}$$

*is called an **optimal refinement operator**.*

The operator guarantees that all rules can be derived from $\top$ in exactly one possible way. So, no rule is generated more than once. Since only literals are added whose upper bound exceeds the threshold, the value of the refinement has a chance of exceeding or matching the current threshold, if $ub(r) \geq \tau$.

In each iteration of the algorithm, the rule body with the highest upper bound is selected for refinement. The support counts x and y_i for the resulting rules are computed and the score and the upper bound is calculated.
Now decisions have to be made about:

1. Raising the threshold τ.
2. Including the current rule in the temporary solution set S.
3. Including the current rule in the set of promising rules P i.e. the set of candidates for future refinement.

Decisions 1. and 2. are actually combined since looking up the threshold is implemented by returning the score of the lowest-ranked rule r_k currently in the solution set. The decision on whether to include the current rule r in the temporary solution set is based on three criteria. First, $\sigma(r)$ has to exceed or match $\sigma(r_k)$ of course. Second, if there already is a rule with the same σ-value

Input: Dataset $\mathcal{E}$, $k \in \mathbb{N}$, $C = \{c_1, ..., c_t\}$
Output: The k rules with highest σ-values on $\mathcal{E}$

$P := \{\top\}$, $S := \emptyset$, $\tau := -\infty$
while $|P| \geq 1$
$\quad p_{mp} = argmax_{p \in P} \, ub_\sigma(p)$
$\quad P := P \setminus \{p_{mp}\}$
$\quad \forall b_i \in \rho(p_{mp})$
$\quad\quad$ **if** $k > |S|$ **then** $S \cup \{b_i\}$
$\quad\quad$ **elsif** $qual(b_i)$ **then**
$\quad\quad\quad S := S \setminus \{s_j \mid argmin_{s_j \in S}\sigma(s_j)\} \cup \{b_i\}$
$\quad\quad\quad \tau = min_{s \in S} \, \sigma(s)$
$\quad\quad$ **endif**
$\quad P := \{p \mid p \in P, ub_\sigma(p) \geq \tau\} \cup \{b_i \mid ub_\sigma(b_i) \geq \tau\}$
return S

Fig. 1. The CorClass Algorithm.

the question is, whether it is a generalization of r. So, r is only included if it has a different support, since otherwise it includes at least one literal giving no additional information. This decision is denoted by the predicate **qual** in the algorithm listing (figure 1). If r is included, r_k is removed and the threshold raised to $\sigma(r_{k-1})$. After processing all refinements the set of promising rules is pruned by removing all rules with $ub_\sigma(r_i) < \sigma(r_k)$. Also, all refinements whose upper bound exceeds the threshold are included in the promising set. In this way, a branch-and-bound algorithm is realized that never traverses parts of the search space known to lead to suboptimal rules.

The algorithm terminates once there is no candidate remaining for refinement. During mining, the rules in the solution set are already ranked by (1) score, (2) generality, and (3) support. The majority class of all instances satisfying a rule is set as the head. Should not all training instances be covered, the majority class of the remaining ones is set as the default class.

4   Combination Strategies

When applying a set of discovered rules for classifying an example, it may be necessary to combine the predictions of multiple rules satisfying the example or to resolve the conflicts between them. Various strategies for realizing this have been considered. We discuss the approaches we evaluate below, following [6].

4.1   Decision List

The most straightforward approach is to use the list of rules created as a decision list (since rules are ranked by quality according to some criterion) and use the first rule satisfied by an example for classification. This approach is chosen by CBA [1]. To compare the quality of the rules found by our solution to CBA's, we will use this approach as well.

4.2 Weighted Combination

The alternative consists of using a weighted combination of all rules being satisfied by an as yet unseen example for classification. The general way to do this is to collect all such rules, assign each one a specific weight and for each c_i predicted by at least one rule sum up the weights of corresponding rules. The class value having the highest value is returned. In the literature on associative classification [6, 2], several approaches for conflict resolution are explored. Among these are:

- Majority Voting [6]: all rules get assigned the weight 1.
- Linear Weight Voting [6]: a rule r is assigned the weight

$$1 - \frac{rank(r)}{rank_{max} + 1},$$

  where $rank(r)$ is the rank a rule has in the list returned by the algorithm and $rank_{max}$ the number of rules returned in total.
- Inverse Weight Voting [6]: a rule r is assigned the weight

$$\frac{1}{rank(r)}$$

- In [2] a heuristic called *weighted-χ^2* is proposed. Since some improvement with regard to the decision list approach is reported, we use this weight as well in our experiments.

Even though other conflict resolution strategies, e.g. *naive Bayes* [12] and *double induction* [13], have been proposed, we do not employ these strategies in our experiments. The main reason for this is that our focus was on comparing our rule discovery approach to established techniques and we use [6] as reference work.

5 Experiments

To compare our approach to existing techniques, we use the results of the experiments by Stefan Mutter, published in [6]. We also compare CorClass to CMAR and for this comparison use the results published in [2]. Our choice of data sets is somewhat limited by the need for comparison with existing work. We used *information gain*, χ^2 and *category utility* [14] as σ. The columns reporting on the classification accuracy of a metric are headed *Gain*,χ^2 and *CU* respectively.

Predictive accuracies in [6] are estimated using stratified 10-fold cross validation followed by discretization of numerical attributes using Fayyad and Irani's MDL method [15]. To facilitate a fair comparison we choose the same evaluation method for CorClass. In the experiments, k was set to 1000 for CorClass.

5.1   Results

The accuracies reported in [6] were obtained using rule sets of size 10–1000. In addition different pruning techniques were evaluated. Due to space restrictions we used the best accuracy *apriori* and *predictive apriori* obtained, regardless of rule set size or pruning method, for comparison. The results for the different combination strategies are shown in tables 3,4,5 and 6. CBA results are considered in the table that reports on the results for decision lists since it uses a (pruned) decision list. The values for CBA are also taken from [6]. Note that the authors report that they have not been able to reproduce the results in [1], neither using the executable obtained from the authors of [1] nor using their own re-implementation.

Table 3. Comparison of Obtained Accuracies using the Inverted Weighting Scheme.

Dataset	*Apriori*	*Apriori$_{pred}$*	*Gain*	χ^2	*CU*
Balance	71.66±5.85	70.55±4.52	70.55±5.06	**74.88±4.71**	74.56±4.49
Breast-w	65.47±0.47	88.94±8.85	94.56±2.84	**94.99±2.05**	**94.99±2.05**
Heart-h	63.95±1.43	**83.01±6.25**	79.95±8.67	79.95±8.67	79.95±8.67
Iris	92±6.89	92±6.89	95.33±5.49	**96±4.66**	94±6.62
Labor	71.67±16.72	79.33±19.55	**82.38±13.70**	77.61±12.28	77.61±12.28
Lenses	66.67±30.43	70±28.11	71.67±31.48	**75±42.49**	71.67±41.61
Pima	65.11±0.36	**74.35±3.95**	74.21±6.02	74.21±6.02	74.08±7.02
Tic-Tac-Toe	65.34±0.43	**96.67±1.67**	75.66±5.15	75.03±4.61	75.03±4.61

Table 4. Comparison of Obtained Accuracies using the Linear Weighting Scheme.

Dataset	*Apriori*	*Apriori$_{pred}$*	*Gain*	χ^2	*CU*
Balance	68.8±6.13	73.92±5.2	70.55±5.06	**74.88±4.71**	74.55±4.49
Breast-w	65.47±0.47	65.47±0.47	**96.99±2.08**	96.85±1.89	96.85±1.89
Heart-h	63.95±1.43	77.17±5.52	81.67±9.58	**81.97±7.71**	**81.97±7.71**
Iris	64±11.42	91.33±6.32	94.67±5.26	**95.33±5.49**	**95.33±5.49**
Labor	64.67±3.22	77±15.19	**82.38±13.70**	**82.38±13.70**	**82.38±13.70**
Lenses	63.33±32.2	70±28.11	**75±32.63**	71.67±41.61	71.67±41.61
Pima	65.11±0.36	66.54±2	74.84±4.28	**75.37±5.43**	**75.37±5.43**
Tic-Tac-Toe	65.34±0.43	**75.78±1.37**	74.69±4.22	75.13±3.92	75.13±3.92

Table 5. Comparison of Obtained Accuracies using Majority Vote.

Dataset	*Apriori*	*Apriori$_{pred}$*	*Gain*	χ^2	*CU*
Balance	**76.46±4.81**	73.92±5.41	71.83±7.76	75.04±4.66	74.72±4.46
Breast-w	88.55±4.03	92.51±9.95	97.28±1.58	**97.42±1.63**	**97.42±1.63**
Heart-h	**83.02±6.35**	82.34±8.54	82.69±7.33	82.01±7.27	82.01±7.27
Iris	92.67±7.34	91.33±9.45	**96±4.66**	**96±4.66**	**96±4.66**
Labor	76±20.05	**86.67±15.32**	84.76±15.26	84.29±11.97	84.28±12.97
Lenses	**68.33±33.75**	**68.33±33.75**	65±38.85	**68.33±43.35**	**68.33±43.35**
Pima	72.14±4.51	**74.49±6.3**	74.45±3.48	73.93±3.87	73.93±3.87
Tic-Tac-Toe	**98.01±1.67**	92.48±3.09	76.18±6.58	74.72±4.79	74.72 ±4.79

Table 6. Comparison of Obtained Accuracies using a Decision List.

Dataset	CBA	Apriori	$Apriori_{pred}$	Gain	χ^2	CU
Balance	71.5±5.97	71.5±5.97	71.5±5.97	70.55±5.06	**74.88±4.71**	74.72±4.71
Breast-w	**95.13±3.03**	88.83±3.73	88.65±8.7	94.13±2.08	94.56±2.42	94.56±2.42
Heart-h	80.63±7.2	82.67±6.43	**83.36±5.26**	78.26±9.39	77.23±10.97	77.23±10.97
Iris	92.67±6.63	91.33±6.32	91.33±6.32	**95.33±5.49**	84±10.04	84±10.04
Labor	79±19.5	79.33±21.07	79.33±21.07	**82.38±13.71**	**82.38±13.71**	**82.38±13.71**
Lenses	66.67±30.43	66.67±30.43	70±28.11	**71.67±31.48**	**71.67±41.61**	**71.67±41.61**
Pima	74.1±4.48	73.83±4.97	**74.22±4.67**	71.99±7.07	73.93±3.87	73.93±3.87
Tic-Tac-Toe	**99.06±1.25**	97.39±1.8	97.28±1.66	76.08±5.76	75.24±4.68	75.24±4.68

The tables list the average predictive accuracy and the standard deviation for each combination of correlation measure and conflict resolution strategy. **Bold** values denote the best accuracy for the respective data set and combination strategy.

As can be seen, CorClass on almost all occasions achieves best accuracy or comes very close. This is most noticable for the Breast-w and Iris data sets on which the different CorClass versions perform very well for all combination strategies. The only set on which CorClass constantly achieves worse accuracy than the other techniques is the Tic-Tac-Toe set. This is caused by the fact that a single literal already restricts coverage quite severely without having any discriminative power. To reliably predict whether x will win, one needs at least three literals. So, the discriminative rules are penalized by correlation measures because their coverage is low.

Results derived using the *weighted-χ^2* heuristic are compared to the accuracies reported in [2] (table 7). It is not entirely clear how the accuracy estimates in [2] were obtained.

The only data set among these on which CMAR performs noticably better than CorClass is again the Tic-Tac-Toe set. It is interesting to see, that using the *weighted-χ^2* heuristic improves CorClass' performance on the Tic-Tac-Toe data set strongly when compared to the other weighting schemes.

Table 7. Comparison of Obtained Accuracies using the *Weighted-χ^2* Scheme.

Dataset	CMAR	Gain	χ^2	CU
Breast-w	**96.4**	96.13±2.44	96.13±2.44	96.13±2.44
Heart-h	82.2	81.33±9.18	**82.3±7.13**	**82.3±7.13**
Iris	94	**96±4.66**	94.67±8.19	94.67±8.19
Labor	**89.7**	84.76±15.26	87.14±16.71	87.14±16.71
Pima	75.1	75.76±4.88	**75.89±5.11**	**75.89±5.11**
Tic-Tac-Toe	**99.2**	86.42±4.51	88.72±4.4	88.72±4.4

Finally, in table 8 we compare the best results obtained by CorClass to several standard machine learning techniques, namely *C4.5*, *PART* and *Ripper*

Table 8. Comparison of CorClass to standard rule learning techniques.

Dataset	Gain	χ^2	CU	C4.5	PART	JRip
Balance	71.83±7.76	75.04±4.66	74.72±4.46	76.65	**83.54**	80.80
Breast-w	97.28±1.58	**97.42±1.63**	**97.42±1.63**	94.69	94.26	94.13
Heart-h	**82.69±7.33**	82.3±7.13	82.3±7.13	81.07	81.02	78.95
Iris	**96±4.66**	**96±4.66**	**96±4.66**	**96**	94	94.67
Labor	84.76±15.26	**87.14±16.71**	**87.14±16.71**	73.67	78.67	77
Lenses	75±32.63	75±42.49	71.67±41.61	**81.67**	**81.67**	75
Pima	74.84±4.28	**75.89±5.11**	**75.89±5.11**	73.83	75.27	75.14
Tic-Tac-Toe	86.42±4.51	88.72±4.4	88.72±4.4	85.07	94.47	**97.81**

as implemented in WEKA [16]. CorClass compares well to those techniques as well. It outperformes them on the Breast-w and Labor data sets, achieves competitive results for the Heart-h, Iris and Pima sets and is outperformed on Balance, Lenses and Tic-Tac-Toe, while still achieving reasonable accuracy.

5.2   Discussion

Generally, CorClass performs well on the data sets considered, and achieves for all but the Tic-Tac-Toe data set repeatedly the best accuracy.

It is interesting to see that there is no single correlation measure that clearly outperforms the other two, even though χ^2 performs slightly better. There is also no ideal combination strategy for combining the mined rules for classification. The majority vote performs surprisingly well (giving rise to the best results for CorClass on five of the eight data sets), considering that it is the least complex of the weighting schemes. This indicates that even rules of relatively low rank still convey important information and do not have to be discounted by one of the more elaborate weighting schemes.

All experiments ran in less than 30 seconds on a 2 GHz Pentium desktop PC with 2 GB main memory, running Linux. The number of *candidate rules* considered by CorClass during mining was on average smaller than the number of *discovered* rules (before pruning) reported in [6]. This difference is probably more pronounced for smaller rule sets since *apriori*-like approaches have to mine all rules satisfying the support and confidence constraints and post-process them while CorClass can stop earlier. We plan to investigate this issue more thoroughly in the future.

6   Conclusion and Future Work

We have introduced CorClass, a new approach to associative classification. Cor-Class differs from existing algorithms for this task insofar, that it does not employ the classical association rule setting in which mining is performed under support and confidence constraints. Instead it directly maximizes correlation measures such as *information gain*, χ^2 and *category utility*. This removes the need for post-processing the rules to obtain the set of actual classification rules.

In the experimental comparison, we evaluated the different correlation measures as well as several strategies for combining the mined rule sets for classification purposes. We compare our algorithm to existing techniques and standard rule learning algorithms. In the experiments, CorClass repeatedly achieves best accuracy, validating our approach.

While the results are promising, there is still room for improvement. As the results on the Tic-Tac-Toe data set show, rules with relatively low coverage but high discriminative power tend to be penalized by our approach. For data sets whose instances are divided in more than two or three classes, this problem will be more pronounced since rules separating, e.g., half of the classes from the other half will achieve a high score but have low classification power. A possible solution to this problem could be to calculate a separate σ-value and upper bound for each such class for a given rule body, thus coming back to the 2-dimensional setting Morishita introduced. During rule mining the best of these values would determine the inclusion in the solution set and promising set respectively. Such an approach would also allow for the usage of heuristics such as *weighted relative accuracy*, cf. [17]. A second approach could lie in maximizing the *weighted-χ^2* criterion directly since the experimental results showed that it offsets the selection bias of the correlation measures somewhat. We plan to extend our technique in these directions.

Since CorClass performs well when compared to the traditional heuristic rule learning algorithms such as Ripper and PART, we hope that CorClass may inspire further work on the use of globally optimal methods in rule learning.

Acknowledgements

This work was partly supported by the EU IST project cInQ (Consortium on Inductive Querying).

This work was strongly inspired by the master's thesis of Stefan Mutter. We sincerely would like to thank him and Mark Hall for allowing us to use their experimental results for comparison purposes. We would also like to thank Andreas Karwath and Kristian Kersting for their helpful suggestions. Finally we thank the anonymous reviewers for their constructive comments on our work.

References

1. Liu, B., Hsu, W., Ma, Y.: Integrating classification and association rule mining. In: Proceedings of the 4th International Conference on Knowledge Discovery and Data Mining, New York, USA, AAAI Press (1998) 80–86
2. Li, W., Han, J., Pei, J.: CMAR: Accurate and efficient classification based on multiple class-association rules. In: Proceedings of the 2001 IEEE International Conference on Data Mining, San Jose, California, USA, IEEE Computer Society (2001) 369–376
3. Cohen, W.W.: Fast effective rule induction. In: Proceedings of the 12th International Conference on Machine Learning, Tahoe City, California, USA, Morgan Kaufman (1995) 115–123

4. Frank, E., Witten, I.H.: Generating accurate rule sets without global optimization. In: Proceedings of the 15th International Conference on Machine Learning, Madison, Wisconsin, USA, Morgan Kaufmann (1998) 144–151
5. Quinlan, J.R.: C4.5: Programs for Machine Learning. Morgan Kaufmann (1993)
6. Mutter, S.: Classification using association rules. Master's thesis, Albert-Ludwigs-Universitä Freiburg/University of Waikato, Freiburg, Germany/Hamilton, New Zealand (2004)
7. Clark, P., Niblett, T.: The CN2 induction algorithm. Machine Learning **3** (1989) 261–283
8. Morishita, S., Sese, J.: Traversing itemset lattices with statistical metric pruning. In: Proceedings of the 19th ACM SIGACT-SIGMOD-SIGART Symposium on Principles of Database Systems, ACM (2000) 226–236
9. Agrawal, R., Srikant, R.: Fast algorithms for mining association rules in large databases. In: Proceedings of the 20th International Conference on Very Large Databases, Santiago de Chile, Chile, Morgan Kaufmann (1994) 487–499
10. Han, J., Pei, J., Yin, Y.: Mining frequent patterns without candidate generation. In: Proceedings of the 2000 ACM SIGMOD International Conference on Management of Data, Dallas, Texas, USA, ACM (2000) 1–12
11. Scheffer, T.: Finding association rules that trade support optimally against confidence. In: Proceedings of the 5th European Conference on Principles and Practice of Knowledge Discovery in Databases. Lecture Notes in Computer Science, Freiburg, Germany, Springer (2001) 424–435
12. Boström, H.: Rule Discovery System User Manual. Compumine AB. (2003)
13. Lindgren, T., Boström, H.: Resolving rule conflicts with double induction. In: Proceedings of the 5th International Symposium on Intelligent Data Analysis. Lecture Notes in Computer Science, Berlin, Germany, Springer (2003) 60–67
14. Gluck, M., Corter, J.: Information, uncertainty, and the utility of categories. In: Proceedings of the 7th Annual Conference of the Cognitive Science Society International Conference on Knowledge Discovery and Data Mining, Irvine, California, USA, Lawrence Erlbaum Associate (1985) 283–287
15. Fayyad, U.M., Irani, K.B.: Multi-interval discretization of continuous-valued attributes for classification learning. In: Proceedings of the 13th International Joint Conference on Artificial Intelligence, Chambéry, France, Morgan Kaufmann (1993) 1022–1029
16. Frank, E., Witten, I.H.: Data Mining: Practical Machine Learning Tools and Techniques with Java Implementations. Morgan Kaufmann (1999)
17. Todorovski, L., Flach, P.A., Lavrac, N.: Predictive performance of weighted relative accuracy. In: Proceedings of the 4th European Conference on Principles and Practice of Knowledge Discovery in Databases. Lecture Notes in Computer Science, Lyon, France, Springer (2000) 255–264

Maximum a Posteriori Tree Augmented Naive Bayes Classifiers

Jesús Cerquides[1] and Ramon Lòpez de Màntaras[2]

[1] Departament de Matemàtica Aplicada i Anàlisi
Universitat de Barcelona
Gran Via 585, Barcelona 08007
[2] Institut d'Investigació en Intel.ligència Artificial (IIIA)
Consejo Superior de Investigaciones Científicas(CSIC)
Campus UAB, Bellaterra 08193

Abstract. Bayesian classifiers such as Naive Bayes or Tree Augmented Naive Bayes (TAN) have shown excellent performance given their simplicity and heavy underlying independence assumptions. In this paper we prove that under suitable conditions it is possible to efficiently compute the maximum a posterior TAN model. Furthermore, we prove that it is also possible to efficiently calculate a weighted set with the k maximum a posteriori TAN models. This allows efficient TAN ensemble learning and accounting for model uncertainty. These results can be used to construct two classifiers. Both classifiers have the advantage of allowing the introduction of prior knowledge about structure or parameters into the learning process. Empirical results show that both classifiers lead to an improvement in error rate and accuracy of the predicted class probabilities over established TAN based classifiers with equivalent complexity.

Keywords: Bayesian networks, Bayesian network classifiers, Naive Bayes, decomposable distributions, Bayesian model averaging.

1 Introduction

Bayesian classifiers as *Naive Bayes* [11] or *Tree Augmented Naive Bayes* (TAN) [7] have shown excellent performance in spite of their simplicity and heavy underlying independence assumptions.

In our opinion, the TAN classifier, as presented in [7], has two weak points: not taking into account model uncertainty and lacking a theoretically well founded explanation for the use of softening of the induced model parameters (see section 2.2).

In [2] an alternative classifier based on empirical local Bayesian model averaging was proposed as a possible improvement for the first weak point. Furthermore, in [4] the fact that decomposable distributions over TANs allow the tractable calculation of the model averaging integral was used to construct SSTB-MATAN, a classifier that takes into account model uncertainty in a theoretically well founded way and that provides improved classification accuracy.

E. Suzuki and S. Arikawa (Eds.): DS 2004, LNAI 3245, pp. 73–88, 2004.

In [2] an alternative softening is proposed with a theoretically more appealing derivation based on multinomial sampling.

In this paper both weak points are addressed. A computationally more efficient alternative to the first weak point is introduced and a well founded softening alternative is proposed that solves the second weak point. More concretely, we show that under the assumption of decomposable distributions over TANs, we can efficiently compute the TAN model with a maximum a posteriori (MAP) probability. This result allows the construction of MAPTAN, a classifier that provides a well founded alternative to the softening proposed in [7] and improves its error rate and the accuracy of the predicted class probabilities. Furthermore, we will also prove that under this assumption we can efficiently compute the k most probable TAN models and their relative probabilities. This result allows the construction of MAPTAN+BMA, a classifier that takes into consideration model uncertainty and improves in time complexity and accuracy over its equivalent presented in [2]. Furthermore, established TAN classifiers do not easily allow the introduction of prior knowledge into the learning process. Being able to compute MAP TAN structures means that we can easily incorporate prior knowledge, whenever our prior knowledge can be represented as a decomposable distribution over TANs.

These results point out the relevance of decomposable distribution over TANs, which are conjugate to TAN models, for the construction of classifiers based on the TAN model.

The paper is structured as follows. In section 2 Tree Augmented Naive Bayes is presented and the notation to be used in the rest of the paper is introduced. In section 3 we present decomposable distributions over TANs. In section 4 we give the main results for finding MAP TAN structures. In section 5 we construct MAPTAN and MAPTAN+BMA, using the previously stated results. In section 6 we provide the empirical results showing that our classifiers improve over established TAN classifiers. We end up with some conclusions and future work in section 7.

2   Tree Augmented Naive Bayes

Tree Augmented Naive Bayes (TAN) appears as a natural extension to the *Naive Bayes* classifier [10, 11, 6]. TAN models are a restricted family of Bayesian networks in which the class variable has no parents, each attribute has as parent the class variable and additionally there is a tree of dependencies between non-class attributes. An example of TAN model can be seen in Figure 1(c).

In this section we start introducing the notation to be used in the rest of the paper. After that we discuss the TAN induction algorithm presented in [7].

2.1   Formalization and Notation

The notation used in the paper is an effort to put together the different notations used in [2, 8, 7, 13] and some conventions in the machine learning literature.

The Discrete Classification Problem. A *discrete attribute* is a finite set, for example we can define attribute *Pressure* as *Pressure* = {*Low, Medium, High*}. A *discrete domain* is a finite set of discrete attributes. We will use $\Omega = \{X_1, \ldots, X_m\}$ for a discrete domain, where $X_1, \ldots, X_m$ are the attributes in the domain. A *classified discrete domain* is a discrete domain where one of the attributes is distinguished as "class". We will use $\Omega_C = \{A_1, \ldots, A_n, C\}$ for a classified discrete domain. In the rest of the paper we will refer to an attribute either as X_i (when it is considered part of a discrete domain), A_i (when it is considered part of a classified discrete domain and it is not the class) and C (when it is the class of a classified discrete domain). We will use $V = \{A_1, \ldots, A_n\}$ for the set of attributes in a classified discrete domain that are not the class.

Given an attribute A, we will use $\#A$ as the number of different values of A. We define $\#\Omega = \prod_{i=1}^{m} \#X_i$ and $\#\Omega_C = \#C \prod_{i=1}^{n} \#A_i$.

An *observation* x in a classified discrete domain Ω_C is an ordered tuple $x = (x_1, \ldots, x_n, x_C) \in A_1 \times \ldots \times A_n \times C$. An *unclassified observation* S in Ω_C is an ordered tuple $S = (s_1, \ldots, s_n) \in A_1 \times \ldots \times A_n$. To be homogeneous we will abuse this notation a bit noting s_C for a possible value of the class for S. A *dataset* $\mathcal{D}$ in Ω_C is a multiset of classified observations in Ω_C.

We will use N for the number of observations in the dataset. We will also note $N_i(x_i)$ for the number of observations in $\mathcal{D}$ where the value for A_i is x_i, $N_{i,j}(x_i, x_j)$ the number of observations in $\mathcal{D}$ where the value for A_i is x_i and the value for A_j is x_j and similarly for $N_{i,j,k}(x_i, x_j, x_k)$ and so on.

A *classifier* in a classified discrete domain Ω_C is a procedure that given a dataset $\mathcal{D}$ in Ω_C and an unclassified observation S in Ω_C assigns a class to S.

Bayesian Networks for Discrete Classification. Bayesian networks offer a solution for the discrete classification problem. The approach is to define a random variable for each attribute in Ω (the class is included but not distinguished at this time). We will use $\mathbf{U} = \{\mathcal{X}_1, \ldots, \mathcal{X}_m\}$ where each $\mathcal{X}_i$ is a random variable over its corresponding attribute X_i. We extend the meaning of this notation to $\mathcal{A}_i$, $\mathcal{C}$ and $\mathcal{V}$. A *Bayesian network* over $\mathbf{U}$ is a pair $B = \langle G, \Theta \rangle$. The first component, G, is a directed acyclic graph whose vertices correspond to the random variables $\mathcal{X}_1, \ldots, \mathcal{X}_m$ and whose edges represent direct dependencies between the variables. The graph G encodes independence assumptions: each variable $\mathcal{X}_i$ is independent of its non-descendants given its parents in G. The second component of the pair, namely Θ, represents the set of parameters that quantifies the network. It contains a parameter $\theta_{i|\Pi_i}(x_i, \Pi_{x_i}) = P_B(x_i|\Pi_{x_i})$ for each $x_i \in X_i$ and $\Pi_{x_i} \in \Pi_{X_i}$, where Π_{X_i} denotes the Cartesian product of every X_j such that $\mathcal{X}_j$ is a parent of $\mathcal{X}_i$ in G. Π_i is the list of parents of $\mathcal{X}_i$ in G. We will use $\overline{\Pi}_i = \mathbf{U} - \{\mathcal{X}_i\} - \Pi_i$. A Bayesian network defines a unique joint probability distribution over $\mathbf{U}$ given by

$$P_B(x_1, \ldots, x_m) = \prod_{i=1}^{m} P_B(x_i|\Pi_{x_i}) = \prod_{i=1}^{m} \theta_{i|\Pi_i}(x_i|\Pi_{x_i}) \tag{1}$$

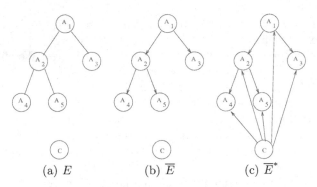

(a) E (b) $\overline{E}$ (c) $\overline{E}^*$

Fig. 1. Notation for learning with trees.

Learning with Trees. Given a classified domain Ω_C we will use $\mathcal{E}$ the set of undirected graphs E over $\{A_1, \ldots, A_n\}$ such that E is a tree (has no cycles). We will use $u, v \in E$ instead of $(A_u, A_v) \in E$ for simplicity. We will use $\overline{E}$ a directed tree for E. Every $\overline{E}$ uniquely determines the structure of a Tree Augmented Naive Bayes classifier, because from $\overline{E}$ we can construct $\overline{E}^* = \overline{E} \cup \{(C, A_i)| 1 \le i \le n\}$ as can be seen in an example in Figure 1. We note the root of a directed tree $\overline{E}$ as $\rho_{\overline{E}}$ (i.e. in Figure 1(b) we have that $\rho_{\overline{E}} = A_1$).

We will use $\Theta_{\overline{E}^*}$ the set of parameters that quantify the Bayesian network $B = \langle \overline{E}^*, \Theta_{\overline{E}^*} \rangle$. More concretely:

$\Theta_{\overline{E}^*} = (\boldsymbol{\theta}_C, \boldsymbol{\theta}_{\rho_{\overline{E}}|C}, \{\boldsymbol{\theta}_{v|u,C}|u, v \in \overline{E}\})$
$\boldsymbol{\theta}_C = \{\theta_C(c)|c \in C\}$ where $\theta_C(c) = P(C = c|B)$
$\boldsymbol{\theta}_{\rho_{\overline{E}}|C} = \{\theta_{\rho_{\overline{E}}|C}(i, c)|i \in A_{\rho_{\overline{E}}}, c \in C\}$ where
$\quad \theta_{\rho_{\overline{E}}|C}(i, c) = P(A_{\rho_{\overline{E}}} = i|C = c, B)$
For each $u, v \in \overline{E}$:
$\quad \boldsymbol{\theta}_{v|u,C} = \{\theta_{v|u,C}(j, i, c)|j \in A_v, i \in A_u, c \in C\}$ where
$\quad \theta_{v|u,C}(j, i, c) = P(A_v = j|A_u = i, C = c, B)$.

2.2   Learning Maximum Likelihood TAN

One of the measures used to learn Bayesian networks is the *log likelihood*:

$$LL(B|\mathcal{D}) = \sum_{x \in \mathcal{D}} \log(P_B(x)) \tag{2}$$

An interesting property of the TAN family is that we have an efficient procedure [7] for identifying the structure of the network which maximizes likelihood. To learn the maximum likelihood TAN we should use the following equation to compute the parameters.

$$\theta_{i|\Pi_i}(x_i, \Pi_{x_i}) = \frac{N_{i,\Pi_i}(x_i, \Pi_{x_i})}{N_{\Pi_i}(\Pi_{x_i})} \tag{3}$$

where $N_{i,\Pi_i}(x_i, \Pi_{x_i})$ stands for the number of times in the dataset that attribute i has value x_i and its parents have values Π_{x_i}. Equivalently, $N_{\Pi_i}(\Pi_{x_i}$ is the number of times in the dataset that the parents of attribute i have values Π_{x_i}.

It has been shown [7] that equation 3 leads to "overfitting" the model. Also in [7] Friedman et al. propose to use the parameters as given by

$$\theta_{i|\Pi_i}(x_i, \Pi_{x_i}) = \frac{N_{i,\Pi_i}(x_i, \Pi_{x_i})}{N_{\Pi_i}(\Pi_{x_i}) + N^0_{i|\Pi_i}} + \frac{N^0_{i|\Pi_i}}{N_{\Pi_i}(\Pi_{x_i}) + N^0_{i|\Pi_i}} \frac{N_i(x_i)}{N} \tag{4}$$

and suggest setting $N^0_{i|\Pi_i} = 5$ based on empirical results. Using equation 4 to fix the parameters improves the accuracy of the classifier. In our opinion, no well founded justification is given for the improvement. In the next section we introduce decomposable distribution over TANs, a family of probability distributions over the space of TAN models that allow to derive a well founded softening alternative.

3   Decomposable Distributions over TANs

Decomposable priors were introduced by Meila and Jaakola in [13] where it was demonstrated for tree belief networks that if we assume a decomposable prior, the posterior probability is also decomposable and can be completely determined analytically in polynomial time.

In this section we introduce decomposable distributions over TANs, which are probability distributions in the space $\mathcal{M}$ of TAN models and an adaptation of decomposable priors, as they appear in [13], to the task of learning TAN.

Decomposable distributions are constructed in two steps. In the first step, a distribution over the set of different undirected tree structures is defined. Every directed tree structure is defined to have the same probability as its undirected equivalent. In the second step, a distribution over the set of parameters is defined so that it is also independent on the structure. In the rest of the paper we will assume ξ implies a decomposable distribution over $\mathcal{M}$ with hyperparameters $\beta, \mathbf{N}'$ (these hyperparameters will be explained along the development). Under this assumption, the probability for a model $B = \langle \overline{E}^*, \Theta_{\overline{E}^*} \rangle$ (a TAN with fixed tree structure $\overline{E}^*$ and fixed parameters $\Theta_{\overline{E}^*}$) is determined by:

$$P(B|\xi) = P(\overline{E}^*, \Theta_{\overline{E}^*}|\xi) = P(\overline{E}^*|\xi)P(\Theta_{\overline{E}^*}|\overline{E}^*, \xi) \tag{5}$$

In the following sections we specify the value of $P(\overline{E}^*|\xi)$ (decomposable distribution over structures) and $P(\Theta_{\overline{E}^*}|\overline{E}^*, \xi)$ (decomposable distribution over parameters).

3.1   Decomposable Distribution over TAN Structures

One of the hyperparameters of a decomposable distribution is an $n \times n$ matrix $\beta = (\beta_{u,v})$ such that $\forall u, v : 1 \leq u, v \leq n : \beta_{u,v} = \beta_{v,u} \geq 0 ; \beta_{v,v} = 0$. We can interpret $\beta_{u,v}$ as a measure of how possible is under ξ that the edge $(\mathcal{A}_u, \mathcal{A}_v)$ is contained in the TAN model underlying the data.

Given ξ, the probability of a TAN structure $\overline{E}^*$ is defined as:

$$P(\overline{E}^*|\xi) = \frac{1}{Z_\beta} \prod_{u,v \in E} \beta_{u,v} \qquad (6)$$

where Z_β is a normalization constant

3.2   Decomposable Distribution over TAN Parameters

Applying equation 1 to the case of TAN we have that

$$P(\Theta_{\overline{E}^*}|\overline{E}^*, \xi) = P(\boldsymbol{\theta}_C|\overline{E}^*, \xi)P(\boldsymbol{\theta}_{\rho_{\overline{E}}|C}|\overline{E}^*, \xi) \times \prod_{u,v \in \overline{E}} P(\boldsymbol{\theta}_{v|u,C}|\overline{E}^*, \xi) \qquad (7)$$

A decomposable distribution has a hyperparameter set $\mathbf{N}'$, which can be understood as the prior observation counts, $\mathbf{N}' = \{N'_{v,u,C}(j,i,c)|1 \le u \ne v \le n \;;\; j \in A_v \;;\; i \in A_u \;;\; c \in C\}$ with the constraint that exist $N'_{u,C}(i,c)$, $N'_C(c)$, N' such that for every u,v:

$$N'_{u,C}(i,c) = \sum_{j \in A_v} N'_{v,u,C}(j,i,c) \qquad (8)$$

$$N'_C(c) = \sum_{i \in A_u} N'_{u,C}(i,c) \qquad (9)$$

$$N' = \sum_{c \in C} N'_C(c) \qquad (10)$$

Given ξ, a decomposable probability distribution over parameters with hyperparameter $\mathbf{N}'$ is defined by equation 7 and the following set of Dirichlet distributions:

$$P(\boldsymbol{\theta}_C|\overline{E}, \xi) = D(\theta_C(.); N'_C(.)) \qquad (11)$$

$$P(\boldsymbol{\theta}_{\rho_{\overline{E}}|C}|\overline{E}, \xi) = \prod_{c \in C} D(\theta_{\rho_{\overline{E}}|C}(.,c); N'_{\rho_{\overline{E}},C}(.,c)) \qquad (12)$$

$$P(\boldsymbol{\theta}_{v|u,C}|\overline{E}, \xi) = \prod_{c \in C} \prod_{i \in A_u} D(\theta_{v|u,C}(.,i,c); N'_{v,u,C}(.,i,c)) \qquad (13)$$

If the conditions in equations 5, 6, 7, 8, 9, 10, 11, 12 and 13 hold, we will say that $P(M|\xi)$ follows a decomposable distribution with hyperparameters $\boldsymbol{\beta}, \mathbf{N}'$.

3.3   Learning with Decomposable Distributions

Assume that the data is generated by a TAN model and that $P(B|\xi)$ follows a decomposable distribution with hyperparameters $\boldsymbol{\beta}$, $\mathbf{N}'$. Then, $P(B|\mathcal{D}, \xi)$, the posterior probability distribution after observing a dataset $\mathcal{D}$, is a decomposable distribution with parameters $\boldsymbol{\beta}^*$, $\mathbf{N}'^*$ given by:

$$\beta^*_{u,v} = \beta_{u,v} W_{u,v} \tag{14}$$

$$N'^*_{u,v,C}(j,i,c) = N'_{u,v,C}(j,i,c) + N_{u,v,C}(j,i,c) \tag{15}$$

where

$$W_{u,v} = \prod_{c \in C} \prod_{i \in A_u} \frac{\Gamma(N'_{u,C}(i,c))}{\Gamma(N'^*_{u,C}(i,c))} \prod_{c \in C} \prod_{j \in A_v} \frac{\Gamma(N'_{v,C}(j,c))}{\Gamma(N'^*_{v,C}(j,c))} \times$$
$$\times \prod_{c \in C} \prod_{i \in A_u} \prod_{j \in A_v} \frac{\Gamma(N'^*_{v,u,C}(j,i,c))}{\Gamma(N'_{v,u,C}(j,i,c))} \tag{16}$$

The proof appears in [5].

3.4 Classifying with Decomposable Distributions Given an Undirected Structure

Assume that the data is generated by a TAN model and that $P(B|\xi)$ follows a decomposable distribution with hyperparameters $\boldsymbol{\beta}$, $\mathbf{N}'$. Then, $P(\mathcal{C} = s_C | \mathcal{V} = S, E, \xi)$, the probability of a class s_C given an unclassified instance S and an undirected TAN structure E, fulfills

$$P(\mathcal{C} = s_C | \mathcal{V} = S, E, \xi) \propto h_0^{S,s_C} \prod_{u,v \in E} h_{u,v}^{S,s_C} \tag{17}$$

where

$$h_0^{S,s_C} = \frac{1}{Z_\beta} \frac{1}{N'} \prod_{A_u \in V} N'_{u,C}(s_u, s_C) \tag{18}$$

$$h_{u,v}^{S,s_C} = \frac{N'_{v,u,C}(s_v, s_u, s_C)}{N'_{u,C}(s_u, s_C) N'_{v,C}(s_v, s_C)} \tag{19}$$

The proof appears in [3].

4 Maximum a Posteriori Results for Decomposable Distributions over TANs

In this section we show that if we assume a decomposable distribution over TANs as prior over the set of models, the undirected tree structure underlying the MAP TAN can be found in $\mathcal{O}((N + r^3) \cdot n^2)$ time where $r = \max(\max_{i \in V} \#A_i, \#C)$. Furthermore, we also show that we can find the k undirected tree structures underlying the k MAP TAN models and their relative weights in $\mathcal{O}((N + r^3 + k) \cdot n^2)$.

Both results are supported by the next result, that shows that computing the most probable undirected tree structure under a decomposable distribution over TANs with hyperparameters $\boldsymbol{\beta}, \mathbf{N}'$ can be reduced to calculating the maximum spanning tree (MST) for the graph with adjacency matrix $\log(\boldsymbol{\beta})$.

4.1 Calculating the Most Probable Undirected Tree Structure Under a Decomposable Distribution over TANs

From equation 6, assuming that $\forall u, v \; u \neq v; \; \beta_{u,v} > 0$, we can characterize the most probable undirected tree given a decomposable distribution over TANs with hyperparameters $\beta, \mathbf{N}'$ is given by

$$MPT(\beta, \mathbf{N}') = \underset{E \in \mathcal{E}}{argmax} \sum_{u,v \in E} \log(\beta_{u,v}) \tag{20}$$

Considering the matrix $\log(\beta)$ as an adjacency matrix, $MPT(\beta, \mathbf{N}')$ is the MST for the graph represented by that adjacency matrix. Hence, if we are given a decomposable distribution over TANs with hyperparameter β, we can find the most probable undirected tree by calculating the logarithm of every element in the matrix and then running any algorithm for finding the MST. The complexity of the MST algorithm for a complete graph is $\mathcal{O}(n^2)$ [15].

4.2 Calculating the MAP TAN Structure Given a Prior Decomposable Distribution over TANs

If we assume a decomposable prior distribution over TANs the posterior distribution after a dataset $\mathcal{D}$ follows a decomposable distribution over TANs (section 3.3). Since the posterior is a decomposable distribution over trees, we can apply the former result for finding the most probable undirected tree over it and we get the MAP tree. We can translate this result into algorithm 1.

4.3 Calculating the k MAP TAN Structures and Their Relative Probability Weights Given a Prior Decomposable Distribution over TANs

The problem of computing the k MST in order is well known and can be solved in $\mathcal{O}(k \cdot n^2)$ for a complete graph [9]. It is easy to see that if in the last step of `MAPTreeStructure` instead of calculating the MST we calculate the k MST and their relative weights as shown in algorithm 2, the algorithm will return the k MAP TANs and their relative probabilities. The time complexity of the new algorithm is $\mathcal{O}((N + r^3 + k) \cdot n^2)$.

5 Constructing the MAPTAN and MAPTAN+BMA Classifiers

From the result in section 3.4 it is easy to see that given an undirected TAN structure E, the probability distribution $P(\mathcal{C} = s_C | \mathcal{V} = S, E, \xi)$ can be represented as a TAN model with structure $\overline{E}^*$, such that its undirected version coincides with E and its parameter set is given by

$$\theta_{u|v,C}(s_u, s_v, s_C) = \frac{N'_{u,v,C}(s_u, s_v, s_C)}{N'_{v,C}(s_v, s_C)}$$
$$\theta_{u|C}(s_u, s_C) = \frac{N'_{u,C}(s_u, s_C)}{N'_C(s_C)} \tag{21}$$
$$\theta_C(s_C) = \frac{N'_C(s_C)}{N'}$$

```
procedure MAPTANStructure (Dataset D,Matrix β,CountingSet N')
    var
        CountingSet N';
        Matrix lβ*;
    begin
        N'* = CalcN'PosteriorTAN(D,N');
        lβ* = CalcLogBetaPosteriorTAN(β,N',N'*);
        return MST(lβ*);

procedure CalcN'PosteriorTAN (Dataset D,CountingSet N')
    var
        CountingSet N'*;
    begin
        foreach attribute u
            foreach attribute v < u
                foreach value xu ∈ Au
                    foreach value xv ∈ Av
                        foreach value c ∈ C
                            N'*u,v,C(xu, xv, c) = N'u,v(xu, xv, c);
        foreach attribute x ∈ D
            foreach attribute u
                foreach attribute v < u
                    N'*u,v,C(xu, xv, xC) = N'*u,v,C(xu, xv, xC) + 1;
        return N'*;

procedure CalcLogBetaPosteriorTAN (Matrix β,CountingSet N', N'*)
    var
        Matrix lβ*;
    begin
        foreach attribute u
            foreach attribute v < u
                lβ*u,v = log βu,v + CalcLogWTAN(N',N'*,u,v);
        return lβ*;

procedure CalcLogWTAN (CountingSet N', N'*, int u, v)
    begin
        w = 0;
        foreach value c ∈ C
            foreach value xu ∈ Au
                w = w + logΓ(N'u,C(xu, c)) - logΓ(N'*u,C(xu, c));
            foreach value xv ∈ Av
                w = w + logΓ(N'v,C(xv, c)) - logΓ(N'*v,C(xv, c));
            foreach value xu ∈ Au
                foreach value xv ∈ Av
                    w = w + logΓ(N'*u,v,C(xu, xv, c))
                          - logΓ(N'u,v,C(xu, xv, c));
        return w;
```

Algorithm 1: Computation of the MAP TAN

A similar result in the case of decomposable distribution over trees can also be found in [14]. Given a decomposable prior we can calculate the decomposable posterior using the result in section 3.3 and then apply the result we have just enunciated to the posterior.

The posterior probability distribution $P(\mathcal{C} = s_C | \mathcal{V} = S, E, \mathcal{D}, \xi)$ can be represented as a TAN model with structure $\overline{E}^*$, such that its undirected version coincides with E and its parameter set is given by

$$\theta_{u|v,C}(s_u, s_v, s_C) = \frac{N'^*_{u,v,C}(s_u,s_v,s_C)}{N'^*_{v,C}(s_v,s_C)}$$

$$\theta_{u|C}(s_u, s_C) = \frac{N'^*_{u,C}(s_u,s_C)}{N'^*_{C}(s_C)} \tag{22}$$

$$\theta_C(s_C) = \frac{N'^*_{C}(s_C)}{N'^*}$$

In [4] we argued that

$$\forall u,v \ ; \ 1 \leq u \neq v \ \leq n \ ; \ \beta_{u,v} = 1 \tag{23}$$

$$\forall u,v; 1 \leq u \neq v \leq n; \forall j \in A_v; \forall i \in A_u; \forall c \in C$$

$$N'_{v,u,C}(j,i,c) = \frac{\lambda}{\#C\#A_u\#A_v} \tag{24}$$

where λ is an equivalent sample size, provide a reasonable choice of the hyper-parameters if no information from the domain is available.

5.1   MAPTAN Classifier

After fixing the prior hyperparameters, the learning step for MAPTAN classifier consists in:

1. Applying algorithm 1 to find the undirected tree E underlying the MAP TAN structure given a dataset $\mathcal{D}$.
2. Randomly choose a root, create a directed tree $\overline{E}$ and from it a directed TAN structure $\overline{E}^*$.
3. Use equation 22 to fix the TAN parameters.

For classifying an unclassified observation, we have to apply the TAN that has been learned for each of the $\#C$ classes to construct a probability distribution over the values of the class C and then choose the most probable class.

This classification algorithm runs in $\mathcal{O}((N+r^3)\cdot n^2)$ learning time and $\mathcal{O}(nr)$ classification time.

```
procedure k-MAPTANs (Dataset D,Matrix β,CountingSet N', int k)
    var
        CountingSet N';
        WeightedTreeSet WTS;
        Matrix lβ*;
    begin
        N'* = CalcN'PosteriorTAN(D,N');
        lβ* = CalcLogBetaPosteriorTAN(β,N',N'*);
        WTS = k-MST(lβ*,k);
        CalcTreeWeights(WTS,lβ*);
        return WTS;
```

Algorithm 2: Computation of the k MAP TANs

5.2   MAPTAN+BMA Classifier

After fixing the prior hyperparameters, the learning stage for MAPTAN+BMA classifier consists in:

1. Applying algorithm 2 to find the k undirected trees underlying the k MAP TAN structures and their relative probability weights given a dataset $\mathcal{D}$.
2. Generate a TAN model for each of the undirected tree structures as we did in MAPTAN.
3. Assign to each TAN model the weight of its corresponding undirected tree.

The resulting probabilistic model will be a mixture of TANs. For classifying an unclassified observation, we have to apply the k TAN models for the $\#C$ classes and calculate the weighted average to construct a probability distribution over the values of the class C and then choose the most probable class.

This classification algorithm runs in $\mathcal{O}((N + r^3 + k) \cdot n^2)$ learning time and $\mathcal{O}(nrk)$ classification time.

5.3   Relevant Characteristics of MAPTAN and MAPTAN+BMA

We have shown that decomposable distributions over TANs can be used to construct two well founded classifiers: MAPTAN and MAPTAN+BMA. In the introduction we highlighted two possible ways in which the TAN classifier, as presented in [7], could be improved: by taking into account model uncertainty and by providing a theoretically well founded explanation for the use of softening.

We have seen that MAPTAN+BMA provides a theoretically well founded way of dealing with model uncertainty. Its learning time complexity regarding N is almost equivalent to that of STAN, and it grows polynomially on k. This is much more efficient than the algorithm for learning k TAN models proposed in [2]. MAPTAN+BMA has a classification time complexity, $\mathcal{O}(nrk)$ reasonably higher than that of STAN. Furthermore, we can use k as an *effort knob*, in the sense of [16], hence providing a useful feature for data mining users that allows them to decide how much computational power they want to spend in the task. In our opinion, MAPTAN+BMA provides a good complexity tradeoff to deal with model uncertainty when learning TAN.

Both MAPTAN and MAPTAN+BMA can be interpreted as using softening in both the structure search and the parameter fixing. This softening appears, in a natural way, as the result of assuming a decomposable distribution over TANs as the prior over the set of models. In our opinion MAPTAN is theoretically more appealing than STAN.

Both MAPTAN and MAPTAN+BMA share the relevant characteristic of allowing the use of some form of prior information if such is available, specially structure related information. For example, if we have expert knowledge that tell us that one of the edges of the tree is much more (equiv. much less) likely than the others it is very easy to incorporate this knowledge when fixing the prior hyperparameter matrix β. Evidently, as was pointed out in [12], decomposable distributions do not allow the expression of some types of prior information such as "if edge (u, v) exists then edge (w, z) is very likely to exist".

6   Empirical Results

We tested four algorithms over 17 datasets from the Irvine repository [1]. To discretize continuous attributes we used equal frequency discretization with 5 intervals. For each dataset and algorithm we tested both error rate and *LogScore*. *LogScore* is calculated by adding the minus logarithm of the probability assigned by the classifier to the correct class and gives an idea of how well the classifier is estimating probabilities (the smaller the score the better the result). If we name the test set $\mathcal{D}'$ we have

$$LogScore(B, \mathcal{D}') = \sum_{(S, s_C) \in \mathcal{D}'} -\log(P(\mathcal{C} = s_C | \mathcal{V} = S, B)) \qquad (25)$$

For the evaluation of both error rate and *LogScore* we used 10 fold cross validation. We tested the algorithm with the 10%, 50% and 100% of the learning data for each fold, in order to get an idea of the influence of the amount of data in the behaviors of both error rate and *LogScore* for the algorithm.

The error rates appear in Tables 1, 3 and 5 with the best method for each dataset boldfaced. *LogScore*'s appear in Tables 2, 4 and 6. The columns of the

Table 1. Averages and standard deviations of error rate using 10% of the learning data.

Dataset	MAPTAN	MAPTAN+BMA	sTAN	sTAN+BMA
ADULT	**17.18 ± 0.68**	17.19 ± 0.71	17.60 ± 0.82	17.60 ± 0.80
AUSTRALIAN	19.91 ± 1.14	**19.62 ± 1.13**	25.39 ± 1.18	24.96 ± 1.13
BREAST	17.23 ± 1.21	16.89 ± 1.28	8.73 ± 0.87	**7.73 ± 0.93**
CAR	17.19 ± 1.04	**16.50 ± 0.84**	19.38 ± 0.95	17.60 ± 0.77
CHESS	9.55 ± 0.80	**9.48 ± 0.86**	10.89 ± 0.56	10.91 ± 0.53
CLEVE	**28.12 ± 1.68**	28.14 ± 1.59	32.37 ± 1.00	31.89 ± 1.27
CRX	19.77 ± 0.91	**19.16 ± 1.00**	25.14 ± 0.87	24.18 ± 0.98
FLARE	23.50 ± 1.09	23.16 ± 1.09	19.94 ± 0.85	**19.92 ± 0.88**
GLASS	47.02 ± 1.66	**45.72 ± 1.59**	59.19 ± 1.78	58.54 ± 1.83
GLASS2	33.69 ± 1.74	**32.87 ± 1.82**	37.75 ± 1.39	36.63 ± 1.37
IRIS	28.67 ± 2.33	26.27 ± 2.30	25.87 ± 3.07	**24.80 ± 2.96**
LETTER	30.22 ± 0.96	**30.19 ± 0.97**	36.11 ± 1.39	34.68 ± 1.37
LIVER	45.52 ± 1.26	44.96 ± 1.06	42.39 ± 0.94	**41.24 ± 1.37**
NURSERY	7.87 ± 1.03	**7.57 ± 1.04**	8.88 ± 1.12	8.50 ± 1.12
PRIMARY-TUMOR	74.52 ± 1.73	74.28 ± 1.66	**71.67 ± 1.54**	71.73 ± 1.44
SOYBEAN	26.53 ± 1.30	**26.51 ± 1.33**	30.79 ± 1.28	30.82 ± 1.33
VOTES	**9.61 ± 0.94**	9.67 ± 0.99	14.14 ± 0.93	14.13 ± 0.71

Table 2. Averages and standard deviations of *LogScore* using 10% of the learning data.

Dataset	MAPTAN	MAPTAN+BMA	sTAN	sTAN+BMA
ADULT	562.25 ± 3.75	**561.39 ± 3.71**	567.09 ± 3.92	567.64 ± 4.00
AUSTRALIAN	18.54 ± 0.95	17.68 ± 0.96	17.85 ± 0.64	**17.06 ± 0.60**
BREAST	23.59 ± 1.67	18.24 ± 1.56	8.12 ± 0.69	**7.56 ± 0.65**
CAR	34.89 ± 1.02	**32.79 ± 0.98**	38.55 ± 0.91	36.52 ± 0.86
CHESS	32.50 ± 0.89	**32.25 ± 0.91**	35.39 ± 0.58	35.40 ± 0.59
CLEVE	11.15 ± 1.06	10.09 ± 0.96	8.49 ± 0.74	**8.23 ± 0.76**
CRX	19.44 ± 1.06	18.30 ± 1.00	17.84 ± 1.05	**16.89 ± 1.00**
FLARE	51.12 ± 1.17	**49.48 ± 1.15**	24332.38 ± 56.59	24332.03 ± 56.59
GLASS	20.49 ± 1.45	**17.14 ± 1.40**	11713.24 ± 72.91	11713.00 ± 72.91
GLASS2	6.45 ± 0.79	5.49 ± 0.64	4.68 ± 0.57	**4.57 ± 0.54**
IRIS	4.58 ± 0.68	4.06 ± 0.69	4.04 ± 0.67	**3.96 ± 0.70**
LETTER	3535.93 ± 12.92	3495.14 ± 13.52	1385.73 ± 8.95	**1300.23 ± 8.38**
LIVER	18.71 ± 0.95	15.87 ± 0.92	12.62 ± 0.79	**11.71 ± 0.65**
NURSERY	112.72 ± 2.47	**111.95 ± 2.47**	3126.39 ± 77.45	3123.62 ± 77.45
PRIMARY-TUMOR	71.74 ± 2.08	**69.08 ± 2.05**	75927.03 ± 123.39	75926.94 ± 123.39
SOYBEAN	68.52 ± 1.77	**65.29 ± 1.55**	41125.59 ± 108.25	41125.46 ± 108.25
VOTES	5.66 ± 0.66	**5.17 ± 0.60**	6.09 ± 0.50	6.03 ± 0.48

Table 3. Averages and standard deviations of error rate using 50% of the learning data.

Dataset	MAPTAN	MAPTAN+BMA	sTAN	sTAN+BMA
ADULT	16.26 ± 0.75	16.28 ± 0.77	16.46 ± 0.78	16.45 ± 0.83
AUSTRALIAN	15.36 ± 0.94	15.13 ± 1.09	18.14 ± 0.91	17.74 ± 0.80
BREAST	5.92 ± 0.74	5.84 ± 0.78	5.26 ± 0.84	4.75 ± 0.72
CAR	7.62 ± 0.75	7.55 ± 0.76	8.68 ± 0.68	8.09 ± 0.58
CHESS	7.87 ± 0.44	7.90 ± 0.44	8.25 ± 0.49	8.15 ± 0.49
CLEVE	19.82 ± 1.30	20.27 ± 1.27	24.01 ± 1.31	23.57 ± 1.28
CRX	15.47 ± 1.01	15.30 ± 1.02	18.12 ± 0.92	17.68 ± 0.85
FLARE	19.83 ± 0.72	19.81 ± 0.65	18.55 ± 0.62	18.54 ± 0.72
GLASS	24.02 ± 1.22	23.31 ± 1.48	33.79 ± 1.14	33.86 ± 0.97
GLASS2	23.69 ± 1.62	22.81 ± 1.61	22.38 ± 1.53	23.40 ± 1.54
IRIS	11.60 ± 1.22	11.07 ± 1.08	8.40 ± 1.00	8.27 ± 0.82
LETTER	14.79 ± 0.78	14.79 ± 0.78	15.62 ± 0.91	15.31 ± 0.83
LIVER	37.33 ± 1.16	36.90 ± 1.15	36.73 ± 1.60	35.17 ± 1.34
NURSERY	6.39 ± 0.77	6.37 ± 0.89	7.09 ± 0.80	6.03 ± 0.97
PRIMARY-TUMOR	59.15 ± 1.67	59.09 ± 1.63	60.23 ± 1.17	59.87 ± 1.33
SOYBEAN	6.64 ± 0.77	6.50 ± 0.85	7.88 ± 0.71	7.80 ± 0.82
VOTES	6.22 ± 0.83	6.25 ± 0.84	7.63 ± 0.93	7.76 ± 0.93

Table 4. Averages and standard deviations of *LogScore* using 50% of the learning data.

Dataset	MAPTAN	MAPTAN+BMA	sTAN	sTAN+BMA
ADULT	507.82 ± 3.82	507.52 ± 3.81	520.03 ± 3.93	518.82 ± 3.91
AUSTRALIAN	12.86 ± 0.82	12.57 ± 0.84	14.79 ± 0.76	14.41 ± 0.59
BREAST	10.95 ± 0.67	9.20 ± 0.69	5.17 ± 0.64	4.40 ± 0.62
CAR	15.96 ± 0.44	15.90 ± 0.40	20.44 ± 0.51	19.73 ± 0.48
CHESS	26.70 ± 0.66	26.66 ± 0.68	27.32 ± 0.73	27.12 ± 0.79
CLEVE	6.83 ± 0.69	6.70 ± 0.66	7.38 ± 0.66	7.15 ± 0.63
CRX	13.28 ± 0.89	12.93 ± 0.85	15.62 ± 1.11	15.21 ± 1.07
FLARE	39.81 ± 1.16	39.45 ± 1.15	4233.42 ± 41.82	4233.31 ± 41.82
GLASS	11.05 ± 0.73	9.37 ± 0.84	309.52 ± 24.49	309.25 ± 24.49
GLASS2	5.06 ± 0.73	4.64 ± 0.64	3.86 ± 0.53	3.68 ± 0.51
IRIS	1.87 ± 0.35	1.77 ± 0.34	1.52 ± 0.37	1.48 ± 0.34
LETTER	1030.65 ± 9.50	1030.65 ± 9.50	574.47 ± 6.13	559.56 ± 6.17
LIVER	13.03 ± 0.89	12.21 ± 0.77	10.78 ± 0.74	10.39 ± 0.71
NURSERY	96.60 ± 2.40	96.52 ± 2.42	1596.96 ± 67.06	1594.32 ± 67.06
PRIMARY-TUMOR	44.24 ± 1.25	43.00 ± 1.23	12028.93 ± 51.79	12028.74 ± 51.79
SOYBEAN	6.79 ± 0.83	6.47 ± 0.74	907.34 ± 42.43	907.27 ± 42.43
VOTES	3.66 ± 0.58	3.54 ± 0.52	5.04 ± 0.80	4.50 ± 0.69

tables are the induction methods and the rows are the datasets. The meaning of the column headers are:

- STAN is the softened TAN induction algorithm as presented in [7].
- STAN+BMA is the classifier resulting from applying local Bayesian model averaging (see [2]) to STAN.
- MAPTAN, is the classifier based on the MAP TAN model described in section 5.
- MAPTAN+BMA is the classifier based on the weighted average of the k MAP TAN models described also in section 5.

Table 5. Averages and standard deviations of error rate using 100% of the learning data.

Dataset	MAPTAN	MAPTAN+BMA	sTAN	sTAN+BMA
ADULT	**16.35 ± 0.73**	16.35 ± 0.73	16.46 ± 0.68	16.42 ± 0.72
AUSTRALIAN	13.68 ± 0.75	**13.65 ± 0.74**	16.49 ± 0.65	16.43 ± 0.72
BREAST	4.75 ± 0.53	4.63 ± 0.48	4.29 ± 0.66	**3.72 ± 0.45**
CAR	**5.76 ± 0.52**	5.78 ± 0.45	6.23 ± 0.55	6.16 ± 0.53
CHESS	7.71 ± 0.25	**7.67 ± 0.21**	7.89 ± 0.38	7.68 ± 0.44
CLEVE	18.74 ± 1.15	**18.53 ± 1.18**	19.99 ± 1.26	19.73 ± 1.18
CRX	13.67 ± 0.53	**13.53 ± 0.58**	15.71 ± 0.66	15.79 ± 0.74
FLARE	19.71 ± 0.49	19.71 ± 0.55	18.46 ± 0.30	**18.31 ± 0.24**
GLASS	**18.46 ± 1.20**	18.74 ± 1.29	26.58 ± 1.22	25.99 ± 1.28
GLASS2	19.81 ± 0.85	20.25 ± 1.39	19.61 ± 1.42	**18.06 ± 1.43**
IRIS	7.73 ± 1.70	7.47 ± 1.66	8.13 ± 1.44	**7.20 ± 1.43**
LETTER	**11.49 ± 0.74**	11.49 ± 0.74	12.69 ± 0.77	12.48 ± 0.83
LIVER	34.35 ± 0.86	33.99 ± 0.77	33.36 ± 0.98	**33.19 ± 1.10**
NURSERY	6.33 ± 0.89	6.26 ± 0.91	6.62 ± 0.75	**4.81 ± 0.76**
PRIMARY-TUMOR	55.09 ± 1.24	**54.68 ± 1.02**	56.74 ± 1.09	56.32 ± 0.93
SOYBEAN	5.47 ± 0.62	**5.27 ± 0.62**	5.97 ± 0.50	5.94 ± 0.49
VOTES	**5.89 ± 0.74**	5.89 ± 0.72	6.26 ± 0.81	6.34 ± 0.56

Table 6. Averages and standard deviations of *LogScore* using 100% of the learning data.

Dataset	MAPTAN	MAPTAN+BMA	sTAN	sTAN+BMA
ADULT	495.88 ± 3.68	**495.70 ± 3.67**	508.10 ± 3.07	508.01 ± 3.07
AUSTRALIAN	10.65 ± 0.46	**10.47 ± 0.44**	12.90 ± 0.65	12.66 ± 0.61
BREAST	8.96 ± 0.87	7.89 ± 0.61	4.85 ± 0.50	**4.28 ± 0.54**
CAR	**14.11 ± 0.40**	14.12 ± 0.40	16.29 ± 0.39	16.31 ± 0.41
CHESS	26.12 ± 0.40	**26.09 ± 0.32**	26.46 ± 0.46	26.22 ± 0.36
CLEVE	6.10 ± 0.43	**6.05 ± 0.38**	6.51 ± 0.44	6.29 ± 0.51
CRX	11.34 ± 0.62	**11.05 ± 0.60**	13.97 ± 0.68	13.76 ± 0.58
FLARE	35.82 ± 0.92	**35.61 ± 0.90**	1532.39 ± 0.62	1532.22 ± 0.65
GLASS	8.53 ± 1.05	7.50 ± 1.03	7.40 ± 0.59	**7.12 ± 0.52**
GLASS2	4.20 ± 0.56	3.91 ± 0.54	3.20 ± 0.39	**3.08 ± 0.37**
IRIS	1.29 ± 0.53	1.22 ± 0.53	1.18 ± 0.44	**1.16 ± 0.44**
LETTER	612.99 ± 7.71	612.99 ± 7.71	441.94 ± 5.61	**433.37 ± 5.84**
LIVER	10.79 ± 0.63	10.61 ± 0.66	**9.59 ± 0.44**	9.72 ± 0.60
NURSERY	94.59 ± 2.45	94.57 ± 2.43	91.52 ± 2.41	**89.41 ± 2.30**
PRIMARY-TUMOR	35.28 ± 0.99	**34.64 ± 0.94**	6327.87 ± 38.33	6327.64 ± 38.33
SOYBEAN	3.45 ± 0.50	**3.38 ± 0.49**	4.49 ± 0.51	4.45 ± 0.48
VOTES	3.74 ± 0.59	**3.57 ± 0.58**	3.96 ± 0.55	3.76 ± 0.46

6.1   Interpretation of the Results

Summarizing the empirical results in the tables, we can conclude that:

- MAPTAN improves STAN error rate for most datasets and has a similar *LogScore*.
- MAPTAN+BMA improves MAPTAN's *LogScore* for most datasets. When little data is available, it also improves its error rate.
- MAPTAN+BMA improves STAN+BMA error rate and *LogScore* for many datasets.

7 Conclusions

We have seen that under a decomposable distribution over TANs it is possible to efficiently determine the MAP undirected TAN structure and the set of k MAP TAN structures and their relative probability weights. We used these results to construct two new classifiers: MAPTAN and MAPTAN+BMA. We have provided empirical results showing that both classifiers improve over established TAN based classifiers (in the case of MAPTAN+BMA reducing also severely the computational complexity). Our results give also a satisfying theoretical explanation for the use of softening in TAN based classifiers (as the result of Bayesian model averaging over parameters). These results highlight the relevance of decomposable distributions over TANs for the construction of TAN based classifiers.

References

1. C. Blake, E. Keogh, and C. J. Merz. UCI repository of machine learning databases, 1998.
2. Jesús Cerquides. Applying General Bayesian Techniques to Improve TAN Induction. In *Proceedings of the International Conference on Knowledge Discovery and Data Mining, KDD99*, 1999.
3. Jesús Cerquides. *Improving Bayesian network classifiers*. PhD thesis, Technical University of Catalonia, 2003.
4. Jesús Cerquides and Ramon López de Màntaras. Tractable bayesian learning of tree augmented naive bayes classifiers. In *Proceedings of the Twentieth International Conference on Machine Learning*, pages 75–82, 2003.
5. Jesús Cerquides and Ramon López de Màntaras. Tractable bayesian learning of tree augmented naive bayes classifiers. long version. Technical Report IIIA-2003-04, Institut d'Investigació en Intel.ligència Artificial, 2003.
6. Pedro Domingos and Michael Pazzani. On the Optimality of the Simple Bayesian Classifier under Zero-One Loss. *Machine Learning*, 29:103–130, 1997.
7. Nir Friedman, Dan Geiger, and Moises Goldszmidt. Bayesian network classifiers. *Machine Learning*, 29:131–163, 1997.
8. D. Heckerman, D. Geiger, and D. Chickering. Learning bayesian networks: The combination of knowledge and statistical data. *Machine Learning*, 20:197–243, 1995.
9. Naoki Katoh, Toshihide Ibaraki, and H. Mine. An algorithm for finding k minimum spanning trees. *SIAM J. Comput.*, 10(2):247–255, 1981.
10. Petri Kontkanen, Petri Myllymaki, Tomi Silander, and Henry Tirri. Bayes Optimal Instance-Based Learning. In C. Nédellec and C. Rouveirol, editors, *Machine Learning: ECML-98, Proceedings of the 10th European Conference*, volume 1398 of *Lecture Notes in Artificial Intelligence*, pages 77–88. Springer-Verlag, 1998.
11. Pat Langley, Wayne Iba, and Kevin Thompson. An Analysis of Bayesian Classifiers. In *Proceedings of the Tenth National Conference on Artificial Intelligence*, pages 223–228. AAAI Press and MIT Press, 1992.
12. M. Meila and T. Jaakkola. Tractable bayesian learning of tree belief networks. In *Proc. of the Sixteenth Conference on Uncertainty in Artificial Intelligence*, 2000.
13. Marina Meila and Tommi Jaakkola. Tractable bayesian learning of tree belief networks. Technical Report CMU-RI-TR-00-15, Robotics Institute, Carnegie Mellon University, Pittsburgh, PA, May 2000.

14. Marina Meila and Michael I. Jordan. Learning with mixtures of trees. *Journal of Machine Learning Research*, 1:1–48, 2000.
15. Seth Pettie and Vijaya Ramachandran. An optimal minimum spanning tree algorithm. *Journal of the ACM (JACM)*, 49(1):16–34, 2002.
16. Kurt Thearling. Some thoughts on the current state of data mining software applications. In *Keys to the Commercial Success of Data Mining, KDD'98 Workshop*, 1998.

Improving Prediction
of Distance-Based Outliers

Fabrizio Angiulli, Stefano Basta, and Clara Pizzuti

ICAR-CNR
Via Pietro Bucci, 41C
Università della Calabria
87036 Rende (CS), Italy
Phone +39 0984 831738/37/24, Fax +39 0984 839054
{angiulli,basta,pizzuti}@icar.cnr.it

Abstract. An unsupervised distance-based outlier detection method that finds the top n outliers of a large and high-dimensional data set D, is presented. The method provides a subset R of the data set, called *robust solving set*, that contains the top n outliers and can be used to predict if a new unseen object p is an outlier or not by computing the distances of p to only the objects in R. Experimental results show that the prediction accuracy of the robust solving set is comparable with that obtained by using the overall data set.

1 Introduction

Unsupervised outlier detection is an active research field that has practical applications in many different domains such as fraud detection [8] and network intrusion detection [14, 7, 13]. Unsupervised methods take as input a data set of unlabelled data and have the task to discriminate between normal and exceptional data. Among the unsupervised approaches, *statistical-based* ones assume that the given data set has a distribution model. Outliers are those points that satisfy a discordancy test, i.e. that are significantly larger (or smaller) in relation to the hypothesized distribution [3, 19]. *Deviation-based* techniques identify outliers by inspecting the characteristics of objects and consider an object that deviates from these features an outlier [2]. *Density-based* methods [5, 10] introduce the notion of *Local Outlier Factor LOF* that measures the degree of an object to be an outlier with respect to the density of the local neighborhood.

Distance-based approaches [11, 12, 17, 1, 7, 4] consider an example exceptional on the base of the distance to its neighboring objects: if an object is isolated, and thus its neighboring points are distant, it is deemed an outlier. These approaches differ in the way the distance measure is defined, however an example can be associated with a weight or score, that is a function of the k nearest neighbors distances, and outliers are those examples whose weight is above a predefined threshold. Knorr and Ng [11, 12] define a point p an outlier if at least k points in the data set lie greater than distance d from p. For Ramaswamy et al. [17] outliers are the top n points p whose distance to their k-th nearest neighbor is

E. Suzuki and S. Arikawa (Eds.): DS 2004, LNAI 3245, pp. 89–100, 2004.
© Springer-Verlag Berlin Heidelberg 2004

greatest, while in [1, 7, 4] outliers are the top n points p for which the sum of distances to their k-th nearest neighbors is greatest.

Distance-based approaches can easily predict if a new example p is an outlier by computing its weight and then comparing it with the weight of the n-th outlier already found. If its weight is greater than that of the n-th outlier then it can be classified as exceptional. In order to compute the weight of p, however, the distances from p to all the objects contained in the data set must be obtained.

In this paper we propose an unsupervised distance-based outlier detection method that finds the top n outliers of a large and high-dimensional data set D, and provides a subset R of data set that can be used to classify a new unseen object p as outlier by computing the distances from p to only the objects in R instead of the overall data set D.

Given a data set D of objects, an object p of D, and a distance on D, let w^* be the n-th greatest weight of an object in D. An outlier w.r.t. D is an object scoring a weight w.r.t. D greater than or equal to w^*. In order to efficiently compute the top n outliers in D, we present an algorithm that applies a pruning rule to avoid to calculate the distances of an object to each other to obtain its k nearest neighbors. More interestingly, the algorithm is able to single out a subset R, called *robust solving set*, of D containing the top n outliers in D, and having the property that the distances among the pairs in $R \times D$ are sufficient to state that R contains the top n outliers. We then show that the solving set allows to effectively classify a new unseen object as outlier or not by approximating its weight w.r.t. D with its weight w.r.t. R. That is, each new object can be classified as an outlier w.r.t. D if its weight w.r.t. R is above w^*. This approach thus allows to sensibly reduce the response time of prediction since the number of distances computed is much lower. Experimental results point out that the prediction accuracy of the robust solving set is comparable with that obtained by using the overall data set.

The paper is organized as follows. In the next section the problems that will be treated are formally defined. In Section 3 the method for computing the top n outliers and the robust solving set is described. Finally, Section 4 reports experimental results.

2   Problem Formulation

In the following we assume that an *object* is represented by a set of d measurements (also called attributes or features).

Definition 1. Given a set D of objects, an object p, a distance d on $D \cup \{p\}$, and a positive integer number i, the *i-th nearest neighbor* $nn_i(p, D)$ of p w.r.t. D is the object q of D such that there exist exactly $i - 1$ objects r of D (if p is in D then p itself must be taken into account) such that $d(p, q) \geq d(p, r)$. Thus, if p is in D then $nn_1(p, D) = p$, otherwise $nn_1(p, D)$ is the object of D closest to p.

Definition 2. Given a set D of N objects, a distance d on D, an object p of D, and an integer number k, $1 \leq k \leq N$, the *weight* $w_k(p, D)$ of p in D (w.r.t. k) is $\sum_{i=1}^{k} d(p, nn_i(p, D))$.

Intuitively, the notion of weight captures the degree of dissimilarity of an object with respect to its neighbors, that is, the lower its weight is, the more similar its neighbors are. We denote by $D_{i,k}$, $1 \leq i \leq N$, the objects of D scoring the i-th largest weight w.r.t. k in D, i.e. $w_k(D_{1,k}, D) \geq w_k(D_{2,k}, D) \geq \ldots \geq w_k(D_{N,k}, D)$.

Definition 3. Given a set D of N objects, a distance d on D, and two integer numbers n and k, $1 \leq n, k \leq N$, the *Outlier Detection Problem ODP*$\langle D, d, n, k \rangle$ is defined as follows: find the n objects of D scoring the greatest weights w.r.t. k, i.e. the set $D_{1,k}, D_{2,k}, \ldots, D_{n,k}$. This set is called the *solution set* of the problem and its elements are called *outliers*, while the remaining objects of D are called *inliers*.

The *Outlier Detection Problem* thus consists in finding the n objects of the data set scoring the greatest values of weight, that is those mostly deviating from their neighborhood.

Definition 4. Given a set U of objects, a subset D of U having size N, an object q of U, called *query object* or simply *query*, a distance d on U, and two integer numbers n and k, $1 \leq n, k \leq N$, the *Outlier Prediction Problem OPP*$\langle D, q, d, n, k \rangle$ is defined as follows: is the ratio $out(q) = \frac{w_k(q,D)}{w_k(D_{n,k},D)}$, called *outlierness* of q, such that $out(q) \geq 1$?

If the answer is "yes", then the weight of q is equal to or greater than the weight of $D_{n,k}$ and q is said to be an *outlier* w.r.t. D (or equivalently a *D-outlier*), otherwise it is said to be an *inlier* w.r.t. D (or equivalently a *D-inlier*). Thus, the outlierness of an object q gives a criterion to classify q as exceptional or not by comparing $w_k(q, D)$ with $w_k(D_{n,k}, D)$.

The ODP can be solved in $\mathcal{O}(|D|^2)$ time by computing all the distances $\{d(p,q) \mid (p,q) \in D \times D\}$, while a comparison of the query object q with all the objects in D, i.e. $\mathcal{O}(|D|)$ time, suffices to solve the *OPP*. Real life applications deal with data sets of hundred thousands or millions of objects, and thus these approaches are both not applicable. We introduce the concept of *solving set* and explain how to exploit it to efficiently solve both the *ODP* and the *OPP*.

Definition 5. Given a set D of N objects, a distance d on D, and two integer numbers n and k, $1 \leq n, k \leq N$, a *solving set* for the *ODP*$\langle D, d, n, k \rangle$ is a subset S of D such that:

1. $|S| \geq \max\{n, k\}$;
2. Let $lb(S)$ denote the n-th greatest element of $\{w_k(p, D) \mid p \in S\}$. Then, for each $q \in (D - S)$, $w_k(q, S) < lb(S)$.

Intuitively, a solving set S is a subset of D such that the distances $\{d(p,q) \mid p \in S, q \in D\}$ are sufficient to state that S contains the solution set of the ODP. Given the $ODP\langle D, d, n, k \rangle$, let n^* denote the positive integer $n^* \geq n$ such that $w_k(D_{n,k}, D) = w_k(D_{n^*,k}, D)$ and $w_k(D_{n^*,k}, D) > w_k(D_{n^*+1,k}, D)$. We call the set $\{D_{1,k}, \ldots, D_{n^*,k}\}$ the *extended solution set* of the $ODP\langle D, d, n, k \rangle$. The following Proposition proves that if S is a solving set, then $S \supseteq \{D_{1,k}, \ldots, D_{n^*,k}\}$ and, hence, $lb(S)$ is $w_k(D_{n,k}, D)$.

Proposition 1. *Let S be a solving set for the ODP $\langle D, d, n, k \rangle$. Then $S \supseteq \{D_{1,k}, \ldots, D_{n^*,k}\}$.*

We propose to use the solving set S to solve the *Outlier Query Problem* in an efficient way by computing the weight of the new objects with respect to the solving set S instead of the complete data set D. Each new object is then classified as an outlier if its weight w.r.t. S is above the lower bound $lb(S)$ associated with S, that corresponds to $w_k(D_{n,k}, D)$.

The usefulness of the notion of solving set is thus twofold. First, computing a solving set for ODP is equivalent to solve ODP. Second, it can be exploited to efficiently answer any OPP. In practice, given $\mathcal{P} = ODP\langle D, d, n, k \rangle$ we first solve $\mathcal{P}$ by finding a solving set S for $\mathcal{P}$, and then we answer any $OPP\langle D, q, d, n, k \rangle$ in the following manner: reply "no" if $w_k(q, S) < lb(S)$ and "yes" otherwise. Furthermore, the outlierness $out(q)$ of q can be approximated with $\frac{w_k(q,S)}{lb(S)}$. We say that q is an *S-outlier* if $\frac{w_k(q,S)}{lb(S)} \geq 1$, and *S-inlier* otherwise.

In order to be effective, a solving set S must be efficiently computable, the outliers detected using the solving set S must be comparable with those obtained by using the entire data set D, and must guarantee to predict the correct label for all the objects in D. In the next section we give a sub-quadratic algorithm that computes a solving set. S, however, does not guarantee to predict the correct label for those objects belonging to $(S - \{D_{1,k}, \ldots, D_{n^*,k}\})$. To overcome this problem, we introduce the concept of *robust solving set*.

Definition 6. *Given a solving set S for the $ODP\langle D, d, n, k \rangle$ and an object $p \in D$, we say that p is S-robust if the following holds: $w_k(p, D) < lb(S)$ iff $w_k(p, S) < lb(S)$. We say that S is robust if, for each $p \in D$, p is S-robust.*

Thus, if S is a robust solving set for the $ODP\langle D, d, n, k \rangle$, then an object p occurring in the data set D is D-outlier for the $OPP\langle D, p, d, n, k \rangle$ iff it is an S-outlier. Hence, from the point of view of the data set objects, the sets D and S are equivalent. The following proposition allows us to verify when a solving set is robust.

Proposition 2. *A solving set S for the $ODP\langle D, d, n, k \rangle$ is robust iff for each $p \in (S - \{D_{1,k}, \ldots, D_{n^*,k}\})$, $w_k(p, S) < lb(S)$.*

In the next section we present the algorithm `SolvingSet` for computing a solving set and how to extend it for obtaining a robust solving set.

3   Solving Set Computation

The algorithm `SolvingSet` for computing a solving set and the top n outliers is based on the idea of repeatedly selecting a small subset of D, called $Cand$, and to compute only the distances $d(p,q)$ among the objects $p \in D - Cand$ and $q \in Cand$. This means that only for the objects of $Cand$ the true k nearest neighbors can be determined in order to have the weight $w_k(q, D)$ actually computed, as regard the others the current weight is an upper bound to their true weight because the k nearest neighbors found so far could not be the true ones. The objects having weight lower than the n-th greatest weight so far calculated are called *non active*, while the others are called *active*. At the beginning $Cand$ contains randomly selected objects from D, while, at each step, it is built selecting, among the active points of the data set not already inserted in $Cand$ during previous steps, a mix of random objects and objects having the maximum current weights. During the execution, if an object becomes non active, then it will not be considered any more for insertion in the set $Cand$, because it can not be an outlier. As the algorithm processes new objects, more accurate weights are computed and the number of non active objects increases more quickly. The algorithm stops when no other objects can be examined, i.e. all the objects not yet inserted in $Cand$ are non active, and thus $Cand$ becomes empty. The solving set is the union of the sets $Cand$ computed during each step.

The algorithm `SolvingSet` shown in Figure 1, receives in input the data set D, containing N objects, the distance d on D, the number k of neighbors to consider for the weight calculation, the number n of top outliers to find, an integer $m \geq k$, and a rational number $r \in [0, 1]$ (the meaning of m and r will be explained later) and returns the solving set $SolvSet$ and the set Top of n couples $\langle p_i, \sigma_i \rangle$ of the n top outliers p_i in D and their weight σ_i. For each object p in D, the algorithm stores a list $NN(p)$ of k couples $\langle q_i, \delta_i \rangle$, where q_i are the k nearest neighbors of p found so far and δ_i are their distances. At the beginning $SolvSet$ and Top are initialized with the empty set while the set $Cand$ is initialized by picking at random m objects from D. At each round, the approximate nearest neighbors $NN(q)$ of the objects q contained in $Cand$ are first computed only with respect to $Cand$ by the function $Nearest(q, k, Cand)$. Thus $weight(NN(p))$, that is the sum of the distances of p from its k current nearest neighbors, is an *upper bound to the weight of the object p*. The objects stored in Top have the property that their weight upper bound coincides with their true weight. Thus, the value $\text{Min}(Top)$ is a *lower bound to the weight of the n-th outlier* $D_{n,k}$ of D. This means that the objects q of D having weight upper bound $\text{weight}(NN(q))$ less than $\text{Min}(Top)$ cannot belong to the solution set. The objects having weight upper bound less than $\text{Min}(Top)$ are called *non active*, while the others are called *active*.

During each main iteration, the active points in $Cand$ are compared with all the points in D to determine their exact weight. In fact, given and object p in D and q in $Cand$, their distance $d(p,q)$ is computed only if either $weight(NN(p))$ or $weight(NN(q))$, is greater than the minimum weight σ in Top (i.e. $Min(Top)$). After calculating the distance from a point p of D to a point

Input: the data set D, containing N objects, the distance d on D, the number
k of neighbors to consider for the weight calculation, the number n
of top outliers to find, an integer $m \geq k$, and a rational number $r \in [0, 1]$.
Ouput: the solving set $SolvSet$ and the top n outliers Top
Let $weight(NN(p))$ return the sum of the distances of p
to its k current nearest neighbors;
Let $Nearest(q, k, Cand)$ return $NN(p)$ computed w.r.t. the set Cand;
Let $Min(Top)$ return the minimum weight in Top;
{
  $SolvSet = \emptyset$;
  $Top = \emptyset$;
  Initialize $Cand$ by picking at random m objects from D;
  while $(Cand \neq \emptyset)$ {
    $SolvSet = SolvSet \cup Cand$;
    $D = D - Cand$;
    for each (q **in** $Cand$) $NN(\text{q}) = Nearest(q, k, Cand)$;
    $NextCand = \emptyset$;
    for each (p **in** D) {
      for each (q **in** $Cand$)
        if $(weight(NN(p)) \geq \texttt{Min}(Top))$ or $(weight(NN(q)) \geq \texttt{Min}(Top))${
          $\delta = \text{d}(p, q)$;
          $\texttt{UpdateNN}(NN(p), \langle q, \delta \rangle)$;
          $\texttt{UpdateNN}(NN(q), \langle p, \delta \rangle)$;
        }
      $\texttt{UpdateMax}(NextCand, \langle p, \texttt{weight}(NN(p)) \rangle)$;
    }
    for each (q **in** $Cand$) $\texttt{UpdateMax}(Top, \langle q, \texttt{weight}(NN(q)) \rangle)$;
    $Cand = \texttt{CandSelect}(NextCand, D - NextCand, r)$;
  }
}

Fig. 1. The algoritm `SolvingSet`.

q of $Cand$, the sets $NN(p)$ and $NN(q)$ are updated by using the procedure $UpdateNN(NN(p_1), \langle q_1, \delta \rangle)$ that inserts $\langle q_1, \delta \rangle$ in $NN(p_1)$ if $\mid NN(p_1) \mid < k$, otherwise substitute the couple $\langle p_1, \bar{\delta} \rangle$) such that $\bar{\delta} = max_{i=1}^{m}\{\delta_i\}$ with $\langle q_1, \delta \rangle$ provided that $\delta < \bar{\delta}$. $NextCand$ contains the m objects of D having the greatest weight upper bound of the current round. After comparing p with all the points in $Cand$, p is inserted in the set $NextCand$ by $UpdateMax(NextCand, \langle p, \sigma \rangle$ where $\sigma = weight(NN[p])$, that inserts $\langle p, \sigma \rangle$) in $NextCand$ if $NextCand$ contains less than m objects, otherwise substitute the couple $\langle q, \bar{\sigma} \rangle$ such that $\bar{\sigma} = min_{i=1}^{h}\{\sigma_i\}$ with $\langle p, \sigma \rangle$ provided that $\sigma > \bar{\sigma}$;

At the end of the inner double cycle, the objects q in $Cand$ have weight upper bound equal to their true weight, and they are inserted in Top using the function $\texttt{UpdateMax}(\text{Top}, \langle q, \texttt{weight}(\text{NN}(\text{q})) \rangle$.

Finally $Cand$ is populated with a new set of objects of D by using the function `CandSelect` that builds the set $A \cup B$ of at most m objects as follows: A is

composed by the at most rm active objects p in $NextCand$ having the greatest
weight($NN(p)$), while B is composed by at most $(1-r)m$ active objects in D
but not in $NextCand$ picked at random. r thus specifies the trade-off between
the number of random objects $((1-r)m$ objects) and the number of objects
having the greatest current weights (rm objects) to select for insertion in $Cand$.
If there are no more active points, then CandSelect returns the empty set and
the algorithm stops: Top contains the solution set and $SolvSet$ is a solving set
for the $ODP\langle Data, \mathrm{d}, n, k\rangle$.

The algorithm SolvingSet has worst case time complexity $\mathcal{O}(|D|^2)$, but
practical complexity $\mathcal{O}(|D|^{1+\beta})$, with $\beta < 1$. Indeed, let S be the solving set
computed by SolvingSet, then the algorithm performed $|D|^{1+\beta} = |D| \cdot |S|$
distance computations, and thus $\beta = \frac{\log|S|}{\log|D|}$.

Now we show how a robust solving set R containing S can be computed with
no additional asymptotic time complexity. By Proposition 2 we have to verify
that, for each object p of S, if the weight of p in D is less than $lb(S)$, then
the weight of p in S remains below the same threshold. Thus, the algorithm
RobustSolvingSet does the following: (i) initialize the robust solving set R to
S; (ii) for each $p \in S$, if $w_k(p, D) < lb(S)$ and $w_k(p, R) \geq lb(S)$, then select a
set C of neighbors of p coming from $D - S$, such that $w_k(p, R \cup C) < lb(S)$, and
set R to $R \cup C$. We note that the objects C added to R are certainly R-robust,
as they are S-robust by definition of solving set and $R \supseteq S$.

If we ignore the time required to find a solving set, then RobustSolvingSet
has worst case time complexity $\mathcal{O}(|R| \cdot |S|)$. Thus, as $|R| \leq |D|$ (and we ex-
pect that $|R| \ll |D|$ in practice), the time complexity of RobustSolvingSet is
dominated by the complexity of the algorithm SolvingSet.

4   Experimental Results

To assess the effectiveness of our approach, we computed the robust solving set
for three real data sets and then compared the error rate for normal query objects
and the detection rate for outlier query objects when using the overall data set
against using the robust solving set to determine the weight of each query. We
considered three labelled real data sets well known in the literature: *Wisconsin
Diagnostic Breast Cancer* [15], *Shuttle* [9], and *KDD Cup 1999* data sets. The first
two data sets are used for classification tasks, thus we considered the examples of
one of the classes as the normal data and the others as the exceptional data. The
last data set comes from the *1998 DARPA Intrusion Detection Evaluation Data*
[6] and has been extensively used to evaluate intrusion detection algorithms.

- *Breast cancer:* The *Wisconsin Diagnostic Breast Cancer* data set is composed
  by instances representing features describing characteristics of the cell nuclei
  present in the digitalized image of a breast mass. Each instance has one of two
  possible classes: *benign*, that we assumed as the normal class, or *malignant*.
- *Shuttle:* The *Shuttle* data set was used in the European StatLog project
  which involves comparing the performances of machine learning, statistical,

and neural network algorithms on data sets from real-world industrial areas [9]. This data set contains 9 attributes all of which are numerical. The data is partitioned in 7 classes, namely, *Rad Flow, Fpv Close, Fpv Open, High, Bypass, Bpv Close*, and *Bpv Open*. Approximately 80% of the data belongs to the class *Rad Flow*, that we assumed as the normal class.

– *KDD Cup:* The KDD Cup 1999 data consists of network connection records of several intrusions simulated in a military network environment. The TCP connections have been elaborated to construct a data set of 23 features, one of which identifying the kind of attack : *DoS, R2L, U2R*, and *PROBING*. We used the TCP connections from 5 weeks of training data (499,467 connection records).

Table 1. The data used in the experiments.

Data set name	Attributes	Normal examples	Normal queries	Exceptional queries
Shuttle	9	40,000	500	1,244
Breast cancer	9	400	44	239
KDD Cup	23	24,051	9,620	18,435

Breast Cancer and *Shuttle* data sets are composed by a training set and a test set. We merged these two sets obtaining a unique labelled data set. From each labelled data set, we extracted three unlabelled sets: a set of *normal examples* and a set of *normal queries*, both containing data from the normal class, and a set of *exceptional queries*, containing all the data from the other classes. Table 1 reports the sizes of these sets. We then used the normal examples to find the robust solving set R for the $ODP\langle D, \mathrm{d}, n, k\rangle$ and the normal and exceptional queries to determine the false positive rate and the detection rate when solving $OPP\langle D, q, \mathrm{d}, n, k\rangle$ and $OPP\langle R, q, \mathrm{d}, n, k\rangle$. We computed a robust solving set for values of the parameter n ranging from $0.01|D|$ to $0.12|D|$ (i.e. from the 1% to the $10\% - 12\%$ of the normal examples set size) and using $r = 0.75$ and $m = k$ in all the experiments and the Euclidean distance as metrics d. The performance of the method was measured by computing the ROC (Receiver Operating Characteristic) curves [16] for both the overall data set of normal examples, denoted by D, and the robust solving set. The ROC curves show how the detection rate changes when specified false positive rate ranges from the 1% to the 12% of the normal examples set size. We have a ROC curve for each possible value of the parameter k. Let Q_I denote the normal queries associated with the normal example set D, and Q_O the associated exceptional queries. For a fixed value of k, the ROC curve is computed as follows. Each point of the curve for the sets D and R resp. is obtained as $(\frac{|I_O|}{|Q_I|}, \frac{|O_O|}{|Q_O|})$, where I_O is the set of queries q of Q_I that are outliers for the $OPP\langle D, q, \mathrm{d}, n, k\rangle$ and $OPP\langle R, q, \mathrm{d}, n, k\rangle$ resp., and O_O is the set of queries of Q_O that are outliers for the same problem.

Figure 2 reports the ROC curves (on the left) together with the curves of the robust data set sizes (on the right), i.e. the curves composed by the points

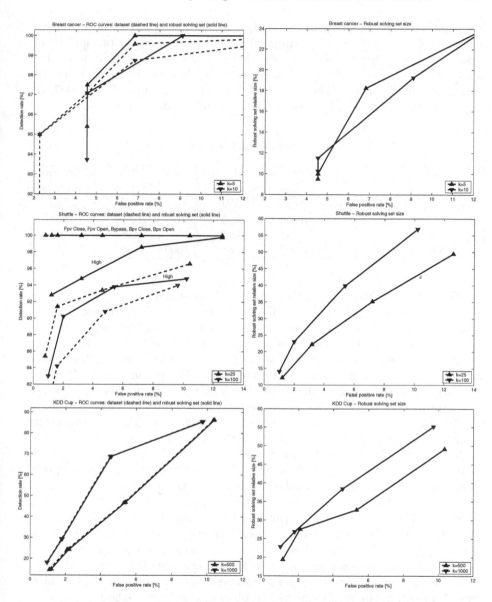

Fig. 2. ROC curves (left) and robust solving set sizes (right).

(false positive rate, $|R|$). For each set we show two curves, associated with two different values of k. Dashed lines are relative to the normal examples set (there called data set), while solid lines are relative to the robust solving set.

The ROC curve shows the tradeoff between the false positive rate and the false negative rate. The closer the curve follows the left and the top border of the unit square, the more accurate the method. Indeed, the area under the curve is a measure of the accuracy of the method in separating the outliers from the inliers. Table 2 reports the approximated areas of all the curves of Figure 2 (in order to compute the area we interpolated the missing points exploiting the fact that the curve starts in $(0,0)$ and stops in $(1,1)$). It is worth to note that the best values obtained using the robust solving set vary from 0.894 to 0.995, i.e. they lye in the range of values characterizing excellent diagnostic tests.

Table 2. The areas of the ROC curves of Figure 2.

Data set name	k	ROC area data set	ROC area solving set	k	ROC area data set	ROC area solving set
Shuttle (*High* class)	25	0.974	0.99	100	0.959	0.964
Shuttle (other classes)	25	0.996	0.994	100	0.996	0.995
Breast cancer	5	0.987	0.976	10	0.986	0.975
KDD Cup	500	0.883	0.883	1000	0.894	0.894

We recall that, by definition, the size $|R|$ of the robust solving set R associated with a data set D is greater than n. Furthermore, in almost all the experiments $n/|D| \simeq$ false positive rate. Thus, the relative size of the robust solving set $|R|/|D|$ is greater than the false positive rate. We note that in the experiments reported, $|R|$ can be roughly expressed as αn, where α is a small constant whose value depends on the data set distribution and on the parameter k. Moreover, the curves of Figure 2 show that using the solving set to predict outliers, not only improves the response time of the prediction over the entire data set, as the size of the solving set is a fraction of the size of the data set, but also guarantees the same or a better response quality than the data set.

Next we briefly discuss the experiments on each data set. As regard the *Breast cancer* data set, we note that the robust solving set for $k = 5$ composed by about the 10% (18% resp.) of the data set, reports a false positive rate 4.5% (6.8% resp.) and detection rate 97.5% (100% resp.). For the *Shuttle* data set, the robust solving set for $k = 25$ composed by about the 12% of the data set, guarantees a false positive rate 1% and a detection rate of the 100% on the classes *Fpv Close*, *Fpv Open*, *Bypass*, *Bpv Close*, and *Bpv Open*. Furthermore, the robust solving set sensibly improves the prediction quality over the data set for the class *High*. These two experiments point out that the method behaves very well also for binary classification problems when one of the two classes is assumed normal and the other abnormal, and for the detection of rare values of the attribute class in a completely unsupervised manner, which differs from the supervised approaches like [18] that search for rare values in a set of labelled data.

As regard the *KDD Cup* data set, a 90% of detection rate is obtained by allowing the 10% of false positive. In this case we note that higher values of the parameter k are needed to obtain good prediction results because of the peculiarity of this data set in which inliers and outliers overlap. In particular, some relatively big clusters of inliers are close to regions of the feature space containing outliers. As a consequence a value of $k \geq 500$ is needed to "erase" the contribution of these clusters to the weight of an outlier query object and, consequently, to improve the detection rate.

5   Conclusions and Future Work

The work is the first proposal of distance-based outlier prediction method based on the use of a subset of the data set. The method, in fact, detects outliers in a data set D and provides a subset R which (1) is able to correctly answer for each object of D if it is an outlier or not, and (2) predicts the outlierness of new objects with an accuracy comparable to that obtained by using the overall data set, but at a much lower computational cost. We are currently studying how to extend the method to *on-line outlier detection* that requires to update the solving set each time a new object is examined.

Acknowledgements

The authors are grateful to Aleksandar Lazarevic for providing the DARPA 1998 data set.

References

1. F. Angiulli and C. Pizzuti. Fast outlier detection in high dimensional spaces. In *Proc. Int. Conf. on Principles of Data Mining and Knowledge Discovery (PKDD'02)*, pages 15–26, 2002.
2. A. Arning, C. Aggarwal, and P. Raghavan. A linear method for deviation detection in large databases. In *Proc. Int. Conf. on Knowledge Discovery and Data Mining (KDD'96)*, pages 164–169, 1996.
3. V. Barnett and T. Lewis. *Outliers in Statistical Data*. John Wiley & Sons, 1994.
4. S. D. Bay and M. Schwabacher. Mining distance-based outliers in near linear time with randomization and a simple pruning rule. In *Proc. Int. Conf. on Knowledge Discovery and Data Mining (KDD'03)*, 2003.
5. M. M. Breunig, H. Kriegel, R.T. Ng, and J. Sander. Lof: Identifying density-based local outliers. In *Proc. Int. Conf. on Managment of Data (SIGMOD'00)*, 2000.
6. Defense Advanced Research Projects Agency DARPA. Intrusion detection evaluation. In *http://www.ll.mit.edu/IST/ideval/index.html*.
7. E. Eskin, A. Arnold, M. Prerau, L. Portnoy, and S. Stolfo. A geometric framework for unsupervised anomaly detection : Detecting intrusions in unlabeled data. In *Applications of Data Mining in Computer Security, Kluwer*, 2002.
8. T. Fawcett and F. Provost. Adaptive fraud detection. *Data Mining and Knowledge Discovery*, 1:291–316, 1997.
9. C. Feng, A. Sutherland, S. King, S. Muggleton, and R. Henery. Comparison of machine learning classifiers to statistics and neural networks. In *AI & Stats Conf. 93*, 1993.
10. W. Jin, A.K.H. Tung, and J. Han. Mining top-n local outliers in large databases. In *Proc. ACM SIGKDD Int. Conf. on Knowledge Discovery and Data Mining (KDD'01)*, 2001.

11. E. Knorr and R. Ng. Algorithms for mining distance-based outliers in large datasets. In *Proc. Int. Conf. on Very Large Databases (VLDB98)*, pages 392–403, 1998.
12. E. Knorr, R. Ng, and V. Tucakov. Distance-based outlier: algorithms and applications. *VLDB Journal*, 8(3-4):237–253, 2000.
13. A. Lazarevic, L. Ertoz, V. Kumar, A. Ozgur, and J. Srivastava. A comparative study of anomaly detection schemes in network intrusion detection. In *Proc. SIAM Int. Conf. on Data Mining (SIAM-03)*, 2003.
14. W. Lee, S.J. Stolfo, and K.W. Mok. Mining audit data to build intrusion detection models. In *Proc. Int. Conf on Knowledge Discovery and Data Mining (KDD-98)*, pages 66–72, 1998.
15. L. Mangasarian and W. H. Wolberg. Cancer diagnosis via linear programming. *SIAM News*, 25(5):1–18, 1990.
16. F. Provost, T. Fawcett, and R. Kohavi. The case against accuracy estimation for comparing induction algorithms. In *Proc. Int. Conf. on Machine Learning (ICML'98)*, 1998.
17. S. Ramaswamy, R. Rastogi, and K. Shim. Efficient algorithms for mining outliers from large data sets. In *Proc. Int. Conf. on Managment of Data (SIGMOD'00)*, pages 427–438, 2000.
18. L. Torgo and R. Ribeiro. Predicting outliers. In *Proc. Int. Conf. on Principles of Data Mining and Knowledge Discovery (PKDD'03)*, pages –, 2003.
19. K. Yamanishi and J. Takeuchi. Discovering outlier filtering rules from unlabeled data. In *Proc. ACM SIGKDD Int. Conf. on Knowledge Discovery and Data Mining (KDD'01)*, pages 389–394, 2001.

Detecting Outliers via Logical Theories and Its Data Complexity

Fabrizio Angiulli[1], Gianluigi Greco[2], and Luigi Palopoli[2]

[1] ICAR-CNR, Via Pietro Bucci 41C, 87030 Rende (CS), Italy
angiulli@icar.cnr.it
[2] DEIS - Università della Calabria, Via P. Bucci 41C, 87030 Rende (CS), Italy
{ggreco,palopoli}@deis.unical.it

Abstract. Detecting anomalous individuals from a given data population, is one major task pursued in knowledge discovery systems. Such exceptional individuals are usually referred to as *outliers* in the literature. Outlier detection has important applications in bioinformatics, fraud detection, network robustness analysis and intrusion detection and several techniques have been developed to obtain it, ranging from clustering-based to proximity-based methods to domain density analysis. Roughly speaking, such techniques models the "normal" behavior of individuals by computing some form of statistics over the given data set.

In this paper we propose a rather different approach to outlier detection that should not be looked at as alternative but, rather, complementary to those statistical-like methods. Our approach consists in modelling what should be "normal" in the form of a logical theory. Then, the given data set is analyzed on the basis of that theory to single out anomalous data elements.

In the paper we first formalize our theory-based approach to outlier detection and then study the cost implied by realizing outlier detection in this setting. As usual with database, we shall concentrate on *data* complexity, that is, the complexity measured assuming the given data set to be the input to the problem, while the underlying theory is considered fixed.

1 Introduction

1.1 Outlier Detection

The development of effective knowledge discovery techniques has become in the recent few years a very active research area due to the important impact it has in several relevant application areas. Knowledge discovery comprises quite diverse tasks and associated methods. One interesting task thereof is that of singling out anomalous individuals from a given population, e.g., to detect rare events in time-series analysis settings, or to identify objects whose behavior is deviant w.r.t. a codified standard set of "social" rules. Such exceptional individuals are usually referred to as *outlier* in the literature. Outlier detection has important applications in bioinformatics, fraud detection, network robustness analysis and

E. Suzuki and S. Arikawa (Eds.): DS 2004, LNAI 3245, pp. 101–113, 2004.
© Springer-Verlag Berlin Heidelberg 2004

intrusion detection. As a consequence, several techniques have been developed to realize outlier detection, ranging from clustering-based to proximity-based methods to domain density analysis (see, e.g., $[6, 12, 17, 8, 1, 5, 13, 7]$). Such techniques models the "normal" behavior of individuals by performing some statistical or probabilistic computation over the given data set (and various methods basically vary on the basis of the way such computation is carried out). In other words, those computations return average settings for considered attribute-value pairs and then outliers are identified roughly speaking as those individuals whose associated attribute-value pairs significantly differ from those "mean" settings.

However, while looking at a data set for discovering outliers, it often happens that we have some "qualitative" description of what the expected normal behavior is. This description might be, for instance, derived by an expert of the domain, and might be formalized by means of a suitable language for knowledge representation. Our claim is that such additional knowledge can be profitably exploited to refine the way outlier detection is carried out, as shown in the following example.

1.2   Example of Outlier Detection

Consider a bank B. The bank approves loan requests put forward by customers on the basis of certain policies. As an example of such policies assume that loan approval policies prescribe that loans for amounts greater than 50K Euro have come along with an endorsement provided as a guarantee by a third party. Information about approved loan requests, are stored into a number of relational tables, that are:

1. REQLOAN(LOAN ID, CUSTOMER, AMOUNT), that records loan requests;
2. ENDORSEMENT(LOAN ID, PARTY), that records the guaranteeing parties for loan requests for more than 50K Euro;
3. APPROVED(LOAN ID), that records approved loan requests.

Moreover, the bank stores information about unreliable customers in the table UNRELIABLE(CUSTOMER) collecting data from an external data source, such as, for example, consulting agencies providing information about the financial records of individuals and companies. In Figure 1, an instance of the bank database is reported.

With this knowledge at hand, the bank policy concerning loan approvals can be easily encoded using a logical rule like the following:

```
Approved(L) ← ReqLoan(L, C, A), A > 50K, Endorsement(L, P),
                                            not Unreliable(P).
```

According to this knowledge, we might notice that there is something strange with the loan l_1 in Figure 1. In fact, since the loan has been approved, the third party p_1 should be reliable. However, this is not the case, as emerges by looking at the database provided by the consulting agency. Then, according to the theory

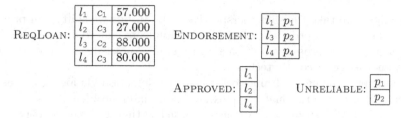

Fig. 1. Example instance of the bank database.

depicted above, we conclude that the unreliable party p_1 is an outlier, witnessed by the fact that the loan l_1 was approved.

Notice that if the tuple APPROVED(l_1) would have dropped, we would have concluded the exact opposite, namely that the loan l_1 is not in the relation APPROVED, i.e. that, according to the normal behavior of the bank, the loan l_1 should not be approved. We call *witnesses* the tuples satisfying this kind of condition. Furthermore, if we drop both the tuples UNRELIABLE(p_1) and APPROVED(l_1), we conclude again that l_1 might be approved. This implies that the loan request l_1 being not approved is a consequence of the fact that p_1 is not reliable, and hence p_1 is an outlier. Thus, in our framework, an *outlier* is a tuple whose removal from the theory defuses some witness.

While in the simple scenario depicted above detecting outliers by exploiting a logical characterization of the domain of interest has been done quite easily, in more general cases (with complex and richer background knowledge) some automatic mechanisms for outlier identification via logic theories is definitively needed. And, in fact, the proposal of a formal framework for implementing this task is the major contribution of this work.

Roughly speaking, our approach relies on the encoding of the general knowledge about the "normal" status of things in a theory, specifically, a (first order) logic program, called the *rule component* in our setting. Then, the data set to be searched for outliers is encoded in its turn in the form of a set of logical facts (using the straightforward and well known correspondence between relational tuples and logical facts), which is called the *observation component* in our setting. Armed with these encodings, outliers are singled out in the observation component looking for those individuals whose *logical properties* differs somehow from what is prescribed within the rule component.

Before describing in some more details the contributions of the paper, we point out that the above depicted approach to outlier detection has been recently investigated as an interesting AI reasoning problem, in the context of *propositional default logic* [2] and *disjunctive propositional logic programming* [3]. However, since these proposal focus on propositional theories only, they would allow the expert of the domain to assert property of individuals (i.e., single tuples) only. Hence, they are not well suited for modelling outlier detection tasks in a classical data mining setting, where first order theories are needed for checking interesting properties over given data sets to be searched for outliers.

1.3   Contribution

In this paper, we take the same perspective of [2, 3], but we shift from propositional theories to first order theories. This is significant exactly because it allows
us to generalize the proposed approach from a pure knowledge representation to
a full a knowledge discovery technique.

Besides introducing the framework for outlier detection via logic theories, we
also study the computational complexity of the main problems arising in the
setting. Specifically, since in a database oriented setting, the "large" component
to be taken care of is the data set at hand, we will be interested in measuring the
data complexity associated with detecting outliers. In fact, in database applications data constitutes indeed the relevant input for the problem, and hence the
investigation of the data complexity is required for understanding the viability of
outlier detection throughout logic theories. Conversely, within the propositional
setting mentioned above ([2, 3]), the *program* complexity of detecting outliers
under various constraints has been quite thoroughly analyzed.

The rest of the paper is organized as follows. Section 2 presents preliminaries
on logic programs and on the main complexity classes we shall deal with in
the paper. Then, Section 3 present the formal framework of outlier detection
problems, while their data complexity is studied in Section 4. Finally, in Section 5
we draw our conclusions.

2   Preliminaries

2.1   Logic Programs

In the following, we assume that the theory describing some interesting properties of a given dataset is encoded by means of a *logic program*. Now we recall
some basic notions of logic programs.

A logic program P^{rls} is a finite set of rules of the form

$$a \leftarrow b_1, \cdots, b_k, \text{not } c_1, \cdots, \text{not } c_n$$

where $k, n \geq 0$, and $a, b_1, \cdots, b_k, c_1, \cdots, c_n$ are atoms; moreover, an *atom* is of
the form $p(t_1, ..., t_k)$ where p is a k-ary predicate symbol and $t_1, ..., t_k$ are terms,
i.e., constants or variables. Finally, a rule with $n = 0$ is called *fact*.

A logic program P is *stratified*, if there is an assignment $s(\cdot)$ of integers to
the atoms in P, such that for each clause r in P the following holds: if p is the
atom in the head of r and q (resp. *not* q) occurs in r, then $s(p) \geq s(q)$ (resp.
$s(p) > s(q)$).

In this paper we will adopt the standard convention of identifying a relational
database instance with a logical theory consisting of ground facts. Thus, a tuple
$\langle a_1, \ldots a_k \rangle$, belonging to relation r, will be identified with the ground atom
$r(a_1, \ldots, a_k)$.

A rule r' is a *ground instance* of a rule r, if r' is obtained from r by replacing
every variable in r with some constant appearing in P. We denote by $ground(P)$
the set of all ground instances of the rules in P.

For a program P, the stable model semantics assigns to P the set $\mathcal{SM}(P)$ of its *stable models* [10]. Let $\mathcal{W}$ be a set of facts.

Then, program P *bravely entails* $\mathcal{W}$ (resp. $\neg\mathcal{W}$), denoted by $P \models_b \mathcal{W}$ (resp. $P \models_b \neg\mathcal{W}$), if there exists $M \in \mathcal{SM}(P)$ such that each fact in $\mathcal{W}$ is evaluated true (resp. false) in M.

Moreover, P *cautiously entails* $\mathcal{W}$ (resp. $\neg\mathcal{W}$), denoted by $P \models_c \mathcal{W}$ (resp. $P \models_c \neg\mathcal{W}$), if for each model $M \in \mathcal{SM}(P)$, each fact in $\mathcal{W}$ is true (resp. false) in M.

2.2   Complexity Classes

We recall some basic definitions about complexity theory, particularly, the polynomial time hierarchy. The reader is referred to [16] for more on this.

The class P is the set of decision problems that can be answered by a deterministic Turing machine in polynomial time.

The classes Σ_k^P and Π_k^P, forming the *polynomial hierarchy*, are defined as follows: $\Sigma_0^P = \Pi_0^P = P$ and for all $k \geq 1$, $\Sigma_k^P = NP^{\Sigma_{k-1}^P}$, and $\Pi_k^P = \text{co-}\Sigma_k^P$. Σ_k^P models computability by a nondeterministic polynomial time Turing machine which may use an oracle, that is, loosely speaking, a subprogram, that can be run with no computational cost, for solving a problem in Σ_{k-1}^P.

The class of decision problems that can be solved by a nondeterministic Turing machine in polynomial time is denoted by NP, while the class of decision problems whose complementary problem is in NP, is denote by co-NP.

The class D_k^P, $k \geq 1$, is defined as the class of problems that consist of the conjunction of two independent problems from Σ_k^P and Π_k^P, respectively. Note that, for all $k \geq 1$, $\Sigma_k^P \subseteq D_k^P \subseteq \Sigma_{k+1}^P$.

3   Defining Outliers

Let P^{rls} be a logic program encoding general knowledge about the world, called *rule program*, and let P^{obs} be a set of facts encoding some *observed* aspects of the current status of the world, called *observation set*. Then, the structure $\mathcal{P} = \langle P^{\text{rls}}, P^{\text{obs}} \rangle$, relating the general knowledge encoded in P^{rls} with the evidence about the world encoded in P^{obs}, is said to be a *rule-observation pair*, and it constitutes the input for the outlier detection problem.

Indeed, given $\mathcal{P}$, we are interested in identifying (if any) a set $\mathcal{O}$ of *observations* (facts in P^{obs}) that are "anomalous" according to the general theory P^{rls} and the other facts in $P^{\text{obs}} \setminus \mathcal{O}$. Quite roughly speaking, the idea underlying the identification of $\mathcal{O}$ is to discover a *witness set* $\mathcal{W} \subseteq P^{\text{obs}}$, that is, a set of facts which would be explained in the theory *if and only if* all the facts in $\mathcal{O}$ were not observed. For instance, in the bank example, the party p_1 is an outlier since the theory is not able to explain why the loan l_1 having p_1 as guarantee party has been approved even though p_1 is not reliable. Such an intuition is formalized in the following definition.

Definition 1 (Outlier). Let $\mathcal{P} = \langle P^{\mathrm{rls}}, P^{\mathrm{obs}} \rangle$ be a rule-observation pair and let $\mathcal{O} \subseteq P^{\mathrm{obs}}$ be a set facts. Then, $\mathcal{O}$ is an *outlier*, under the cautious (resp. brave) semantics, in $\mathcal{P}$ if there exists a non empty set $\mathcal{W} \subseteq P^{\mathrm{obs}}$, called *outlier witness* for $\mathcal{O}$ in $\mathcal{P}$, such that:

1. $P(\mathcal{P})_{\mathcal{W}} \models_c \neg \mathcal{W}$ (resp. $P(\mathcal{P})_{\mathcal{W}} \models_b \neg \mathcal{W}$), and
2. $P(\mathcal{P})_{\mathcal{W},\mathcal{O}} \not\models_c \neg \mathcal{W}$ (resp. $P(\mathcal{P})_{\mathcal{W},\mathcal{O}} \not\models_b \neg \mathcal{W}$),

where $P(\mathcal{P})$ is the logic program $P^{\mathrm{rls}} \cup P^{\mathrm{obs}}$, $P(\mathcal{P})_{\mathcal{W}} = P(\mathcal{P}) \backslash \mathcal{W}$ and $P(\mathcal{P})_{\mathcal{W},\mathcal{O}} = P(\mathcal{P})_{\mathcal{W}} \backslash \mathcal{O}$. □

Notice that, in the above definition, we have distinguished between brave and cautious semantics. Indeed, the semantics is part of the input, since it is provided by the designer of the rules encoding the general knowledge of the world. Obviously, if P^{rls} has a unique stable model (for instance, in the case it is positive or stratified), the two semantics coincide. In the rest of the paper, for stratified or positive programs we do not distinguish among the semantics - for instance, we shall simply say that P entails a set $\mathcal{W}$.

Now that the framework has been introduced, given a rule-observation pair $\mathcal{P} = \langle P^{\mathrm{rls}}, P^{\mathrm{obs}} \rangle$, we shall look at the following problems related to outlier detection:

- EXISTENCE: does $\mathcal{P}$ have an outlier?
- OUTLIER − CHECKING: given $\mathcal{O} \subseteq P^{\mathrm{obs}}$, is $\mathcal{O}$ an outlier for any witness set $\mathcal{W}$?
- WITNESS − CHECKING: given $\mathcal{W} \subseteq P^{\mathrm{obs}}$, is $\mathcal{W}$ a witness for any outlier $\mathcal{O}$ in T?
- OW − CHECKING: given $\mathcal{O}, \mathcal{W} \subseteq P^{\mathrm{obs}}$, is $\mathcal{O}$ an outlier in $\mathcal{P}$ with witness $\mathcal{W}$?

The complexity of problem EXISTENCE is related to the complexity of finding at least an outlier in the database at hand.

Problem OUTLIER − CHECKING consists in deciding if a given set of observations is an outlier in the input rule-observation pair, while problem WITNESS − CHECKING consists in deciding if a given set of observations is a witness for some outlier in the input rule-observation pair.

Finally, problem OW − CHECKING is relevant as it may constitute the basic operator to be implemented in a system of outlier detection. Its complexity states the computational effort that must be taken into account in order to check that a given pair $\mathcal{O}, \mathcal{W}$ actually represents an outlier and its associated witness.

4   Complexity of Outlier Detection Problems

4.1   Overview of the Results

In this section, we study the data complexity of the introduced outlier detection problems. Table 1 reports a summary of the results for both general and stratified logic programs.

We next provide the detailed proofs for the most basic problem arising in this setting, i.e., the EXISTENCE problem of deciding the existence of an outlier.

Table 1. Data complexity of outlier detection.

	EXISTENCE	WITNESS − CHECKING	OUTLIER − CHECKING	OW − CHECKING
General LP	Σ_2^P-complete	Σ_2^P / D^P-complete	Σ_2^P-complete	D^P-complete
Stratified LP	NP-complete	NP-complete	NP-complete	P-complete

Theorem 1 states that the data complexity of the EXISTENCE problem when stratified logic programs are considered is NP-complete, while Theorems 2 and 3 state that the data complexity of the same problem when general logic programs under cautious and brave semantics are considered is Σ_2^P-complete. The result for the other problems can be obtained with similar constructions, and are therefore omitted due to space limits.

From Table 1 it is clear that, except for problem OW − CHECKING on stratified logic programs, these problems are unlikely to be solved in polynomial time (unless P=NP). It is worth to point out that these results do not imply that outlier detection is not practicable as a data mining technique. Indeed, although several relevant knowledge discovery tasks are computationally intractable (see, for example, [9,11,15,4] for some complexity analysis pertaining clustering, decision tree and association rule induction), researchers developed a number of efficient algorithms providing solutions that are "good in practice" (e.g., K-Means, C4.5, A-Priori). Rather, computational results on outlier detection indicate that algorithmic efforts must concern the design of efficient heuristics for practically solving detection problems in real cases.

4.2 Complexity Results

We preliminary introduce some basic definitions that will be exploited in the proofs. Let L be a consistent set of literals. We denote with σ_L the truth assignment on the set of letters occurring in L such that, for each positive literal $p \in L$, $\sigma_L(p) = \textbf{true}$, and for each negative literal $\neg p \in L$, $\sigma_L(p) = \textbf{false}$.

Let L be a set of literals. Then we denote with L^+ the set of positive literals occurring in L, and with L^- the set of negative literals occurring in L.

Let σ be a truth assignment of the set $\{x_1, \ldots, x_n\}$ of boolean variables. Then we denote with $\text{Lit}(\sigma)$ the set of literals $\{\ell_1, \ldots, \ell_n\}$, such that ℓ_i is x_i if $\sigma(x_i) = \textbf{true}$ and is $\neg x_i$ if $\sigma(x_i) = \textbf{false}$, for $i = 1, \ldots, n$.

Theorem 1. *Let $\mathcal{P} = \langle P^{\text{rls}}, P^{\text{obs}} \rangle$ be a rule-observation pair such that P^{rls} is a fixed stratified logic program. Then* EXISTENCE *is NP-complete.*

Proof. (Membership) Given a fixed stratified logic program P^{rls} and a set of ground facts P^{obs}, we must show that there exist two sets $\mathcal{W}, \mathcal{O} \subseteq P^{\text{obs}}$ such that $P(\mathcal{P})_\mathcal{W} \models \neg \mathcal{W}$ (query q') and $P(\mathcal{P})_{\mathcal{W}, \mathcal{O}} \not\models \neg \mathcal{W}$ (query q''). $P(\mathcal{P})$ is stratified and, hence, admits a unique minimal model. Since the size of both programs $ground(P(\mathcal{P})_\mathcal{W})$ and $ground(P(\mathcal{P})_{\mathcal{W}, \mathcal{O}})$ is polynomially related to the size of P^{obs}, then query q' and q'' can be answered in polynomial time. Thus, we can build a polynomial-time nondeterministic Turing machine solving EXISTENCE as follows: the machine guesses both the sets $\mathcal{W}$ and $\mathcal{O}$ and then solves queries q' and q'' in polynomial time.

(Hardness) Recall that deciding whether a Boolean formula in conjunctive normal form $\Phi = c_1 \wedge \ldots \wedge c_m$ over the variables $x_1, \ldots, x_n$ is satisfiable, i.e., deciding whether there exists truth assignments to the variables making each clause c_j true, is an NP-hard problem, even if each clause contains at most three distinct (positive or negated) variables, and each variable occurs in at most three clauses [16]. W.l.o.g, assume Φ contains at least one clause and one variable. We associate with Φ the following set of facts $P^{obs}(\Phi)$:

$$o_1 : sat.$$
$$o_2 : disabled.$$
$$o_3 : variable(x_i). \qquad\qquad 1 \le i \le n$$
$$o_4 : \begin{array}{l} clause(c_j, c_{(j+1) \bmod (m+1)}, \wp(t_{j,1}), \ell(t_{j,1}), \\ \wp(t_{j,2}), \ell(t_{j,2}), \wp(t_{j,3}), \ell(t_{j,3})). \end{array} \qquad 1 \le j \le m$$

where $c_j = t_{j,1} \vee t_{j,2} \vee t_{j,3}$, $1 \le j \le m$, $\ell(t)$ denotes the atom occurring in the literal t, and $\wp(t)$ is the constant p, if t is a positive literal, and the constant n, if t is a negative literal. Consider the following stratified logic program P^{rls}:

$$r_0 : clauseTrue(C) \leftarrow clause(C, \_, p, X, \_, \_, \_, \_), variable(X), not\ disabled.$$
$$r_1 : clauseTrue(C) \leftarrow clause(C, \_, n, X, \_, \_, \_, \_), not\ variable(X), not\ disabled.$$
$$r_2 : clauseTrue(C) \leftarrow clause(C, \_, \_, \_, p, X, \_, \_), variable(X), not\ disabled.$$
$$r_3 : clauseTrue(C) \leftarrow clause(C, \_, \_, \_, n, X, \_, \_), not\ variable(X), not\ disabled.$$
$$r_4 : clauseTrue(C) \leftarrow clause(C, \_, \_, \_, \_, \_, p, X), variable(X), not\ disabled.$$
$$r_5 : clauseTrue(C) \leftarrow clause(C, \_, \_, \_, \_, \_, n, X), not\ variable(X), not\ disabled.$$

$$r_6 : unsat \leftarrow clause(C, \_, \_, \_, \_, \_, \_, \_), not\ clauseTrue(C, \_, \_, \_, \_, \_, \_, \_).$$
$$r_7 : clause(c_0, c_1, p, x_0, p, x_0, p, x_0).$$
$$r_8 : variable(x_0).$$
$$r_9 : unsound \leftarrow clause(C1, C2, \_, \_, \_, \_, \_, \_), not\ clause(C2, \_, \_, \_, \_, \_, \_, \_).$$
$$r_{10} : sound \leftarrow not\ unsound.$$
$$r_{11} : sat \leftarrow sound, not\ unsat.$$

Now we show that Φ is satisfiable $\Leftrightarrow$ there exists an outlier in $\mathcal{P}(\Phi) = \langle P^{rls}, P^{obs}(\Phi) \rangle$.

($\Rightarrow$) Suppose that Φ is satisfiable, and take one of such satisfying truth assignments, say σ. Consider the set $\mathcal{W}$ containing the fact sat, and the set $\mathcal{O}$ containing all the facts $variable(x_i)$ associated to the variables x_i that are false in σ plus the fact $disabled$. Obviously, the program $P(\mathcal{P}(\Phi))_{\mathcal{W}}$ has a unique minimal model, in which each fact $clauseTrue(c_j)$ is false since $disabled$ is true, for it being not removed from $P^{obs}(\Phi)$. Hence, $\neg sat$ is entailed in $P(\mathcal{P}(\Phi))_{\mathcal{W}}$. Moreover, the program $P(\mathcal{P}(\Phi))_{\mathcal{W},\mathcal{O}}$ has the effect of evaluating the truth value of Φ. Since σ is a satisfying assignment, the unique model of $P(\mathcal{P}(\Phi))_{\mathcal{W},\mathcal{O}}$ will contain sat. Hence, $\mathcal{O}$ is an outlier in $\mathcal{P}(\Phi)$, and $\mathcal{W}$ is a witness for it.

($\Leftarrow$) Suppose that there exists an outlier $\mathcal{O}$ with witness $\mathcal{W}$ in $\mathcal{P}(\varPhi)$. Notice that only the fact sat can be entailed by the program among the facts in P^{obs}. Assume by contradiction that $sat \notin \mathcal{W}$, then $P(\varPhi)_{\mathcal{W},\mathcal{O}} \models \neg\mathcal{W}$ and $\mathcal{W}$ is not an outlier witness. Thus, it must be the case that $sat \in \mathcal{W}$, and hence that $P(\varPhi)_{\mathcal{W}} \models \neg sat$. Moreover, if some other fact f is in $\mathcal{W}$, then we have both $P(\varPhi)_{\mathcal{W}} \models \neg f$ and $P(\varPhi)_{\mathcal{W},\mathcal{O}} \models \neg f$, i.e., f cannot be exploited for satisfying the point 2 of Definition 1. It follows that in order to have $P(\varPhi)_{\mathcal{W},\mathcal{O}} \not\models \neg\mathcal{W}$, the only possibility is that $P(\varPhi)_{\mathcal{W},\mathcal{O}}$ entails sat. From what above stated, it must be the case that $\mathcal{W} \subseteq \{sat\}$ and $\mathcal{O} \supseteq \{disabled\}$. Furthermore, it cannot be the case that $\mathcal{W} \cup \mathcal{O}$ contains one of the facts in the set o_4. Indeed, in this case the rule r_9 makes the fact $unsound$ true, and sat cannot be entailed by $P(\varPhi)_{\mathcal{W},\mathcal{O}}$. Thus, we can conclude that $P(\varPhi)_{\mathcal{W},\mathcal{O}}$ entails sat only if the formula $\varPhi$ is satisfiable. $\square$

Theorem 2. *Let* $\mathcal{P} = \langle P^{\mathrm{rls}}, P^{\mathrm{obs}} \rangle$ *be a rule-observation pair such that* P^{rls} *is a fixed general logic program. Then* **EXISTENCE** *under the cautious semantics is* $\varSigma_2^P$-*complete.*

Proof. (Membership) Given a fixed general logic program P^{rls} and a set of ground facts P^{obs}, we must show that there exist two sets $\mathcal{W}, \mathcal{O} \subseteq P^{\mathrm{obs}}$ such that $P(\mathcal{P})_{\mathcal{W}} \models \neg\mathcal{W}$ (query q') and $P(\mathcal{P})_{\mathcal{W},\mathcal{O}} \not\models \neg\mathcal{W}$ (query q'') We recall that the complexity of the entailment problem for general propositional logic programs is co-NP-complete. Thus, we can build a polynomial-time nondeterministic Turing machine with an NP oracle solving **EXISTENCE** as follows: the machine guesses both the sets $\mathcal{W}$ and $\mathcal{O}$, computes the propositional logic programs $ground(P(\mathcal{P})_{\mathcal{W}})$ and $ground(P(\mathcal{P})_{\mathcal{W},\mathcal{O}})$ – this task can be done in polynomial time since the size of these programs is polynomially related to the size of P^{obs}, and then solves queries q' and q'' by two calls to the oracle.

(Hardness) Let $\varPhi = \exists \mathbf{X} \forall \mathbf{Y} f$ be a quantified Boolean formula in disjunctive normal form, i.e., f is a Boolean formula of the form $D_1 \vee \ldots \vee D_m$, over the variables $\mathbf{X} = x_1, \ldots x_n$, and $\mathbf{Y} = y_1, \ldots y_q$. We associate with $\varPhi$ the following set of facts $P^{\mathrm{obs}}(\varPhi)$:

$$o_1 : unsat.$$
$$o_2 : disabled.$$

$o_3 : variable\exists(x_k).$	$1 \le k \le n$
$o_4 : variable\forall(y_i, y_{(i+1) \bmod (q+1)}).$	$1 \le i \le q$
$o_5 : \begin{array}{l} clause(c_j, c_{(j+1) \bmod (m+1)}, \wp(t_{j,1}), \ell(t_{j,1}), \\ \quad \wp(t_{j,2}), \ell(t_{j,2}), \wp(t_{j,3}), \ell(t_{j,3})). \end{array}$	$1 \le j \le m$

where $c_j = t_{j,1} \wedge t_{j,2} \wedge t_{j,3}$, $1 \le j \le m$, $\ell(t)$ denotes the atom occurring in the literal t, and $\wp(t)$ is the constant p, if t is a positive literal, and the constant n, if t is a negative literal. Consider the following general logic program P^{rls}:

$r_0 : clauseTrue \leftarrow clause(\_, \_, p, X_1, p, X_2, p, X_3),$
$\qquad variable\exists(X_1), variable\exists(X_2), variable\exists(X_3).$
$r_1 : clauseTrue \leftarrow clause(\_, \_, n, X_1, p, X_2, p, X_3),$
$\qquad not\ variable\exists(X_1), variable\exists(X_2), variable\exists(X_3).$

$\vdots$

$r_7 : clauseTrue \leftarrow clause(\_, \_, n, X_1, n, X_2, n, X_3),$
$\qquad not\ variable\exists(X_1), not\ variable\exists(X_2), not\ variable\exists(X_3).$
$r_8 : clauseTrue \leftarrow clause(\_, \_, p, Y_1, p, X_2, p, X_3),$
$\qquad variable\forall True(Y_1), variable\exists(X_2), variable\exists(X_3).$

$\vdots$

$r_{63} : clauseTrue \leftarrow clause(\_, \_, n, Y_1, n, Y_2, n, Y_3), not\ variable\forall True(Y_1, \_),$
$\qquad not\ variable\forall True(Y_2, \_), not\ variable\forall True(Y_3, \_).$
$r_{64} : variable\forall True(Y) \leftarrow variable\forall(Y, \_), not\ variable\forall False(Y).$
$r_{65} : variable\forall False(Y) \leftarrow variable\forall(Y, \_), not\ variable\forall True(Y).$
$r_{66} : clause(c_0, c_1, p, x_0, p, x_0, p, x_0).$
$r_{67} : variable\exists(x_0).$
$r_{68} : variable\forall(y_0, y_1).$
$r_{69} : unsound \leftarrow clause(C1, C2, \_, \_, \_, \_, \_, \_), not\ clause(C2, \_, \_, \_, \_, \_, \_, \_).$
$r_{70} : unsound \leftarrow variable\forall(Y1, Y2), not\ variable\forall(Y2, \_).$
$r_{71} : sound \leftarrow not\ unsound.$
$r_{72} : sat \leftarrow sound, clauseTrue.$
$r_{72} : unsat \leftarrow not\ sat.$
$r_{73} : unsat \leftarrow not\ disabled.$

Now we show that Φ is valid $\Leftrightarrow$ there exists an outlier in $\mathcal{P}(\Phi) = \langle P^{\text{rls}}, P^{\text{obs}} \rangle$.

($\Rightarrow$) Suppose that Φ is valid, and let σ^X be a truth value assignment for the existentially quantified variables $\mathbf{X}$ that satisfies f. Consider the set $\mathcal{W}$ composed by the fact $unsat$ plus all the facts $variable\exists(x_i)$ associated to the variables that are false in σ^X, that is the set $\{unsat\} \cup \{variable\exists(x) \mid x \in \text{Lit}(\sigma^X)^-\}$, and consider the set $\mathcal{O}$ composed only by the fact $disabled$. We note that the stable models $\mathcal{M}_Y$ of the program $P(\mathcal{P}(\Phi))_{\mathcal{W}}$ are in one-to-one correspondence with the truth assignments σ_Y of the universally quantified variables (consider rules r_{64} and r_{65}). Indeed, since the formula is satisfied by σ^X, for each $\mathcal{M}_Y$, $sat \in \mathcal{M}_Y$ and $unsat \notin \mathcal{M}_Y$. Hence, $P(\mathcal{P}(\Phi))_{\mathcal{W}} \models_c \neg \mathcal{W}$. Conversely, the program $P(\mathcal{P}(\Phi))_{\mathcal{W},\mathcal{O}}$ in which $disabled$ is false, trivially derives $unsat$. We can conclude that $\mathcal{O}$ is an outlier in $\mathcal{P}(\Phi)$, and $\mathcal{W}$ is a witness for it.

($\Leftarrow$) Suppose that there exists an outlier $\mathcal{O}$ with witness $\mathcal{W}$ in $\mathcal{P}(\Phi)$. As $unsat$ is the unique fact in $P^{\text{obs}}(\Phi)$ that can be derived by $P(\mathcal{P}(\Phi))_{\mathcal{W},\mathcal{O}}$, then in order to satisfy condition (2) of Definition 1, it is the case that $\mathcal{W}$ contains $unsat$. Furthermore, in order to satisfy condition (1) of Definition 1, $disabled$ does not belong to $\mathcal{W}$. Now, we show that $\{unsat\} \subseteq \mathcal{W} \subseteq \{unsat, variable\exists(x_1), \ldots, variable\exists(x_n)\}$. By contradiction, assume that a fact of the set o_4 (o_5 resp.) belongs to $\mathcal{W}$, then $P(\Phi)_{\mathcal{W}} \models unsat$ by rule r_{70} (r_{69} resp.).

Let $\mathcal{X}$ be the subset $\{variable\exists(x) \mid x \in (\mathcal{W} \setminus \{unsat\})\}$ and let σ^X be the truth value assignment $\sigma_{(\{x_1,\dots,x_n\}\setminus\mathcal{X})\cup\neg\mathcal{X}}$ to the set of variables $\mathbf{X}$. Clearly, $P(\mathcal{P}(\Phi))_{\mathcal{W}} \models_c (\{variable\exists(x_1),\dots,variable\exists(x_n)\} \setminus \mathcal{X}) \cup \neg\mathcal{X}$. Furthermore, as $P(\mathcal{P}(\Phi))_{\mathcal{W}} \models_c \neg unsat$, then it is the case that for each subset Y of $\mathbf{Y}$, the stable model $\mathcal{M}_Y$ of $P(\mathcal{P}(\Phi))_{\mathcal{W}}$ associated to Y, that is the model $\mathcal{M}_Y$ containing $\{variable\forall True(y) \mid y \in Y\}$ and no other fact of the same predicate, is such that $sat \in \mathcal{M}_Y$. That is, for each truth value assignment σ^Y to the variables in the set Y, there exists at least a disjunct such that $\sigma^X \circ \sigma^Y$ makes the formula f true. As a consequence, Φ is valid. To conclude the proof, note that $\mathcal{O} = \{disabled\}$ is always an outlier having such a witness. $\qquad\square$

Theorem 3. *Let $\mathcal{P} = \langle P^{\mathrm{rls}}, P^{\mathrm{obs}} \rangle$ be a rule-observation pair such that P^{rls} is a fixed general logic program. Then* **EXISTENCE** *under the brave semantics is Σ_2^P-complete.*

Proof. (Membership) Analogous to that of Theorem 2.

(Hardness) Let $\Phi = \exists\mathbf{X}\forall\mathbf{Y} f$ be a quantified Boolean formula in disjunctive normal form, i.e., f is a Boolean formula of the form $d_1 \vee \dots \vee d_m$, over the variables $\mathbf{X} = x_1,\dots x_n$, and $\mathbf{Y} = y_1,\dots y_q$. Deciding the validity of such formulas is a well-known Σ_2^P-complete problem. W.l.o.g., assume that each disjunct d_j contains three literals at most.

We associate with Φ the following set of facts $P^{\mathrm{obs}}(\Phi)$:

$o_1 : sat.$
$o_2 : disabled.$
$o_3 : variable\exists(x_k).$ $\hspace{4cm}$ $1 \le k \le n$
$o_4 : variable\forall(y_i, y_{(i+1) \bmod (q+1)}).$ $\hspace{1.8cm}$ $1 \le i \le q$
$o_5 : \begin{array}{l} clause(c_j, c_{(j+1) \bmod (m+1)}, \wp(t_{j,1}), \ell(t_{j,1}), \\ \quad \wp(t_{j,2}), \ell(t_{j,2}), \wp(t_{j,3}), \ell(t_{j,3})). \end{array}$ $\hspace{1cm}$ $1 \le j \le m$

where $c_j = t_{j,1} \wedge t_{j,2} \wedge t_{j,3}$, $1 \le j \le m$, $\ell(t)$ denotes the atom occurring in the literal t, and $\wp(t)$ is the constant p, if t is a positive literal, and the constant n, if t is a negative literal. Consider the following general logic program P^{rls}:

$r_0 : clauseTrue \leftarrow not\ disabled, clause(\_, \_, p, X_1, p, X_2, p, X_3),$
$\qquad variable\exists(X_1), variable\exists(X_2), variable\exists(X_3).$
$r_1 : clauseTrue \leftarrow not\ disabled, clause(\_, \_, n, X_1, p, X_2, p, X_3),$
$\qquad not\ variable\exists(X_1), variable\exists(X_2), variable\exists(X_3).$
$\vdots$
$r_7 : clauseTrue \leftarrow not\ disabled, clause(\_, \_, n, X_1, n, X_2, n, X_3),$
$\qquad not\ variable\exists(X_1), not\ variable\exists(X_2), not\ variable\exists(X_3).$
$r_8 : clauseTrue \leftarrow not\ disabled, clause(\_, \_, p, Y_1, p, X_2, p, X_3),$
$\qquad variable\forall True(Y_1), variable\exists(X_2), variable\exists(X_3).$
$\vdots$

r_{63} : $clauseTrue \leftarrow not\ disabled, clause(\_, \_, n, Y_1, n, Y_2, n, Y_3),$
  $not\ variable\forall True(Y_1, \_), not\ variable\forall True(Y_2, \_),$
  $not\ variable\forall True(Y_3, \_).$
r_{64} : $variable\forall True(Y) \leftarrow variable\forall(Y, \_), not\ variable\forall False(Y).$
r_{65} : $variable\forall False(Y) \leftarrow variable\forall(Y, \_), not\ variable\forall True(Y).$
r_{66} : $clause(c_0, c_1, p, x_0, p, x_0, p, x_0).$
r_{67} : $variable\exists(x_0).$
r_{68} : $variable\forall(y_0, y_1).$
r_{69} : $unsound \leftarrow clause(C1, C2, \_, \_, \_, \_, \_, \_), not\ clause(C2, \_, \_, \_, \_, \_, \_, \_).$
r_{70} : $unsound \leftarrow variable\forall(Y1, Y2), not\ variable\forall(Y2, \_).$
r_{71} : $sound \leftarrow not\ unsound.$
r_{72} : $sat \leftarrow sound, clauseTrue.$

Now we show that Φ is valid $\Leftrightarrow$ there exists an outlier in $\mathcal{P}(\Phi) = \langle P^{\mathrm{rls}}, P^{\mathrm{obs}} \rangle$.

Following the same line of reasoning of Theorem 2, it is easy to see that Φ is valid $\Leftrightarrow \mathcal{O} = \{disabled\} \cup \mathcal{X}'$ is an outlier with witness $\mathcal{W} = \{sat\} \cup (\mathcal{X} \setminus \mathcal{X}')$ in $\mathcal{P}(\Phi)$, where $\mathcal{X} = \{variable\exists(x) \mid x \in \mathrm{Lit}(\sigma^X)^-\}$, σ^X is a truth value assignment for the existentially quantified variables $\mathbf{X}$ that satisfies f, and $\mathcal{X}'$ is a generic subset of $\mathcal{X}$.   □

5 Conclusions and Discussion

In this paper, we introduced a novel approach to outlier detection that can be considered complementary to existing statistical-like methods.

Roughly speaking, classical outlier detection methods examine the input database in order to find individuals having associated attribute-value pairs significantly deviating from an average value. On the contrary, the approach proposed exploits background knowledge, encoded in the form of a logical theory, in order to single out individuals of the input database having a behavior for which no logical justification can be found in the theory at hand. We believe that this approach is of interest in several fields, such as bioinformatics, fraud detection, network robustness analysis and intrusion detection, where a domain expert might exploit his background knowledge for improving the effectiveness of traditional outlier mining algorithms.

In the framework we have presented, the background knowledge is encoded in the form of a first order logic program, while the observations are encoded in the form of a set of logical facts representing the status of a database at hand.

Besides discussing the novel approach, we studied the data complexity of detecting outliers. It turned out that most of the problems are not feasible in polynomial time (unless P=NP) and as such outlier detection via logical theories is "intrinsically" intractable, as well as many other relevant data mining tasks (e.g. mining of association rules, decision tree induction, etc. [15, 11, 4]). As a further research we plan to investigate some practical algorithms based on suitable heuristics for singling out outliers in some applicative scenarios.

References

1. C. C. Aggarwal and P.S. Yu. Outlier detection for high dimensional data. In *Proc. ACM Int. Conference on Managment of Data*, 37–46, 2001.
2. F. Angiulli, R. Ben-Eliyahu-Zohary, and L. Palopoli. Outlier detection using default logic. In *Proc. Int. Joint Conf. on Artificial Intelligence*, 833–838, 2003.
3. F. Angiulli, R. Ben-Eliyahu-Zohary, and L. Palopoli. Outlier detection using disjunctive logic programming. In *Proc. of the European Conf. on Artificial Intelligence*, 2004.
4. F. Angiulli, G. Ianni, and L. Palopoli. On the complexity of inducing categorical and quantitative association rules. *Theoretical Computer Science*, (1-2)314, 217–249, 2004
5. F. Angiulli and C. Pizzuti. Outlier detection in high dimensional spaces. In *Proc. of the European Conf. on Principles and Practice of Know. Discovey in Databases*, 15–26, 2002.
6. A. Arning, R. Aggarwal, and P. Raghavan. A linear method for deviation detection in large databases. In *Proc. Int. Conf. on Know. Discovery and Data Mining*, 164–169, 1996.
7. S.D. Bay and M. Schwabacher. Mining Distance-Based Outliers in Near Linear Time with Randomization and a Simple Pruning Rule. In *Proc. Int. Conf. on Knowledge Discovery and Data Mining*, 2003.
8. M.M. Breunig, H. Kriegel, R.T. Ng, and J. Sander. LOF: Identifying density-based local outliers. In *Proc. ACM Int. Conf. on Managment of Data*, 93–104, 2000.
9. P. Brucker. On the complexity of some clustering problems. *Optimization and Operations Research*, 45–54, Springer-Verlag, 1978.
10. M. Gelfond and V. Lifschitz. The stable model semantics for logic programming. In *Fifth Int'l Conf.Symp. on Logic Programming*, 1070–1080, 1988.
11. D. Gunopulos, H. Mannila, and S. Saluja. Discovering all most specific sentences by randomized algorithms. In *Proc. Int. Conf. on Database Theory*, 215–229, 1997.
12. E. Knorr and R. Ng. Algorithms for mining distance-based outliers in large datasets. In *Proc. Int. Conf. on Very Large Databases*, 392–403, 1998.
13. A. Lazarevic and L. Ertoz and V. Kumar and A. Ozgur and J. Srivastava. A Comparative Study of Anomaly Detection Schemes in Network Intrusion Detection. In *Proc. SIAM Int. Conf. on Data Mining*, 2003.
14. N. Leone, G. Pfeifer, W. Faber, T. Eiter, G. Gottlob, S. Perri, and F. Scarcello. The dlv system for knowledge representation and reasoning. *ACM Transactions on Computational Logic. To Appear.*
15. S. Morishita. On classification and regression. In *Proc. of the 1st Int. Conf. on Discovery Science*, 40–57, 1998.
16. C.H. Papadimitriou. *Computatational Complexity*. Addison-Wesley, Reading, Mass., 1994.
17. S. Ramaswamy, R. Rastogi, and K. Shim. Efficient algorithms for mining outliers from large data sets. In *Proc. ACM Int. Conf. on Managment of Data*, 427–438, 2000.

Fast Hierarchical Clustering Algorithm Using Locality-Sensitive Hashing

Hisashi Koga, Tetsuo Ishibashi, and Toshinori Watanabe

Graduate Schools of Information Systems, University of Electro-Communications
{koga,watanabe}@sd.is.uec.ac.jp

Abstract. A hierarchical clustering is a clustering method in which each point is regarded as a single cluster initially and then the clustering algorithm repeats connecting the nearest two clusters until only one cluster remains. Because the result is presented as a dendrogram, one can easily figure out the distance and the inclusion relation between clusters.

One drawback of the agglomerative hierarchical clustering is its large time complexity of $O(n^2)$, which would make this method infeasible against large data, where n expresses the number of the points in the data.

This paper proposes a fast approximation algorithm for the single linkage clustering algorithm that is a well-known agglomerative hierarchical clustering algorithm. Our algorithm reduces its time complexity to $O(nB)$ by finding quickly the near clusters to be connected by use of Locality-Sensitive Hashing known as a fast algorithm for the approximated nearest neighbor search. Here B expresses the maximum number of points thrown into a single hash entry and practically grows a simple constant compared to n for sufficiently large hash tables.

By experiment, we show that (1) the proposed algorithm obtains similar clustering results to the single linkage algorithm and that (2) it runs faster for large data than the single linkage algorithm.

1 Introduction

Clustering is a powerful tool for data analysis which gives an insight into characteristics of the given data by classifying them into several groups. Because of its usefulness, clustering has been used in various application areas such as biology, medicine and information science. The conventional clustering methods are categorized into two classes, that is, hierarchical clustering and non-hierarchical clustering. The hierarchical clustering is a kind of unsupervised classification method which excels in that no information from the outside such as the number of clusters and various threshold parameters need to be specified.

The agglomerative hierarchical clustering approach regards each point as a single cluster initially and connects the most similar pair of clusters step by step until the number of clusters becomes one finally. In the the agglomeration step, the clustering algorithm first searches the most similar pair of clusters under the condition that the distances between all of the existing clusters are known.

E. Suzuki and S. Arikawa (Eds.): DS 2004, LNAI 3245, pp. 114–128, 2004.

Then, the most similar pair of clusters are merged into a new single cluster. Finally the distances between this new cluster and other unchanged clusters are updated. The definition of distance between clusters depends on the clustering algorithm, which in turn leads to the variation in the agglomeration process. The clustering result is typically drawn as a dendrogram which illustrates the distance and the inclusion relation between clusters and helps the analyzers to acquire a proper understanding of the data. By contrast, a notorious drawback of the agglomerative hierarchical clustering is its awful time complexity mainly caused by the calculation of the distances between all clusters which becomes more serious as the data size grows large.

In this paper, we focus on the single linkage algorithm [9] known as an agglomerative hierarchical clustering algorithm which suffers from large time complexity as we mentioned. Especially we propose its fast randomized approximation algorithm.

In the single linkage algorithm, the distance between two clusters is defined as the distance between the two nearest points each of which is a member of each cluster. This algorithm produces clusters such that every member of a cluster is more closely related to at least one member of its cluster than to any point outside it. This algorithm is apt to gather a chain of points into one cluster and can discover clusters in various shapes besides the common ovals. Its time complexity reaches $O(n^2)$ where n represents the number of points in the data [7]. We remark here that this bound of $O(n^2)$ originates in computing the distances between all pairs of these n points.

Our approximation algorithm utilizes the *Locality-Sensitive Hashing* [5] (abbreviated as LSH hereafter) that is an probabilistic approximated algorithm for the nearest neighbor search. We rely on LSH to find efficiently the close clusters to be merged. For this reason, our algorithm is named LSH-link. The hierarchical structure is formed by combining the multiple clustering results derived with different parameter values. LSH-link achieves a time complexity of $O(nB)$, where B is the maximum number of points in a single hash entry and may be treated as a small constant compared to n by preparing large hash tables usually. This algorithm is useful especially when the analyzers want to discover the coarse-grained hierarchical structure from the given data in a short period. We show by experiment that (1) LSH-link obtains similar clustering results to the single linkage algorithm and that (2) it runs faster for large data than the single linkage algorithm.

The structure of this paper is summarized as follows. In Sect. 2, we introduce LSH in detail. Sect. 3 explains our LSH-link algorithm. There we discussion its time complexity. In Sect. 4 the experimental result of LSH-link is reported. Sect. 5 mentions the conclusion and the future work.

2   Locality-Sensitive Hashing(LSH)

LSH [5] is a randomized algorithm for searching approximated nearest neighbor points for a given query point q from a set of points P in an Euclidean space.

LSH uses a hash function satisfying the property that near points are stored in the same entry (bucket) with high probability, while remote points are stored in the same bucket with low probability. Thus it narrows down the search range to the points kept in the same bucket as q, accelerating the nearest neighbor search. In addition, LSH prepares multiple hash functions to reduce the probability that a near point to q is not noticed. LSH is proved both in theory and by experiment to run efficiently even for high-dimensional data [2]. We explain its mechanism from now on.

2.1   Hash Functions

Let $p = (x_1, x_2, \ldots, x_d)$ be a point in a d-dimensional space where the maximal coordinate value of any point is less than a constant C. We can transform p to a Cd-dimensional vector $v(p) = \text{Unary}_C(x_1)\text{Unary}_C(x_2) \cdots \text{Unary}_C(x_d)$ by concatenating unary expressions for every coordinate. Here $\text{Unary}_C(x)$ is a sequence of x ones followed by $C - x$ zeroes as shown in Equation (1). For example, when $p = (3, 2)$ and $C = 5$, $v(p) = 1110011000$.

$$\text{Unary}_C(x) = \underbrace{11 \ldots 11}_{x}\underbrace{00 \ldots 00}_{C-x}. \tag{1}$$

In LSH, a hash function is defined as follows. Let I be a set consisting of k distinct elements chosen uniformly randomly from the set of Cd arithmetic values $\{1, 2, \cdots Cd\}$, where k is a parameter of LSH. A hash function for LSH computes a hash value for a point p by concatenating the k bits from $v(p)$ corresponding to the set I.

LSH prepares l distinct hash functions which differ from one another in the selection of I. k and l are important parameters for LSH whose effects are discussed in Sect. 2.3. k is called the number of sampled bits and l is called the number of hash functions throughout this paper.

2.2   Approximated Nearest Neighbor Search

The procedure to find the nearest point to q from the set of points P follows the next two steps.

Step 1: The generation of hash tables
l hash functions $h_1, h_2, \cdots .h_l$ are constructed from l sets of sampled bits $I_1, I_2, \cdots, I_l$. Then, for each point in P, l hash values are computed. As the result, l hash tables are built.

Step 2: The nearest neighbor search for the query q

1. l hash values are computed for q.
2. Let P_i be the set of points in P classified into the same bucket as q on the hash table for h_i. In LSH, the points in $P_1 \cup P_2 \cup \cdots \cup P_l$ become the candidates for the nearest point to q. Among the candidates, the nearest point to q is determined by measuring the actual distance to q.

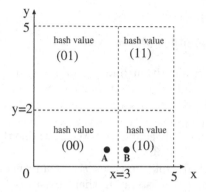

Fig. 1. Space partition by a hash function in LSH.

2.3 Parameters in LSH

This section mentions the intuitive significance of parameters in LSH.

The Number of Sampled Bits k: In a word, a hash function in LSH cuts the d-dimensional space into cells by k hyperplanes. Fig. 1 illustrates a two-dimensional case when $C = 5$ and $k = 2$. The hash value of each cell is written in it, provided that the third bit (i.e. the line $x = 3$) and the seventh bit (i.e. the line $y = 2$) are selected from the total $2 \times 5 = 10$ bits. This figure tells that near points tend to take the same hash value. The sizes of generated cells are expectedly large for small values of k, while they are expectedly small for large values of k. Thus, k determines how far points are likely to take the same hash value. Our algorithm in Sect. 3 exploits this characteristic of LSH.

The Number of Hash Functions l: The reason why LSH utilizes multiple hash functions is that two points may take different hash values depending on the result of the probabilistic space division, even if they are sufficiently close to each other. For example, in Fig. 1, though the two points A and B are closely located, the cell containing A does not cover B.

However, by increasing the number of the hash functions, A and B will enter the same hash bucket at least for one hash function. This reduces the probability of the oversight of the near neighbor points. On the other hand, one should be aware of the danger that, for large values of l, far points might be contained in the candidates for the nearest neighbors (that is, false positive). LSH premises the existence of false positives and, therefore, inspects the points in the candidate set for the nearest neighbors by measuring the real distances.

3   LSH-Link Algorithm

This section explains our new clustering algorithm LSH-link that exploits the hash tables generated by LSH. This algorithm outputs clustering results that approximates the single linkage algorithm.

3.1   Design Concepts

Our algorithm works in multiple phases. A single phase corresponds to a single layer in the hierarchy.

In a single phase, our algorithm finds the clusters within a distance of r for each cluster by LSH and combines them into a single cluster. Here r is a parameter which should be adjusted to the number of the sampled bits k in LSH. The analyzer only has to specify r for the first phase. In the phases except the first phase, r is automatically configured. We postpone the explanation about how to adjust r to k until Sect. 3.3.

After a phase ends, LSH-link increases r and advances to the next phase if there remain more than 2 clusters (If there exists only one cluster, LSH-link terminates). By increasing r, we anticipate that clusters not merged in the previous phase will have a chance to be unified in this phase. Repeating multiple phases creates the hierarchical structure. We say that a phase i is *higher* than a phase j, if i is performed later than j in the execution of LSH in the sense that the phase i corresponds to a higher layer in the final hierarchy than the phase j.

Note that the single linkage algorithm merges the closest pair of clusters step by step. Hence, we can consider that a single phase of LSH-link puts together several steps in the single linkage algorithm. LSH-link differs greatly from the single linkage algorithm in that it discovers clusters to be merged very fast with omitting the calculation of the distances between all clusters, which contributes to the reduction of time complexity.

3.2   Details of the Algorithm

The action of LSH-link is described below. At the beginning, each point is viewed as n clusters containing a single point only.

LSH-Link Algorithm:
Step 1: For each point p, l hash values $h_1(p)$, $h_2(p)$, $\cdots$, $h_l(p)$ are computed. On the i-th hash table, p is thrown into the hash bucket with the index $h_i(p)$. However, if another point belonging to the same cluster as p has already been saved in the very bucket, p is not stored in it.
Step 2: For each point p, from the set of points which enter the same bucket as p at least on one hash table, LSH-link finds those whose distances from p are less than r.
Step 3: The pairs of clusters each of which corresponds to a pair of points yielded in Step 2 are connected.
Step 4: If the number of clusters is bigger than 1 after Step 3, r is set to a new larger value and LSH-link advances to Step 5. Otherwise the algorithm finishes.
.**Step 5:** LSH-link diminishes k in accordance with the increase of r. It goes back to Step 1, after generating new l hash functions by using this new k.

In Step 1, any points belonging to the same cluster are assured not to enter the same bucket. This feature is essential in order to avoid the performance degradation of hashing, which tends to take place in higher phases in which many points take the same hash value against small values of k. Note Step 1

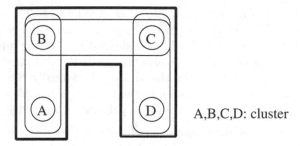

Fig. 2. Process of merging.

and Step 2 are equivalent to the original LSH except that only one point per cluster is allowed to stay in one hash bucket. Thus LSH-link examines at most l hash buckets for each point. In Step 2, LSH-link removes the point pairs whose distances are greater than r from the set of candidates for near point pairs.

Step 3 is responsible for merging clusters. Let us give an example of this step in Fig. 2. In this picture, the cluster A and the cluster B, the cluster B and the cluster C, and the cluster C and the cluster D form three pairs of clusters to be merged respectively (i.e the three areas surrounded by thin lines). In this case, all of the four clusters compose one large cluster surrounded by a thick line. Thus, like the single linkage algorithm, LSH-link is able to discover clusters in various shapes besides the common ovals.

3.3   Parameters in LSH-Link

This section discusses important parameters in LSH-link which influence the processing speed and the quality of the clustering results.

- R that is the initial value of r: LSH-link connects the pairs of points whose distances are less than R in the first phase. This means that, in order to identify point pairs whose distances are small, R should be set to a small value also.
- The increase ratio A of r: Let r_c be the value of r in the current phase and r_n be the one in the next phase. When the current phase ends, LSH-link increases r such that $r_n = Ar_c$ $(A > 1)$. A is a constant consistently used in all phases. For large values of A, the number of executed phases decreases and the execution time shrinks. By contrast, for small values of A, though the execution time increases, the clustering result gets detailed and more similar to the single linkage method.

In this way, R and A controls the fineness of the clustering result.

- the number of hash functions l: For the reason claimed in Sect. 2.3 this parameter is important. For high dimensional data, l need to be somewhat big, so that all pairs of points within a distance of r may be found without fail.

- the relation between r and the number of sampled bits k: Though the sampled bits are selected randomly, the sizes of cells are roughly determined, when k is fixed. Therefore, r should be balanced with k. We adapt as r the diagonal length of a cell under the assumption that the Euclidean space is partitioned at an equal interval by the sampled bits ideally. Since k sampled bits are distributed to the d dimension, the number of sampled bits per one dimension is equal to $\frac{k}{d}$. As the result, hypercubes whose edges are $\frac{C}{\frac{k}{d}+1}$ in length are made. Thus, the diagonal length of a hypercube becomes $\frac{dC}{k+d}\sqrt{d}$. Therefore, if p, q are the two points that have entered the same bucket, the associated two clusters are merged when

$$||p - q|| \leq r = \frac{dC}{k+d}\sqrt{d}. \tag{2}$$

3.4   Time Complexity

This section proves the time complexity of LSH-link does not go beyond $O(nB)$, where B is the maximum number of points that one hash bucket holds over the execution of LSH-link. Strictly speaking, B is influenced by n in the sense that n points are distributed in the hash tables. On the other hand, B depends on the occupancy ratio of hash tables and can be regarded as a constant by far smaller than n if we prepare sufficiently large hash tables. Though the hash table size decreases in the running of LSH-link as the phase gets higher and k gets smaller, LSH-link keeps the occupancy ratio small by prohibiting multiple points from the same cluster entering the identical hash bucket.

Our method repeats the following three operations until only one cluster is left. The number of this repetition is denoted by s. We clarify the time complexity incurred in all of the three operations. Let K be the value of k for the first phase hereafter.

1. Generation of hash tables
2. Search for clusters to be merged
3. Updating clusters by connecting them

Lemma 1. *The time complexity in generating hash tables is $O(\frac{A}{A-1}nlK)$.*

Proof. The generation of hash tables is almost the same as LSH. However, when a point is saved in a hash bucket, an extra judgment must be performed which checks if another point from the same cluster has already entered the hash bucket. This judgment finishes in $O(1)$ time by accompanying each hash bucket with an array showing from which cluster the hash bucket is holding the points currently. This computational overhead is smaller than the calculation of hash functions mentioned below and may be ignored.

Now let us evaluate the overhead in calculating hash functions. Regarding to the first phase, it takes $O(K)$ time to compute a hash function once, because the K sampled bits are looked into. Since we need to acquire l hash values for all the n points, a time of $O(nlK)$ is spent for the generation of hash tables in the first phase. From the formula (2), k is inversely proportional to r. Hence, k is reduced to $\frac{K}{A^{i-1}}$ in the i-th phase in which r increases up to $A^{i-1}R$. Therefore, the total time of calculating hash functions over all the phases is at most

$$O(nlK + \frac{1}{A}nlK + \frac{1}{A^2}nlK + \cdots + \frac{1}{A^{s-1}}nlK) \leq O(\sum_{i=0}^{\infty} \frac{1}{A^i}nlK) \leq O(\frac{A}{A-1}nlK).$$

Interestingly, since k decreases exponentially, the time complexity over all phases is not more than constant times that in the first phase.

Because all of A, l and K are constants independent from n, the generation of hash tables is complete in $O(n)$ time. $\square$

Lemma 2. *The time complexity in searching for clusters to be merged is $O(nlB)$.*

The proof resembles to Lemma 3 and is omitted here.

Lemma 3. *The time complexity in updating clusters is $O(n + nlB)$.*

Proof. First we consider the time complexity per phase. Consider the graph in which each vertex expresses a single cluster and a edge runs between two vertices such that these two vertices are a pair of clusters to be merged. Then, constructing new large clusters is equivalent to the decomposition of the graph into connected components such that a connected component represents a new large cluster. Here, the decomposition of a graph into connected components is realized by depth-first search in $O(N + M)$ time by using the list expression of the graph, where N is the number of vertices and M is the number of edges. In our instance, $N \leq n$ evidently.

From now on, we show $M \leq nlB$. Note that each edge of the graph represents a pair of clusters to be merged which is selected from the set (denoted by SH) of point pairs entering the same hash bucket. Therefore M is smaller than the cardinality of SH. Next, as we have l hash tables and one hash bucket contains at most B points, the cardinality of SH is bounded from above by nlB. Hence $M \leq nlB$.

By summarizing the above discussion, it is proved that the time complexity of constructing new cluster grows $O(n + nlB)$. After merging clusters, we need to make an array that indicates to which cluster each point is assigned. This array is useful in the next phase, when the algorithm checks whether a hash bucket does not have multiple points from the identical cluster. A time of $O(n)$ is necessary for the arrangement of this array.

As there are s phases in total, the total time in updating clusters grows $O(s(n + nlB))$. We conclude the proof by showing that s is a constant of $O(1)$ size.

In the i-th phase, pairs of clusters within a distance of $A^{i-1}R$ are connected. Let r' be the distance between the farthest point pair in the data and s' be the smallest integer satisfying $A^{s'-1}R \geq r'$. Namely, $s' \geq 1 + \log_A\left(\frac{r'}{R}\right)$. Because s' is the smallest integer satisfying this inequality, we have $s' \leq 2 + \log_A\left(\frac{r'}{R}\right)$. From the definition of r' and s', the number of phases s is smaller than s'.

Hence, we obtain the next formula: $s \leq s' \leq 2 + \log_A \left(\frac{r'}{R} \right)$. Here r' depends on the distribution of points but has nothing to do with the data size. Because A and R are also independent from n, s is proved to be a constant of $O(1)$ size independent from n.

As above, since the time complexity per phase is $O(n+nlB)$ and the number of phases s is $O(1)$, the operation of updating clusters over all the phases finishes in a time of $O(n + nlB)$. Because both l is a constant independent from n, the update of clusters is complete in $O(nB)$ time. $\square$.

From Lemma 1, 2 and 3, we derive Theorem 1 regarding to the total time complexity of LSH-link.

Theorem 1. *The time complexity of LSH-link is $O(nB)$.*

3.5   Related Works

Now that we have described our algorithm, we briefly review the related works of our approach. The approach to apply LSH to clustering is previously attempted by [3] in the context of web-page clustering. Though our paper coincides with theirs in that near pairs of points are derived by LSH, their clustering algorithm is complete linkage method and different from our LSH-link. Our paper is superior to their paper in that they neither handle the hierarchical clustering issue nor discuss the time complexity.

In terms of low time complexity, BIRCH [10] achieves an $O(n)$ I/O cost. BIRCH is a incremental hierarchical clustering algorithm and constructs a data structure called a CF (cluster feature) tree by scanning the data only once. Each node in the CF tree maintains a data structure named CF which represents a cluster. The size of the CF is so small that the cluster operation may be manipulated in the fast memory. On the other hand, LSH-link is an agglomerative hierarchical clustering algorithm. While LSH-link need to scan the data multiple times until the final one cluster is obtained, LSH-link is better than BIRCH in that BIRCH performs poorly, if the clusters are not spherical in shape as reported in [8]. Recently a new agglomerative clustering algorithm CHAMELEON [6] is suggested whose time complexity is $O(n \log n)$. This algorithm uses two evaluation measures to judge whether to connect two clusters together, that is, relative inter-connectivity and relative closeness. LSH-link is based on distances solely and does not direct to this approach.

Our approach divides an Euclidean space randomly into grids that we have called cells so far. As for the grid-based clustering, an approach that treats only high-density cells as being valid and neglects low-density cells as outliers is known, for example, in CLIQUE [1] and in DENCLUE [4]. This approach is effective especially for high-dimensional data which often contain noise data points. However this approach cannot handle clusters with different densities and does not consider hierarchical structures between clusters. Because LSH-link inherits the single linkage algorithm, it is free from these difficulties. With respect to the noise exclusion, we suppose that it is easy for LSH-link to adopt the density-based approach by investigating the densities of the hash buckets as its preprocessing.

4   Experimental Results

This section confirms the practicality of LSH-link by experiments from the three viewpoints below.

- The similarity of the dendrogram to the single linkage method.
- The ability to extract clusters in various shapes besides the common ovals.
- The execution time for a large dataset.

4.1   Similarity of Dendrogram

For evaluating a pattern recognition algorithm, a dataset *Iris* is used very frequently. We compare the dendrogram by LSH-link with that by the single linkage algorithm for Iris. Iris consists of 150 points in a 5-dimensional space. The five dimensions express the length and the width of the petal, the length and the width of the cup, and the kind of the iris. In the experiment we use only the four dimensional data except the kind of the iris.

The parameters of LSH-link are set as follows: the number of hash functions $l = 20$, the number of the sampled bits in the initial phase $K = 200$. Regarding to A (the increase ratio of r), we test the two values (i)$A = 1.4$ and (ii)$A = 2.0$. About R, we decide it from K, obeying the formula (2).

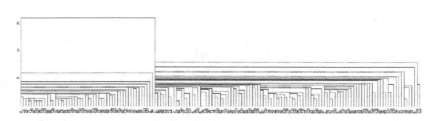

Fig. 3. Dendrogram by single linkage method.

From Fig. 3 to Fig. 5, the clustering results are illustrated as dendrograms which show the process of mering clusters. The vertical axis represents the distance between a pair of merged clusters. In our web page [11], one can acquire the large images of the dendrograms. In the page, the dendrogram for $A = 1.2$ is also exhibited.

From these dendrograms, one can see that LSH-link yields a dendrogram that approximates the single linkage method both for $A = 1.4$ and for $A = 2.0$. Especially, the data are classified into the two large cluster at the top layer in all of the three dendrograms. We confirmed that the members of these two clusters accord exactly between the single linkage method and our LSH-link. Be conscious that the clustering result for $A = 1.4$ resembles more to the single linkage algorithm than that for $A = 2.0$. This phenomenon claims that an analyzer can specify the grain of the clustering result arbitrarily by adjusting A.

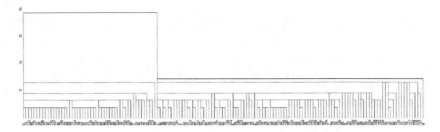

Fig. 4. Dendrogram by LSH-link($A = 1.4$).

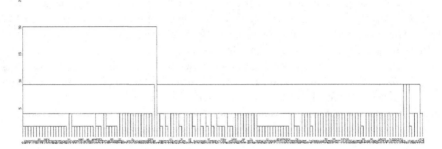

Fig. 5. Dendrogram by LSH-link($A = 2.0$).

4.2 Ability to Extract Clusters in Various Shapes

Next, we apply LSH-link to the two dimensional data in Fig. 6 which contains clusters in various shapes such as a bar and a ring. In the data, 913 points are plotted in the area of 100x100 size and there are 6 clusters. The minimum distance between two clusters that corresponds to the one between the ring cluster and the sphere cluster inside the ring cluster is $\sqrt{53} \approx 7.28$ and the maximum distance between a pair of points belonging to the identical cluster is $\sqrt{17} \approx 4.12$. As these two values are rather close, in order to extract the 6 clusters exactly a clustering algorithm is requested to feature the high accuracy.

We repeat the experiments 10 times under the condition that $l = 10$, $K = 100$ and $A = 1.3$. As the result, LSH-link succeeds in discovering the 6 clusters with a probability of $\frac{10}{10} = 100\%$ when the number of remaining clusters equals 6 as shown in Fig. 7. Thus, the ability of LSH-link to extract clusters in various shapes besides the common oval is verified. This result also demonstrates the stability of our randomized algorithm.

4.3 Execution Time for a Large Dataset

Finally, LSH-link is applied to a 20-dimensional large dataset and its execution time is compared with the single linkage algorithm.

Fig. 6. Experimental data.

Fig. 7. The clustering result by LSH-link.

The experimental data consists of 6 clusters each of which is generated from a Gaussian distribution whose variance is equal to 1. The central points of these 6 clusters are listed as follows: $(10, 10, \ldots, 10), (20, 20, \ldots, 20), (30, 30, \ldots, 30),$ $(50, 50, \ldots, 50), (60, 60, \ldots, 60), (70, 70, \ldots, 70)$. Note that there exist two lumps of three clusters.

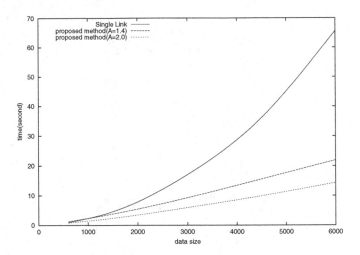

Fig. 8. Comparison of execution time.

In the experiment we change the number of points from 600 (100 per cluster) to 6000 (1000 per cluster) and see how the execution time varies as the number of points increases. The single linkage method is realized with the Partial Maximum Array [7]. As for the parameters in LSH-link, we set K to 200 and l to 20. Then we examine the two values for A again, i.e., 1.4 and 2.0.

Before the discussion of the length of the running time, we briefly refer to the accuracy of LSH-link. LSH-link could extract 6 clusters correctly both for $A = 1.4$ and $A = 2.0$ without any fault. In addition, when we set the number of the produced clusters to two, it was confirmed that the 6 clusters are divided into two groups of three clusters reliably. As for the dendrogram, visit our web page [11].

The execution times averaged over ten trials of the single linkage algorithm and LSH-link are compared in Fig. 8. The horizontal axis indicates the data size and the vertical axis indicates the execution time in second. The experimental environment is build on a PC (CPU:Pentium IV 2.4GHz, memory size 512MB) with the Redhat Linux OS(kernel 2.4.20-8).

From this graph, we conclude that, whereas the single linkage algorithm requires a time of $O(n^2)$, LSH-link runs in $O(n)$ time almost. Because the expectedly same hash table size determined by K is used throughout this experiment, B increases for large value of n, though B is still even smaller than n. However, its effect does not appear in the graph, because the computation of hash functions is the dominant factor in the execution time of the LSH-link. As the graph indicates, the performance of LSH-link is by far superior to the single linkage algorithm for large data sizes.

The averaged number of phases amounts 9.2 for $A = 1.4$ and 5.2 for $A = 2.0$ when the data size is 6000. This difference appears as the gap of the execution time for the two parameter values.

5   Conclusion

This paper proposes a new fast approximation algorithm named LSH-link for the single linkage clustering algorithm. LSH-link attains a time complexity of $O(nB)$ by utilizing an approximated nearest neighbor search algorithm LSH. For practical use, B becomes a small constant compared to n if sufficiently large hash tables are prepared. By experiment, we show that (1) the proposed algorithm obtains similar clustering results to the single linkage algorithm and that (2) it runs faster for large data than the single linkage algorithm. This algorithm suits for the fast discovery of the coarse-grained hierarchical structure from the given data in particular.

One limitation of this algorithm is that it demands the analyzers to specify the two parameters, i.e. R (the initial value of r) and A (the increase ratio of r). However, we suppose LSH-link is still practical. This is because, even if the analyzers could not get the quality as they desired owing to the poor parameter choice, they can try another set of parameters again in a reasonable time.

Hence, one future direction of this work is to devise a mechanism to select parameters automatically. The random sampling technique is promising for this goal. In addition to this theme, evaluating LSH-link with more challenging experimental data will be essential. Finally it is important to seek application areas in which LSH-link exhibits its full ability, though we propose LSH-link as a generic clustering algorithm like the single linkage algorithm without premising specific applications in this paper.

References

1. R. Agrawal, J. Gehekr, D. Gunopulos and P. Raghavan. Automatic Subspace Clustering of High-Dimensional Data for Data Mining Applications., *Proc. of ACM SIGMOD Conference*, pages 94–105, 1998.
2. A. Gionis, P. Indyk and R. Motwani. Similarity Search in High Dimensions via Hashing. *Proc. of the 25th VLDB Conference*, pages 518–528, 1999.
3. T.H. Haveliwala, A. Gionis and P. Indyk. Scalable Techniques for Clustering the Web. *Proc. of the Third International Workshop on the Web and Databases* pages 129–134, 2000.
4. A. Hinneburg and D.A. Keim. An Efficient Approach to Clustering in Large Multimedia Databases with Noise., *Proc. of 4th International Conferences on Knowledge Discovery and Data Mining*, pages 58–65, 1998.
5. P. Indyk and R. Motwani. Approximate Nearest Neighbors: Towards Removing the Curse of Dimensionality., *Proc. of 30th ACM Symposium on Theory of Computing*, pages 604–613, 1998.
6. G. Karypis, E. Han and V. Kumar. CHAMELEON: A Hierarchical Clustering Algorithm Using Dynamic Modeling. *IEEE Computer*, Vol.32 No.8, pages 68–75, 1999.
7. S.Y. Jung and T. Kim. An Agglomerative Hierarchical Clustering using Partial Maximum Array and Incremental Similarity Computation Method. *Proc. of the 2001 IEEE International Conference on Data Mining*, pages 265–272, 2001.

8. G. Sheikholeslami, S. Chatterjee, and A. Zhang. WaveCluster: A Multi-Resolution Clustering Approach for Very Large Spatial Databases. *Proc. of the 24th VLDB Conference*, pages 428–439, 1998.
9. R. Sibson. SLINK: An Optimally Efficient Algorithm for the Single Link Cluster Method. *Computer Journal*, 16, pages 30–34, 1973.
10. T. Zhang, R. Ramakrishnan and M. Livny. BIRCH: An Efficient Data Clustering Model for Very Large Databases. *Proc. of the 1996 ACM SIGMOD International Conference on Management of Data*, pages 103–114, 1996.
11. http://sd.is.uec.ac.jp/~koga/DSdata.html

Measuring the Similarity for Heterogenous Data: An Ordered Probability-Based Approach

SiQuang Le and TuBao Ho

Japan Advanced Institute of Science and Technology
Tatsunokuchi, Ishikawa 923-1292 Japan
{quang,bao}@jaist.ac.jp

Abstract. In this paper we propose a solution to the similarity measuring for heterogenous data. The key idea is to consider the similarity of a given attribute-value pair as the probability of picking randomly a value pair that is less similar than or equally similar in terms of order relations defined appropriately for data types. Similarities of attribute value pairs are then integrated into similarities between data objects using a statistical method. Applying our method in combination with distance-based clustering to real data shows the merit of our proposed method.

Keywords: data mining, similarity measures, heterogenous data, order relations, probability integration.

1 Introduction

Measuring similarities between data objects is one of primary objectives in data mining systems. It is essential for many tasks, such as finding patterns or ranking data objects in databases with respect to queries. Moreover, measuring similarities between data objects affects significantly effectiveness of distance-based data mining methods, e.g., distance-based clustering methods and nearest neighbor techniques.

To measure the similarity for heterogenous data that comprise different data types, quantitative data, qualitative data, item set data, etc., has been a challenging problem in data mining due to natural differences among data types. Two primary tasks of the problem are: (1) determining the same essential (dis)similarity measures for different data types, and (2) integrating properly (dis)similarities of attribute value pairs into similarities between data objects. The first task is rather difficult because each data type has its own natural properties that leads to itself particular similarity measures. For example, the similarity between two continuous values are often considered as their absolute difference meanwhile the similarity between two categorical values is simply identity or non-identity of these two values. Thus, it is hard to define one proper similarity measure for all data types. Further, similarity measures for some data types are so poor due to the poorness of data structure (e.g., categorial, item set).

To date, a few similarity measuring methods for heterogeneous data have been proposed [1–6]. Their approach is to apply the same similarity measure scheme for different data types, and then, dissimilarity between two data objects

E. Suzuki and S. Arikawa (Eds.): DS 2004, LNAI 3245, pp. 129–141, 2004.

is assigned by adding linearly dissimilarities of their attribute value pairs or by distance Minkowski. In [1–3], the authors measure the similarity between two values based on three factors: *position*, *span*, and *content* where *position* indicates the relative position of two attribute values, *span* indicates the relative sizes of attribute values without referring to common parts, and *content* is a measure of the common parts between two values. It is obvious that similarities of value pairs of different data types have the same meaning since they are all based on these three factors. However, it is not hard to see that these three factors are not always applicable or suitable for all data types. For example, *position* arises only when the values are quantitative. Similar to that, methods of [4–6] take sizes of union (the joint operation $\otimes$) and intersection (the meet operation $\oplus$) of two values into account of measuring their similarity. Obviously, these two operators are not always suitable for all data types. For example, the intersection is not suitable for continuous data since continuous values are often different and therefore, the intersection of two continuous values seems to be always empty. In short, the methods [1–6] are based on factors or operators that are required to be suitable for all data types. However, due to the nature difference of data types, the factors or operators are hard to exist or not discovered yet.

In this paper, we address the similarity measuring problem for heterogeneous data by a probability-based approach. Our intuition in similarity for a value pair is that the more number of pairs that are less or equally similar, the greater their similarity. Based on the idea, we define the similarity of one value pair as the probability of picking randomly a value pair that is less similar than or equally similar in terms of order relations defined appropriately for each data types. By this way, we can obtain the same meaning similarities between values of different data types meanwhile each of the similarities is still based on particular properties of the corresponding data type. After that, similarities of attribute value pairs of two objects are then integrated using a statistical method to assign the similarity between them.

This paper is organized as follows. The ordered probability-based similarity measuring method for single attributes is described in Section 2. The integration methods and an example are given in Section 3 and Section 4. In section 5, we investigate characteristics of the proposed method. Next, complexity evaluation and experiment evaluations in combination with clustering methods to real data sets is shown in Section 6. Conclusions and further works are discussed in the last section.

2   Ordered Probability-Based Similarity Measure for an Attribute

Let $A_1, \ldots, A_m$ be m data attributes where A_i can be any data type such as quantitative data, qualitative data, interval data, item set data, etc. Let $D \subseteq A_1 \times \ldots \times A_m$ be a data set and $\mathbf{x} = (x_1, \ldots, x_m), x_i \in A_i$ be a data object of D. For each attribute A_i, denote $\preceq_i$ an order relation on A_i^2 where $(x_i', y_i') \preceq_i (x_i, y_i)$ implies that value pair (x_i', y_i') is less similar than or equally similar to value pair (x_i, y_i).

2.1 Ordered Probability-Based Similarity Measure

The first task of measuring similarity for heterogeneous data is to determine similarity measures for value pairs of each attribute. We define the ordered probability-based similarity for value pair (x_i, y_i) of attribute A_i as follows:

Definition 1. *The ordered probability-based similarity between two values x_i and y_i of attribute A_i with respect to order relation $\preceq_i$, denoted by $S_{\preceq_i}(x_i, y_i)$, is the probability of picking randomly a value pair of A_i that is less similar than or equally similar to (x_i, y_i)*

$$S_{\preceq_i}(x_i, y_i) = \sum_{(x_i', y_i') \preceq_i (x_i, y_i)} p(x_i', y_i')$$

*where $p(x_i', y_i')$ is the probability of **picking** value pair (x_i', y_i') of A_i.*

Definition 1 implies that the similarity of one value pair depends on both the number of value pairs that are less similar than or equally similar and probabilities of picking them. It is obvious that the more number of less than or equally similar value pair one pair has, the more similar they are.

As it can be induced from Definition 1, similarities of value pairs do not depend on data types. They are based only on order relations and probability distributions of value pairs. Hence, similarities of value pairs have the same meaning regardless of their data types. In other hand, each similarity is based on an order relation built properly for each data type. Thus, the similarity measure still reserved particular properties of this data type.

2.2 Order Relations for Real Data

In the following we define order relations of some common real data types, e.g. continuous data, interval data, ordinal data, categorical data, and item set data.

- **Continuous data:** A value pair is less similar or equally similar to another value pair if and only if the absolute difference of the first pair is greater than or equal to that of the second pair.

$$(x', y') \preceq (x, y) \Leftrightarrow |x' - y'| \geq |x - y|$$

- **Interval data:** A value pair is less similar than or equally similar to another value pair if and only if the proportion between the intersection interval and the union interval of the first pair is smaller than or equal to that of the second pair.

$$(x', y') \preceq (x, y) \Leftrightarrow \frac{|x' \cap y'|}{|x' \cup y'|} \leq \frac{|x \cap y|}{|x \cup y|}$$

- **Ordinal data:** A value pair is less similar than or equally similar to another value pair if and only if the interval between two values of the first pair contains that of the second pair:

$$(x', y') \preceq (x, y) \Leftrightarrow [x'..y'] \supseteq [x..y]$$

- **Categorical data:** A value pair is less similar than or equally similar to another value pair if and only if either they are identical or values of the first pair are not identical meanwhile those of the second pair are:

$$(x', y') \preceq (x, y) \Leftrightarrow \begin{cases} x' = x, y' = y \\ x' \neq y', x = y \end{cases}.$$

- **Item set data:** Following the idea of Geist [7], the order relation for item set value pairs that come from item set M is defined as follows:

$$(X, Y), (X', Y') \in M^2 : (X', Y') \preceq (X, Y) \Leftrightarrow \begin{cases} X' \cap Y' \subseteq X \cap Y \\ \overline{X}' \cap \overline{Y}' \subseteq \overline{X} \cap \overline{Y} \\ X' \cap \overline{Y} \supseteq X \cap \overline{Y} \\ \overline{X}' \cap Y' \supseteq \overline{X} \cap Y \end{cases}$$

It is easy to see that these order relations are transitive.

2.3   Probability Approximation

Now we present a simple method to estimate the probability of picking randomly a value pair. Assuming that values of each attribute are independent, the probability of picking a value pair (x_i, y_i) of A_i is approximately estimated as:

$$p(x_i, y_i) = \frac{\delta(x_i)\delta(y_i)}{n^2}$$

where $\delta(x_i)$ and $\delta(y_i)$ are the numbers of objects that have attribute value x_i, y_i respectively, and n is the number of data objects.

3   Integration Methods

The similarity between two data objects consisting of m attributes is measured by a combination of m similarities of their attribute value pairs. Taking advantage of measuring similarities of attribute value pairs in terms of probability, we consider integrating similarities of m attribute value pairs as the problem of integrating m probabilities.

Denote $S(\mathbf{x}, \mathbf{y}) = f(S_1, \ldots, S_m)$ the similarity between two data objects $\mathbf{x}$ and $\mathbf{y}$ where S_i is the similarity between values x_i and y_i of attribute A_i, and $f(.)$ is a function for integrating m probabilities $S_1, \ldots, S_m$. Here we describe some popular methods for integrating probabilities [8–10].

The most popular method is due to Fisher's transformation [8], which uses the test statistic

$$T_F = -2 \sum_{i=1}^{m} \ln S_i$$

and compares this to the χ^2 distribution with $2m$ degrees of freedom.

In [9], **Stouffer et al.** defined

$$T_s = \sum_{i=1}^{m} \frac{\Phi^{-1}(1 - S_i)}{\sqrt{m}}$$

where Φ^{-1} is the inverse normal cumulative distribution function. The value T_s is compared to the standard normal distribution.

Another P-value method was proposed by Mudholkar and Geore [10]

$$T_M = -c \sum_{i=1}^{m} \log \frac{S_i}{1 - S_i}$$

where

$$c = \sqrt{\frac{3(5m + 4)}{m\pi^2(5m + 2)}}$$

The combination value of $S_1, \ldots, S_m$ is referenced to the t distribution with $5m + 4$ degrees of freedom.

In practice, probability integrating functions are often non-decreasing functions. It means that the greater $S_1, \ldots, S_m$ are, the greater $S(\mathbf{x}, \mathbf{y})$ is. In particular, it is easy to prove that the mentioned probability integrating functions are non-decreasing functions.

4 Example

To illustrate how the similarity between two data objects is measured using our method, consider the simple data set given in Table 1 that was obtained from an user internet survey. This data set contains 10 data objects comprising 3 different attributes e.g. age (continuous data), connecting speed (ordinal data), and time on internet (interval data). Consider the first data object ({26, 128k, [6..10]} and the second one {55, 56k, [7..15]}, the similarity between them is measured as follows:

$$S_{age}(26, 55) = p(23, 55) + p(55, 23) + p(25, 55) + p(55, 25) + \ldots + p(57, 26)$$
$$= \frac{1 \times 1}{10^2} + \frac{1 \times 1}{10^2} + \frac{1 \times 1}{10^2} + \frac{1 \times 1}{10^2} + \ldots + \frac{1 \times 1}{10^2}$$
$$= 0.18$$

$$S_{speed}(128k, 56k) = p(14k, 128k) + p(128k, 14k) + p(28k, 128k)$$
$$+ p(128k, 28k) + \ldots + p(56k, > 128k) + p(> 128k, 56k)$$
$$= \frac{2 \times 1}{10^2} + \frac{2 \times 2}{10^2} + \frac{2 \times 1}{10^2} + \frac{2 \times 2}{10^2} + \ldots + \frac{2 \times 1}{10^2} + \frac{2 \times 2}{10^2}$$
$$= 0.42$$

Table 1. An example: a data set obtained from an user internet survey includes 10 data objects, comprising 3 different attributes e.g., age (continuous data), connecting speed (ordinal data) and time on internet (interval data).

No.	Age (year)	Connecting Speed (k)	Time on Internet (hour)
1	26	128	[6..10]
2	55	56	[7..15]
3	23	14	[5..10]
4	25	36	[20..30]
5	56	> 128	[12..20]
6	45	56	[15..18]
7	34	28	[3..4]
8	57	28	[3..7]
9	48	14	[8..12]
10	34	> 128	[5..10]

$$S_{time}([6..10],[7..15]) = p([5..10],[20..30]) + p([20..30],[5..10]) + p([5..10],[12..20])$$
$$+ p([12..20],[5..10]) + \ldots + p([3..7],[5..12])$$
$$= \frac{1 \times 1}{10^2} + \frac{1 \times 1}{10^2} + \frac{1 \times 1}{10^2} + \frac{1 \times 1}{10^2} + \ldots + \frac{1 \times 1}{10^2}$$
$$= 0.76$$

Now we use Fisher's transformation test statistic [8] to integrate S_{age}, S_{speed} and S_{time}:

$$T_F = -2(\ln S_{age} + \ln S_{speed} + \ln S_{time})$$
$$= -2(\ln(0.18) + \ln(0.42) + \ln(0.76))$$
$$= 5.71$$

The value of the χ^2 distribution with 6 degrees of freedom at point 5.71 is 0.456. Thus, the similarity between the first and the second objects, $S(\{26, 128k, [6..10]\}, \{55, 56k, [7..15]\})$, is 0.456.

5   Characteristics

In this subsection, we investigate characteristics and properties of our proposed method. For convenience let us recall an important required property of similarity measures that was proposed by Geist et al. [7].

Definition 2. *Similarity measure* $\rho : \Gamma^2 \rightarrow R^+$ *is called an order-preserving similarity measure with respect to order relation* $\preceq$ *if and only if it holds true for:*

$$\forall (\mathbf{x}, \mathbf{y}), (\mathbf{x}', \mathbf{y}') \in \Gamma^2, (\mathbf{x}', \mathbf{y}') \preceq (\mathbf{x}, \mathbf{y}) \Rightarrow \rho(\mathbf{x}', \mathbf{y}') \leq \rho(\mathbf{x}, \mathbf{y})$$

Since order-preserving measures play important roles in practice, most common similarity measures (e.g., Euclidean, Hamming, Russel and Rao, Jaccard and Needham) possess the property with respect to reasonable order relations.

Theorem 1. *Similarity measure* $S_{\preceq_i} : A_i^2 \to R^+$ *is an order-preserving similarity measure with respect to order relation* $\preceq_i$ *if order relation* $\preceq_i$ *is transitive.*

Proof. Denote $\Lambda(x_i, y_i)$ the set of pairs which are smaller than or **equally** (x_i, y_i)

$$\Lambda(x_i, y_i) = \{(x_i', y_i') : (x_i', y_i') \preceq_i (x_i, y_i)\}$$

Since $\preceq_i$ is a transitive relation, for any two value pairs (x_{i_1}, y_{i_1}) and (x_{i_2}, y_{i_2}), when $(x_{i_1}, y_{i_1}) \preceq_i (x_2, y_2)$ we have $\forall (x_i, y_i) \in \Lambda(x_{i_1}, y_{i_1}) : (x_i, y_i) \preceq (x_{i_1}, y_{i_1})$ implies $(x_i, y_i) \preceq_i (x_{i_2}, y_{i_2})$. This means $(x_i, y_i) \in \Lambda(x_{i_2}, y_{i_2})$, and thus

$$\Lambda(x_{i_1}, y_{i_1}) \subseteq \Lambda(x_{i_2}, y_{i_2}) \tag{1}$$

In other hand, we have

$$S_{\preceq_i}(x_i, y_i) = \sum_{(x_i', y_i') \preceq_i (x_i, y_i)} p(x_i', y_i') = \sum_{(x_i', y_i') \in \Lambda(x_i, y_i)} p(x_i', y_i') \tag{2}$$

From (1) and (2),

$$S_{\preceq_i}(x_{i_1}, y_{i_1}) = \sum_{(x_i, y_i) \in \Lambda(x_{i_1}, y_{i_1})} p(x_i, y_i) \le \sum_{(x_i, y_i) \in \Lambda(x_{i_2}, y_{i_2})} p(x_i, y_i) = S_{\preceq_i}(x_{i_1}, y_{i_1})$$

Thus, $S_{\preceq_i}(.,.)$ is an order-preserving measure. $\square$

In practice, order relation $\preceq_i$ are often transitive. Thus, the ordered probability-based similarity measures for attributes are also order-preserving similarity measures.

Denote $\mathbb{A} = A_1 \times \ldots \times A_m$ the product space of m attributes $A_1, \ldots, A_m$. We define the product of order relation $\preceq_1, \ldots, \preceq_m$ as follows:

Definition 3. *The product of order relations* $\preceq_1, \ldots, \preceq_m$*, denoted by* $\prod_{i=1}^m \preceq_i$*, is an order relation* $\preceq$ *on* $\mathbb{A}^2$*, for which one data object pair is said to be less similar than or* **equally similar** *to another data object pair with respect to* $\prod_{i=1}^m \preceq_i$ *if and only if attribute value pairs of the first data object pair are less similar than or equally similar to those of the second data object pair*

$$\forall (\mathbf{x}, \mathbf{y}), (\mathbf{x}', \mathbf{y}') \in \mathbb{A}^2 : (\mathbf{x}', \mathbf{y}') \preceq (\mathbf{x}, \mathbf{y}) \Leftrightarrow (x_i', y_i') \preceq_i (x_i, y_i), i = 1, \ldots, m$$

Proposition 1. *The* **product** *of order relations* $\preceq_1, \ldots, \preceq_m$ *is transitive when order relations* $\preceq_1, \ldots, \preceq_m$ *are transitive.*

Proof. Denote $\preceq = \prod_i^m \preceq_i$. For any triple data object pairs $(\mathbf{x_1}, \mathbf{y_1}), (\mathbf{x_2}, \mathbf{y_2})$, and $(\mathbf{x_3}, \mathbf{y_3})$. if $(\mathbf{x_1}, \mathbf{y_1}) \preceq (\mathbf{x_2}, \mathbf{y_2})$, and $(\mathbf{x_2}, \mathbf{y_2}) \preceq (\mathbf{x_3}, \mathbf{y_3})$, we have

$$(\mathbf{x_1}, \mathbf{y_1}) \preceq (\mathbf{x_2}, \mathbf{y_2}) \Leftrightarrow (x_{i_1}, y_{i_1}) \preceq_i (x_{i_2}, y_{i_2}) \; \forall i = 1 \ldots m$$
$$(\mathbf{x_2}, \mathbf{y_2}) \preceq (\mathbf{x_3}, \mathbf{y_3}) \Leftrightarrow (x_{i_2}, y_{i_2}) \preceq_i (x_{i_3}, y_{i_3}) \; \forall i = 1 \ldots m$$

Since $\preceq_i$ is transitive for $i = 1 \ldots m$, $(x_{i_1}, y_{i_1}) \preceq_i (x_{i_2}, y_{i_2})$ and $(x_{i_2}, y_{i_2}) \preceq_i (x_{i_3}, y_{i_3})$ implies $(x_{i_1}, y_{i_1}) \preceq_i (x_{i_3}, y_{i_3})$. Hence $(\mathbf{x_1}, \mathbf{y_1}) \preceq (\mathbf{x_3}, \mathbf{y_3})$.

Thus, $\prod_i^m \preceq_i$ is transitive. $\square$

Theorem 2. *Similarity measure* $S : \mathbb{A}^2 \to R^+$ *is an order-preserving similarity measure with respect to* $\prod_{i=1}^m \preceq_i$ *when order relations* $\preceq_1, \ldots, \preceq_m$ *are transitive and probability integrating function* f *is non-decreasing.*

Proof. Denote $\preceq = \prod_i^m \preceq_i$, $(\mathbf{x}', \mathbf{y}')$ and $(\mathbf{x_2}, \mathbf{y_2})$ two data object pairs. We have

$$(\mathbf{x}', \mathbf{y}') \preceq (\mathbf{x}, \mathbf{y}) \Leftrightarrow (x_i', y_i') \preceq (x_i, y_i) \; \forall i = 1, \ldots, m;$$

Since $\preceq_i$ is transitive for $i = 1, \ldots, m$, following Theorem 1,

$$S_i' = S_{\preceq_i}(x_i', y_i') \leq S_{\preceq_i}(x_i, y_i) = S_i \; \forall i = 1, \ldots, m$$

Since f is a non-decreasing function,

$$S(\mathbf{x}', \mathbf{y}') = f(S_1', \ldots, S_m') \leq f(S_1, \ldots, S_m) = S(\mathbf{x}, \mathbf{y})$$

Since $(\mathbf{x}', \mathbf{y}') \preceq (\mathbf{x}, \mathbf{y}) \Rightarrow S(\mathbf{x}', \mathbf{y}') \leq S(\mathbf{x}, \mathbf{y})$, $S(.,.)$ is an order-preserving similarity measure with respect to $\prod_i^m \preceq_i$. $\square$

Theorem 2 says that if attribute value pairs of a object pair are less similar than or **equally similar** to those of another object pair, the similarity of the first object pair is smaller than **or equally** the similarity of the second object pair in conditions that order-relations $\preceq_1, \ldots, \preceq_m$ are transitive and probability integrating function f is non-decreasing.

6   Evaluation

6.1   Complexity Evaluation

In this subsection, we analysis the complexity for computing similarity for a value pair and for a two data objects described by m attributes.

The simplest way two measure the similarity between two values of attribute A_k is to scan all value pairs of this attribute. By this way, the complexity of measuring similarity for a value pair is obviously $O(n_k^2)$ where n_k is the number of values of attribute A_k. In practice, n_k is often small (from dozens to a hundred) and therefore the complexity is absolutely acceptable. However, n_k may be large (up to n) when A_k is continuous data. In this case, we design two especial methods for computing similarity between two continuous values in $O(\log_2 n_k)$ or $O(n_k)$ depending on memory space requirements.

Let denote A_k a continuous attribute with n_k values $a_1, \ldots, a_{n_k}$. Assuming that $a_1 < \ldots < a_{n_k}$.

Computing Similarity for Continuous Data in $O(\log_2 n_k)$. In this methods, we first sort n_k^2 value pairs. Then the similarity of value pair (v, v') at index i is simply the similarity of pair (u, u') at index $i-1$ plus the probability of getting (v, v') and stored in vector S. After that the similarity between any value pair can be referred from vector S in $O(\log_2 n_k)$ by the binary search technique [11]. The method is rather convenient in sense of complexity since $O(\log_2 n_k)$ is so small even when n_k is very large. However, it requires $O(n_k^2)$ memory space.

Computing Similarity for Continuous Data in $O(\log n_k)$. Since $O(n_k^2)$ memory requirement is out of today computer's ability when the number of values is up to hundred thousands or millions, the method of computing similarity for value pairs in $O(\log_2 n_k)$ seems to be unrealistic when facing with data sets describing by continuous attributes with large numbers of values. In this part, we introduce the method required $O(n_k)$ memory space and gives the similarity between two values in $O(n_k)$.

Theorem 3. *Given* $a_1, ..., a_{n_k}$ *be a ordered values. For any value pair* (v, v') *and* (a_i, a_j) *with* $i \leq j$, *it holds true*

1. *if* $(a_i, a_j) \preceq (v, v')$, *then* $(a_i, a_t) \preceq (v, v')$ *when* $t \geq j$.
2. *if* $(a_i, a_j) \npreceq (v, v')$, *then* $(a_i, a_t) \npreceq (v, v')$ *when* $i \leq t \leq j$.

Proof.   1. We have $(a_i, a_j) \preceq (v, v') \Leftrightarrow a_j - a_i \geq |v - v'|$. Since $a_t \geq a_j$ when $t \geq j$, $a_t - a_i \geq |v - v'|$. Thus $(a_i, a_j) \preceq (v, v')$.□

2. We have $(a_i, a_j) \npreceq (v, v') \Leftrightarrow a_j - a_i \ngeq |v - v'|$. Since $a_j \leq a_t \leq a_i$ when $i \leq t \leq j$, $a_t - a_i \ngeq |v - v'|$. Thus, $(a_i, a_t) \npreceq (v, v')$.□

From Theorem 3, it is easy to see that the similarity between two values v and v' can be computed as

$$Sim(v, v') = \sum_{i=1}^{n_k} p(a_i) \sum_{j=t_i}^{n_k} p(a_j) \tag{3}$$

where t_i is the smallest number that is greater than i and satisfies $(a_i, a_{l_i}) \preceq (v, v')$.

Based on equation 3, we build an algorithm for determining similarity between two value (v, v') (see Figure 1). It is not hard to prove that the complexity for computing similarity between two values is $O(n_k)$ and required memory store is $O(n_k)$.

After obtaining similarities for m attribute value pairs of two data objects, it is not hard to prove that integrating these m similarities requires $O(m)$.

It is obvious that the complexity to measure a value pair of the proposed measure is higher than $O(m)$. However, in real applications, the complexity is acceptable as the value number of each attribute is often small or using the especial methods to reduce the complexity.

Procedure Sim_Determine
IN: two values v and v'.
OUT: Similarity of (v, v')
BEGIN

1: $i = 1, j = 1, sp = 1, Sim = 0$
2: **for** $i = 1$ to n_k **do**
3:      **while** $((v, v') \preceq (a_i, a_j))$ and $(j \leq n_i)$ **do**
4:          $sp = sp - p(a_j)$
5:          $j = j + 1$
6:      **end while**
7:      $Sim = Sim + p(a_i) * sp$
8: **end for**
9: return Sim

END

Fig. 1. Algorithm for computing similarity between two continuous values in $O(n_k)$.

6.2   Evaluation with Real Data

In the following we analyze real data sets using our similarity measuring approach in conjunction with clustering methods. We try to mine group of users with particular properties from internet survey data.

Data Set. The Cultural Issues in Web Design data set was obtained from the GVU's 8th WWW User Survey (http://www.cc.gatech.edu/gvu/user_surveys/survey-1997-10/). The data set is a collection of users's opinions on influences of languages, colors, culture, etc. on web designs. The data set includes 1097 respondents, which are described by 3 item set attributes, 10 categorical attributes, and 41 ordinal attributes.

Methodology

- *Similarity measure method*
  We apply the proposed method to measure similarities between respondents of the Cultural Issues in Web Design data set. We use the order relations and the probability approximation method as mentioned in Section 2. We choose the Fisher's transformation to integrate similarities of attribute value pairs.
- *Clustering method*
  A clustering method can be categorized into either partitioning approaches (e.g., K-means [12], Kmedoid [13]) or hierarchical approaches (e.g., single linkage [14], complete linkage [15], group average linkage [16, 17]). Since partitioning approaches are not proper for noncontinuous data, we choose agglomeration hierarchical average linkage clustering method, which overcomes the *chain* problem of single linkage methods and discover more *balanced* clusters than complete linkage methods do.

Table 2. Characteristics of three discovered clusters.

Cluster 1

No.	Att. Names	Value	P_a	Value	P_a	Value	P_a	Value	P_a
1	Unfamiliar sites	Can't write	92					Other	8
2	Read Arabic	None	100						
3	Read Hebrew	None	100						
4	Speak Bengali	None	100						
5	Speak Hebrew	None	100						
6	Primary same as Native	Yes	98	No	2				
7	Important problem	Can't write	92	None	2			Other	6
8	American images	None	79					Other	21
9	Native Language	English	79	Chinese	4	German	4	Other	12
10	Read German	None	71	Basic phrases	19	Native	8	Other	2
11	Software	Yes both	73	Yes get	25	No	2		
12	Speak English	Native	69	Conver.	17	None	14		
13	Provide native sites	Agree strongly	69	Agree somewhat	23	Disag. somewhat	4	Other	4

Cluster 2

No.	Att. Names	Value	P_a	Value	P_a	Value	P_a	Value	P_a
1	Unfamiliar sites	Can't translate	81					Other	19
2	Read Chinese	None	100						
3	Read Hindi	None	100						
4	Read Japanese	None	100						
5	Speak Hindi	None	100						
6	Due to culture	No	93	Yes-both	8				
7	Sites in non-fluent	Few	89	None	9	Most	2		
8	Non-English sites	Few	89	None	8	Half	4		
9	Translations	Yes-useful	87					Other	13
10	Read German	None	83	Basic phrases	9	Literate	8		
11	Native Language	English	81	Spanish	8	Arabic	2	Other	9
12	Speak German	None	81	Basic phrases	11	Conver.	8		
13	Designed culture	Yes	70	No	28	Don't know	2		

Cluster 3

No.	Att. Names	Value	P_a	Value	P_a	Value	P_a	Value	P_a
1	Unfamiliar sites	Can't read	84					Other	16
2	Read Arabic	None	100						
3	Read Chinese	None	100						
4	Read Hindi	None	100						
5	Speak Arabic	None	100						
6	Speak Bengali	None	100						
7	Speak Hindi	None	100						
8	Read Italian	None	93	Basic phrases	4	Native	2	Other	2
9	Speak Italian	None	93	Basic phrases	7				
10	Speak Spanish	None	84	Basic phrases	14	Conver.	2		
11	Read Spanish	None	82	Basic phrases	18				
12	Sites designed for culture	Yes	68	No	29	Dontknow	4		
13	Sites in non-fluent	Few	77	All	11	None	7	Other	5
14	Software	Yes get	77	Yesboth	18	No	5		
15	Non-English sites	Few	68	None	21	Half	9	Other	2

Clustering Results. The Cultural Issues in Web Design data set was clustered into 10 clusters. However, we present characteristics of only three clusters due to space limitation, see Table 2. A characteristic of a cluster is presented as an attribute value that majority of respondents of the cluster answered. For example, value *can't write* of attribute *Unfamiliar site* is considered as a characteristic of the first cluster because 92% respondents of this cluster answered the value.

Discussion. As it can be seen from Table 2, the clusters have many characteristics, e.g. the first cluster has 13 characteristics, the third has 15 characteristics. Moreover, characteristics are different from cluster to cluster. In particular, when visiting an *unfamiliar site*, the problem of 92% respondents of the first cluster is *cannot write*, while 81% respondents of the second cluster is *cannot translate*, and 84% respondents of the third cluster is *cannot read*. Moreover, answers of respondents in the same clusters are somehow similar. For example, all respondents of the first cluster can neither read *Rabic* and *Hebrew* nor speak *Bengari* and *Hebrew*. In short, almost respondents in the same cluster have the same answers but they are different from answers of respondents of different clusters. The analysis of characteristics from these clusters shows that our similarity measuring method in combination with the agglomeration hierarchical average linkage clustering method discovers valuable clusters of real data sets.

7   Conclusions and Further Works

We introduced a method to measure the similarity for heterogeneous data in the statistics and probability framework. The main idea is to define the similarity of one value pair as the probability of picking randomly a value pair that is less similar than or equally similar in terms of order relations defined appropriately for data types. Similarities of attribute value pairs of two objects are then integrated using a statistical method to assign the similarity between them.

The measure possess the order-preserving similarity property. Moreover, applying our approach in combination with clustering methods to real data shows the merit of our proposed method.

However, the proposed method is designed for data sets whose data objects have the same number of attribute values. In future works, we will adapt this method for more complex data sets whose data objects may have different numbers of attribute values.

Acknowledgments

We appreciate professor Gunter Weiss and Le Sy Vinh at the Heinrich-Heine University of Duesseldorf, Germany for helpful comments on the manuscript.

References

1. Gowda K. C. and Diday E. Symbolic clustering using a new dissimilarity measure. *In Pattern Recognition*, 24(6):567–578, 1991.
2. Gowda K. C. and Diday E. Unsuppervised learning throught symbolic clustering. *In Pattern Recognition lett.*, 12:259–264, 1991.

3. Gowda K. C. and Diday E. Symbolic clustering using a new similarity measure. *IEEE Trans. Syst. Man Cybernet*, 22(2):368–378, 1992.
4. Ichino M. and Yaguchi H. Generalized minkowski metrics for mixed feature-type data analysis. *IEEE Transactions on Systems Man, and Cybernetics*, 24(4), 1994.
5. de Carvalho F.A.T. Proximity coefficients between boolean symbolic objects. In E. et al Diday, editor, *New Approaches in Classification and Data Analysis*, volume 5 of *Studies in Classification, DataAnalysis, and Knowledge Organisation*, pages 387–394, Berlin, 1994. Springer-Verlag.
6. de Carvalho F.A.T. Extension based proximity coefficients between constrained boolean symbolicobjects. In Hayashi C. et al., editor, *IFCS96*, pages 370–378, Berlin, 1998. Springer.
7. Geist S., Lengnink K., and Wille R. An order-theoretic foundation for similarity measures. In Diday E. and Lechevallier Y., editors, *Ordinal and symbolic data analysis, studies in classification, data analysis, and knowledge organization*, volume 8, pages 225–237, Berlin, Heidelberg, 1996. Springer.
8. Fisher R.A. *Statistical methods for research workers*. Oliver and Boyd, 11th edition, 1950.
9. Stouffer S.A, Suchman E.A, Devinney L.C, and Williams R.M. Adjustment during army life. *The American Solder*, 1, 1949.
10. Mudholkar G.s and George E.O. The logit method for combining probabilities. In J. Rustagi, editor, *Symposium on Optimizing methods in statistics*, pages 345–366. Academic press, NewYork, 1979.
11. Thomas H. Cormen, Charles E. Leiserson, Ronald L. Rivest, and Clifford Stein. *Introduction to Algorithms*. MIT Press and McGraw-Hill., the third edition, 2002.
12. MacQueen J. Some methods for classification and analysis of multivariate observation. In *Proceedings 5th Berkeley Symposium on Mathematical Statistics and Probability*, pages 281–297, 1967.
13. Kaufmann L. and Rousseeuw P.J. Clustering by means of medoids. *Statistical Data Analysis based on the L1 Norm*, pages 405–416, 1987.
14. Sneath P.H.A. The application of computers to taxonomy. *Journal of general microbiology*, 17:201–226, 1957.
15. McQuitty L.L. Hierarchical linkage analysis for the isolation of types. *Education and Psychological measurements*, 20:55–67, 1960.
16. Sokal R.R. and Michener C.D. Statistical method for evaluating systematic relationships. *University of Kansas scicnce bulletin*, 38:1409–1438, 1958.
17. McQuitty L.L. Expansion of similarity analysis by reciprocal pairs for discrete and continuous data. *Education and Psychological measurements*, 27:253–255, 1967.

Constructive Inductive Learning
Based on Meta-attributes

Kouzou Ohara[1], Yukio Onishi[1], Noboru Babaguchi[2], and Hiroshi Motoda[1]

[1] I.S.I.R., Osaka University, 8-1 Mihogaoka, Ibaraki, Osaka 567-0047, Japan
{ohara,motoda}@ar.sanken.osaka-u.ac.jp
[2] Graduate School of Engineering, Osaka University,
2-1 Yamadaoka, Suita, Osaka 565-0871, Japan
babaguchi@comm.eng.osaka-u.ac.jp

Abstract. Constructive Inductive Learning, CIL, aims at learning more accurate or comprehensive concept descriptions by generating new features from the basic features initially given. Most of the existing CIL systems restrict the kinds of functions that can be applied to construct new features, because the search space of feature candidates can be very large. However, so far, no constraint has been applied to combining the basic features. This leads to generating many new but meaningless features. To avoid generating such meaningless features, in this paper, we introduce *meta-attributes* into CIL, which represent domain knowledge about basic features and allow to eliminate meaningless features. We also propose a Constructive Inductive learning system using Meta-Attributes, CIMA, and experimentally show it can significantly reduce the number of feature candidates.

1 Introduction

Most of inductive learning methods such as decision tree learning[11] and Inductive Logic Programming[10,12] are *selective* because they selectively use predefined features to learn a concept description[9], which is used as a classifier in a domain in most cases. Such a selective inductive learning method, or simply a *selective learner* would fail in learning a useful classifier if the learner is not adequate with respect to features initially given, or the *basic features*. *Constructive Inductive Learning*, CIL, aims at solving this problem. Namely, for a selective learner, a CIL system constructs classifiers while simultaneously generating new useful features likely to improve the classification accuracy and comprehensibility of resulting classifiers[8]. In the last decade, several researchers have proposed various CIL systems[1, 4, 6, 7].

There are similar statistical approaches such as Support Vector Machine and basis expansions[3]. Although these statistical methods can construct more accurate classifiers by transforming basic features into new features, the resulting classifiers are difficult to interpret, while the results by a CIL system based on selective learning are comprehensive. Propositionalization in Inductive Logic Programming is a special case of CIL[5]. It is regarded as a representation change

E. Suzuki and S. Arikawa (Eds.): DS 2004, LNAI 3245, pp. 142–154, 2004.
© Springer-Verlag Berlin Heidelberg 2004

from a relational representation to a propositional one. In other words, it generates new features that can be accepted by an arbitrary propositional learner from relational background knowledge and relational basic features.

The problem in CIL is the size of the search space consisting of new feature candidates, which exponentially increases as the number of available functions to construct new features becomes larger. Thus most of CIL systems are designed to work well only for some restricted kinds of functions such as logical ones[1, 4]. FICUS[7], however, is a more flexible CIL system, which provides an efficient search algorithm that is independent of available functions. The user can specify available functions for each domain. These existing systems evaluate a new feature candidate based on a heuristic function such as information gain measured *after its construction*. In other words, so far, no constraint has been imposed on the basic features to be combined into a new candidate except for simple rules concerning the data types. This leads to the generation of many new but meaningless candidates. Such meaningless candidates worsen the efficiency of a system. Furthermore, a system might generate a classifier that is not applicable to unseen instances or not understandable for the user if such meaningless features happen to fit to training examples.

To avoid generating such meaningless candidates, in this paper, we introduce *meta-attributes* into CIL, which represent the domain knowledge about the basic features a user possesses and are used to eliminate meaningless candidates during the search for new useful features. We also propose *Feature Definition Grammar*, FDG, which provides means to construct new features based on meta-attributes. Using both meta-attributes and FDG allows us to reduce the search space and to use many more functions to construct new features. Furthermore, we present a Constructive Induction system using Meta-Attributes, CIMA, and experimentally show that CIMA can improve both the classification accuracy and the description length of resulting classifiers, while reducing the number of candidates to be evaluated.

2 Constructive Inductive Learning

2.1 A Motivating Example

The performance of a classifier learned by a selective learner is highly dependent on the basic features initially given. If the basic features are not adequate to describe the underlying characteristics of a target class, a selective learner would fail in learning a classifier that can be widely applicable to instances of the target class. For example, let us consider the scale balancing problem whose purpose is finding out the condition to balance a scale from examples described by the four basic features: the distance from its center to a sinker in the left-hand side, the same corresponding distance in the right-hand side, and the weight of each sinker. In this paper, we denote these features by D_L, D_R, W_L, and W_R, respectively. Then a selective learner might obtain a rule, something like the following one based on the actual values of the basic features:

if   ($W_L > 4$ and $D_L \leq 2$ and $W_R \leq 2$ and $D_R > 4$)
then it is balanced.

This rule is correct for the given examples, but would not generalize to unknown data.

CIL aims at overcoming this problem by generating new features such as the following one:

$$\frac{W_R * D_R}{W_L * D_L} \tag{1}$$

The above feature fully represents the underlying characteristic that the momentum of a scale is determined based on the product of the weight of each side's sinker and its distance from the center of a scale. Thus, if by f_{new} we mean the above new feature, a selective learner could obtain the following concise and more general rule:

if ($f_{new} = 1$) **then** it is balanced.

The operators to construct new features from existing ones such as $*$ are called *constructor functions* in CIL.

2.2   Existing CIL Systems

GALA[4] and GCI[1] are designed to deal with only logical functions. GALA finds out the best new feature from all possible feature candidates based on its evaluation function. GCI adopts a genetic algorithm for traversing a search space. Restricting kinds of constructor functions allows us to generate new features efficiently. However domains to which this approach is applicable are limited. For example, it is obvious that these algorithms never generate new features such as f_{new} in the scale balancing problem because they can use only logical functions.

On the other hand, FICUS[7] is a more flexible CIL system which is capable of generating features with arbitrary constructor functions specified by a user. In FICUS, one can specify a set of constructor functions customized for each domain by a description called the *feature space specification*, FSS. FICUS generates new features based on the information about the basic features and the constraints on constructor functions in FSS, by means of only four primitive operations: *Compose, Insert, Replace,* and *Interval.* These operations generate new features from one or two existing features by applying a constructor function to them or by slightly modifying their definition.

These systems apply constructor functions either to randomly selected basic features or to all possible combinations of them, and then evaluate resulting features using a heuristic function such as information gain. FICUS imposes a constraint on the data type of basic features to be combined. But there is no other constraint on combining the basic features. This may generate meaningless features. For example, in the scale balancing problem, $D_L - W_R$ can be a new feature to be evaluated although its meaning is not understandable. The purpose of this paper is to provide a framework to eliminate such meaningless feature candidates as much as possible.

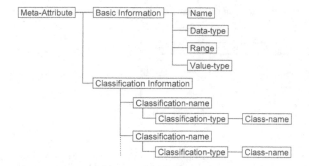

Fig. 1. Definition of the meta-attribute.

Table 1. Available data-types and value-types in the basic information.

Data-type	Value-type
binary	boolean
char	nominal, orderedNominal
string	nominal, orderedNominal
int	nominal, orderedNominal, continuous
float	nominal, orderedNominal, continuous

3   Feature Construction Using Meta-attributes

To avoid generating meaningless combinations of the basic features, in this section, we introduce a method of combining features based on *meta-attributes* and *feature definition grammar*.

3.1   Meta-attributes

Meta-attributes are the additional information about the basic features, and consist of the *basic information* and the *classification information* as shown in Fig. 1. The basic information contains *name, range, data-type* and *value-type* of a basic feature, and is used for selecting constructor functions applicable to it. Values of *name* and *range* are determined by a user, whereas values of *data-type* and *value-type* have to be selected from the predefined ones shown in Table 1, where *int* and *orderedNominal* mean integer and nominals with an order, respectively.

On the other hand, the classification information is used to group the basic features that are related to each other, and has three kinds of information: *classification-name, classification-type,* and *class-name*. One can classify the basic features into disjoint classes from multiple viewpoints. A unique classification-name has to be assigned to each viewpoint. The classification-type expresses a fundamental criterion for such classification, and it is defined as either *function, position,* or *scale*. Table 2 shows meta-attributes for the basic features for the scale balancing problem. In this example, the basic features D_L and W_L are

Table 2. Meta-attributes for the balance scale problem.

Classification Name

Basic Feature	Data Type	Range	Value Type	position1	scale1
D_L	int	1-5	continuous	left	length
W_L	int	1-5	continuous	left	weight
D_R	int	1-5	continuous	right	length
W_R	int	1-5	continuous	right	weight

Class Name

Classification Name	Classification Type
position1	position
scale1	scale

Basic Information   Classification Information

$\langle$NewF$\rangle$	::=	$\langle$binarizeF$\rangle$ $\mid$ $\langle$countF$\rangle$ $\mid$ $\langle$aggregateF$\rangle$ $\mid$ $\langle$relativeF$\rangle$ $\mid$ $\langle$statisticF$\rangle$ $\mid$ $\langle$conversionF$\rangle$
$\langle$binarizeF$\rangle$	::=	'binaryF('$\langle\langle$basicF$\rangle\rangle$ ',' $\langle$binaryFunc$\rangle$ ',' $\langle\langle$CONST$\rangle\rangle$ ').'
$\langle$countF$\rangle$	::=	'countF('$\langle$F_set$\rangle$ ',' $\langle$binaryFunc$\rangle$ ',' $\langle\langle$CONST$\rangle\rangle$ ').'
$\langle$aggregateF$\rangle$	::=	'aggregateF('$\langle$aggregateFunc_set$\rangle$ ',' $\langle$F_set$\rangle$ ').'
$\langle$relativeF$\rangle$	::=	'relativeF('$\langle$relativeFunc$\rangle$ ',' {$\langle$transferFunc$\rangle$} ',' $\langle\langle$basicF$\rangle\rangle$ $\langle\langle$basicF$\rangle\rangle$ ').'
$\langle$statisticF$\rangle$	::=	'statisticF('$\langle$statisticalFunc$\rangle$ ',' $\langle\langle$basicF$\rangle\rangle$ ').'
$\langle$conversionF$\rangle$	::=	'conversionF('$\langle$convertFunc$\rangle$ ',' $\langle\langle$basicF$\rangle\rangle$ ').'
$\langle$F_set$\rangle$	::=	$\langle\langle$C_Name$\rangle\rangle$ = $\langle\langle$ClassName$\rangle\rangle$ {';' $\langle\langle$C_Name$\rangle\rangle$ ('=' $\mid$ '$\neq$') $\langle\langle$ClassName$\rangle\rangle$ }
$\langle$binaryFunc$\rangle$	::=	'=' $\mid$ '$\geq$'
$\langle$aggregateFunc$\rangle$	::=	'sum' $\mid$ 'ave' $\mid$ 'max' $\mid$ 'min' $\mid$ 'stdev' $\mid$ 'product'
$\langle$relativeFunc$\rangle$	::=	'diff' $\mid$ 'incRatio'
$\langle$statisticalFunc$\rangle$	::=	'ratio' $\mid$ 'asds' $\mid$ 'rank'
$\langle$convertFunc$\rangle$	::=	'log' $\mid$ 'sqrt'

Fig. 2. Feature Definition Grammar.

classified into the same class of *left* from the viewpoint of the position, while they are classified into different classes from the other viewpoint, scale.

Thanks to the classification information in meta-attributes, we can create new and reasonable features from those which belong to the same class value.

3.2   Feature Definition Grammar

Feature Definition Grammar, FDG, is a description that defines kinds of new features, as well as constructor functions to construct them. In this paper, we employ FDG shown in Fig. 2. In this FDG, as defined in its first line, 6 kinds of new features can be constructed: *binary, count, aggregate, relative, statistic, and conversion* features. To construct them, 15 constructor functions are available. In FDG, the value of a keyword placed between $\langle\langle$ and $\rangle\rangle$ is instantiated using the values that appear in the given meta-attributes.

The binarize feature converts the value of a basic feature into a truth value based on a binary function, which is either = or $\geq$. For example, = returns *True* if the feature value is equal to a constant $\langle\langle$CONST$\rangle\rangle$; otherwise returns *False*. For a set of basic features denoted by $\langle$F_Set$\rangle$, the count feature returns the number of features in it whose value satisfy a selected binary function. The aggregate feature applies an aggregate function to one or more feature values of an instance, while the relative feature applies a relative function to only two feature values. The set of basic features to which an aggregate function is applied is also defined by $\langle$F_Set$\rangle$. Given a basic feature, the value of the statistic feature is calculated by a statistical function, which regards all existing values for that

Table 3. Valid combinations of features based on their data-type and value-type.

	Basic Feature		Resulting Feature	
	Data Type	Value Type	Data Type	Value
⟨binarizeF⟩	any	any	binary	boolean
⟨countF⟩	any	any	int	continuous
⟨aggregateF⟩	int, float	orderedNominal, continuous	int, float	continuous
⟨conversionF⟩	int, float	orderedNominal, continuous	int, float	continuous
⟨relativeF⟩	int, float	orderedNominal, continuous	int, float	continuous
⟨statisticF⟩	int, float	orderedNominal, continuous	int, float	continuous

feature as its population. The conversion feature converts the value of a basic feature into either its log or square root.

A set of basic features ⟨F_Set⟩ in the FDG plays an important role in our framework because it allows us to consider only meaningful or reasonable combinations of basic features. It consists of features satisfying a condition based on the classification information in meta-attributes. The condition is a conjunction of one or more equations in the form of either $C = g$ or $C \neq g$, where C means a classification-name and g a class-name, denoted by ⟨⟨C_Name⟩⟩ and ⟨⟨ClassName⟩⟩, respectively in Fig. 2. $C = g$ means "the class in a classification C is g", whereas $C \neq g$ means "the class in C is other than g." The first element of the conjunction must be in the form of $C = g$. For example, given the meta-attributes shown in Table 2 for the scale balancing problem, we can construct a new feature, $aggregateF(sum, scale1 = weight)$, according to the FDG in Fig. 2. This feature returns as its value the summation of values of W_L and W_R because their class-name in the classification-name $scale1$ is $weight$. Note that a feature that returns summation of values of W_L and D_R is never constructed because they have no common class-name.

In addition to this constraint on ⟨F_Set⟩, we further introduce the following constraints on constructor functions.

- The basic features used to construct a new feature must have appropriate data and value types specified in Table 3.
- Two basic features used to construct a relative feature must belong to the same class whose classification-type is scale.

In Table 3, *any* means an arbitrary type. By the first constraint, we can avoid that each function receives an invalid input. The second one prevents a relative function from applying to the basic features with different scales.

4   CIL System Using Meta-attributes: CIMA

4.1   Outline of CIMA

Fig. 3 shows the conceptual design of CIMA. CIMA generates new features using a builtin FDG and meta-attributes given by a user, and has two internal modules that are called *Generator* and *Evaluator*, respectively. Generator produces a set of new features likely to be useful for constructing a classifier, while Evaluator

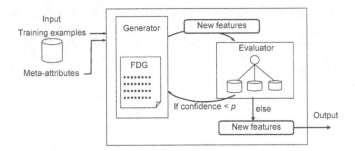

Fig. 3. Conceptual design of CIMA.

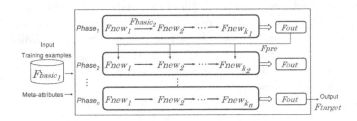

Fig. 4. Outline of Generator.

tests whether the new features produced by Generator are adequate to classify given examples. In fact, Evaluator creates a classifier with an internal learner L_e using extended training examples that have values of those generated features. If the *confidence*, or the ratio of the number of examples correctly classified over the total number of examples, is less than a predefined threshold p, CIMA invokes Generator again and gives as its input examples belonging to a class with the highest misclassification rate. CIMA iterates this process until the threshold p is satisfied.

Generator, as shown in Fig. 4, accepts as its input training examples described by the basic features and meta-attributes for them. Once Generator is invoked, it repeatedly activates its internal procedure, *Phase*, which progressively generates new features according to a list called the *Phase List*, PL. Fig. 5 shows an example of PL. Each line in a PL is processed by one execution of Phase. Thus Generator preforms Phase n times if a given PL includes n lines. In other words, the j-th Phase generates new features according to the j-th line in the PL. For example, given the PL in Fig. 5, the first Phase generates aggregate features from basic ones. After that, it constructs relative features using the new aggregate features, and then terminates. The maximum number of features that appear in a line of a PL is specified by a parameter d, which is given by a user. Once the value of d is determined, a PL is automatically generated by CIMA based on a *seed phase list* that is empirically predefined. This step-by-step execution of Phase reduces irrelevant features that could affect the succeeding construction of new features by pruning them in each step.

⟨aggregateF⟩;⟨relativeF⟩
⟨aggregateF⟩;⟨binaryF⟩
⟨aggregateF⟩;⟨binaryF⟩; ⟨countF⟩
⟨aggregateF⟩;⟨binaryF⟩; ⟨countF⟩;⟨relativeF⟩

Fig. 5. An example of Phase List.

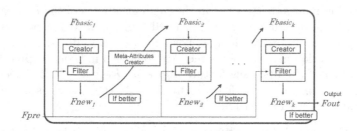

Fig. 6. Outline of Phase.

4.2 Phase

The outline of Phase is shown in Fig. 6. Phase consists of the three internal modules: *Creator, Filter, and Meta-Attributes Creator*. The $Phase_j$ that is the j-th invoked Phase receives as its input original training examples, their meta-attributes, and the output of $Phase_{j-1}$ denoted by F_{pre}, if $j > 1$. The set of original basic features is regarded as the *first basic feature set* F_{basic_1}. Suppose that $T_1, \cdots, T_k$ are feature types included in the j-th line of a PL and they appear in this order. Then $Phase_j$ generates the *first new feature set* F_{new_1} in which the type of each feature is T_1, using Creator and Filter: Creator generates new feature candidates, and Filter selects useful ones from them. Next, Meta-Attributes Creator is invoked to convert F_{new_1} into the second basic feature set F_{basic_2} by creating meta-attributes for features in F_{new_1}. After that, similarly $Phase_j$ generates the second new feature set F_{new_2} from F_{basic_2}. $Phase_j$ iterates this process until it generates F_{new_k}, and finally outputs it as its output F_{out}.

The details of these three modules are as follows:

Creator: based on the PL, generates at most N_{new} new feature candidates from the i-th basic feature set $(1 \leq i \leq k)$, F_{basic_i}, using FDG and meta-attributes. Creator randomly selects available features from F_{basic_i} to construct new features according to the definition of T_i in the FDG. N_{new} is a parameter given by a user.

Filter: performs feature selection. Filter receives as its input the set of feature candidates F_{cand} created by Generator and F_{pre}, and selects features that are likely to be useful for generating a better classifier. For that purpose, Filter creates a classifier H_i with its internal learner L_f using *extended training examples* that have additional feature values for both F_{cand} and F_{pre}. Then it selects features that appear in H_i, and outputs them as the i-th new feature set, together with the description length and confidence of H_i, denoted by $S(H_i)$ and $C(H_i)$, respectively.

Meta-attributes Creator: receives F_{new_i} from Filter, and creates meta-attributes of each feature in it in order to generate $F_{basic_{i+1}}$ that is the input of the next execution of Creator if $i < k$. As for the basic information of a new feature f in F_{new_i}, its data-type, range, and value-type are decided based on actual values f_i can take, while a unique name such as $newAttr33$ is automatically assigned to its name. For its classification information, all classification-names that appear in the meta-attributes of the i-th basic features used for constructing f are inherited from them to f. The class-name is also inherited if possible. If those basic features have different class-names for a classification-name, those names are combined in lexicographical order. For example, if the first new feature $aggregateF(sum, scale1 = weight)$ is constructed from the basic features in Table 2, then its class-name in the classification $part1$ becomes $left\_right$.

We say that the i-th new feature set F_{new_i} is *better* if and only if it satisfies either of the following conditions:

$$Size(H_i) < Size(H_{i-1}) \text{ and } Conf(H_{i-1}) \leq Conf(H_i), \quad (2)$$
$$Size(H_i) = Size(H_{i-1}) \text{ and } Conf(H_{i-1}) < Conf(H_i). \quad (3)$$

When $i = 1$, H_0 means the classifier generated with the basic features. Intuitively a *better* new feature set improves at least either the description length or the confidence of a classifier generated from the existing features. These conditions are similar to the minimum description length principle, but slightly different because they require that neither the size nor confidence of a classifier increase in any case. If F_{new_i} is no *better*, Phase outputs the $(i - 1)$-th new feature set $F_{new_{i-1}}$ as its final output F_{out} even if it has not constructed all kinds of features specified in the i-th line, in order to eliminate features that are no *better* and could affect the succeeding feature construction process.

5   Experimental Evaluation

To evaluate the usefulness of CIMA, we compared the performance of classifiers produced by a standard learner using both new features generated by CIMA and original basic ones with the performance of classifiers produced by the same learner using only the basic ones with respect to their classification accuracy and description length. We adopted C4.5[11] as the standard learner, and avoided using its pruning function since our purpose is to exactly know the differences incurred by generated features. It is noted that CIMA can adopt any standard, or external learner other than C4.5. We adopted C4.5 in the following experiments in order to compare CIMA with FICUS that adopted a decision tree learner called DT as its external learner[7]. DT was originally developed as an internal module of FICUS and does not perform pruning. Comparing the results by CIMA with those by FICUS is helpful to evaluate the performance of CIMA although we can not compare them directly because of the difference of their external learners.

Table 4. Summary of three datasets.

Dataset	balance	heart	tic-tac-toe
Number of classes	3	2	2
Number of basic features	continuous:4	continuous:5, nominal:8	nominal:9
Number of examples	625	303	958

Table 5. Meta-attributes for the heart dataset.

Basic Feature	Data_Type	Range	Value_Type	position1	scale1
age	int	0-77	continuous	non	year
sex	binary	0-1	nominal	non	non
cp	int	1-4	nominal	heart	non
trestbps	int	0-200	continuous	blood	pa
chol	int	0-603	continuous	blood	g
fbs	binary	0-1	boolean	blood	g
restecg	int	0-2	nominal	heart	non
thalach	int	0-202	continuous	heart	count
exang	binary	0-1	boolean	heart	non
oldpeak	float	−2.6-6.2	continuous	heart	mv
slope	int	1-3	nominal	heart	non
ca	int	0-3	orderedNominal	blood	count
thal	int	3-7	nominal	heart	non

The classification-types of position1 and scale1 are position and scale, respectively.

Table 6. Meta-attributes for the tic-tac-toe dataset.

Basic Feature	Data_Type	Value_Type	row	column
Top_Left	char	nominal	top	left
Top_Mid	char	nominal	top	mid
Top_Right	char	nominal	top	right
Middle_Left	char	nominal	mid	left
Middle_Mid	char	nominal	mid	mid
Middle_Right	char	nominal	mid	right
Bottom_Left	char	nominal	bot	left
Bottom_Mid	char	nominal	bot	mid
Bottom_Right	char	nominal	bot	right

The classification-type of both row and column is position.

In addition, in order to directly evaluate the effect of meta-attributes, we also examined the performance of a variant of CIMA, denoted by NOMA that is the abbreviation of No Meta-Attributes, in which Creator generates new feature candidates by randomly selecting the basic features without using meta-attributes. Actually we compared them with respect to the total number of candidates generated by Creator, which is equal to the number of candidates evaluated by Filter.

A prototype of CIMA was implemented employing C4.5 as its internal learner both in Evaluator and in Filter, and tested on three datasets, *balance, heart,* and *tic-tac-toe* taken from the UCI machine learning repository[2], which were used for evaluation of FICUS in [7]. The summary of these datasets is shown in Table 4, and the meta-attributes for the heart and tic-tac-toe datasets are shown in Tables 5 and 6, respectively. In Table 6, we omitted the row for "range" because the data-type of all the basic features is "char" for which "range" is not

Table 7. Experimental results: comparison in the tree size and classification accuracy of resulting tree and in the number of evaluated candidates.

Dataset		C4.5	C4.5+CIMA	C4.5+NOMA	DT+FICUS
balance	Size	115	6	7	5
	Accuracy(%)	78.60	99.84	99.80	99.84
	# of candidates		2,787	4,808	5,485
tic-tac-toe	Size	187	39	20	64.5
	Accuracy(%)	84.70	98.10	97.00	96.45
	# of candidates		1,262	3,730	21,625
heart	Size	75	68	52	67.2
	Accuracy(%)	74.90	76.20	73.7	77.3
	# of candidates		1,831	2,869	6,770

defined. As for the balance dataset, we used the meta-attributes shown in Table 2. We adopted the default configuration of CIMA where $d = 2$, $p = 0.9$, and $N_{new} = 1,000$ in all experiments, except for the cases of testing the sensitivity of CIMA to the parameters d and N_{new}.

Table 7 shows the details of experimental results, which also include the results by DT+FICUS taken from [7]. Each value in Table 7 is an average of the results of 10-fold cross-validation. Compared with C4.5, it is obvious that CIMA is successful in significantly improving both the classification accuracy and the size of classifiers for the balance and tic-tac-toe datasets. Also for the heart dataset, CIMA improves them, but the differences are less significant. This means that there is a room for refinement of the meta-attributes and FDG we adopted for this experiment. However this result for the heart dataset is not so pessimistic because the accuracy achieved by using CIMA is still comparable to that by FICUS.

Compared with NOMA, CIMA is successful in reducing the number of evaluated candidates by about half. Thus it is said that the meta-attributes function served as a good bias to decrease meaningless combinations of the basic features in Creator. Furthermore, the number of evaluated candidates of CIMA is much less than that of FICUS in spite of the fact that the number of available constructor functions in CIMA is 15 whereas that of FICUS is at most 5. This means that using meta-attributes along with the search algorithm based on a PL is very useful to reduce the size of the search space, or to concentrate on meaningful or useful candidates.

On the other hand, the accuracy by NOMA is slightly worse than that by CIMA, even though NOMA further reduces the size of classifiers. Especially for the heart dataset, the accuracy achieved by NOMA is lower than that by C4.5. This is because the resulting classifiers were overfitted to training examples due to newly generated features. From this result, it is said that constructing new features with meta-attributes also can reduce the possibility of such overfitting.

Furthermore, we investigated the sensitivity of CIMA to the parameters d and N_{new}. The complexity of a new feature depends on the value of d because d means the maximum depth of a nested feature. On the other hand, N_{new} is the upperbound of the number of feature candidates generated by Creator in Phase. Thus it is expected that N_{new} has an effect on the diversity of new features. We

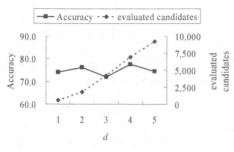

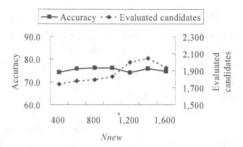

Fig. 7. Change of the classification accuracy and the number of evaluated candidates for the parameter d.

Fig. 8. Change of the classification accuracy and the number of evaluated candidates for the parameter N_{new}.

used the heart dataset for this investigation exploring for a further improvement of the accuracy (the other two datasets have already achieved high accuracy).

The results are shown in Fig. 7 and Fig. 8. The number of evaluated candidates increases proportionally to the value of d, but there is no significant difference among the accuracies. In case that $d = 3$, the resulting tree was overfitted to the training examples in some trials. The reason for such increase in the number of evaluated candidates is that the number of times Creator is invoked increases proportionally to the size of PL, in which both the number of lines and that of features in each line increase as d becomes larger. Consequently we can say excessively complex feature is not necessary to improve the accuracy of a classifier. This claim is supported also by the fact that for the other two datasets we achieved extensive improvement in the accuracy by setting d to 2.

From Fig. 8, we can not observe any significant difference in the accuracy even if N_{new} becomes larger. The number of evaluated candidates slightly increases and has a peak at $N_{new} = 1,400$, but the difference is not significant. We observed the same tendency when we set d to a larger value. These results show that by using meta-attributes the number of candidates generated by Creator is effectively restrained. Namely we do not have to be sensitive to setting value of the parameter N_{new}, as long as we use meta-attributes.

6   Conclusions

In this paper, we introduced the meta-attributes and Feature Generation Grammar into Constructive Inductive Learning in order to exclude meaningless combinations of basic features being attempted. We designed and implemented a constructive inductive learning system, CIMA, under this framework. Our experimental results show that, compared with the existing system, CIMA can significantly reduce the number of evaluated candidates while improving both the classification accuracy and the description length of classifiers. Smaller number of candidates could allow us to make use of many more constructor functions. In fact, the number of constructor functions that are available in CIMA is much larger than that of the existing systems.

One of the immediate future work is analyzing the scalability of CIMA by applying it to a more realistic dataset with many more basic features. For that purpose, we are optimizing the implementation of CIMA. Current prototype system is written in Visual Basic for Application on MS Access in order to use its useful data handling functions, but it is computationally expensive. We also plan to make a more deep analysis of generated features. Using meta-attributes allows us to focus on meaningful combinations of the basic features. We, thus, can expect that the understandability or transparency of the resulting features could be improved. In addition, we will investigate the usefulness of CIMA for external learners other than C4.5 we adopted in this paper.

References

1. Bensusan, H. and Kuscu, I.: Constructive Induction using Genetic Programming, In Proceedings of the ICML'96 Workshop on Evolutionary computing and Machine Learning (1996)
2. Blake, C.L. and Merz, C.J.: UCI Repository of machine learning databases, http://www.ics.uci.edu/~mlearn/MLRepository.html, Irvine, CA: University of California, Department of Information and Computer Science. (1998)
3. Hastie, T., Tibshirani, R. and Friedman, J.: The Elements of Statistical Learning, Springer-Verlag(2001)
4. Hu, Y. and Kibler, D.: Generation of Attributes for Learning Algorithms, Proc.of AAAI-1996 (1996) 806–811
5. Kramer, S., Lavrač, N. and Flach, P.: Propositionalization Approaches to Relational Data Mining, Džeroski, S. and Lavrač, N. ed., "Relational Data Mining", Springer-Verlag (2001), 262–291
6. Krawiec, K.: Constructive Induction in Learning of Image Representation, Institute of Computing Science Poznan University of Technology Research Report RA-006/2000 (2000)
7. Markovitch, R. and Rosenstein, D.: Feature Generation Using General Constructor Functions, Machine Learning, Vol.49 (2002) 59–98
8. Michalski, R.: Pattern Recognition as Knowledge-Guided Computer Induction, Department of Computer Science, University of Illinois, Urbana, Report No.927 (1978)
9. Michalski, R.: A Theory and Methodology of Inductive Learning, Artificial Intelligence, Vol.20, No.2 (1983) 111–161
10. Muggleton, S.: Inverse Entailment and Progol, New Generation Computing, Vol.13 (1995) 245–286
11. Quinlan, J., C4.5: Programs for Machine Learning, Morgan Kaufmann, San Mateo, California (1992)
12. Quinlan, J. and Camereon-Jones, R.M.: Induction of Logic Programs:FOIL and Related Systems, New Generation Computing, Vol.13 (1995) 287–312

Resemblance Coefficient
and a Quantum Genetic Algorithm for Feature Selection*

Gexiang Zhang[1,2], Laizhao Hu[1], and Weidong Jin[2]

[1] National EW Laboratory, Chengdu 610031 Sichuan, China
dylan7237@sina.com
[2] School of Electrical Engineering, Southwest Jiaotong University,
Chengdu 610031 Sichuan, China

Abstract. Feature selection is always an important and difficult issue in pattern recognition, machine learning and data mining. In this paper, a novel approach called resemblance coefficient feature selection (RCFS) is proposed. Definition, properties of resemblance coefficient (RC) and the evaluation criterion of the optimal feature subset are given firstly. Feature selection algorithm using RC criterion and a quantum genetic algorithm is described in detail. RCFS can decide automatically the minimal dimension of good feature vector and can select the optimal feature subset reliably and effectively. Then the efficient classifiers are designed using neural network. Finally, to bring into comparison, 3 methods, including RCFS, sequential forward selection using distance criterion (SFSDC) and a new method of feature selection (NMFS) presented by Tiejun Lü are used respectively to select the optimal feature subset from original feature set (OFS) composed of 16 features of radar emitter signals. The feature subsets, obtained from RCFS, SFSDC and NMFS, and OFS are employed respectively to recognize 10 typical radar emitter signals in a wide range of signal-to-noise rate. Experiment results show that RCFS not only lowers the dimension of feature vector greatly and simplifies the classifier design, but also achieves higher accurate recognition rate than SFSDC, NMFS and OFS, respectively.

1 Introduction

Feature selection is an important problem in pattern recognition, data mining and machine learning. The main task of feature selection is to select the most discriminatory features from original feature set to lower the dimension of pattern space in terms of internal information of feature samples. [1-5] Although it is very good that error probability of a classifier is chosen as the criterion of feature selection, it is very complicated to compute the error probability of a classifier even if class-condition probability density function is known. Moreover, the function is usually unknown in

---

* This work was supported by the National Defence Foundation (No.51435030101ZS0502 No.00JSOS.2.1.ZS0501), by the National Natural Science Foundation (No.69574026), by the Doctoral Innovation Foundation of SWJTU and by the Main Teacher Sponsor Program of Education Department of China.

E. Suzuki and S. Arikawa (Eds.): DS 2004, LNAI 3245, pp. 155–168, 2004.

factual situation, which brings many difficulties to analyzing the validity of a feature using directly the classification standard based on error probability. So a good solution to solve this problem is that some practical criterions must be found to decide the separability of different classes. [1-2,5] Up to now, many researchers have done much work in feature selection and have presented multiple class separability criterions and algorithms for finding the best feature subset, such as distance criterion [5], information entropy criterion [2,5], feature importance criterion [1], linear programming [3], independent feature selection [4], sequential forward selection [5], feature selection based on genetic algorithm [6], unsupervised feature selection [7], scalable feature selection [8], correlation-based feature selection method [9], feature selection based on analysis of class regions [10], feature selection with genetic algorithms [6,11-13], mutual information feature selection and Taguchi feature selection method [14]. Although some feature selection methods have been presented, the methods cannot arrive at completely satisfactory results. Feature selection is still a hot topic and is still paid much attention to by lots of researchers. [1-14]

In radar emitter signal recognition, the most significant and valid feature is not found easily because of many uncertain reasons. Some empirical methods and heuristic approaches are often used to extract features from radar emitter signals. Thus, subjectivity and guess are usually brought into feature extraction process. What is more, radar emitter signals are always interfered with by plenty of noise in the process of transmission and processing in scouts. Signal-to-noise rate (SNR) of radar emitter signals received by radar scouts varies in a large range from several dB to several tens of dB. These factors result in out-of-order distribution of feature vector and much overlapping among the features of different radar emitter signals in feature space so as to lower accurate recognition rate greatly. To eliminate the subjectivity and to enhance accurate recognition rate, multiple features must be extracted from radar emitter signals using different methods and good feature selection approach based on good evaluation criterion and efficient search algorithm must be explored to select the optimal feature subset. [5,6]

In this paper, a novel approach called resemblance coefficient feature selection (RCFS) is proposed. The main ideas of RCFS are that resemblance coefficient is chosen as class separability criterion and quantum genetic algorithm (QGA) [15,16] with rapid convergence and good global search capability is used to search the optimal feature subset from original feature set (OFS) composed of 16 features. First of all, definition and properties of resemblance coefficient and evaluation criterion based on resemblance coefficient are given. The detailed feature selection algorithm based on resemblance coefficient criterion and QGA is described. Then efficient classifiers are designed using neural network. Finally, to bring into comparison, RCFS, sequential forward selection based on distance criterion (SFSDC) [5] and a new method of feature selection (NMFS) [6] are used to make the experiment of feature selection. Dispensing with designating the dimension of feature vector, RCFS can select the best feature subset reliably and effectively. Furthermore, the feature subset obtained by RCFS achieves much higher accurate recognition rate than that of SFSDC, NMFS and OFS.

The rest of this paper is organized as follows. Section 2 describes feature selection algorithm in details. Section 3 discusses classifier design using neural network. Experiment is made in section 4 to demonstrate that RCFS is effective, which is then followed by the conclusions in Section 5.

2   Resemblance Coefficient Feature Selection Algorithm

2.1   Definition and Property of RC

Definition 1. Suppose that one-dimensional functions $f(x)$ and $g(x)$ are continuous, positive and real, i.e.

$$f(x) \geq 0, g(x) \geq 0 \tag{1}$$

Resemblance coefficient of function $f(x)$ and $g(x)$ is defined as

$$C_r = \frac{\int f(x)g(x)dx}{\sqrt{\int f^2(x)dx} \cdot \sqrt{\int g^2(x)dx}} \tag{2}$$

In equation (2), the integral domains of $f(x)$ and $g(x)$ are their definable domains of the variable x. Moreover, when x is within its definable domain, the value of function $f(x)$ or $g(x)$ cannot be always equal to 0.

Property 1 The value domain of resemblance coefficient C_r is

$$0 \leq C_r \leq 1 \tag{3}$$

Because $f(x)$ and $g(x)$ are positive functions, according to the famous *Cauchy Schwartz* inequality, we can obtain

$$0 \leq \int f(x)g(x)dx \leq \sqrt{\int f^2(x)dx} \cdot \sqrt{\int g^2(x)dx} \tag{4}$$

$$0 \leq \frac{\int f(x)g(x)dx}{\sqrt{\int f^2(x)dx} \cdot \sqrt{\int g^2(x)dx}} \leq 1 \tag{5}$$

Obviously, we can get $0 \leq C_r \leq 1$. According to the conditions of *Cauchy Schwartz* inequality, if $f(x)$ equals to $g(x)$, resemblance coefficient C_r of $f(x)$ and $g(x)$ gets the maximal value 1. In fact, if and only if the $f(x)$-to-$g(x)$ ratio in every point is constant, resemblance coefficient C_r equals to 1. If and only if the integral of product of $f(x)$ and $g(x)$ is zero, i.e. for arbitrary x, $f(x)=0$ or $g(x)=0$, resemblance coefficient C_r equals to the minimal value 0.

From definition 1, computing resemblance coefficient of two functions corresponds to computing the correlation of the two functions. The value of resemblance coefficient mainly depends on the characteristics of two functions. If $f(x)$ is in proportion to $g(x)$, i.e. $f(x)=kg(x)$, $k>0$, the value of resemblance coefficient C_r equals to 1, which indicates function $f(x)$ resembles $g(x)$ completely. As the overlapping of the two functions decreases gradually, resemblance coefficient C_r will increase gradually, which indicates that $f(x)$ and $g(x)$ are resemblant partly. When $f(x)$ and $g(x)$ are completely separable, C_r gets to the minimal value 0, which implies $f(x)$ does not resemble $g(x)$ at all.

2.2   Class Separability Criterion

Class separability criterion based on probability distribution must satisfy three conditions [5]: (i) the criterion function value is not negative; (ii) if there is not overlapping part of distribution functions of two classes, the criterion function value gets to the maximal value; (iii) if distribution functions of two classes are identical, the criterion function value is 0.

Class separability criterion function based on resemblance coefficient is defined as

$$J = 1 - \frac{\int f(x)g(x)dx}{\sqrt{\int f^2(x)dx} \cdot \sqrt{\int g^2(x)dx}} = 1 - C_r \qquad (6)$$

According to the definition and property of resemblance coefficient, the value of J is always equal to or more than zero. For any x, if $f(x) \neq 0$ and $g(x)=0$ or if $g(x) \neq 0$ and $f(x)=0$, J arrives at the maximal value. If $f(x)$ is the same as $g(x)$, $J=0$. So the criterion function J given in equation (6) satisfies the three class separability conditions and can be used as a standard to decide whether two classes are separable or not.

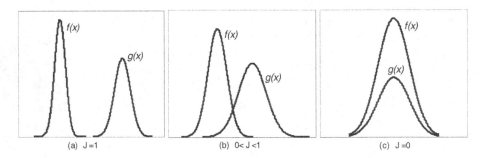

Fig. 1. Three separability cases of function $f(x)$ and $g(x)$.

When the two functions $f(x)$ and $g(x)$ in equation (6) are regarded respectively as probability distribution functions of feature samples of two classes A and B, several separability cases of A and B are shown in figure 1. For all x, if one of $f(x)$ and $g(x)$ is zero at least, which is shown in figure 1(a), A and B are completely separable and the criterion function J arrives at the maximal value 1. If there are some points of x that make $f(x)$ and $g(x)$ not equal to 0 simultaneously, which is shown in figure 1(b), A and B are partly separable and the criterion function J lies in the range between 0 and 1. For all x, if $f(x)=kg(x)$, which is shown in figure 1(c), $k=2$, A and B are not completely separable and the criterion function J arrives at the minimal value 0. Therefore, it is reasonable and feasible that the criterion function in equation (6) is used to compute separability of two classes. An additional explanation is that any function satisfied the conditions in definition 1 can be used as $f(x)$ or $g(x)$ in equation (6).

2.3 Feature Selection Algorithm

In feature extraction of radar emitter signals, all feature samples always vary in the neighboring area of expectation value because of plenty of noise and measurement errors. If occurrences of all samples are computed in statistical way, a feature probability distribution function can be obtained. The function can be considered approximately as a Gaussian distribution function with the parameters of expectation and variance of feature samples. Thus, according to the above criterion function, the feature selection algorithm of radar emitter signals is given as follows in detail.

Step 1 For a certain feature F_1, computing the class separability criterion function values of n radar emitter signals and constructing a matrix S called class separability matrix. S is

$$S = \begin{bmatrix} s_{11} & s_{12} & \cdots & s_{1n} \\ s_{21} & s_{22} & \cdots & s_{2n} \\ \vdots & \vdots & \vdots & \vdots \\ s_{n1} & s_{n2} & \cdots & s_{nn} \end{bmatrix} \quad (7)$$

where $s_{ij}\,(i, j = 1, 2, \cdots, n)$ obtained according to equation (6) is separability of the ith radar emitter signal and the jth radar emitter signal. In the process of computing S, the feature samples of each radar emitter signal are considered as a Gaussian function with the parameters of expectation μ and variance σ of feature samples, where μ and σ are respectively

$$u = E(Z) \quad (8)$$

$$\sigma = D(Z) \quad (9)$$

Where Z is a vector and Z is

$$Z = [z_1, z_2, \cdots, z_M] \quad z_p(p = 1, 2, \cdots, M) \quad (10)$$

Where z_p is the pth feature value and M is the number of feature values. According to definition and properties of resemblance coefficient, we can see easily that $s_{ii} = 1, (i = 1, 2, \cdots, n)$ and $s_{ij} = s_{ji}, (i, j = 1, 2, \cdots, n)$, i.e. class separability matrix S is a diagonal matrix. So it is enough to compute all elements above diagonal.

Step 2 Choosing a threshold value r of separable degree of two signals. If the element s_{ij} in matrix S is more than r, the ith radar emitter signal and the jth radar emitter signal are separable and we set its corresponding value to zero. Otherwise, If the element s_{ij} of matrix S is less than r, the ith radar emitter signal and the jth radar emitter signal are not separable and we set its corresponding value to one. Thus, we obtain another matrix P called class separability reduction matrix. Matrix P is composed of "0" and "1".

Step 3 Computing class separability matrix S^l of the lth feature and class separability reduction matrix $P^l (l = 1, 2, \cdots L)$ in the same way in step 1 and step 2, where L is the number of features. That is $P^l = \{p_{ij}^l\}, i, j = 1, 2, \cdots L$.

After step 1, 2 and 3 are finished, there are l class separability reduction matrices because every feature has a class separability reduction matrix. Thus, a problem appears subsequently that which separability reduction matrices should be chosen to form the best feature subset and how the matrices are chosen. Obviously, the problem is a combinatorial problem with T_c combinations and T_c is

$$T_c = C_l^1 + C_l^2 + \cdots + C_l^l \tag{11}$$

Genetic algorithm (GA) is a global optimization method and GA has strong robustness and general applicability. Because GA cannot be restricted by the nature of optimization problems and GA can deal with very complicated problems that cannot be solved by using traditional optimization methods, GA has become an attractive optimization approach and has been used generally in many fields [6,11-13,17-19]. Especially, quantum genetic algorithm (QGA), a new probability optimization method, is paid much attention to in recent years [15-16,20-22]. QGA is based on the concepts and principles of quantum computing. QGA uses a novel quantum bit (qubit) chromosome representation instead of binary, numeric, or symbol representation. The characteristic of the representation is that any linear superposition of solutions can be represented. Different from conventional GAs in which crossover and mutation operations are used to maintain the diversity of population, the evolutionary operation in QGA is implemented by updating the probability amplitudes of basic quantum states using quantum logic gates so as to maintain the diversity of population. QGA has good characteristics of rapid convergence, good global search capability, simultaneous exploration and exploitation, small population instead of degrading the performances of algorithm. [15,16, 20-22] So in this paper, QGA is used to solve the described combinatorial optimization problem.

Step 4 QGA [15,16] is used to search automatically the optimal feature subset from original feature set. The detailed steps using QGA are as follows.

(1) Initialization of QGA includes the process of choosing population size h and choosing the number m of qubits. The population containing h individuals is represented as $P=\{p_1,p_2,\cdots,p_h\}$, where p_j ($j=1,2,\cdots,h$) is the jth individual and p_j is

$$p_j = \begin{bmatrix} \alpha_1 & \alpha_2 & \cdots & \alpha_m \\ \beta_1 & \beta_2 & \cdots & \beta_m \end{bmatrix} \tag{12}$$

In equation (12), $\alpha_i, \beta_i (i = 1, 2, \cdots, m)$ are respectively probability amplitudes of quantum basic states $|1\rangle$ and $|0\rangle$ of the ith qubit. In the beginning of QGA, all α_i, β_i

(i=1,2,$\cdots$, m) =1/$\sqrt{2}$, which indicates that all quantum basic states are superposed by the same probability. Evolutionary generation *gen* is set to 0.

(2) According to the probability amplitudes of all individuals in population P, observation state R of quantum superposition is constructed. R={a_1, a_2,$\cdots$, a_h}, where a_j (j=1, 2,$\cdots$,h) is observation state of each individual, i.e. a binary string composed of "0" and "1".

(3) Fitness function is used to evaluate all individuals in population P. If the dimension of feature vector of an individual in P is d and the binary string $b_1 b_2 \cdots b_K$ (b_k='0' or '1', $k = 1, 2, \cdots, m$) is quantum observation state in QGA, the fitness function is defined as

$$f = d + \sum_{i=1}^{n-1} \sum_{j=i+1}^{n} q_{bij}$$

(13)

$$q_{bij} = (b_1 q_{ij}^1) \& (b_2 q_{ij}^2) \cdots \& (b_m q_{ij}^m)$$

where the symbol '&' stands for 'AND' operation in Boolean algebra. q_{bij} ($i, j = 1, 2, \cdots, n$) is the element of the ith row and the jth column in matrix Q_b that is class separability reduction matrix of an individual in population. Obviously, the smaller the function f is, the better the feature subset obtained is.

(4) Maintaining the optimal individual in population P and judging terminal condition of QGA. If terminal condition is satisfied, the algorithm ends. Otherwise, the algorithm continues.

(5) Quantum rotation angles of quantum rotation gates are obtained using the method in reference [16]. Quantum rotation gates obtained operate on probability amplitudes of all individuals, i.e. the probability amplitudes of all individuals in population P are updated.

(6) Evolutionary generation *gen* increases 1. The algorithm goes to (2) and continues.

3   Classifier Design

Feature selection can be considered as a transformation that transforms the radar emitter signal feature from high dimensional feature space into low dimensional feature space and extracts the most discriminatory information and removes the redundant information. Although feature selection is an important process, it is not the final step in radar emitter signal recognition. The recognition task is to be finished only by the classifier. So classifier design is also an important process subsequent to feature extraction and feature selection in radar emitter signal recognition.

The recent vast research activities in neural classification have established that neural networks are a promising alternative to various conventional classification methods. Neural networks have become an important tool for classification because neural networks have the following advantages in theoretical aspects. [23,24] First, neural networks are data driven self-adaptive methods in that they can adjust themselves to the data without any explicit specification of functional or distributional form for the underlying model. Second, they are universal functional approximators in that neural networks can approximate any function with arbitrary accuracy. Third, neural networks are nonlinear models, which makes them flexible in modeling real world complex relationships. Finally, neural networks are able to estimate the posterior probabilities, which provide the basis for establishing classification rule and performing statistical analysis. So neural network classifers are used generally in signal recognition.

The structure of neural network classifier is shown in Fig.2. In Fig.2, L_1 is the input layer that has L neurons corresponding to radar emitter signal features. L_2 is the hidden layer and '*tansig*' is chosen as the transfer functions. L_3 is output layer that has the same number of neurons as radar emitter signals to be recognized. Transfer function in output layer is '*logsig*'. We choose RPROP algorithm [25] as the training algorithm of the neural network. The ideal outputs of neural network are "1". Output tolerance is 0.05 and output error is 0.001.

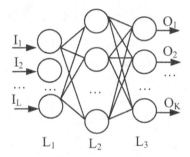

Fig. 2. The structure of neural network classifier.

4   Experimental Results

To demonstrate the feasibility and effectiveness of the proposed approach, 10 typical radar emitter signals are chosen to make the simulation experiment. They are CW, BPSK, QPSK, MPSK, LFM, NLFM, FD, FSK, IPFE and CSF, respectively. In our prior work [26-30], 16 features of 10 radar emitter signals were studied. The original feature set is composed of 16 features that are labeled as 1, 2, $\cdots$, 16. The features are respectively fractal dimensions including information dimension [26], box dimension [26] and correlation dimension [27], two resemblance coefficient features [28], Lempel-Ziv complexity [27], approximate entropy [29], wavelet entropy and eight energy distribution features based on wavelet packet decomposition [30]. In the experiment, for every radar emitter signal, 150 feature samples are generated in each SNR point of 5dB, 10dB, 15dB and 20dB. Thus, 600 samples of each radar emitter signal in total are generated when SNR varies from 5dB to 20dB. The samples are classified into two groups: training group and testing group. Training group, one third of the total samples generated, is applied to make the simulation experiment of feature selection and neural network classifer training. Testing group, two thirds of the total samples generated, is used to test trained neural network classifers.

In resemblance coefficient feature selection (RCFS), threshold value r is set to 0.99, and population size in QGA is set to 20 and the number of qubits is set to 16. To bring into comparison, sequential forward selection using distance criterion (SFSDC) [5] and a new method of feature selection (NMFS) [6] are also used to select the best feature subset.

Distance criterion function in SFSDC is

$$G_q = \sum_i \sum_j \frac{(m_i - m_j)^2}{\sigma_i^2 + \sigma_j^2}, \quad q = 1, 2, \cdots, n \tag{14}$$

where m_i and m_j are the mean values of all samples of the ith signal and the jth signal, respectively. σ_i and σ_j are the variance values of all samples of the ith signal and the jth signal, respectively. n is the number of features.

The criterion function for evaluating the best feature subset in NMFS is described as follows. Suppose the distance d_{ij} between class ω_i and class ω_j is

$$d_{ij} = \left(\sum_n^N \omega_{ij}^n \cdot \left| m_i^n - m_j^n \right|^p \right)^{\frac{1}{p}}, \quad p \geq 1 \tag{15}$$

In equation (15), m_i and m_j are respectively average feature vectors of class ω_i and class ω_j and they are respectively

$$m_i = (m_i^1, m_i^1, \cdots, m_i^N) \tag{16}$$

$$m_j = (m_j^1, m_j^1, \cdots, m_j^N) \tag{17}$$

In equation (15), ω_{ij}^n is the weighted value of class ω_i and class ω_j and is defined as

$$\omega_{ij}^n = e^{-a(\sigma_i^n + \sigma_j^n)}, \quad a > 0 \tag{18}$$

In equation (18), σ_i^n and σ_j^n are respectively variances of feature vectors of class ω_i and class ω_j and they are respectively

$$\sigma_i = (\sigma_i^1, \sigma_i^1, \cdots, \sigma_i^N) \tag{19}$$

$$\sigma_j = (\sigma_j^1, \sigma_j^1, \cdots, \sigma_j^N) \tag{20}$$

In NMFS, genetic algorithm is used to search the best feature subset from the original feature set. If there are R classes , the fitness function is

$$f = \sum_{i=1}^{R-1} \sum_{j>1}^{R} d_{ij} \tag{21}$$

Because SFSDC and NMFS cannot decide automatically the dimension of the best feature subset selected, the dimension of feature subset obtained by RCFS is chosen as that of SFSDC and NMFS so as to draw a comparison of recognition results of three methods.

First of all, the original feature set (OFS) is used to train neural network classifiers (NNC) whose structure is 16-25-10. The samples in testing group are employed to test the trained neural network classifiers in a wide range of SNR. After 20 experiments, the recognition results are shown in table 1. The average recognition error rate is 4.83% in table 1. Then, RCFS is applied to select the best feature subset from the original feature set. The feature subset selected is composed of feature 5 and 10. The result is identical in 30 experiments. After neural network classifiers that have the structure of 2-15-10 are trained, testing results are shown in table 2, in which the average recognition error rate is only 1.34%. Finally, the dimension of feature subset in SFSDC and NMFS is designated as 2 and the two methods are respectively used to make the experiment of feature selection. The feature subset obtained by SFSDC is composed of feature 4 and 5. Feature 6 and 7 constitute the optimal feature subset in NMFS. Obviously, the structures in SFSDC and NMFS are the same as that in RCFS.

Table 1. Average recognition error rates obtained using original feature set.

Types	5 dB	10 dB	15 dB	20 dB	Average
BPSK	0.00	0.00	0.00	0.00	0.00
QPSK	66.67%	33.33%	0.00	0.00	24.75%
MPSK	4.00%	0.20%	0.00	0.00	1.41%
LFM	0.00	0.00	0.00	0.00	0.00
NLFM	0.00	0.00	0.00	0.00	0.00
CW	0.00	0.00	0.00	0.00	0.00
FD	0.10%	0.00	0.00	0.00	0.03%
FSK	0.00	0.00	0.00	0.00	0.00
IPFE	0.00	0.00	0.00	0.00	0.00
CSF	23.00	33.33%	0.00	0.00	14.08%

Table 2. Average recognition error rates using the feature subset obtained by RCFS.

Types	5 dB	10 dB	15 dB	20 dB	Average
BPSK	0.00	5.68%	0.00	0.00	1.42%
QPSK	4.57%	42.47%	0.82%	0.00	11.96%
MPSK	0.00	0.00	0.00	0.00	0.00
LFM	0.00	0.00	0.00	0.00	0.00
NLFM	0.00	0.00	0.00	0.00	0.00
CW	0.00	0.00	0.00	0.00	0.00
FD	0.00	0.00	0.00	0.00	0.00
FSK	0.00	0.00	0.00	0.00	0.00
IPFE	0.00	0.00	0.00	0.00	0.00
CSF	0.00	0.00	0.00	0.00	0.00

When SNR varies from 5dB to 20dB, recognition results of SFSDC and NMFS are shown in table 3 and table 4, respectively. The results in table 1, 2, 3 and 4 are statistical results of 20 experiments and all values are recognition error rates. Table 5 shows comparison results of OFS, RCFS, SFSDC and NMFS. In table 5, ATG an ARR are the abbreviations of average training generation of NNC and accurate recognition rate, respectively.

From table 1 to table 5, several conclusions can be drawn: (i) in comparison with OFS, RCFS not only lowers the dimension of feature vector and the cost of feature extraction greatly, but also simplifies classifier design and enhances recognition efficiency and accurate recognition rate; (ii) in comparison with SFSDC and GADC, RCFS selects better features and achieves higher accurate recognition rate because

Table 3. Recognition results using the feature subset obtained by SFSDC.

Types	5 dB	10 dB	15 dB	20 dB	Average
BPSK	9.20%	8.00%	13.53%	10.90%	10.41%
QPSK	43.00%	34.87%	45.02%	42.15%	41.0%
MPSK	24.87%	18.45%	17.22%	26.53%	21.77%
LFM	15.90%	0.00	0.00	0.00	3.97%
NLFM	0.00	0.00	0.00	0.00	0.00
CW	0.00	0.00	0.00	0.00	0.00
FD	0.00	0.00	0.00	0.00	0.00
FSK	0.00	0.00	0.00	0.00	0.00
IPFE	2.75%	0.57%	8.22%	0.00	2.88%
CSF	0.75%	0.00	1.75%	1.97%	1.12%

Table 4. Recognition results using the feature subset obtained by NMFS.

Types	5 dB	10 dB	15 dB	20 dB	Average
BPSK	32.30%	0.00	0.00	0.00	8.07%
QPSK	7.20%	0.50%	0.00	0.00	1.92%
MPSK	58.10%	0.00	0.00	0.00	14.53%
LFM	0.00	0.00	0.00	0.00	0.00
NLFM	55.54%	10.50%	0.00	0.00	16.47%
CW	0.00	0.00	0.00	0.00	0.00
FD	0.00	0.00	0.00	0.00	0.00
FSK	0.00	0.00	0.00	0.00	0.00
IPFE	23.70%	0.00	0.00	0.00	5.92%
CSF	19.30%	0.00	0.00	0.00	4.82%

Table 5. Comparison results of 4 feature sets.

Methods	Feature set	Structure of NNC	ATG of NNC	Average ARR
RCFS	5,10	2-15-10	248.10	98.66%
OFS	1~16	2-25-10	324.67	95.17%
SFSDC	4,5	2-15-10	4971.80	91.88%
NMFS	6,7	2-15-10	1146.50	94.83%

the training generation of NNC using the feature subset selected by RCFS is much less than that of SFSDC and GADC, and the average recognition error rate of RCFS is only 1.34% which is less 6.78%, 3.83% and 3.49% than that of SFSDC, GADC and OFS, respectively.

5 Concluding Remarks

This paper proposes a novel feature selection called resemblance coefficient feature selection approach. The main points of the introduced method are as follows:

(1) An effective class separability criterion function is defined with resemblance coefficient. The definition of resemblance coefficient is given and the properties of resemblance coefficient are analyzed. Using resemblance coefficient, a novel class separability criterion function is presented. The criterion function satisfies the three conditions [5] that any class separability criterion based on probability distribution must satisfy. Only using the internal information of feature samples, the presented evaluation criterion can decide the dimension of the best feature subset automatically. Therefore, the class separability criterion can overcome the problems that most existing feature selection methods need choose the dimension of the feature subset before feature selection is made and multiple tries of different dimensions must be done.

(2) An efficient optimization algorithm called quantum genetic algorithm is introduced to select the best feature subset from the original feature set composed of a large number of features. In QGA, a novel chromosome representation called qubit representation is used to represent more individuals than conventional genetic algorithm with the same population size and a novel evolutionary operation called quantum rotation gate update procedure is applied to generate the next population. Thus, QGA can maintain the population diversity in the process of searching the optimal solution and avoid the problem of selection pressure of conventional genetic algorithm. Also, there are little relations between the individuals in the child population in QGA. So QGA has good characteristics of rapid convergence, good global search capability, simultaneous exploration and exploitation, small population instead of degrading the performances of algorithm.

In order to bring into comparison, 3 methods including RCFS, SFSDC and NMFS are used respectively to select the optimal feature subset from original feature set (OFS) composed of 16 features of radar emitter signals in this paper. In the simulation experiments, the feature subsets, obtained from RCFS, SFSDC and NMFS, and OFS are employed respectively to recognize 10 typical radar emitter signals in a wide range of signal-to-noise rate. Experimental results show that RCFS not only lowers the dimension of feature vector greatly and simplifies the classifier design, but also achieves 98.66% accurate recognition rate, which is higher 6.78%, 3.83% and 3.49% than that of SFSDC, NMFS and OFS, respectively.

Although RCFS is used only in radar emitter signal feature selection in this paper, in fact, from the above analysis, RCFS has generality and can be applied in many other fields, such as data mining and machine learning. Moreover, the distribution function of feature samples in computing resemblance coefficients is not limited to Gaussian distribution.

References

1. Zhao J., Wang G.Y., Wu Z.F., etc.: The Study on Technologies for Feature Selection. Proc. of 1th Int. Conf. on Machine Learning and Cybernetics. (2002) 689-693
2. Molina L.C., Belanche L., Nebot A.: Feature Selection Algorithms: A Survey and Experimental Evaluation. Proc. of Int. Conf. on Data Mining. (2002) 306-313
3. Guo G.D., Dyer C.R.. Simultaneous Selection and Classifier Training Via Linear Programming: A Case Study for Face Expression Recognition. Proc. of IEEE Computer Society Conf. on Computer Vision and Pattern Recognition. (2003) 346-352
4. Bressan M., Vitria` J.: On the Selection and Classification of Independent Features. IEEE Trans. on Pattern Analysis and Machine Intelligence. Vol.25, No.10, (2003) 1312-1317
5. Bian Z.Q., Zhang X.G.: Pattern recognition (Second edition). Beijing: Tsinghua University Press, 2000
6. Lü T.J., Wang H., Xiao X.C.:. Recognition of Modulation Signal Based on a New Method of Feature Selection. Journal of Electronics and Information Technology. Vol.24, No.5, (2002) 661-666
7. Rhee F.C.H. and Lee Y.J.: Unsupervised feature selection using a fuzzy genetic algorithm. Proceedings of 1999 IEEE International Fuzzy Systems Conference. (1999) Vol.3, 1266-1269
8. Chakrabarti S., Dom B., Agrawal R and Raghavan P.: Scalable feature selection, classification and signature generation for organizing large text databases into hierarchical topic taxonomies. The VLDB Jounal. Springer-Verlag. (1998) Vol.7, 163-178
9. Liu H.Q., Li J.Y., Wong L S.: A comparative study on feature selection and classification methods using gene expression profiles. Genome Informatics. (2002) Vol.13, 51-60
10. Thawonmas R. and Abe S.: A novel approach to feature selection based on analysis of class regions. IEEE Transactions on Systems, Man, and Cybernetics----Part B: Cybernetics.(1997) Vol.27, No.2, 196-207
11. Garrett D., Peterson D.A., Anderson C W and Thaut M H.: Comparison of linear, nonlinear, and feature selection methods for EEG signal classification. IEEE Transactions on Neural Systems and Rehabilitation Engineering. (2003) Vol.11, No.2, 141-144
12. Haydar A., Demirekler and Yurtseven.: Speaker identification through use of features selected using genetic algorithm. Electronics Letters. (1998) Vol.34, No.1, 39-40
13. Jack L.B. and Nandi A.K.: Genetic algorithms for feature selection in machine condition monitoring with vibration signals. IEE Proceedings on Vision Image Signal Processing. (2000) Vol.147, No.3, 205-212
14. Kwak N. and Choi C.H.: Input feature selection for classification problems. IEEE Transactions on Neural Networks. (2002) Vol.13, No.1, 143-159
15. Zhang G.X., Jin W.D., Li N.: An improved quantum genetic algorithm and its application, Lecture Notes in Artificial Intelligence. (2003) Vol.2639, 449-452
16. Zhang G.X., Hu L.Z., Jin W.D.: Quantum Computing Based Machine Learning Method and Its Application in Radar Emitter Signal Recognition. Lecture Notes in Computer Science (LNCS), 2004.8 (to appear)
17. Youssef H., Sait S.M., Adiche H.: Evolutionary algorithms, simulated annealing and tabu search: a comparative study. Engineering Application of Artificial Intellegence. (2001) Vol.14, 167-181
18. Srinivas M., Patnaik L.M.: Genetic algorithms: a survey. Computer. (1994) Vol.27, No.6, 17-26

19. Koza J.R.: Survey of genetic algorithm and genetic programming. Proceedings of the 1995 Microelectronics Communications Technology Producing Quality Products Mobile and Portable Power Emerging Technologies. (1995) 589-594
20. Han K.H., Kim J.H.: Genetic quantum algorithm and its application to combinatorial optimization problems. Proceedings of the 2000 IEEE Conference on Evolutionary Computation. (2000) 1354-1360
21. Yang J.A., Li B., Zhuang Z.Q.: Research of quantum genetic algorithm and its application in blind source separation. Journal of Electronics. (2003) Vol.20, No.1, 62-68
22. Li Y., Jiao L.C.: An effective method of image edge detection based on parallel quantum evolutionary algorithm. Signal Processing. (2003) Vol.19, No.1, 69-74
23. Zhang G.P.: Neural Networks for Classification: a Survey. IEEE Trans. on System, Man, and Cybernetics-Part C: Application and Reviews. Vol.30, No.4, (2000) 451-462
24. Kavalov D., Kalinin V.: Neural Network Surface Acoustic Wave RF Signal Processor for Digital Modulation Recognition. IEEE Trans. on Ultrasonics, Ferroelectrics, and Frequency Control. Vol.49, No.9, (2002) 1280-1290
25. Riedmiller M., Braun H.: A Direct Adaptive Method for Faster Back Propagation Learning: The RPROP Algorithm. Proc. of IEEE Int. Conf. on Neural Networks. (1993) 586-591
26. Zhang G.X., Jin W.D., Hu L.Z.: Fractal Feature Extraction of Radar Emitter Signals. Proc. of the third Asia-Pacific conf. on Environmental Electromagnetics. (2003) 161-164
27. Zhang G.X., Hu L.Z., Jin W.D.: Complexity Feature Extraction of Radar Emitter Signals. Proc. of the third Asia-Pacific Conf. on Environmental Electromagnetics. (2003) 495-498
28. Zhang G.X., Rong H.N., Jin W.D., Hu L.Z.: Radar emitter signal recognition based on resemblance coefficient features, Lecture Notes in Computer Science (LNCS). (2004) Vol.3066, 665-670
29. Zhang G.X., Rong H.N., Hu L.Z., Jin W.D.: Entropy Feature Extraction Approach of Radar Emitter Signals. Proceedings of International Conference on Intelligent Mechatronics and Automation. 2004.8 (to appear)
30. Zhang G.X., Jin W.D., Hu L.Z.: Application of Wavelet Packet Transform to Signal Recognition. Proceedings of International Conference on Intelligent Mechatronics and Automation. 2004.8 (to appear)

Extracting Positive Attributions from Scientific Papers

Son Bao Pham and Achim Hoffmann

School of Computer Science and Engineering
University of New South Wales, Australia
{sonp,achim}@cse.unsw.edu.au

Abstract. The aim of our work is to provide support for reading (or skimming) scientific papers. In this paper we report on the task to identify concepts or terms with positive attributions in scientific papers. This task is challenging as it requires the analysis of the relationship between a concept or term and its sentiment expression. Furthermore, the context of the expression needs to be inspected. We propose an incremental knowledge acquisition framework to tackle these challenges. With our framework we could rapidly (within 2 days of an expert's time) develop a prototype system to identify positive attributions in scientific papers. The resulting system achieves high precision (above 74%) and high recall rates (above 88%) in our initial experiments on corpora of scientific papers. It also drastically outperforms baseline machine learning algorithms trained on the same data.

1 Introduction

Knowing the advantages and disadvantages of a particular concept or algorithm is important for every researcher. It helps researchers in learning a new field or even experienced researchers in keeping up to date. Unfortunately, such information is usually scattered across many papers. Survey papers are generally written on an irregular basis, and hence up-to-date surveys may not be available. Furthermore, in new and emerging fields, often survey papers do not exist at all. Having a tool that could collect all the relevant information for a concept of interest would therefore be of tremendous value.

For example, suppose we want to check if a particular algorithm is suitable for our task at hand, such a tool could go through available papers and extract sentences together with the contexts that mention the advantages and disadvantages of the algorithm. This would make our task much simpler. We only have to look at those extracted sentences rather than going through a large number of entire papers.

Another useful scenario is the following: before reading a paper, we could quickly have a look at what the advantages and disadvantages of the things discussed in the paper are to make a decision whether the paper is relevant to our interest.

In this paper, we introduce a new framework for acquiring rules to classify text segments, such as sentences, as well as extracting information relevant to the classification from the text segments. We apply our framework to extract advantages of concepts or actions, i.e. positive attributions of the concepts/actions, in technical papers. An advantage is detected when a positive sentiment is expressed towards the concept or action. For example, given the following sentences:

E. Suzuki and S. Arikawa (Eds.): DS 2004, LNAI 3245, pp. 169–182, 2004.

There is some evidence that Randomizing is better than Bagging in low noise settings.
It is more efficient to use Knowledge Acquisition to solve the task.

We would like to detect that the algorithm *Randomizing* and the action *to use Knowledge Acquisition to solve the task* have been mentioned with a positive sentiment. Analysis of positive sentiments towards a concept is a challenging task that requires deep understanding of the textual context, drawing on common sense, domain knowledge and linguistic knowledge. A concept could be mentioned with a positive sentiment in a local context but not in a wider context. For example,

We do not think that X is very efficient.

If we just look at the phrase *"X is very efficient"*, we could say that X is of positive sentiment, but considering its wider context it is not.

In this paper, we will first describe the underlying methodology of our framework in section 2- 4. Section 5 illustrates the process by giving examples on how the knowledge base evolves. In section 6, we present experimental results. Section 7 discusses related work and our conclusions are found in the last section.

2   Methodology

In this section we present the basic idea behind Ripple-Down Rules upon which our approach is based.

Knowledge Acquisition with Ripple Down Rules: Ripple Down Rules (RDR) is an unorthodox approach to knowledge acquisition. RDR does not follow the traditional approach to knowledge based systems (KBS) where a knowledge engineer together with a domain expert perform a thorough domain analysis in order to come up with a knowledge base. Instead a KBS is built with RDR incrementally, while the system is already in use. No knowledge engineer is required as it is the domain expert who repairs the KBS as soon as an unsatisfactory system response is encountered. The expert is merely required to provide an explanation for why in the given case, the classification should be different from the system's classification.

This approach resulted in the expert system PEIRS used for interpreting chemical pathology results [4]. PEIRS appears to have been the most comprehensive medical expert system yet in routine use, but all the rules were added by a pathology expert without programming or knowledge engineering support or skill whilst the system was in routine use. Ripple-Down Rules and some further developments are now successfully exploited commercially by a number of companies.

Single Classification Ripple Down Rules: A single classification ripple down rule (SCRDR) tree is a finite binary tree with two distinct types of edges. These edges are typically called *except* and *if not* edges. See Figure 1. Associated with each node in a tree is a *rule*. A rule has the form: *if α then β* where α is called the *condition* and β the *conclusion*.

Cases in SCRDR are evaluated by passing a case to the root of the tree. At any node in the tree, if the condition of a node N's rule is satisfied by the case, the case is passed on to the exception child of N. Otherwise, the case is passed on the N's if-not child.

The conclusion given by this process is the conclusion from the last node in the RDR tree which fired. To ensure that a conclusion is always given, the root node typically contains a trivial condition which is always satisfied. This node is called the *default* node.

A new node is added to an SCRDR tree when the evaluation process returns a wrong conclusion. The new node is attached to the last node evaluated in the tree. If the node has no exception link, the new node is attached using an exception link, otherwise an *if not* link is used. To determine the rule for the new node, the expert formulates a rule which is satisfied by the case at hand. Importantly, new node is added only when its rule is consistent with the knowledge base i.e. all cases that have been correctly classified by existing rules will not be classified differently by the new rule.

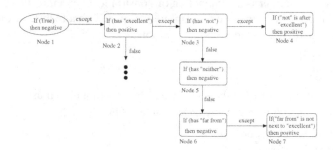

Fig. 1. An example SCRDR tree with simple rule language to classify a text into positive or negative class. Node 1 is the default node. A text that contains *excellent* is classified as *positive* by Node 2 as long as none of its exception rules fires, i.e., the text does neither contain *not, neither* nor *far from* so Node 3,5,6 would not fire. A text that has *not excellent* is classified as *negative* by Node 3 while it is classified as *positive* by Node 4, if it contains *excellent but not*. If it contains *far from excellent* then it is classified as *negative* by Node 6.

3   Our Approach

While the process of incrementally developing knowledge bases will eventually lead to a reasonably accurate knowledge base, provided the domain does not drift and the experts are making the correct judgements, the time it takes to develop a good knowledge base depends heavily on the appropriateness of the used language in which conditions can be expressed by the expert.

Some levels of abstraction in the rule's condition is desirable to make the rule expressive enough in generalizing to unseen cases. To realize this, we use the idea of annotation where phrases that have similar roles are deemed to belong to the same annotation type.

3.1   Rule Description

A rule is composed of a condition part and a conclusion part. A condition has an annotation pattern and an annotation qualifier. An annotation is an abstraction over string tokens. Conceptually, string tokens covered by annotations of the same type are considered to represent the same concept. An annotation contains the character locations of the beginning and ending positions of the annotated text in the document along with the type of annotation and a list of feature value pairs.

The annotation pattern is a simplified regular expression over annotations. It can also post new annotations over matched phrases of the pattern's sub-components. The following is an example of a pattern which posts an annotation over the matched phrase:

({Noun} {VG} {Noun}):MATCH

This pattern would match phrases starting with a Noun annotation followed by a VG followed by another Noun annotation. When applying this pattern on a piece of text, MATCH annotations would be posted over phrases that match this pattern.

The annotation qualifier is a conjunction of constraints over annotations, including newly posted ones. An annotation constraint may require that a feature of that annotation must have a particular value as in this example:

VG.voice==active
Token.string=increase

A constraint can also require that the text covered by an annotation must contain (or not contain) another annotation or a string of text, such as here:

NP.hasAnno == LexGoodAdj
VG.has == outperform
VG.hasnot == not

A rule condition is satisfied by a phrase, if the phrase matches the pattern and satisfies the annotation qualifier. For example we have the following rule condition:

((({NP}):Noun1 {VG.voice==active}
({NP.hasAnno == LexGoodAdj}):Noun2):MATCH

This pattern would match phrases starting with a NP annotation followed by a VG annotation (with feature *voice* having value *active*) followed by another NP annotation (Noun2), which must also contain a LexGoodAdj annotation for the annotation qualifier to be satisfied. When a phrase satisfies the above rule condition, a MATCH annotation would be posted over the whole phrase and Noun1, Noun2 annotations will be posted over the first and second NP in the pattern respectively. Note that Noun1 is not used in the condition part but it could be used later in the conclusion part or in the exception of the current rule.

A piece of text is said to satisfy the rule condition if it has a substring that satisfies the condition. The following sentence matches the above rule condition as *useful* is annotated by the LexGoodAdj annotation, being a purpose built lexicon containing terms indicating a positive sentiment:

[NP Parallelism NP][VG is VG][NP a useful way NP] to speed up computation.
This sentence triggers the posting of the following new annotations:

[MATCH Parallelism is a useful way MATCH]
[Noun1 Parallelism Noun1]
[Noun2 a useful way Noun2]

However, the following sentences do not match:

(1) [NP Parallelism NP] [VG is VG] [NP a method NP] used in our approach.
(2) [NP Parallelism NP] [VG has been shown VG] [VG to be VG] very useful.

Sentence (1) matches the pattern, but it does not satisfy the annotation constraints. Sentence (2) does not match the pattern.

The rule's conclusion contains the classification of the input text. In our task, it is *true* if the text mentions an advantage or a positive aspect of a concept/term and *false* otherwise.

Besides classification, our framework also offers an easy way to do information extraction. Since a rule's pattern can post annotations over components of the matched phrase, extracting those components is just a matter of selecting appropriate annotations. In this work, we extract the concept/term of interest whenever the case is classified as containing a positive aspect by specifying the target annotation. A conclusion of the rule with the condition shown above could be:

Conclusion: true
Concept Annotation: Noun1

The rule's conclusion contains a classification and an annotation to be extracted. In regards to whether a new exception rule needs to be added to the KB, a conclusion is deemed to be incorrect if either part of the conclusion is incorrect.

3.2   Annotations and Features

Built-in Annotations: As our rules use patterns over annotations, the decision on what annotations and their corresponding features should be are important for the expressiveness of rules. We experimentally tested the expressiveness of rules on technical papers and found that the following annotations and features make patterns expressive enough to capture all rules we want to specify for our tasks.

We have Token annotations that cover every token with the *string* feature holding the actual string, the *category* feature holding the part-of-speech and the *lemma* feature holding the token's lemma form.

As a result of the Shallow Parser module, we have several forms of noun phrase annotations ranging from simple to complex noun phrases, e.g., NP (simple noun phrase), NPList (list of NPs) etc. All forms of noun phrase annotations are covered by a general Noun annotation. There is also a VG (verb group) annotation with *type, voice* features, several annotations for clauses, e.g., PP (prepositional phrase), SUB (subject), OBJ (object).

An important annotation that makes rules more general is Pair which annotates phrases that are bound by commas or brackets. With this annotation, the following sentences:

The EM algorithm (Dempster, Laird, & Rubin, 1977) is effective.....
...the algorithm, in noisy domains, outperformed .....

could be covered by the following patterns respectively:

{Noun}({Pair})?{Token.lemma==be}{LexGoodAdj}
{Noun}({Pair})?{Token.lemma==outperform}

Every rule that has a non-empty pattern would post at least one annotation covering the entire matched phrase. Because rules in our knowledge base are stored in an exception structure, we want to be able to identify which annotations are posted by which rule. To facilitate that, we number every rule and enforce that all annotations posted by rule number x should have the prefix RDRx_. Therefore, if a rule is an exception of rule number x, it could use all annotations with the prefix RDRx_ in its condition pattern or annotation qualifier.

Custom Annotations: In our current implementation we manually created a list of about 50 *good* adjectives and adverbs as a seed lexicon. We post LexGoodAdj annotations over words in that lexicon. In fact, users could form new named lexicons during the knowledge acquisition process. The system would then post a corresponding annotation over every word in such a lexicon. Doing this makes the effort of generalizing the rule quite easy and keeps the rules compact.

4   Implementation

We built our framework using GATE [2]. A set of reusable modules known as AN-NIE is provided with GATE. These are able to perform basic language processing tasks such as part-of-speech (POS) tagging and semantic tagging. We use *Tokenizer, Sentence Splitter, Part-of-Speech Tagger and Semantic Tagger* processing resources from ANNIE. *Semantic Tagger* is a JAPE finite state transducer that annotates text based on JAPE grammars. Our rule's annotation pattern is implemented as a JAPE grammar with some additional features.

We also developed additional processing resources for our task:

Lemmatizer: a processing resource that automatically puts a *lemma* feature into every Token annotation containing the lemma form of the token's string [11]. Lemmatizer uses information from WordNet [5] and the result from the POS Tagger module.

Shallow Parser: a processing resource using JAPE finite state transducer. The shallow parser module consists of cascaded JAPE grammars recognizing noun groups, verb groups, propositional phrases, different types of clauses, subjects and objects. These constituents are displayed hierarchically in a tree structure to help experts formulate patterns, see e.g. Figure 2.

All these processing resources are run on the input text in a pipeline fashion. This is a pre-processing step which produces all necessary annotations before the knowledge base is applied on the text.

5   Examples of How to Build a Knowledge Base

The following examples are taken from the actual KB as discussed in section 6. Suppose we start with an empty knowledge base (KB) for recognizing advantages. I.e. the KB would only contain a default rule which always produces a *'false'* conclusion. When the following sentence is encountered:

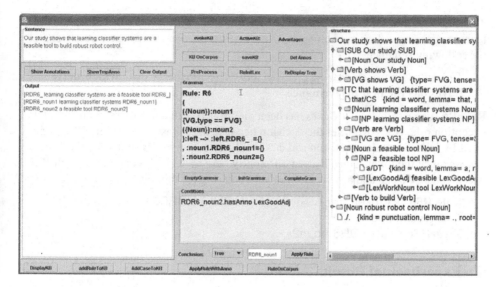

Fig. 2. The interface to enter a new rule where the rule is automatically checked for consistency with the existing KB before it gets committed. Annotations including those created by the shallow parser module are shown in the tree in the *structure* box.

Our study shows that learning classifier systems are a feasible tool to build robust robot control.

our empty KB would initially use the default rule to suggest it does not belong to the *Advantages* class. This can be corrected by adding the following rule to the KB:

Rule:R6
(({Noun}):RDR6_noun1 {VG.type==FVG}
({Noun.hasAnno == LexGoodAdj}):RDR6_noun2):RDR6_
Conclusion: true
Target Concept: RDR6_noun1

This rule would match phrases starting with a Noun annotation, followed by a VG annotation (with feature *type* equal to *FVG*) followed by a Noun annotation. Furthermore, the second Noun annotation must contain a LexGoodAdj annotation covering its substring. As there is a LexGoodAdj annotation covering the token *feasible*, the phrase *learning classifier systems are a feasible tool* is matched by **Rule6** and *learning classifier systems* is extracted as the concept of interest. When we encounter this sentence:

*Given a data set, it is often not clear beforehand which **algorithm will yield the best performance**.*

Rule **R6** suggests that the sentence mentions *algorithm* with a positive sentiment (the matched phrase is highlighted in boldface) which is not correct. The following exception rule is added to fix that:

Rule:R32   ({Token.lemma==which}{RDR6_}):RDR32_
Conclusion: false

This rule says that if the phrase matched by **Rule6** follows a *which* token, then the sentence containing it does not belong to *Advantages* class. However, when we encounter the following sentence

> *The latter approach searches for the subset of attributes over* **which naive Bayes has the best performance***.*

Rule **R6** suggests that *naive Bayes* has been mentioned with a positive sentiment but its exception rule, **R32**, overrules the decision because the phrase that matches **R6** (annotated by RDR6-) follows a token *which*. Obviously, *naive Bayes* should be the correct answer since the token *which* is used differently here than in the context in which **R32** was created. We can add an exception to **R32** catering for this case:

> **Rule:R56** ({Token.string==over} {RDR32-}):RDR56-
> **Conclusion:** true
> **Target Concept:** RDR6-noun2

6   Experimental Results

A corpus was collected consisting of 140 machine learning conference and journal papers downloaded from citeseer, and converted from PDF into text. Even though these papers are from the same domain, we have to stress that the topics they cover are quite diverse including most subtopics in machine learning.

We have applied our framework to build a knowledge base (KB) for recognizing sentences that contain advantages of a concept/term as well as extracting that concept/term. A sentence is considered to mention an advantage of a concept/term if the author expresses a positive sentiment towards that concept/term. Given a sentence, the fired rule from the KB would give the *true* conclusion if the sentence is considered to be of the *Advantages* class and *false* otherwise.

We randomly selected 16 documents from 16 different authors and grouped them into 2 corpora. The first corpus has 3672 sentences from 9 documents called the training corpus. The second corpus contains 4713 sentences from 7 documents called the test corpus.

Using the training corpus, we have built a KB consisting of 61 rules. Applying the knowledge base on the test corpus, it classified 178 sentences as belonging to *Advantages* class. Checking the accuracy, 132 cases are correct, resulting in a precision of 74% (132/178). A case is deemed correct, if in fact it mentions a concept/term with an advantage and that concept/term must be at least partly extracted. This flexibility allows for the imperfection of our shallow parser. The KB missed 18 cases resulting in a recall rate of 88% and an F-measure of 80.4%. Examples of sentences with the extracted concept in bold face are:

> *(3) Again,* **EM** *improves accuracy significantly.*
> *(4) In this low dimensional problem it was more computationally efficient* **to consider a random candidate** *set.*
> *(5)* **The topology of a KBANN-net** *is better suited to learning in a particular domain than a standard ANN.*

There are now 445 sentences in the corpus that contain at least one word from the seed lexicon. Our KB returned 178 cases out of which many do not contain any word from the seed lexicon. For example, sentence (3) above is returned because of the verb *improve*. This is to show that naively selecting sentences containing *good* adjectives or adverbs is not enough.

6.1 Comparison with Machine Learning Approaches

As a baseline with which to compare our approach, we used decision tree C4.5, Naive Bayes and Support Vector Machine algorithms using the bag-of-word representation on the task of classifying a sentence into the *true* or *false* class depending on whether the sentence mentions an advantage of a concept. Notice that this task is simpler than our task of extracting positive attributions as we do not ask the learner to extract the concept.

We prepared four different datasets which are variants of the training corpus (3672 sentences) described in the previous section:

- DataSet 1: It contains all full sentences from the training corpus.
- DataSet 2: It contains all sentences from the training corpus but sentences belonging to the *Advantages* class (i.e. containing an advantage of a concept/term) are duplicated 15 times. This is to give a balanced number of positive and negative examples in the training data.
- DataSet 3: Only the relevant phrase within a sentence is used in the bag-of-word representation for those sentences that are classified into the *Advantages* class. I.e. in building the KB, when an expert identifies a sentence as belonging to the *Advantages* class, s/he has to select a relevant phrase of the sentence and formulates a rule that would match the phrase. The phrase can be thought of as the most relevant part in the sentence which indicates that the sentence mentions an advantage. All other sentences are used as in the previous two datasets. This is done to give the learner more expert knowledge.
- DataSet 4: It contains all cases in DataSet 3 except that the *true* class cases are duplicated 15 times as done in Dataset 2.

Because there are 216 positive examples and 3456 negative examples in the original training corpus, Dataset 2 and 4 duplicate positive examples to give a balanced training data set.

We use J48, NaiveBayes and SMO in Weka [14] as implementations for C4.5, Naive Bayes and Support Vector Machine algorithms respectively. The 3 learners are trained on 4 datasets and are tested against the same test corpus described in the previous section. Their results are shown in table 1.

When the learners were trained on the data containing only the class label of the entire sentences, as in dataset 1 and 2, they performed very poorly giving the best F-measure of 27.6%. In dataset 3 and 4, more expert knowledge is provided in the form of relevant phrases of the sentences boosting the best F-measure to 40.5%, which is still significantly lower compared to the performance of our manually created KB of 80.4% based on the same data.

Table 1. **P**recision, **R**ecall and **F**-measure of the three algorithms trained on different training datasets but tested on the same test corpus.

	C4.5(J48)			Naive Bayes			SVM(SMO)		
	P	R	F	P	R	F	P	R	F
DataSet 1 *full sentence*	37	6.7	**11.3**	12.1	38	**18.3**	20	10	**13.3**
DataSet 2 *full sentence, balanced*	23.9	32.7	**27.6**	7.6	74	**13.8**	22.6	16	**18.8**
DataSet 3 *relevant phrase if available*	0	0	**0**	30.6	7.3	**11.8**	43.8	35.3	**39.1**
DataSet 4 *relevant phrase if available, balanced*	31.8	42.7	**36.5**	6.4	82.7	**11.8**	33.8	50.7	**40.5**

It is possible that the bag-of-word approach and the number of training examples given to those learners are not large enough to achieve a high precision and recall. It however supports our approach of utilizing more expert knowledge instead of requiring much more training data given that it takes only about 2 minutes for the expert to formulate and enter a rule into the KB. If the expert has to classify the sentences anyway, it does not take much more time to also provide some explanations on some of the sentences for why they should be classified differently to how the current knowledge base classified the sentence. In general, we believe that our approach leads to a significantly higher accuracy based on a given set of sentences than could be achieved by machine learning approaches which would only exploit a boolean class label as expert information for each sentence. As a consequence, the total number of sentences an expert needs to read and classify in order to achieve a certain classification accuracy can be expected to be much less with our approach than a machine learning approach.

6.2   Can the KB's Performance Be Improved?

We fixed the errors made by the KB on the test corpus by adding exception rules to the KB. The resulting KB2 had 74 rules. We randomly selected 5 documents (containing 1961 sentences) from 5 different authors which are all different from the authors in the training and test corpus described in previous sections. When applied to this new test corpus, the KB2 achieved a precision of 79%(60/78) and recall of 90%(60/67). This is an indication that as we add more rules to the knowledge base, both precision and recall are improved.

6.3   Extracting Advantages

To verify the quality of our approach, we looked at an application scenario of finding advantages for a particular concept/term e.g. *decision tree*. The literal string *decision tree* appears at least 720 times in the corpus. We only considered those sentences, which were classified as *Advantages* by the KB and had the string *decision tree* in the extracted *target annotation* to be of interest. Clearly, this simple analysis would miss cases where the actual string is not in the sentence but is mentioned via an anaphora. Future work will address this point.

Our analysis results in 22 cases that have *decision tree* in the target annotation. These 22 cases were from 7 documents of 5 different authors. Some examples are:

> **Decision trees** *are a particularly useful tool in this context because they perform classification by a sequence of simpler, easy-to-understand tests whose semantics is intuitively clear to domain experts.*
>
> *Results clearly indicate that* **decision trees** *can be used to improve the performance of CBL systems and do so without reliance on potentially expensive expert knowledge.*

Out of the suggested sentences, 12 are correct giving a precision of 55%.

6.4   Analysis of KB's Errors

Inspecting misclassified sentences of the above experiment reveals that 6 of them are from the same document (accounting for 60% of the error) and are of the same type:

> *...what is the probability that [*RDR1_ *[*RDR1_noun2 *the smaller decision tree* RDR1_noun2*] is more accurate* RDR1_ *].*

which does not say that the decision tree is more accurate if we consider the sentential context. This misclassification is due to the fact that we have not seen this type of pattern during training which could easily be overcome by adding a single exception rule.

A number of misclassifications is due to errors in modules we used, such as the POS tagger or the Shallow Parser, that generate annotations and features used in the rule's condition. For example, in

> *Comparing classificational accuracy alone, assistant performed better than cn 2 in these particular domains.*

performed is tagged as VBN (past participle), rather than VBD (past tense), causing our Shallow Parser not to recognize *performed* as a proper VG. Consequently, our rule base did not classify this case correctly. We also noticed that several errors came from noise in the text we get from pdf2text program.

Overall, errors appear to come from the fact that rules tend to either be overly general or too specific and, thus, do not cover exactly what they should cover. While the particular choice of generality or specificity of an exception rule affects to some degree the convergence speed of the knowledge base towards complete accuracy, it is not that crucial in our approach. This is because suboptimal early choices are naturally patched up by further rules as the knowledge acquisition process progresses. Where a too general rule was entered a subsequent exception rule will be entered, while a subsequent if-not rule patches a too specific rule.

6.5   Structure of Exception Rules

Apart from the default rule, every rule is an exception rule which is created to correct an error of another rule. An SCRDR KB could be viewed as consisting of layers of exception rules where every rule in layer n is an exception of a rule in layer $n-1$ [1].

---

[1] Rules in Figure 1 are presented in layers structure.

The default rule is the only rule in layer 0. A conventional method that stores rules in a flat list is equivalent to layer 1 in our SCRDR KB. Our latest KB2 (from section 6.2) had 25, 43 and 5 rules in layers 1, 2 and 3 respectively. Having more than one level of exceptions indicates the necessity of the exception structure in storing rules. An example of 3 levels of exception rules is shown in section 5.

7   Related Work

Most of the related work on sentiment analysis has focused on news and review genres [9, 13, 7, 10]. Our work instead looks at technical papers as it appears more challenging. However, our framework is domain independent.

A majority of existing approaches only analyze co-occurrences of simple phrases (unigrams or bigrams) within a short distance [7, 10] or indicative adjectives and adverbs [13]. In contrast to that, we look at more complex patterns in relation to the subject matter to determine its polarity. One reason for this difference might be that those applications tend to classify the polarity of the whole article assuming that all sentiment expressions in the article contribute towards the article's subject classification. In this work, we identify sentiment expressions to extract the subject of the expression as well as its sentiment classification. [9] has a similar goal as ours but they do not consider the surrounding context of the sentiment expression which could affect the subject's polarity. The sentential context containing the sentiment expression is in fact analyzed in our work for sentiment classification.

Our approach takes a knowledge acquisition approach where experts create rules manually. This differs from machine learning approaches that automatically learn rule patterns [12, 6], or learn implicit rules (not expressed as patterns) to classify sentiment expressions [10, 13]. We strongly believe that our approach is superior to automatic approaches since experts need to spend their time to classify the sentences even where machine learning approaches are taken. This has been proved true on the task of classifying sentences into *Advantages* category described in Section 6.1 where our manually build KB outperformed several machine learning algorithms trained on the same data. Similar findings for medical knowledge bases have been obtained in systematic studies for Ripple Down Rules [1].

Many researchers [9, 7] share the same view by manually creating patterns and lexicons in their approaches. We take one step further by helping experts to incrementally acquire rules and lexicons as well as controlling their interactions in a systematic manner.

Our work could also be considered as a kind of information extraction. Allowing the pattern to be a regular expression over annotations, the expressiveness of our rule language is at least as good as existing IE systems (e.g. [6, 12], see [8] for a survey). Our framework combines the following valuable features of those IE systems:

- it allows syntactic and semantic constraints on all components of the patterns including the extracted fields.
- it can accommodate both single-slot and multi-slot rules.
- it can be applied to a wide variety of documents ranging from rigidly formatted text to free text.

8   Conclusion and Future Work

In this paper, we have presented a domain independent framework to easily build a knowledge base for text classification as well as information extraction. We applied the framework to the task of identifying the mention of advantages of a concept/term in a scientific paper. The objective being to support scientists in digesting the ever increasing literature available on a subject.

Initial experiments on this task have shown that the knowledge base built using our framework achieved precision of at least 74% and recall rates of more than 88% on unseen corpora.

We compared the performance of our knowledge base on a simpler task, i.e. sentence classification only (without extracting any information), with that of classifiers constructed using three different machine learning techniques, often used for text mining. Our knowledge base proved to be far superior to those machine learning approaches, based on the same text. Furthermore, it should be noted that for both approaches, our knowledge acquisition approach as well as supervised machine learning approaches, a human is required who classifies the individual sentences. The additional time our human expert required to enter rules into the knowledge base was only a fraction of the time it took to classify the sentences in the first place. The reason being that many more sentences need to be classified than the number of rules that was entered. This suggests that our approach uses the expert's time much more economically than supervised machine learning approaches which only utilize the class label as the indication of what to do with a sentence.

It should be noted that all documents in those corpora were from different authors covering different topics. Furthermore, building the KB of the above quality was reasonably quick. This suggests that by using our framework, it is feasible to quickly, i.e. within a few days, build new knowledge bases for different tasks in new domains.

Although it appears easy for experts to create rules, it would be desirable if the experts are presented with possible candidate rules to choose from. Even when the suggested rules are not correct to be used as-is, using them as a starting point to create the final rules should be helpful. We are working towards using machine learning techniques to automatically propose candidate rules to be used in a *mixed initiative* style [3].

References

1. P. Compton, P. Preston, and B. Kang. The use of simulated experts in evaluating knowledge acquisition. In *Proceedings of the Banff KA workshop on Knowledge Acquisition for Knowledge-Based Systems*. 1995.
2. H. Cunningham, D. Maynard, K. Bontcheva, and V. Tablan. Gate: An architecture for development of robust hlt applications. In *Proceedings of the 40th Annual Meeting of the Association for Computational Linguistics(ACL)*, Philadelphia, PA, 2002.
3. D. Day, J. Aberdeen, L. Hirschman, R. Kozierok, P. Robinson, and M. Vilain. Mixed-initiative development of language processing systems. In *Fifth ACL Conference on Applied Natural Language Processing*, Washington, DC, 1997.

4. G. Edwards, P. Compton, R. Malor, A. Srinivasan, and L. Lazarus. PEIRS: a pathologist maintained expert system for the interpretation of chemical pathology reports. *Pathology*, 25:27–34, 1993.
5. C. Fellbaum, editor. *WordNet - An electronic lexical database*. MIT PRESS, Cambridge, MA, 1998.
6. J. Kim and D. Moldovan. Acquisition of linguistic patterns for knowledge-based information extraction. *IEEE Transactions on Knowledge and Data Engineering*, 7(5):713–724, 1995.
7. S. Morinaga, K. Yamanishi, K. Tateishi, and T. Fukushima. Mining product reputations on the web. In *Proceedings of the Eighth ACM International Conference on Knowledge Discovery and Data Mining(KDD)*, pages 341–349, 2002.
8. I. Muslea. Extraction patterns for information extraction tasks: A survey. In *The AAAI Workshop on Machine Learning for Information Extraction*, 1999.
9. T. Nasukawa and J. Yi. Sentiment analysis: Capturing favorability using natural language processing. In *Proceedings of the 2nd International Conference on Knowledge Capture(K-Cap)*, Florida, 2003.
10. B. Pang and L. Lee. Thumbs up? sentiment classification using machine learning techniques. In *Proceedings of the Conference on Empirical Methods in Natural Language Processing(EMNLP)*, pages 79–86, 2002.
11. M. Porter. An algorithm for suffix stripping. *Program*, 14(3):130–137, 1980.
12. S. Soderland. Learning information extraction rules for semi-structured and free text. *Machine Learning*, 34(1-3):233–272, 1999.
13. P. Turney. Thumbs up or thumbs down? semantic orientation applied to unsupervised classification of reviews. In *Proceedings of the 40th Annual Meeting of the Association for Computational Linguistics(ACL)*, pages 417–424, 2002.
14. I. H. Witten and E. Frank. *Data Mining: Practical machine learning tools with Java implementations*. Morgan Kaufmann, 2000.

Enhancing SVM with Visualization

Thanh-Nghi Do and François Poulet

ESIEA Recherche
38, rue des Docteurs Calmette et Guérin
Parc Universitaire de Laval-Changé
53000 Laval - France
{dothanh,poulet}@esiea-ouest.fr

Abstract. Understanding the result produced by a data-mining algorithm is as important as the accuracy. Unfortunately, support vector machine (SVM) algorithms provide only the support vectors used as "black box" to efficiently classify the data with a good accuracy. This paper presents a cooperative approach using SVM algorithms and visualization methods to gain insight into a model construction task with SVM algorithms. We show how the user can interactively use cooperative tools to support the construction of SVM models and interpret them. A pre-processing step is also used for dealing with large datasets. The experimental results on Delve, Statlog, UCI and bio-medical datasets show that our cooperative tool is comparable to the automatic LibSVM algorithm, but the user has a better understanding of the obtained model.

1 Introduction

The SVM algorithms proposed by Vapnik [22] are a well-known class of data mining algorithms using the idea of kernel substitution. SVM and kernel related methods have shown to build accurate models but the support vectors found by the algorithms provide limited information. Most of the time, the user only obtains information regarding the support vectors and the accuracy. It is impossible to explain or even understand why a model constructed by SVM performs a better prediction than many other algorithms. Understanding the model obtained by the algorithm is as important as the accuracy. A good comprehension of the knowledge discovered can help the user to reduce the risk of wrong decisions. Very few papers have been published about methods trying to explain SVM results ([3], [20]). Our investigation aims at using visualization methods to try to involve more intensively the user in the construction of the SVM model and to try to explain their results. A new cooperative method based on a set of different visualization techniques and large scale Mangasarian SVM algorithms [10], [16] gives an insight into the classification task with SVM. We will illustrate how to combine some strength of different visualization methods with automatic SVM algorithms to help the user and improve the comprehensibility of SVM models. The experimental performance of this approach is evaluated on Delve [8], Statlog [18], UCI [2] and bio-medical [13] data sets. The results show that our cooperative method is comparable with LibSVM (a high performance automatic SVM algorithm [4]). We also use a pre-processing step to deal with very large data-

E. Suzuki and S. Arikawa (Eds.): DS 2004, LNAI 3245, pp. 183–194, 2004.
© Springer-Verlag Berlin Heidelberg 2004

sets. The feature selection with 1-norm SVM [11] can select a subset from an entire large number of dimensions (thousands of dimensions). With very large number of data points, we sample the data points from the clusters created by SOM [15] or K-means [17] algorithms to reduce the dataset size. And thus, our tool works only on this reduced dataset. A case study on the UCI Forest cover type dataset shows that this approach provides an accurate model.

In section 2, we briefly introduce supervised classification with SVM algorithms. In section 3, we present the cooperative algorithm using visualization methods and SVM algorithms for classification tasks. In section 4, we propose to use a multiple view approach based on different visualization methods to interpret SVM results. We present experimental results in section 5 before the conclusion and future work.

2 Support Vector Machines

Let us consider a linear binary classification task, as depicted in figure 1, with m data points in the n-dimensional input space R^n, denoted by the x_i $(i=1, ..., m)$, having corresponding labels $y_i = \pm 1$.

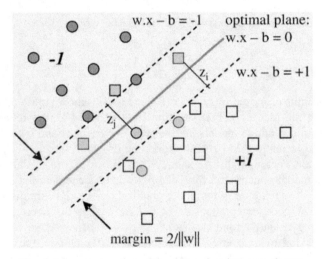

Fig. 1. Linear separation of the data points into two classes.

For this problem, the SVM try to find the best separating plane, i.e. furthest from both class $+1$ and class -1. It can simply maximize the distance or margin between the support planes for each class ($x.w - b = +1$ for class $+1$, $x.w - b = -1$ for class -1). The margin between these supporting planes is $2/||w||$ (where $||w||$ is the 2-norm of the vector w). Any point falling on the wrong side of its supporting plane is considered to be an error. Therefore, the SVM has to simultaneously maximize the margin and minimize the error. The standard SVM formulation with linear kernel is given by the following quadratic program (1):

$$Min \quad \Psi(w, b, z) = (1/2) \, ||w||^2 + C \sum_{i=1}^{m} z_i$$

$$\text{s.t.} \quad y_i(w.x_i - b) + z_i \geq 1 \tag{1}$$
$$z_i \geq 0 \quad (i=1, \ldots, m)$$

where slack variable $z_i \geq 0$, constant $C > 0$ is used to tune errors and margin size.

The plane (w,b) is obtained by the solution of the quadratic program (1). And then, the classification function of a new data point x based on the plane is: $f(x) = sign \, (w.x - b)$

SVM can use some other classification functions, for example a polynomial function of degree d, a RBF (Radial Basis Function) or a sigmoid function. To change from a linear to non-linear classifier, one must only substitute a kernel evaluation in (1) instead of the original dot product. More details about SVM and others kernel-based learning methods can be found in [1].

Recent developments for massive linear SVM algorithms proposed by Mangasar-ian [10], [16] reformulate the classification as an unconstraint optimization and these algorithms require thus only solution of linear equations of (w, b) instead of quadratic programming. If the dimensional input space is small enough (less than 10^4), even if there are millions data points, the new SVM algorithms are able to classify them in minutes on a Pentium. The algorithms can deal with non-linear classification tasks; however the m^2 kernel matrix size requires very large memory size and execution time. Reduced support vector machine (RSVM) [16] creates rectangular mxs kernel matrix of size $(s<<m)$ by sampling, the small random data points S being a represen-tative sample of the entire dataset (and uses it as a set of support vectors). RSVM reduces the problem size. Our investigation aims at using visualization methods to try to involve the user in the construction of the SVM model. The user gains insights of datasets by visualizing them. This can provide him with some ideas to choose repre-sentative dataset sample S, kernel type and kernel parameters in input of RSVM.

3   Cooperation Between Visualization and SVM for Classification

Data-mining is intended to extract hidden useful knowledge from large datasets in a given application. This usefulness relates to the user goal, in other words only the user can determine whether the resulting knowledge answers his goal. Therefore, data mining tool should be highly interactive and participatory. The idea here is to increase the human participation through interactive visualization techniques in a data-mining environment. The effective cooperation can bring out some progress towards reaching advantages [9], [14], and [19] such as:

- the user can be an expert of the data domain and can use this domain knowledge during the whole model construction,
- the confidence and comprehensibility of the obtained model are improved because the user was involved at least partially in its construction,
- we can use the human pattern recognition capabilities.

3.1   Visualization Methods

Over the last decade, a large number of visualization methods developed in different domains have been used in data exploration and knowledge discovery process. The visualization methods are used for data selection (pre-processing step) and viewing mining results (post-processing step). Some recent visual data mining methods try to involve more intensively the user in the data-mining step through visualization. We only present two geometrical techniques: the 2D scatter-plot matrices [5] and the parallel coordinates [12] significantly used for data exploration.

3.1.1   2D Scatter-Plot Matrices
The data points are displayed in all possible pair wise combinations of dimensions in 2D scatter plot matrices. For *n-dimensional* data, this method visualizes *n(n-1)/2* matrices. Figure 2 (left) displays the Iris dataset from the UCI repository which contains *150* data points in *4*-dimensional input space with *3* classes (corresponding to *3* colors).

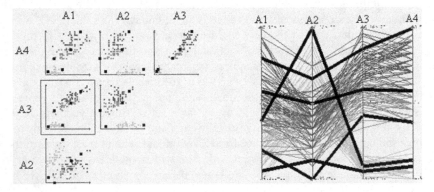

Fig. 2. 2D scatter-plot matrices and parallel coordinates with the Iris dataset.

3.1.2 Parallel Coordinates
The parallel axes represent the data dimensions. A data point corresponds to a poly-line intersecting the vertical axes at the position corresponding to the data value. Figure 2 (right) displays the Iris dataset with parallel coordinates.

3.1.3 Multiple Linked Views
No single visualization tool is the best for high dimensional data exploration: some visualization methods are the best for showing clusters or outliers, some other visualization methods can handle very large dataset. In all cases, we would like to combine different visualization techniques to overcome the single one. The same information is displayed in different views with different visualization techniques providing useful information to the user. Furthermore, interactive linking and brushing can be also applied to multiple views: the user can select points in one view and these points are automatically selected (highlighted) in the other available views (figure 2). Thus, the linked multiple views provide more information than the single one.

3.2   Cooperative Method

The cooperative method tries to involve the user in the construction of SVM model with multiple linked views. The starting point of the cooperation here is the multiple views used to visualize the same dataset. The user can choose appropriate visualization methods to gain insight of data. The interactive graphical methods provide some utilities for example brushing, zoom, rotation, linking, etc. that can help the user to select data points (S) being near the separating boundary. These points are used as support vectors by the RSVM algorithm.

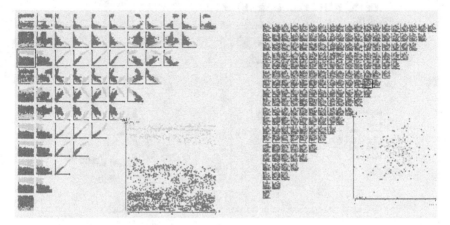

Fig. 3. Visualization of linearly and non-linearly separable datasets.

Furthermore, the selection of user-defined kernel is not an easy task with an automatic SVM algorithm. The dataset visualization also provides an overview of the separating boundary complexity; this brings some ideas for choosing a kernel type. For example, if the visualization of data shows that the dataset is linearly separable (like in the left part of figure 3) then the user can use a linear kernel, or if the frontier between two classes (figure 3 right) is complex the user can use a non-linear kernel type (RBF or polynomial).

And thus, RSVM classifies the dataset and gives the resulting accuracy. If the user is not satisfied with this result he can repeat the previous steps until he gets the expected model.

Figure 4 is a data view of the UCI Segment dataset with a set of 2D scatter-plot matrices. This dataset consists of *2310* data points in a *19*-dimensional space with *7* classes. Once the visualization method has displayed the data on the screen, the interactive selection of support vectors starts. We suppose the user tries to separate the class six (considered as *+1* class with light grey points) from the other ones (considered as *-1* class with dark grey points). With some graphical utilities and the human pattern recognition capabilities, the user can very easily select data points near the boundary (black colored). In this example, the visualization shows the boundary between class six and other ones is non-linear and not complex, so this gives the user a good idea of choosing an appropriate kernel type and tuning parameters too. By using RBF kernel (i.e. $K(x,y) = exp(-\gamma||x-y||^2)$) and setting γ parameter to small value (based

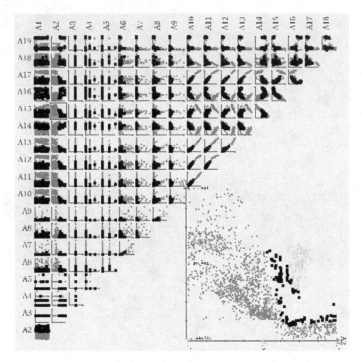

Fig. 4. Interactive selection of support vectors (in black).

on data separating boundary complexity provided by the visualization), for example γ = 0.0001 (we choose a small γ value if the frontier is nearly linear and increase the γ value if we want to obtain a boundary very close to +1 class). RSVM classifies the dataset (*6-against-all*) with 99.96 % accuracy. The user is able to repeat this process for improving results or he goes on to separate another class against all. It is possible to simultaneously work on multiple graphical representations and thus the user can see useful information given by appropriate visualization techniques.

3.3 Pre-processing for the Cooperative Method with Large Datasets

Recent advances in new technologies such as satellite systems or telephone networks are generating massive datasets with large number of data points and even large number of dimensions. These datasets are challenging the ability of data mining tools to analyze and visualize them.

Most of the visualization techniques work well with small datasets (with about thousand data points in less than fifty dimensions). For large datasets, a limitation is the resolution of visualization systems.

To deal with datasets having large number of data points, we have proposed a method to reduce dataset sizes. First of all, we use SOM or k-means algorithms to create the clusters and then we sample the data points from the clusters. Our cooperative method is used to display multiple linked views of the reduced data. Then the user can interactively work with the cooperative method. We illustrate our approach on the UCI Forest cover type dataset with *581012* data points in *54*-dimensions and *7* classes. This is a well known difficult classification problem for SVM algorithms. Collobert and his colleagues [7] trained the models (class two against all) with SVMTorch [6] and a RBF kernel using *100000* training data points and *50000* testing data points on a PC. The learning task needed more than *2* days and *5* hours with an accuracy being *83.24 %*. We have also tried to classify the class two against all on a PC; we have used *500000* data points for training and *50000* data points to test. The LibSVM has not finished the learning task after several days. To adapt our cooperative method to this dataset, we needed about *1* hour as a pre-processing step for creating *200* clusters (*100* for each class) and sampling *10000* data points (sampling *2 %* data points of each cluster). Then, we have interactively selected support vectors from the reduced dataset views as shown in figure 5. A rectangular RBF kernel was created with a γ parameter being *2* in input of RSVM. The learning task needed about *8* hours for constructing the model (i.e. about *150* times faster). We have obtained *84.322 %* accuracy on the testing set. This is a first promising result of our method with large datasets.

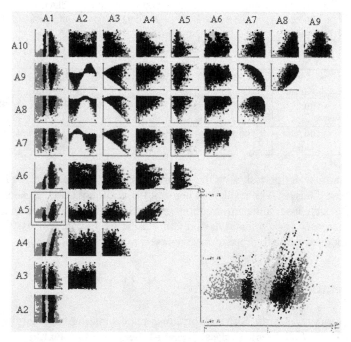

Fig. 5. Interactive selection of support vectors from reduced data view.

Most of the bio-data analysis problems process datasets with a very large number of dimensions and few training data. To deal with these datasets, we use feature selection with Mangasarian 1-norm linear SVM algorithm [11] as data preprocessing. The 1-norm linear SVM algorithm maximizes the margin by minimizing 1-norm (instead of 2-norm with standard SVM) of plane coefficients (w). This algorithm provides results having many null coefficients. This can efficiently select few dimensions corresponding to non-null coefficients without losing too much information. We have evaluated the performances on the bio-medical datasets described in table 1.

Table 1. Bio-medical dataset description.

	Classes	Points	Attributes	Evaluation method
AML-ALL Leukemia	2	72	7129	38 trn – 34 tst
Breast Cancer	2	97	24481	78 trn – 19 tst
Colon Tumor	2	62	2000	Leave-1-out
Lung Cancer	2	181	12533	32 trn – 149 tst
Ovarian Cancer	2	253	15154	Leave-1-out

Table 2. Results on bio-medical datasets.

	+1 accuracy		-1 accuracy		Accuracy	
	Feature Selection	No selection	Feature Selection	No selection	Feature selection	No Selection
AML-ALL Leukemia	**100 %** 5-dim	95 %	85.71 % 5-dim	**92.86 %**	**94.12 %** 5-dim	**94.12 %**
Breast Cancer	**91.67 %** 10-dim	83.33 %	**57.14 %** 10-dim	**57.14 %**	**78.95 %** 10-dim	73.68 %
Colon Tumor	**95.45 %** 19-dim	86.36 %	**97.5 %** 19-dim	92.5 %	**96.77 %** 19-dim	90.32 %
Lung Cancer	**100 %** 9-dim	**100 %**	96.27 % 9-dim	**98.51 %**	96.64 % 9-dim	**98.66 %**
Ovarian Cancer	**100 %** 13-dim	**100 %**	**100 %** 10-dim	**100 %**	**100 %** 13-dim	**100 %**

After a feature selection task with the 1-norm linear SVM, we have used LibSVM to classify these datasets. The results concerning the accuracy (table 2) show that the pre-processing step lose some information in one case (Lung Cancer dataset); the accuracy is increased for two datasets and is the same for the other ones. Thus, visualization methods are able to process these datasets to interpret the results.

4   Interpretation of SVM Results

For classification tasks with SVM algorithms, understanding the margin (furthest distance between +1 class and -1 class) is one of the most important keys of the support vector classification. It is necessary to see the points near the separating plane between the two classes. These points naturally represent the boundary between the two classes.

For achieving this goal, we propose to use the data distribution according to the distance to the separating surface. While the classification task is processed (based on the support vectors), we also compute the distribution of the data points according to the distance to the separating surface. For each class, the positive distribution is the set of correctly classified data points, and the negative distribution is the set of mis-classified data points. When the bar charts corresponding to the points near the frontier are selected, the data points are also selected in the other views (visualization methods) by using the brushing and linking technique. The user can see approximately the boundary between classes, and the margin width. This helps the user to evaluate the robustness of the model obtained by support vector classification algorithms. He can also know the interesting dimensions (corresponding to the projections providing a clear boundary between the two classes) in the obtained model.

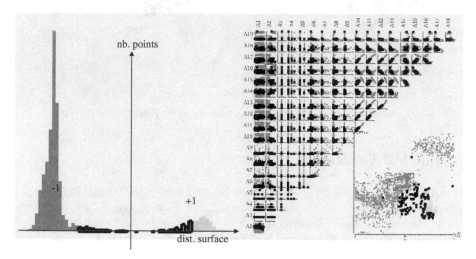

Fig. 6. Visualization of SVM classification results, Segment dataset (6-versus-all).

We have used SVM classification result on the Segment dataset (class six against all) obtained by the cooperative method described in section 3.2. In figure 6, we start with the data distribution tool. By brushing data points being near the separating surface, we visualize them in a 2D scatter-plot matrix. And then, this gives us an approximate idea of the separating boundary. The dimensions corresponding to the projections provide a clear boundary between the two classes and are interesting in the model obtained.

This idea is extended for interpreting SVM classification results on bio-medical datasets having very large number of dimensions (thousands dimensions). It is necessary to select the "interesting" dimensions by using 1-norm SVM for binary classification tasks.

For example, the Lung Cancer dataset has *181* data points in *12533*-dimensional input space with *2* classes. 1-norm SVM needs only *9* dimensions for classification. And then, it is easy to visualize the classification result on this dataset as shown in figure 7 with linking data distribution and a 3D visualization method. It shows that three dimensions are interesting in the model obtained because they clearly show the boundary between the two classes. The model comprehensibility can be improved.

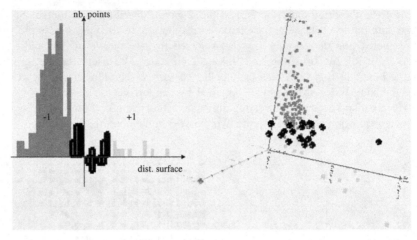

Fig. 7. Visualization of SVM classification results on the Lung Cancer dataset.

5 Numerical Test Results

The software program is written in C/C++ on SGI-O2 station (IRIX) and PC (Linux). It consists of about ten visualization methods and five SVM algorithms for very large datasets. To validate the performances of our cooperative method (Viz-RSVM), we have classified the Delve, Statlog and UCI datasets described in table 3.

Table 3. Dataset description.

	Classes	Points	Attributes	Evaluation method
Bupa	2	345	6	10-fold
Pima	2	768	8	10-fold
Twonorm	2	7400	20	300 trn – 7100 tst
Ringnorm	2	7400	20	300 trn – 7100 tst
Segment	7	2310	19	10-fold
Satimage	6	6435	36	4435 trn – 2000 tst
Forest	7	581012	54	500000 trn – 50000 tst

Thus, we have obtained the results concerning accuracy shown in table 4 on a personal computer Pentium-4 (2.4 GHz, 512 MB RAM, Linux Redhat 7.2). The one-against-all approach has been used in our cooperative method to classify multi-class datasets (more than 2 classes). LibSVM uses the one-against-one approach for classifying multi-class datasets and there is no overfitting as with the one-against-all approach, but its training time is more important. We have used RBF kernel for SVM algorithms. The best result is in bold face for each dataset. With the UCI Forest cover type dataset, we have only classified the class two against all.

As we can see in table 4, our cooperative method Viz-RSVM has a good classification accuracy compared to automatic LibSVM algorithm, the accuracy is not significantly different. The parameter tuning and kernel choice with an automatic SVM algorithm is a well-known high cost task based on cross validation tests. With our cooperative method, the visualization of dataset also brings some ideas for choosing a kernel type and its parameters. The major advantage of the cooperative approach is to increase the human role in the construction of SVM model through appropriate visualization methods. And thus, the user gains insight into data and model. This can help him to avoid the risk of wrong decisions because he gets more comprehensibility and confidence in the model constructed at least partially by himself.

Table 4. Results on Statlog and UCI datasets.

	Viz-RSVM	LibSVM
Bupa	**76.18 %**	73.62 %
Pima	**78.86 %**	77.34 %
Twonorm	97.28 %	**97.35 %**
Ringnorm	97.15 %	**97.28 %**
Segment	96.02 %	**97.10 %**
Satimage	91.70 %	**92.05%**
Forest	**84.32 %**	N/A

6 Conclusion and Future Work

We have presented a new cooperative approach of SVM algorithms and visualization method increasing the human role in model construction tasks of SVM algorithms by using appropriate visualization methods and human pattern recognition capabilities. We have shown how the user can interactively use cooperative method to choose support vectors and provide some ideas for the user-defined kernel in input of automatic SVM algorithms. A pre-processing step is also used for dealing with large datasets. The performances on Delve, Statlog and UCI datasets show that our approach gives promising results compared with the automatic LibSVM algorithm.

We have also proposed a way to interpret SVM results. The distribution of the data points according to the distance to the separating surface linked with other visualization methods in multiple views help us to naturally see the separating boundary between two classes. The graphical representation shows the interesting dimensions in the obtained model.

A first forthcoming improvement will be to combine our method with other graphical techniques to construct another kind of cooperation between SVM and visualization tools for a better use of the user domain knowledge in the mining task.

Another improvement will be to extend our approach combining visualization methods and automatic algorithms to some other data mining techniques. We had already developed an interactive decision tree construction algorithm able to deal with very large data sets [21], here we have worked with SVM but it is always supervised classification tasks. We are working on the same kind of approach for unsupervised classification and outlier detection in high dimensional data sets.

References

1. Bennett, K., and Campbell, C.: Support Vector Machines: Hype or Hallelujah ?. in *SIGKDD Explorations*, 2(2), 2000, pp. 1-13.
2. Blake, C. and Merz, C.: UCI Machine Learning Repository. 1998. http://www.ics.uci.edu/~mlearn/MLRepository.html
3. Caragea, D., Cook, D., and Honavar, V.: Towards Simple, Easy-to-Understand, yet Accurate Classifiers. in proc. of VDM@ICDM'03, the 3rd Int. Workshop on Visual Data Mining, Florida, USA, 2003, pp. 19-31.   .
4. Chang C-C., and Lin, C-J.: LIBSVM – A Library for Support Vector Machines. 2003. http://www.csie.ntu.edu.tw/~cjlin/libsvm/
5. Cleveland, W-S.: *Visualizing Data*. Hobart Press, Summit NJ, 1993.
6. Collobert, R., and Bengio, S.: SVMTorch: Support Vector Machines for Large-Scale Regression Problems. *Journal of Machine Learning Research*, Vol. 1, pp. 143-160, 2001. ftp://ftp.idiap.ch/pub/learning/SVMTorch.tgz
7. Collobert, R., Bengio, S., and Bengio, Y.: A parallel mixture of SVMs for very large scale problems. in *Advances in Neural Information Processing Systems*, NIPS'02, MIT Press, 2002, Vol. 14, pp. 633-640.
8. Delve – Data for Evaluating Learning in Valid Experiments. 1996. http://www.cs.toronto.edu/~delve
9. Fayyad, U., Grinstein, G., and Wierse, A.: *Information Visualization in Data Mining and Knowledge Discovery*. Morgan Kaufmann Publishers, 2001.
10. Fung, G., and Mangasarian, O.: Proximal Support Vector Machine Classifiers. in proc. of ACM SIGKDD'01, the 7th Int. Conf. on KDD'01, San Francisco, USA, 2001, pp. 77-86.
11. Fung, G., and Mangasarian, O.: A Feature Selection Newton Method for Support Vector Machine Classification. Data Mining Institute Technical Report 02-03, Computer Sciences Department, University of Wisconsin, Madison, USA, 2002.
12. Inselberg, A.: The Plane with Parallel Coordinates. in *Special Issue on the Computational Geometry of The Visual Computer*, 1(2), 1985, pp. 69-97.
13. Jinyan, L., and Huiqing, L.: Kent Ridge Bio-medical Data Set Repository. 2002. http://sdmc.lit.org.sg/GEDatasets/Datasets.html
14. Keim, D.: Information Visualization and Visual Data Mining. in *IEEE Transactions on Visualization and Computer Graphics*, 8(1), 2002, pp. 1-8.
15. Kohonen, T.: *Self-Organizing Maps*. Springer, Berlin, Heidelberg, New York, 1995.
16. Lee, Y-J., and Mangasarian, O.: RSVM: Reduced Support Vector Machines. Data Mining Institute Technical Report 00-07, Computer Sciences Department, University of Wisconsin, Madison, USA, 2000.
17. MacQueen, J.: Some Methods for classification and Analysis of Multivariate Observations. in proc. of 5th *Berkeley Symposium on Mathematical Statistics and Probability*, Berkeley, University of California Press, 1967, Vol. 1, pp. 281-297.
18. Michie, D., Spiegelhalter, D., and Taylor, C.: *Machine Learning, Neural and Statistical Classification*. 1999. http://www.amsta.leeds.ac.uk/~charles/statlog/
19. Poulet, F.: Full-View: A Visual Data Mining Environment. in *Int. Journal of Image and Graphics*, 2(1), 2002, pp. 127-143.
20. Poulet, F.: Cooperation between Automatic Algorithms, Interactive Algorithms and Visualization for Visual Data Mining. in proc. of VDM@ECML/PKDD'02, 2nd International Workshop on Visual Data Mining, Helsinki, Finland, Aug.2002, 67-79.
21. Poulet, F.: Towards Visual Data Mining. in proc. of ICEIS'04, the 6th Int. Conf. on Enterprise Information Systems, Porto, 2004, Vol. 2, pp. 349-356.
22. Vapnik, V.: *The Nature of Statistical Learning Theory*. Springer-Verlag, New York, 1995.

An Associative Information Retrieval Based on the Dependency of Term Co-occurrence

Mina Akaishi[1], Ken Satoh[2], and Yuzuru Tanaka[3]

[1] RCAST, University of Tokyo,
4-6-1 Komaba Meguro-ku, Tokyo 153-8904, Japan
mina@ai.rcast.u-tokyo.ac.jp
[2] National Institute of Informatics,
2-1-2 Hitotsubashi, Chiyoda-ku, Tokyo 101-8430, Japan
ksatoh@nii.ac.jp
[3] Meme Media Laboratory, Hokkaido University,
N13 W8, Kita-ku, Sapporo 060-8628, Japan
tanaka@meme.hokudai.ac.jp

Abstract. This paper proposes a method to lead users to appropriate information based on associative information access. People of nowadays can access Web resources easily. However it is difficult to reach the information that satisfies users' requests. The results obtained from search engines usually contain large amount of unnecessary information. Our target users are the persons who could not express appropriate keywords for information retrieval. To support such users, we clarified the several requirements of an associative information access support system. According to these requirements, we propose *Word Colony* that visualizes the overview of the contents of results and suggests unexpected words as new keywords for the next retrieval. To design *Word Colony*, we introduced the notion of the dependency of term co-occurrence (*DTC*) in a document. In this paper, we describe the details of *DTC*, *Word Colony* and its demonstration.

1 Introduction

The rapid proliferation of information sources in recent years and the advent of the Internet have created a world wide web of interconnected information resources. Today, the Web represents the largest collection of information resources that an individual has ever been able to access - and it is continuously growing at accelerating paces.

Several kinds of tools have been developed in recent years to help individual users to access Web resources. The so-called search engines are the most popular among these tools, as they allow users to access the resources they index using a very simple search mechanism, namely keywords or combinations of keywords. However, the answers obtained from search engines usually contain large amounts of information that is not related to what the user had in mind when asking the question. Moreover, users can not always specify the appropriate keywords to get necessary information.

E. Suzuki and S. Arikawa (Eds.): DS 2004, LNAI 3245, pp. 195–206, 2004.
© Springer-Verlag Berlin Heidelberg 2004

Generally, the person who knows a part of content of target information can describe his/her requests concretely. However, for the person who knows nothing about content of necessary information, it is difficult to declare a query. The former is looking for already-known information, which means he/she has accessed the target information before or he/she has convinced the existence of the target information in some way. On the other hand, the latter is looking for really new information, namely unknown information or unexpected information at that moment. In other words, the latter person would like to discover new/unknown/unexpected relations to reach the target information.

To solve this problem, we propose an associative information access support system that suggests several possible relations to reach the target objects. This system provides users with two main functions. The first one is to visualize overviews of search results to grasp the content of retrieved information at glance and to see whether the result satisfies user's requests. The second function is to suggest several keywords that are associated with the previous keywords, which express user's retrieval purpose at that moment. When the search result does not satisfy user's requirement, the second function helps users to choose appropriate keywords for the next retrieval. Especially, the words that are not popular in the specific domain but closely related with the users' requirement are very useful. It is difficult for users to notice such words by themselves. By adding/changing keywords according to the system suggestions, the user would see the overview of the new retrieved result. This process is continued until the user finds the necessary information. Finally, the users clarify their search goal with the process of information access.

The goal of this paper is to propose a method to lead users to appropriate information based on associative information access. For such purpose, we introduced the notion of the dependency of term co-occurrence, called *DTC*. The DTC analyzes a document to calculate not only the importance of each term but also the strength of the associative relation among terms. The analysis result is visualized by a directed graph based on the spring model; it is called *Word Colony*. Word Colony shows the overview of document content, which is associative relation among terms and the importance of each term in the document. By using Word Colony with search engines, the system suggests the direction to choose other keywords for the next retrieval to change topics or to narrow/expand an argument.

The remainder of this paper is organized as follows. In section 2, we discuss requirements of an associative information access support system and examine whether the existing visualization tools can be used for such purpose. In section 3, we propose the notion of the dependency of term co-occurrence (*DTC*) and the visualization tool, *Word Colony*. In section 4, we show an example of associative information access with Word Colony. Finally, in section 5 we offer some concluding remarks.

2   Associative Information Access Support System

When we start information access, we sometimes do not know exactly what we want to get. In the case that a user knows a part of target information, it is easy to specify the query. For example, when a user is looking for the book written by Goethe, he can just ask a query to a database in order to select books whose author is Goethe. However, in the case that a person who wants to improve his/her English conversation ability, would like to know more efficient and effective way to study English, that image of result is ambiguous even the retrieval purpose is clear. Namely, the user has to find new information which he/she cannot specify completely.

Generally, the person who knows a part of content of target information can describe his requests as a query. However the person who knows nothing about content of necessary information, it is difficult to declare a query. The former is looking for already-known information, which means he/she has accessed the target information before or he/she has convinced the existence of the target information in some way. On the other hand, the latter is looking for really new information. The aim of associative information access support system is to suggest several possible relations to reach the target objects.

For such purpose, we assume that the basic two functions required for associative information access support system in following section 2.1. Then in the section 2.2, we discuss how to support an associative information access from the viewpoint of the types of association among terms in documents.

2.1   Basic Requirements
for Associative Information Access Support System

To support associative information access, we consider that the following two functions are basic requirements. The first one is to visualize overviews of search results. The second one is to suggest several keywords associated with user's retrieval purpose.

When a user asks a query to databases or search engines, the user needs to see immediately whether the retrieval succeeds or not. To see the retrieved result, the function to visualize overviews of search results is important to grasp the content of retrieved information at glance. It helps user's judgments to continue the retrievals more or not. In the case that the search result does not satisfy user's requirement, the user has to think of other suitable keywords to get the revised results. To provide candidate words for the next retrieval helps users to reach the correct goal efficiently. Especially, it is very useful to suggest the words that are not popular in the specific domain but closely related with the users' requirement. It is difficult for users to notice such words by themselves.

According to these two basic requirements, we examine the existing visualization tools to see if they can be used for an associative information access support system. (They are summarized in table 1.)

Nowadays, software is emerging to analyze search results, sort them automatically into categories and present them visually to provide far more information

Table 1. The basic requirements for an associative information access support system.

	Visualization of overview for results	Suggestion of associative terms for the next retrieval
Vivisimo	○	×
Grokker	○	×
DualNAVI	○	○
KeyGraph	○	○

than the typical textual list. For example, Vivisimo [1] Clustering Engine automatically categorizes search results into hierarchical folders. End-users can find information from the types of available information. Vivisimo does not use predefined categories. Its software determines them, depending on the search results. The filing is done through a combination of linguistic and statistical analysis. Another example is Grokker [2], that not only sorts search results into categories but also "maps" the results in a holistic way, showing each category as a colorful circle. Within each circle, subcategories appear as more circles that can be clicked on and zoomed in. These tools show all the possible categories of information the Internet offers. Such results help users to understand the overviews of retrieved results at a glance. However, when users ask questions, users can not always specify the appropriate keywords to get necessary information. Sometimes, users do not know how to express keywords exactly to get what they have in mind. To support it, we proposed the second basic function, that is to suggest the terms associated with the retrieved purpose as candidates for the next retrieval keywords. Vivisimo and Grokker do not supply the second function.

DualNAVI [4–6] is an engine which could improve the efficiency of searches on collections of documents. DualNAVI has dual windows. Hits for a keyword are listed on the left-hand side of the window. On the right of the window, called 'topic word graph', a set of related keywords by analyzing the retrieved documents are arranged roughly in the order of their frequencies. It enables users to execute a new search using the selected topic words as search key. This often yields articles that the initial query missed. DualNAVI is very powerful. It is not only a search tool but also a tool to support thinking.

KeyGraph [3] is an algorithm for extracting keywords representing the asserted main point in a document, without relying on external devices such as natural language processing tools or a document corpus. It is based on the co-occurrence between terms in a document. KeyGraph does not use each term's average frequency in a corpus. KeyGraph is a content-sensitive, domain independent device of indexing. This system is not designed to aim to suggest associative terms. However, we consider it would be used for this purpose.

However, DualNAVI and KeyGraph do not provide all the necessary associative words, even though they satisfied the basic requirements for associative information access. In the next section, we describe the types of associative terms to bring out the hidden important words.

2.2   Types of Association Between Terms

In this section, we describe a classification of association between terms. Let us assume that a term t is given and the term t expresses a user's interest at that moment. Then let a document set D the retrieved documents with the keyword t. Of course, the frequency of the term t in the retrieved documents is high. Then other terms in the result documents are grouped into four categories in the table 2 based on the term frequency and on the relation with the term t.

A term frequency is one of the indicators to show the popularity of a term in the selected documents. The terms are grouped into the high frequency terms and the low frequency terms. Besides, the strength of associative relation between terms are given by the location where terms co-occur, because we consider that the physical distances between terms in text are one of the indicators of the relations of terms. The terms co-occurring in each sentence have close relationships than the terms co-occurred in each document. So the terms co-occurring with the term t in each sentence are grouped into the category "strong relation with the term t", while the terms co-occurring with the term t in each document are grouped into the category "weak relation with the term t". By two axes, the terms are grouped into four areas I to IV in the table 2.

A term in the category I or II is popular term because of its high frequency. Such term is useful to find trend or main topic in document. Generally, high frequent terms are conspicuous, even though a term in category II has less relation with the term t than a term in category I. A term in the category III has low frequency but it is closely associated with the term t. Such term is useful to find hidden or minor relation. We consider that it is necessary to show the terms in the categories I, II, III for support associative information access. Especially the terms in the category III suggest unexpected terms for users.

DualNAVI gives the documents list and topic word graph in which a set of related keywords are arranged roughly in the order of their frequencies. This topic word graph visualizes the terms in the category I and II without distinction between two types of terms. The topic word graph gives priority to showing the order of frequencies over showing the co-occurrence relation. KeyGraph extracts keywords representing the main point in a document based on the co-occurrence between terms in a document. It gives the terms in the category I, because the original purpose of KeyGraph is to extract keywords.

As a consequence of the previous discussion, the existing systems are not enough for support associative information access. We propose the Word Colony to visualize the terms in the categories I to III, that are required for associative information access support. In the next section, we describe the details of the notion of the dependency of term co-occurrence and the visualization tool, Word Colony. Then we propose that using Word Colony with databases or search engines. That satisfies conditions to support associative information access in table 1 and 2.

Table 2. The four groups of terms associated with the term t.

	High Frequency (HF)	Low Frequency (LF)
Strong Relation with the term t (SR)	I	III
Weak Relation with the term t (WR)	II	IV

3   Word Colony

Text has a rich structure, in which sentences are grouped and related to one another in a variety of ways. Each sentence consists of terms. In this paper, we regard the dependency of co-occurrence as cohesion of each term in a document. We focus on the characteristics of the dependency of co-occurrence among terms which is divided into three groups. The first one is the group of interdependent terms. These terms appear close in a document. The second one is dependent terms. One term depends on another but not the reverse. The last group is independent terms.

To visualize such relationships, important topic terms which are surrounded by dependent terms, gives an outline of document content. At the same time, it indicates some direction to narrow the issues in user's mind down or to arouse user's interest in the relevant information.

In the following subsections, we describe how to calculate the cohesion and how to visualize it. Then in the next section 4, we mention how each visualized component corresponds to each of the categories in the table 2.

3.1   Dependency of Term Co-occurrence

In this section, we describe the method to calculate the cohesion among terms. Interdependent terms are related with each other very strong. It is easy to guess their relation. While dependent terms suggest weak associable relations between main terms and subordinate terms. It can be regarded that the main terms are the important words that are surrounded by subordinate terms to suggest several aspects concerning about topic words. Except for them, there are independent terms. It seems difficult to guess the relation with other words but those terms appear in the same document.

To extract associable relations among terms mentioned before, we propose to use of the dependency between two co-occurred terms.

The dependence of a term t on a term t' in a document d, denoted by $dtc_d(t, t')$, is given by the conditional probability as follows.

$$dtc_d(t, t') = \frac{sentences_d(t \cap t')}{sentences_d(t)}, \tag{1}$$

where $sentences_d(t \cap t')$ is the number of sentences that includes the terms t and t' in the document d, and $sentences_d(t)$ is the number of sentences that includes the term t in the document d.

Let us assume that the term t is given. Based on $dtc_d(t, t')$, any other term t' is grouped into the following types.

1. Interdependent terms
   If $dtc_d(t, t') \geq \delta$ and $dtc_d(t', t) \geq \delta$, then the term t and t' are interdependent terms. These are close relevant terms in the context of the original document. Such terms suggest some notion vaguely by working together.
2. Dependent terms
   The dependent terms are divided into the following two group based on the direction of the dependency between the terms t and t'.
   (a) Superordinate terms
       If $dtc_d(t, t') > \sigma \geq \mu \geq dtc_d(t', t)$, then term t depends on the term t'. The term t emphasizes one aspect of the terms t' in the context.
   (b) Subordinate terms
       If $dtc_d(t', t) > \sigma \geq \mu \geq dtc_d(t, t')$, then term t' depends on the term t. The term t' emphasizes one aspect of the terms t in the context.
3. Independent terms
   If $dtc_d(t, t')$ and $dtc_d(t', t)$ do not satisfy the previous conditions, then the terms t and t' are independent terms.

3.2   Word Colony Graph

The dependency relationship among terms co-occurrence forms the directed graph, called Word Colony. In Word Colony, the strength of co-occurrence relation between terms is mapped to the distance between nodes of terms and the frequency is regarded as the size of nodes. To visualize this relation, we use a spring model graph. This graph visualizes the overview of a document and the topic transition of documents. It also provides users an associative way to find the next step for searching information.

The interdependent terms are embedded into the same node, because such terms have close relationships among them. They suggest some notion according to the context of the document.

The size of node is proportional to the function $attr_d(t)$. It is the sum of dependency of any term t'' in a document on the term t. It is calculated as follows:

$$attr_d(t) = \sum_{t'' \in T} dtc_d(t'', t) \tag{2}$$

The directed edge links from the node of the term t to the node of the term t', if the term t depends on the term t'. The length of edge is inverse proportion to $dtc_d(t, t')$. The dependent terms indicate several keywords for the relevant topics.

Figure 1 shows a simple example of Word Colony. The right side of the interface is the document window, while the left side is Word Colony graph window. The text appeared in the document window is analyzed by DTC. Word Colony graph window visualizes the analysis result, that is the dependency relation among co-occurrence terms by Touch Graph [8] based on the spring model.

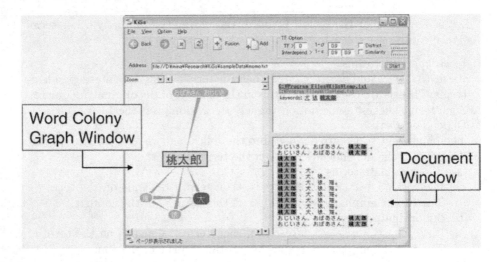

Fig. 1. An example of Word Colony.

4 Associative Information Access

By combining search engine results with Word Colony, we provide the function of the associative information access. In this section we use Google as a search engine to combine with Word Colony, because Google is one of the popular search engines. The Google search engine is set at the document window. When a user inputs keywords to Google, the document window (Google) returns a list of web pages that contain top 10 pages' title with short description texts. The DTC analyzes the relationships among terms in a top ten list page and visualizes them as Word Colony. It shows the overview of the relation among not only terms with high frequency but also the terms with low frequency. Let us assume that an interesting term be the term t. Then, a term in category I in the table 2 appears close to the term t and its node size is big. A term in category II appears far from the term t, even though its node size is big. A term in category III appears close to the term t and its node size is smaller. A term in category IV appears far from the term t and its node size is smaller. This visualization suggests some ranking of keywords for the next retrieval to find more appropriate keywords for user's intention.

Let us assume that we are looking for *something* to improve English conversation ability. At this moment, nobody knows the exact search results. The table 3 shows a process of information search.

Before associative information access, the user did not have any exact image of the result. At the end, the user gets the scenario books of English movies. They are useful to check the speech words in movies before/after watching movies. The important point is that the user did not imagine the term "Scenario" at the beginning of the retrieval. It is suggested during the retrieval process. Moreover, by the comparison of the Word Colony at step 1 (figure 2) and the one at step 4 (figure 4), the topic transition is observed.

Table 3. An example of an associative information access process with Word Colony.

[Step 1]
Asking Google with the keywords "**English Conversation**" and "**Improvement**" in Japanese.
To improve English, a user specified the keywords "English Conversation" and "Improvement". Then DTC returns Word Colony which is shown in figure 2. Around the term "English Conversation", there are a group of terms concerning about English Conversation School, a group of studying materials (books, softwares, etc.) and others. Then the user found the term "**Movie**" and chose it for the next search keywords, because the user likes movies.
[Step 2]
Asking Google with the keywords "**Movie**", "English Conversation" and "Improvement" in Japanese.
In the same way as Step 1, the user saw the Word Colony and chose the term "**DVD**", because there is the term "practical use" between the term "DVD" and "Movie". The user wanted to know how to use DVD efficiently for improving English Conversation ability.
[Step 3]
Asking Google with the keywords "**DVD**", "Movie", "English Conversation" and "Improvement" in Japanese.
This time, the user found the term "**Scenario**" near "Movie" and "DVD" (See figure 3). It is completely unexpected term at the beginning.
[Step 4]
Asking Google with the keywords "**Scenario**", "DVD", "Movie", "English Conversation" and "Improvement" in Japanese.
The terms "Scenario" "Movie" and "DVD" are related each other extremely. (See figure 4). Then the user checked the Google search results and found that the first and second entries in the result list are the page about the Scenario books for English Movies. That seems the one of answers that the user wants.

Table 4 shows (i) the ranking of the selected terms that is calculated by the term frequency and the formula $attr_d$ and (ii) the distance of selected terms to other keywords that is estimated by the minimal length between two terms' nodes. Based on the combination of the ranking and the distance, each selected term t_3, t_4, t_5 is categorized in III, I, III, respectively. It shows that the user selected terms from the category III and I. In other words, the values of ranking of the selected terms in each step show that the ranking is not influential with keywords selection. It means that Word Colony's term arrangement is effective and helpful to select new keywords. Table 4 shows also the rank of the final result page at each step. At step 1, it is impossible to find that page from Google result list. By selecting appropriate keywords, that page comes up at the end. It also shows the effectiveness of Word Colony.

Table 4. The ranking of selected term, and the distance between selected term and other keywords.

	Keywords	Selected term (and category)	The ranking of the selected term	Minimal length of the path between two terms	Final result page rank by Google
Step 1	t_1, t_2	t_3 (III)	$R_{tf}(t_3)$=134 * $R_a(t_3)$=42 **	1 ($t_3 \leftrightarrow t_2$) 1 ($t_3 \leftrightarrow t_1$)	under 200th
Step 2	t_1, t_2, t_3	t_4 (I)	$R_{tf}(t_4)$=9 * $R_a(t_4)$=4 **	1 ($t_4 \leftrightarrow t_3$) 1 ($t_4 \leftrightarrow t_2$) 1 ($t_4 \leftrightarrow t_1$)	26th
Step 3	t_1, t_2, t_3, t_4	t_5 (III)	$R_{tf}(t_5)$=43 * $R_a(t_5)$=30 **	2 ($t_5 \leftrightarrow t_4$) 1 ($t_5 \leftrightarrow t_3$) 3 ($t_5 \leftrightarrow t_2$) 2 ($t_5 \leftrightarrow t_1$)	4th

t_1= "English Conversation", t_2= "Improvement", t_3= "Movie", t_4= "DVD", t_5= "Scenario"

*R_{tr} is calculated by the term frequency.
**R_{attr} is calculated by the function $attr_d$.

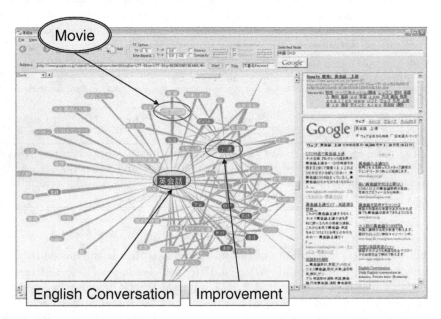

Fig. 2. Word Colony and Google Search result with the keywords "English Conversation" and "Improvement" (Step 1).

5   Conclusion

In this paper, we proposed the use of the dependency of term co-occurrence and a system supporting an associative information access called *Word Colony*. The

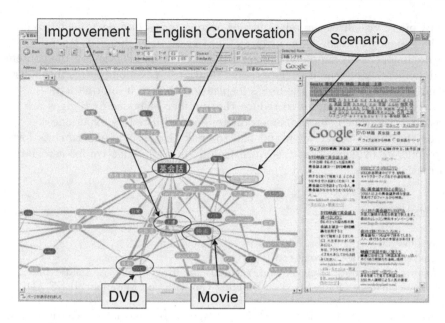

Fig. 3. Word Colony and Google Search result with the keywords "DVD", "Movie", "English Conversation" and "Improvement".

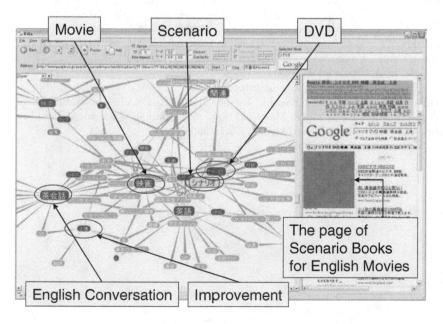

Fig. 4. Word Colony and Google Search result with the keywords "**Scenario**", "DVD", "Movie", "English Conversation" and "Improvement" (Step 4).

Word Colony visualizes the relationships among terms in a document based on the dependency of term co-occurrence. That relation is changed depending on each document. This simple visualization tool gives an overview of the content of a document. Moreover it suggests some relationships among terms that a user was not aware of. It draws the user's inspiration to search information.

However, the techniques to reduce the number of nodes and edges are required. For the time being, some parameters are adjusted by experiences.

For the future work, we would like to make large-scale experiments for a more concrete evidence that the Word Colony is really effective for associative information access.

References

1. Vivisimo clustering engine, (2001)http://vivisimo.com
2. Grokker, http://www.groxis.com/
3. Ohsawa, Y., Benson, Nels E. and Yachida, M.: KeyGraph: Automatic Indexing by Co-occurrence Graph based on Building Construction Metaphor, IEEE ADL'98 (1998)
4. Takano, A., Niwa, Y., Nishioka, S., Iwayama, M., Hisamatsu, T., Imaichi, O. and Sakurai, H.: Information Access Based on Associative Calculation. SOFSEM 2000, LNCS vol. 1963, Springer-Verlag (2000) 187–201
5. Takano, A., Niwa, Y., Nishioka, S., Iwayama, M., Hisamitsu, T., Imaichi, O. and Sakurai, H.: Associative Information Access Using DualNAVI. Kyoto International Conference on Digital Libraries (ICDL'00), (2000) 285-289
6. Takano, A.: Association Computation for Information Access, *Proc. of Discovery Science 2003*, LNCS 2843, (2003) 33-44
7. Google, http://www.google.com
8. TouchGraph, http://www.touchgraph.com/
9. Yuji Matsumoto, Akira Kitauchi, Tatsuo Yamashita, Yoshitaka Hirano, Hiroshi Matsuda, Kazuma Takaoka, Masayuki Asahara: Manual of Japanese Morphological Analysis System ChaSen version 2.2.1 (2000)

On the Convergence
of Incremental Knowledge Base Construction

Tri M. Cao[1], Eric Martin[1,2], and Paul Compton[1]

[1] School of Computer Science and Engineering
University of New South Wales
Sydney 2052, Australia
{tmc,emartin,compton}@cse.unsw.edu.au
[2] National ICT Australia Limited*

Abstract. Ripple Down Rules is a practical methodology to build knowledge-based systems, which has proved successful in a wide range of commercial applications. However, little work has been done on its theoretical foundations. In this paper, we formalise the key features of the method. We present the process of building a correct knowledge base as a discovery scenario involving a user, an expert, and a system. The user provides data for classification. The expert helps the system to build its knowledge base incrementally, using the output of the latter in response to the last datum provided by the user. In case the system's output is not satisfactory, the expert guides the system to improve its future performance while not affecting its ability to properly classify past data. We examine under which conditions the sequence of knowledge bases constructed by the system eventually converges to a knowledge base that faithfully represents the target classification function. Our results are in accordance with the observed behaviour of real-life systems.

1 Introduction

An expert can be defined as someone who is able to make excellent judgements in some specific domain; that is, someone who is able to draw appropriate conclusions from the data available. A central problem for building knowledge-based systems is that although an ability to draw conclusions from data generally implies some ability to indicate features relevant to the conclusion, it does not imply an ability to provide a general model of the whole domain. That is, the expert will only indicate some distinguishing features and never all the features that distinguish a particular conclusion from all other conclusions in the domain. Compton and Jansen [5] and Richards and Compton [14] argue that people cannot give a comprehensive explanation for their decision making, but at most justify why a decision is preferable to the other alternate decisions under consideration in the context.

Ripple-Down Rules (RDR) is a knowledge acquisition methodology developed to deal with this contextual nature of knowledge. It requires experts only to deal with

---

* National ICT Australia is funded by the Australian Government's Department of Communications, Information Technology and the Arts and the Australian Research Council through Backing Australia's Ability and the ICT Centre of Excellence Program.

E. Suzuki and S. Arikawa (Eds.): DS 2004, LNAI 3245, pp. 207–218, 2004.

specific cases and to justify why a specific conclusion applies to a case rather than some other conclusion. The expert has to identify features in the case that distinguish it from other specific cases that have the alternate conclusion. The approach is further grounded in an expert's concrete experience, by building the knowledge base while it is in use processing real cases. The expert supervises the output of the system and corrects it when it makes an inappropriate decision, by identifying the distinguishing features of the case. The expert only deals with cases and makes no attempt to generalise or structure the knowledge; the RDR system is responsible for organising the knowledge. This approach contrasts significantly with conventional knowledge acquisition where generally the expert and knowledge engineer collaborate to design a knowledge model of the domain, and it is hoped that this knowledge is effectively complete before being put into use.

RDR systems have been developed for a range of application areas and tasks. The first industrial demonstration of this approach was the PEIRS system, which provided clinical interpretations for reports of pathology testing and had almost 2000 rules built by pathologist [5, 13]. The approach has also been adapted to a range of tasks: control [16], heuristic search [2], document management using multiple classification [9], and configuration [6]. The level of evaluation in these studies varies, but overall they clearly demonstrate very simple and highly efficient knowledge acquisition. There is now significant commercial experience of RDR confirming the efficiency of the approach. Following the PEIRS example, one company, Pacific Knowledge Systems supplies tools for pathologist to build systems to provide interpretative comments for medical Chemical Pathology reports. One of their customers now processes up to 14,000 patient reports per day through their 23 RDR knowledge bases with a total of about 10,000 rules, giving very highly patient-specific comments. They have a high level of satisfaction from their general practitioner clients and from the pathologists who keep on building more rules – or rather who keep on identifying distinguishing features to provide subtle and clinically valuable comments. A pathologist generally requires less than one day's training and rule addition is a minor addition to their normal duties of checking reports; it takes at most a few minutes per rule (Pacific Knowledge Systems, personal communication[1]).

Given the success of the knowledge representation scheme and the knowledge revision procedure, it is of interest to investigate the properties of RDR to account for its success and shape its future developments. We examine the relevance of the approach in the perspective of formal learning theory (Inductive Inference). A paradigm of formal learning theory investigates under which conditions the sequence of hypotheses output by an agent, in response to longer and longer finite initial segments of a potentially infinite stream of data, converges to an accurate and finite description of a target function. A simple learning scenario together with the corresponding concepts of "convergence" and "accuracy" were first defined by Gold in his model of learning in the limit [8]. This model is well adapted to address the kind of issues we are interested in. However, further investigation in other learning frameworks, such as PAC learning or statistical query models [10], would be necessary to provide complementary insight into the issues.

---

[1] Dr. L. Peters http://www.pks.com.au/main.htm

In this paper, we formalise the process of building a correct knowledge base as a learning scenario where success is expressed in terms of convergence in the limit. More precisely, we examine important conditions that do or do not guarantee whether the sequence of knowledge bases built by the system eventually converges to a knowledge base that faithfully represents the target classification function.

In Section 2, we give an informal representation of Ripple Down Rules. In Section 3, we formalise the process of incremental knowledge base construction. The convergence of the process is investigated in Section 4. We conclude in Section 5.

2   Ripple Down Rules

In this section, we present a knowledge representation and reasoning scheme of RDR systems. There seems to be two reasonable approaches for translating an RDR knowledge base into a logical theory. The first approach would use prioritised logic, which would involve a second-order predicate to capture the refinement semantics. The second approach is presented in more detail since it is our preferred choice.

A binary Ripple Down Rules knowledge base is a finite binary tree with two distinct types of edges, labelled with either *except* or *if-not*, whose nodes are labelled with rules of the form *"if α then C"* or *"if α then not C"*, where α is a formula in a propositional language and C is a fixed atom outside this language. Figure 1 depicts an example of a binary RDR knowledge base. We call α the condition and C the conclusion of the rule.

Fig. 1. An RDR knowledge base.

Note that we only consider binary RDR knowledge bases. (In full generality, knowledge bases can have n branches and conclusions can be built from different atoms. It is not difficult to reduce such knowledge bases to the kind of knowledge base we consider here; details are omitted.)

In the above example, we can identify the following elements of the knowledge base:

- attributes, namely: bird, young, airplane, penguin, in-plane,...;
- conclusions, namely: fly and not fly.

The binary RDR knowledge base is used as follows. A data case d is passed to the tree starting from the root. If the data case entails the condition of the current node, the data case is said to fire the rule and the conclusion is temporarily recorded (and overrides the previous conclusion). The data case is then passed to the next node in the *except* branch. If the case does not entail the condition of the rule, it is passed to the next node in the *if-not* branch. The process continues until there is no suitable node to pass the case to. This algorithm defines a path from the root of the tree to a node for each case.

Procedures to translate a binary RDR tree into a set of logical rules have been proposed in [15, 11, 4]. One of the results is a representation such that, for any given data case d, there is a unique rule from the translated set which can be applied to the data case. For example, the above RDR rule tree can be translated into the following set:

$$\text{bird} \land \text{not young} \land \text{not penguin} \rightarrow \text{fly}$$
$$\text{not bird} \land \text{airplane} \rightarrow \text{fly}$$
$$\text{bird} \land \text{young} \rightarrow \text{not fly}$$
$$\text{bird} \land \text{not young} \land \text{penguin} \land \text{not in-plane} \rightarrow \text{not fly}$$
$$\text{bird} \land \text{not young} \land \text{penguin} \land \text{in-plane} \rightarrow \text{fly}$$
$$\text{not bird} \land \text{not airplane} \rightarrow \text{not fly}$$

One of the strengths of the RDR framework is that rule bases are easy to revise. When an expert spots a rule which gives a wrong conclusion, she only needs to create a new exception to that rule. A new condition will be required from the expert to distinguish between the current data case and the past data cases which have been correctly classified by the rule. In the rule set, this action will break the rule into two rules. The conditions of the new rules will be the condition of the old rule in conjunction with the differentiating condition (from the expert) or its negation. For example, if we feed the RDR knowledge base in Figure 1 with the data case bird $\land$ kiwi, the answer from the system will then be fly, which is undesired, and the rule which gives the wrong conclusion is

$$\text{bird} \land \text{not young} \land \text{not penguin} \rightarrow \text{fly}$$

Therefore, if the differentiating condition given by the expert is kiwi, then the RDR knowledge base becomes as represented in Figure 2.

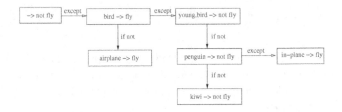

Fig. 2. The revised RDR knowledge base.

The new translated rule set is

$$\text{not bird} \land \text{airplane} \rightarrow \text{fly}$$
$$\text{bird} \land \text{young} \rightarrow \text{not fly}$$
$$\text{bird} \land \text{not young} \land \text{penguin} \land \text{not in-plane} \rightarrow \text{not fly}$$
$$\text{bird} \land \text{not young} \land \text{penguin} \land \text{in-plane} \rightarrow \text{fly}$$

bird $\wedge$ not young $\wedge$ not penguin $\wedge$ not kiwi $\rightarrow$ fly

bird $\wedge$ not young $\wedge$ not penguin $\wedge$ kiwi $\rightarrow$ not fly

not bird $\wedge$ not airplane $\rightarrow$ not fly

In the next section, we will formalise this kind of translated set of rules and use the formalisation as the knowledge representation scheme of the system.

3   The Formal Discovery Framework

Our framework can be intuitively described as a scenario which involves: an agent or *user*, a *system*, an *expert*, and a *classification function*. The expert knows whether a given datum is negatively or positively classified by the classification function ϕ, though he does not know ϕ. The user presents the system with a stream of data. The aim of the system is to correctly classify all data in the stream with the help of the expert. We will formalise this scenario in Section 3.2 after we have introduced the necessary definitions.

3.1   Basic Concepts

We denote by $\mathcal{L}$ the set of propositional formulas built from a fixed countable set of propositional atoms $\{a_0, a_1, \ldots\}$. We choose an arbitrary tautology and refer to it by $\top$. Depending on the context, we will refer to members of $\mathcal{L}$ either as *formulas* or as *classification functions*.

Definition 1. *A* data case *is a finite set of natural numbers.*

A data case d can be seen as a model in which finitely many atoms, namely $\{a_i : i \in d\}$, have the value true. Hence given $\alpha \in \mathcal{L}$, we will write $d \models \alpha$ to mean that d is a model of α. We denote by Truth the binary function that maps every pair of the form (α, d), where α is a formula and d is a data case, to 1 if $d \models \alpha$, and to 0 otherwise.

Definition 2. *Let a finite set $F = \{\alpha_1, \alpha_2, \ldots, \alpha_n\}$ of formulas be given.*

F is said to be independent *iff for all data cases d and for all distinct members j, k of $\{1 \ldots n\}$, if $d \models \alpha_j$ then $d \models \neg\alpha_k$.*

F is said to be a coverage of the domain *iff for all data cases d, there exists a member j of $\{1 \ldots n\}$ such that $d \models \alpha_j$.*

F is said to be a partition of the domain *iff F is both independent and a coverage of the domain.*

We abstract an RDR knowledge base as a finite set of rules whose antecedents are formulas that represent conditions to be tested against data cases and whose conclusions are either 1 or 0 depending on whether the data case is positively or negatively classified by the system. A key condition is that the set of premises of the rules forms a partition of the domain.

Definition 3. *A knowledge base is a finite set of the form:*

$$\{(\alpha_1, c_1), (\alpha_2, c_2), \ldots, (\alpha_n, c_n)\}$$

where $\alpha_1, \ldots, \alpha_n$ are formulas such that $\{\alpha_1, \alpha_2, \ldots, \alpha_n\}$ is a partition of the domain and $c_1, \ldots, c_n$ are members of $\{0, 1\}$.

Intuitively, if a pair of the form $(\alpha, 1)$ (respect., $(\alpha, 0)$) belongs to a knowledge base K, then for every data case d that is a model of α, K positively (respect., negatively) classifies d.

Our scenario involves infinite sequences of knowledge bases. Two issues have to be considered: convergence of the sequence and, in case of convergence, correctness of the limit knowledge base. The next two definitions capture these concepts.

Definition 4. *Let a sequence $(K_i)_{i\in\mathbb{N}}$ of knowledge bases and a knowledge base K be given. We say that $(K_i)_{i\in\mathbb{N}}$ converges to K iff there exists $n \in \mathbb{N}$ such that for all $i \geq n$, $K_i = K$.*

Definition 5. *Let $K = \{(\alpha_1, c_1), (\alpha_2, c_2), \ldots, (\alpha_k, c_k)\}$ be a knowledge base. K is said to be* correct *with respect to a classification function ϕ iff:*

$$\models \bigvee_{(\alpha,1)\in K} \alpha \rightarrow \phi \quad and \quad \models \bigvee_{(\alpha,0)\in K} \alpha \rightarrow \neg\phi.$$

In the previous definition, implications can be replaced by equivalences.

Proposition 1. *Let $K = \{(\alpha_1, c_1), (\alpha_2, c_2), \ldots, (\alpha_k, c_k)\}$ be a knowledge base. K is correct with respect to a classification function ϕ iff*

$$\models \bigvee_{(\alpha,1)\in K} \alpha \leftrightarrow \phi \quad and \quad \models \bigvee_{(\alpha,0)\in K} \alpha \leftrightarrow \neg\phi.$$

Proof. Let d be a data case and assume that $d \models \phi$. By the definition of K, there exists a unique $(\gamma, c) \in K$ with $d \models \gamma$. It suffices to show that $c = 1$. By way of contradiction, assume that $c = 0$. Since $\models \bigvee_{(\alpha,0)\in K} \rightarrow \neg\phi$, $\gamma \in \{\alpha : (\alpha, 0) \in K\}$ and $d \models \gamma$, it follows that $d \models \neg\phi$. So $c = 1$.

The previous proposition expresses that we can get the normal form of the target classification function by combining the antecedents of the rules in the correct knowledge base. The next definition abstracts the behaviour of the system which, for a given data case d, will automatically determine the unique rule in its knowledge base that applies to d.

Definition 6. *The* system function *is the (unique) function, denoted by S, that maps any pair of the form (K, d), where K is a knowledge base and d is a data case, to the unique member (α, c) of K such that $d \models \alpha$; we denote α by $S_1(K, d)$ and c by $S_2(K, d)$.*

Two knowledge bases may represent distinct classification functions but from the user's point of view, both knowledge bases are equivalent if they agree on that user's (finite) set of data. The next definition formalises this notion.

Definition 7. *Let a set D of data cases and two knowledge bases K, K' be given. We say that K and K' are D-equivalent iff for all members d of D, $S_2(K, d) = S_2(K', d)$.*

3.2 The Discovery Scenario

Before we formalise the scenario that has been outlined at the beginning of the previous section, we describe it in more detail, using the concepts that have been introduced.

1. The user presents the system with some datum d.
2. The system returns $S(K, d) = (\alpha, c)$ where K is the current knowledge base in the system.
3. The expert analyses (α, c). More precisely, if $c = 1$ (respect., $c = 0$) and the expert considers that d should be positively (respect., negatively) classified, then go back to step 1; otherwise, go to step 4.
4. The expert E determines a formula δ such that if (α, c) is removed from K and replaced by the more specific rules $(\alpha \wedge \delta, c)$ and $(\alpha \wedge \neg\delta, 1 - c)$, then the resulting knowledge base K' has the following properties:
    - if D is the set of data received by the system before d, then K and K' are D-equivalent;
    - If the expert positively (respect., negatively) classifies d, then $S_2(K', d) = 1$ (respect., $S_2(K', d) = 0$).
    Go back to step 1.

To fix the knowledge base, the expert essentially breaks the rule that gives wrong classification on the last datum into two rules by adding a differentiating condition. The first rule will take care of all the previous data cases that apply to the old rule. The second rule will apply to the new data case only.

The next lemma is important because it guarantees that the expert can always find a differentiating condition and thus refine the knowledge base in an incremental manner. This is one of the strengths of RDR in building knowledge-based systems. Note that the lemma is just an existence statement. It is a theoretical result that does not impose any restriction on how an expert might come up with a differentiating condition. In practice, such a condition is likely to be more complex than the one given here.

Lemma 1. *Let a knowledge base K, a classification function ϕ, and a finite set D of data cases be such that for all members d of D, $S_2(K, d) = \text{Truth}(\phi, d)$. Let a data case d that does not belong to D be given. Let $(\alpha, c) = S(K, d)$. Then there exists a formula δ with the following properties. Set:*

$$K' = K \backslash \{(\alpha, c)\} \cup \{(\alpha \wedge \delta, c), (\alpha \wedge \neg\delta, 1 - c)\}.$$

Then:

- *K' is a knowledge base;*
- *K and K' are D-equivalent;*
- *$\text{Truth}(\delta, d) = c$.*

Proof. Define $D_\alpha = \{d' \in D : d' \models \alpha\}$ to be the support data of the rule (α, c). We will construct a formula δ such that $d \models \alpha \wedge \delta$ and for all $d' \in D_\alpha$, $d' \models \alpha \wedge \neg\delta$. Given a data case d', define:

$$\text{Diff}(d, d') = \{a \in \mathcal{L} : d \models a \wedge d' \models \neg a\}.$$

In case d' is a data case that is distinct from d then the set $\mathrm{Diff}(d, d') \cup \mathrm{Diff}(d', d)$ is not empty. For all $d' \in D_\alpha$, let

$$\delta_{d'} = \bigwedge_{a \in \mathrm{Diff}(d,d')} a \wedge \bigwedge_{a \in \mathrm{Diff}(d',d)} \neg a.$$

Let $\delta = \bigwedge_{d' \in D_\alpha} \delta_{d'}$. Then δ satisfies the claims of the lemma.

We are now ready to formalise the learning scenario outlined at the beginning of this section. For a given sequence of data cases and a given target classification function, it defines:

- a sequence of formulas that represent the expert's strategy in response to the data received so far, and the way these data are classified by the system;
- a sequence of knowledge bases that are built incrementally by the system following the expert's guidance.

Definition 8. *Let a sequence* $(d_n)_{n \in \mathbb{N} \setminus \{0\}}$ *of data cases and a classification function* ϕ *be given. An* admissible scenario *for* $(d_n)_{n \in \mathbb{N} \setminus \{0\}}$ *and* ϕ *is a pair*

$$((\delta_n)_{n \in \mathbb{N} \setminus \{0\}}, (K_n)_{n \in \mathbb{N}})$$

where $(\delta_n)_{n \in \mathbb{N} \setminus \{0\}}$ *is a sequence of formulas and* $(K_n)_{n \in \mathbb{N}}$ *is a sequence of knowledge bases such that* $K_0 = \{(\top, 1)\}$ *and for all* $n \in \mathbb{N}$, *the following holds. Let* $(\alpha, c) = S(d_{n+1}, K_n)$ *and* $D = \{d_1, \dots, d_n\}$. *Then:*

- *if* $\mathrm{Truth}(\phi, d_{n+1}) = c$ *then* $\delta_{n+1} = \top$;
- $K_{n+1} = K_n \setminus \{(\alpha, c)\} \cup \{(\alpha \wedge \delta_{n+1}, c), (\alpha \wedge \neg \delta_{n+1}, 1 - c)\}$;
- K_n *and* K_{n+1} *are* D-*equivalent;*
- $\mathrm{Truth}(\delta_{n+1}, d_{n+1}) = c$.

Let a sequence $(d_n)_{n \in \mathbb{N} \setminus \{0\}}$ of data cases and a classification function ϕ be given. In Definition 8, the existence of at least one admissible scenario for $(d_n)_{n \in \mathbb{N} \setminus \{0\}}$ and ϕ is justified by Lemma 1. Intuitively, δ_{n+1} is the expert's opinion in response to d_{n+1}, α and c. In the above definition, if the expert agrees with the system on the classification of data case d_n then δ is assigned $\top$ (the tautology) and as a result, K_{n+1} is the same as K_n.

Technically, it is convenient to be able to refer to the sequence of knowledge bases, with no mention of the sequence of formulas put forward by the expert:

Definition 9. *Let a sequence* $(d_n)_{n \in \mathbb{N} \setminus \{0\}}$ *of data cases and a classification function* ϕ *be given. A* knowledge base refinement *for* $(d_n)_{n \in \mathbb{N} \setminus \{0\}}$ *and* ϕ *is a sequence* $(K_n)_{n \in \mathbb{N}}$ *of knowledge bases for which there exists a sequence* $(\delta_n)_{n \in \mathbb{N} \setminus \{0\}}$ *of formulas such that* $((\delta_n)_{n \in \mathbb{N} \setminus \{0\}}, (K_n)_{n \in \mathbb{N}})$ *is an admissible scenario for* $(d_n)_{n \in \mathbb{N} \setminus \{0\}}$ *and* ϕ.

4   Convergence Properties

To explore the scenario we described in the previous section, we need to introduce an additional notion. As in standard learning paradigms, we assume that data are noise-free and complete in the following sense.

Definition 10. *We say an enumeration of data cases is* complete *iff it contains at least one occurrence of every data case.*

Though enumerations of data might not be complete in practice, completeness is one of the main assumptions in all basic paradigms of formal learning theory. A good understanding of scenarios based on this assumption paves the way to more realistic descriptions where data can be noisy or incomplete. So we apply the scenario to the case where $(d_n)_{n\in\mathbb{N}\setminus\{0\}}$ is a complete enumeration of data cases. Since the knowledge base correctly classifies all the observed data cases, it performs no worse than the expert on those cases. Hence even if the stream of data is not complete, the sequence of generated knowledge bases is practically valuable; the completeness assumption is not necessary to guarantee the consistency of the knowledge bases with past data.

We now investigate the conditions under which the sequence of knowledge bases built by the system converges. The reason we are interested in the convergence is that the limit knowledge base correctly classifies all data cases with respect to the target classification function, as expressed in the next proposition.

Proposition 2. *Let a complete sequence $(d_n)_{n\in\mathbb{N}\setminus\{0\}}$ of data cases and a classification function ϕ be given. Let $(K_n)_{n\in\mathbb{N}}$ be a knowledge base refinement for $(d_n)_{n\in\mathbb{N}\setminus\{0\}}$ and ϕ. If $(K_n)_{n\in\mathbb{N}}$ converges to some knowledge base K then K is correct w.r.t. ϕ.*

Proof. For all $n \in \mathbb{N}$, if $K_{n+1} = K_n$ then K_n correctly classifies d_{n+1}. Moreover, for all $n \in \mathbb{N}$, K_n correctly classifies $d_0, \ldots, d_n$ by construction. The proposition follows. The next proposition shows that some real expertise is needed in order to help the system to converge to some knowledge base.

Proposition 3. *For every classification function ϕ, there exists uncountably many complete enumerations of data $(d_n)_{n\in\mathbb{N}\setminus\{0\}}$ for which some knowledge base refinement for $(d_n)_{n\in\mathbb{N}\setminus\{0\}}$ and ϕ does not converge to any knowledge base.*

Proof. We show that there exists an enumeration of data on which the knowledge base is not guaranteed to converge. Let's assume without loss of generality that the classification function ϕ is a_0. Let $\{n_i\}_{i\in\mathbb{N}}$ be an increasing sequence of numbers with $n_0 > 0$. Let

$$p_{n_i} = \mathrm{Odd}(i)\,\overbrace{0\ldots01}^{n_0-1}\,\overbrace{0\ldots0}^{n_1-n_0-1}\,1\ldots$$

denote the data case where the 0th attribute is classified as positive (resp. classified as negative) if i is odd (resp. even) and only the $n_0^{th}, \ldots, n_i^{th}$ attributes are positive. We construct a complete enumeration $(d_j)_{j\in\mathbb{N}\setminus\{0\}}$ of data cases from the sequence $(p_{n_i})_{i\in\mathbb{N}}$ as follows:

$$d_1 = p_{n_0} = \underset{A_0}{00\ldots010\ldots}$$

$$d_{j_1} = p_{n_1} = \underset{A_1}{10\ldots010\ldots10\ldots}$$

$$\ldots$$

where for all $i \in \mathbb{N}$, A_i is an enumeration of all data cases d such that for all $k \geq n_i$, $d \models \neg a_k$. Remember from Definition 8 that $K_0 = \{(\top, 1)\}$. When the user presents

d_1 to the system, the system misclassifies d_1 (positively instead of negatively). In order to correctly classify d_1, the expert can choose the formula $\neg a_{n_0}$ as the differentiating condition, in which case the system updates its knowledge base to $K_1 = \{(\neg a_{n_0}, 1), (a_{n_0}, 0)\}$. Some of the member of A_0 will be misclassified by K_1. In order to remedy these misclassifications, the expert will need to break the first rule of K_1 (namely, $(\neg a_{n_0}, 1)$) and subsequently break some of the derived rules; indeed, the expert cannot break the rule $(a_{n_0}, 0)$ since none of the members of A_0 is a model of a_{n_0}. When d_{j_1} is presented to the system by the user, the faulty rule $(a_{n_0}, 0)$ has to be broken into two rules so that the system correctly classifies d_{j_1}. Suppose that the expert chooses the formula $\neg a_{n_1}$. The two new rules are $(a_{n_0} \wedge \neg a_{n_1}, 0)$ and $(a_{n_0} \wedge a_{n_1}, 1)$. Then the pattern described before repeats itself: the rule $(a_{n_0} \wedge a_{n_1}, 1)$ is left untouched until d_{j_2} (which is p_{n_2}) shows up, etc. For all $i \in \mathbb{N} \setminus \{0\}$, the knowledge base will be changed in view of d_{j_i} and the sequence of knowledge bases does not converge. Since there are uncountably many increasing sequences $\{n_i\}_{i \in \mathbb{N}}$, it follows that there are uncountably many data enumerations that can fail the expert.

The proof of Proposition 3 shows that experts who can correctly classify a single data case, but who are unable to select the relevant attributes, will induce the system to produce infinitely many knowledge bases. At any given time, the number of attributes occurring in the data cases observed so far is finite. When the expert has to revise the current knowledge base, he could take into consideration only the finite set of observed attributes, and converge to a correct knowledge base. The problem with the expert defined in the proof of the previous proposition is not that the number of attributes is potentially infinite (with more and more of them being observed as more and more data come in), but the fact that he chooses the wrong attribute from the finite set of available ones. It should also be noted that the behaviour of experts described in the proof is in accordance with practical RDR based systems.

Theorem 1. *Let a complete enumeration of data $(d_n)_{n \in \mathbb{N} \setminus \{0\}}$, a classification function ϕ, a sequence $(\delta_n)_{n \in \mathbb{N} \setminus \{0\}}$ of formulas, and a sequence $(K_n)_{n \in \mathbb{N}}$ of knowledge bases be such that:*

- *$((\delta_n)_{n \in \mathbb{N} \setminus \{0\}}, (K_n)_{n \in \mathbb{N}})$ is an admissible scenario for $(d_n)_{n \in \mathbb{N} \setminus \{0\}}$ and ϕ;*
- *there exists a finite set A of propositional atoms such that for all $n \in \mathbb{N}$, δ_n is built from members of A only.*

Then $(K_n)_{n \in \mathbb{N}}$ converges to some knowledge base.

Proof. Let $n \in \mathbb{N}$ be given. By Definition 3, for all distinct members $(\alpha_1, c_1), (\alpha_2, c_2)$ of K_n, α_1 and α_2 are not logically equivalent. Moreover, it follows from Definition 8 that for all $(\alpha, c) \in K_n$, α is built from members of A only. Hence there exists $N \in \mathbb{N}$ such that for all $n \in \mathbb{N}$, the number of rules in K_n is bounded by N. Also from Definition 8, the number of rules in K_{n+1} is greater than or equal to the number of rules in K_n, for all $n \in \mathbb{N}$. This immediately implies the sequence $(K_n)_{n \in \mathbb{N}}$ converges.

The theorem states that if the expert bases his judgement of the classification on a finite set of attributes, then the knowledge base will eventually converge to some limiting knowledge base. Proposition 2 guarantees that the limit knowledge base is correct with respect to the target classification function.

One of the possible implications of the previous theorem can be applied to the domain of web page classification. If the keywords are taken as attributes, then it is possible to assume that the number of distinct attributes in the domain is large (infinite). On the other hand, the number of attributes that occur in a given web page is small (finite). Hence a web page can be identified as a data case in the sense of our framework. Therefore, if we assume that the classification function can be defined in terms of a finite number of attributes (keywords), then a good expert, namely, an expert who considers new attributes only when she is forced to, will eventually construct a good classifier.

There are alternative approaches for building knowledge bases from a stream of cases, some of these approaches might be conceptually simpler, and enjoy a straightforward convergence result. It is beyond the scope of this paper to compare RDR with competing approaches. RDR has turned out to be very valuable in practice, in particular because it not only classifies cases but also encodes knowledge that supports classification. Hence RDR is a useful tool for explanation and teaching. For this reason, it is important that the convergence property of the RDR approach is guaranteed.

Feature selection is the problem of removing irrelevant or redundant features from a given set [3] and is often formalised as a search problem. It is considered as a hard problem (even if the number of attributes is small) that has attracted lots of attention recently [7, 12]. Our work is related to this problem in the sense that it analyses how expertise can be used in the selection process. RDR has proved useful in acquiring search knowledge [2] and could provide some insight to the feature selection problem.

5   Conclusion and Further Work

Ripple Down Rules has proved successful in practice as a methodology to build knowledge based systems. However, there has been so far no systematic analysis of the fundamental features that make it successful. In this paper, we have presented a crude but accurate formalisation of RDR methodology. We have shown that concepts from learning theory can be fruitfully applied and capture a desirable property of practical systems, namely that they eventually stabilise to an accurate representation of a target classification function. In practice, the number of potential attributes is very large and neither the expert nor the system knows the small number of those that are relevant. The number of attributes that appear in a data case as well as the number of attributes on the basis of which the target classification function can be expressed is small. We have examined the consequences of this asymmetry and shown that it might prevent the system from converging to a correct knowledge base, but that experts can apply some strategy to ensure convergence. We also showed that convergence guarantees correctness.

Our aim is to come up with formalisations of general classes of strategies, and evaluate their chance of success. This paper represents a first step in this direction, and illustrates the potential of the approach. It paves the way to more fine-grained models where the behaviour of the agents involved (user, expert, system) can be described more realistically. In particular, further models will incorporate possibly noisy or incomplete data, and consider less stringent classification criteria. We will also consider under which conditions the convergence of the sequence of knowledge bases is subject to an ordinal mind change bound [1]. Finally, alternative models should be investigated to shed light on other issues, e.g., speed of convergence.

References

1. A. Ambainis, R. Freivalds, and C. Smith. Inductive inference with procrastination: back to definitions. *Fundam. Inf.*, 40(1):1–16, 1999.
2. G. Beydoun and A. Hoffmann. Incremental acquisition of search knowledge. *Journal of Human-Computer Studies*, 52:493–530, 2000.
3. A. Blum and P. Langley. Selection of relevant features and examples in machine learning. *Artificial Intelligence*, 97(1-2):245–271, 1997.
4. R. Colomb. Representation of propositional expert systems as partial functions. *Artificial Intelligence*, 109(1-2):187–209, 1999.
5. P. Compton and G. Edwards. A philosophical basis for knowledge acquisition. *Knowledge Acquisition*, 2:241–257, 1990.
6. P. Compton, Z. Ramadan, P. Preston, T. Le-Gia, V. Chellen, and M. Mullholland. A trade-off between domain knowledge and problem solving method power. In B. Gaines and M. Musen, editors, *11th Banff KAW Proceeding*, pages 1–19, 1998.
7. M. Dash and H. Liu. Consistency-based search in feature selection. *Artificial Intelligence*, 151(1-2):155–176, 2003.
8. M. E. Gold. Language identification in the limit. *Information and Control*, 10:447–474, 1967.
9. B. Kang, K. Yoshida, H. Motoda, and P. Compton. A help desk system with intelligence interface. *Applied Artificial Intelligence*, 11:611–631, 1997.
10. M. J. Kearns. Efficient noise-tolerant learning from statistical queries. In *Proceedings of the 25th ACM Symposium on the Theory of Computing*, pages 392–401. ACM Press, 1993.
11. R. Kwok. Translation of ripple down rules into logic formalisms. In Rose Dieng and Olivier Corby, editors, *EKAW 2000*, volume 1937 of *Lecture Notes in Computer Science*. Springer, 2000.
12. H. Motoda and H. Liu. Data reduction: feature aggregation. In *Handbook of data mining and knowledge discovery*, pages 214–218. Oxford University Press, Inc., 2002.
13. P. Preston, G. Edwards, and P. Compton. A 2000 rule expert system without a knowledge engineer. In B. Gaines and M. Musen, editors, *8th Banff KAW Proceeding*, 1994.
14. D. Richards and P. Compton. Taking up the situated cognition challenge with ripple down rules. *Journal of Human-Computer Studies*, 49:895–926, 1998.
15. T. Scheffer. Algebraic foundation and improved methods of induction of ripple down rules. In *Pacific Rim Workshop on Knowledge Acquisition Proceeding*, 1996.
16. G. Shiraz and C. Sammut. Combining knowledge acquisition and machine learning to control dynamic systems. In *Proceedings of the 15th International Joint Conference in Artificial Intelligence (IJCAI'97)*, pages 908–913. Morgan Kaufmann, 1997.

Privacy Problems
with Anonymized Transaction Databases

Taneli Mielikäinen

HIIT Basic Research Unit
Department of Computer Science
University of Helsinki, Finland
Taneli.Mielikainen@cs.Helsinki.FI

Abstract. In this paper we consider privacy problems with anonymized
transaction databases, i.e., transaction databases where the items are
renamed in order to hide sensitive information. In particular, we show
how an anonymized transaction database can be deanonymized using
non-anonymized frequent itemsets. We describe how the problem can
be formulated as an integer programming task, study the computational
complexity of the problem, discuss how the computations could be done
more efficiently in practice and experimentally examine the feasibility of
the proposed approach.

1 Introduction

A *transaction database* is a bag (i.e., a multi-set) of *itemsets* (called *transactions*
in the transaction database) that are finite subsets of the set of items. The
prominent example of transaction data is market basket data. In that case each
transaction corresponds to one purchase event and each item is a salable product.
Also the popular representation of text documents as a set of words contained
in the document can be seen as transaction data: items are possible words in
documents and each transaction represents one document. A collection of web
pages is yet another example of transaction databases. For each web page in the
collection there is one transaction containing its out-going (or alternatively in-
coming) links. Thus, the items in this case are the links. Transaction databases
have important role in data mining. For example, association rules were first
defined for transaction databases [1].

Many times the owner of the transaction database is not willing to reveal the
database to others due to confidentiality or potential valuability of the database
but might still be interested to share the data if the privacy problems can be
avoided. For example, the data owner might be interested to compute and possibly also to distribute some summaries computed from the data. Hiding sensitive
information about the data has been recognized to be an important aspect of
data mining and it has recently been studied actively, see e.g. [2, 3].

One particularly simple approach to hide sensitive information is to rename
the set $\mathcal{I}$ of items e.g. by a random bijection $f : \mathcal{I} \rightarrow \mathcal{J}$ where $\mathcal{J}$ is a set with
same cardinality as $\mathcal{I}$.

E. Suzuki and S. Arikawa (Eds.): DS 2004, LNAI 3245, pp. 219–229, 2004.

A great advantage of this approach to anonymize transaction databases is that the existing methods to manipulate transaction data can be directly applied to the anonymized version of the data if the methods depend only on the syntactical structure of data. Examples of such methods are e.g. decision tree induction [4], learning Bayesian networks [5] and association rule mining [6].

A downside of the method is that the data mining results computed from the anonymized transaction database are not always very informative. Sometimes it is not necessary to remove the anonymization from the data mining results. For example, in the case of classification, the data owner can provide the anonymized data to a data miner that constructs the classifier for the anonymized data and gives that to the data owner. As the data owner knows the mapping to anonymize the data, she can classify the transactions of other parties on demand by anonymizing the transactions, classifying them by the classifier and deanonymizing the classification result, i.e., the class of the transaction. The knowledge about the data represented by the classifier is not shared by sharing the anonymized version of the classifier.

Even more difficult problems have to be faced when the data mining result is inherently representing some knowledge about the original data and the purpose of the knowledge is to be shared with other parties. This is the case e.g. with association rules. They are a summary of the data as conditional empirical probabilities over sets of items. The data owner might be willing to share (or more probably sell) this kind of summary information about her database although she does not want to reveal the actual database.

It has recently been shown that finding a database compatible with given collection of association rules or frequent itemsets is NP-hard [7, 8]. Thus, at least some privacy is preserved when releasing the frequent itemsets of the transaction database although it is known that frequent itemsets tell much also about other itemsets [9]. That is, releasing frequent itemsets might be an acceptable privacy risk for the data owner.

If the data owner can compute the association rules by herself then there is no need for anonymizing the data. Sometimes, however, the data owner is not interested or capable of conducting the data mining by herself but is willing to use the services of some data miner. In that case it is quite reasonable to assume that the data owner is not interested or willing to do very advanced obscuration of the database.

There are methods for hiding sensitive association rules if the sensitive rules are known [10–13]. (The problem of hiding all sensitive rules optimally has been shown to be NP-hard [14].) This requires, however, that the association rules that should not be revealed are known in advance and modifying the transactions changes also the frequencies of the itemsets. When the data miner mines association rules from the anonymized database, the data owner can interactively control which itemsets she is willing to reveal and which must be hidden. (There are methods for hiding such rules that tolerate certain attacks [15].)

In this paper we show that it is very dangerous to share the anonymized data and non-anonymized data mining results. In particular, we show that it is possible to estimate the original names of frequent items from anonymized data and non-anonymized frequent itemsets in non-deterministic polynomial time even in the case when the frequencies are independently perturbed by noise, some frequent itemsets are omitted and some itemsets are falsely claimed to be frequent. Furthermore, we show that even the simplest variants of the problem are at least as hard as graph isomorphism problem, but the exact maximum likelihood solution can be feasible in practice and that more efficiently solvable relaxed versions of the problems give reasonable solutions for the exact problem.

The rest of the paper is organized as follows. In Section 2 we describe how the problem of finding the original item names can be formulated as an integer programming task of reasonable size. In Section 3 we study the computational issues of the problem: we show that it is at least as difficult as graph isomorphism and discuss how the integer programming formulations could be relaxed and simplified without losing too much of the accuracy of the solution. The feasibility of the integer programming approach and its relaxations are evaluated experimentally in Section 4. Section 5 is a short conclusion.

2   Deanonymizing Transactions

To be able to consider anonymized transaction database, it is useful to define transaction databases and anonymizations of the transaction databases:

Definition 1 (transaction databases). *Let the collection of all items be denoted by $\mathcal{I}$. A set of items is called an* itemset. *A transaction database D is a* bag *of* transactions, *i.e., a multiset of subsets $D_1, \ldots, D_n$ of $\mathcal{I}$.*

Definition 2 (anonymization). *An anonymization of a transaction database D over $\mathcal{I}$ is a random mapping $f : \mathcal{I} \to \{1, \ldots, |\mathcal{I}|\}$. We use the shorthands $f(X) = \{f(A) : A \in X\}$ for itemsets $X \subseteq \mathcal{I}$, $f(\mathcal{F}) = \{f(X) : X \in \mathcal{F}\}$ for itemset collections $\mathcal{F} \subseteq 2^{\mathcal{I}}$, and $f(D) = \{f(D_i) : 1 \le i \le n\}$ for transaction databases.*

The most prominent data mining task for transaction databases is finding all σ-frequent itemsets in the database.

Problem 1 (discovery of σ-frequent itemsets). Given a transaction database D and minimum frequency threshold $\sigma \in [0, 1]$, find all σ-frequent itemsets, i.e., find the collection

$$\mathcal{F}(\sigma, D) = \{X \subseteq \mathcal{I} : fr(X, D) \ge \sigma\}.$$

where the *frequency $fr(X, D)$* of an itemset X in D is the fraction of transactions containing X, i.e.,

$$fr(X, D) = \frac{|\{i : X \subseteq D_i \in D\}|}{|D|}.$$

There exist highly optimized methods for finding the σ-frequent itemsets but still the resources needed can be too high for an ordinary user [16]. Thus, the

data owner might buy this service from a specialized data mining company if
the privacy issues are not of the highest priority. It would be desirable that the
privacy of the database would not be sacrificed completely. A very simple and
computationally cheap solution is to first anonymize the database and give that
to the data miner and deanonymize the data mining result. More specifically we
consider the following procedure:

Anonymization of the database. The data owner chooses a random anonymization function f and sends $f(D)$ and a minimum frequency threshold
$\sigma \in [0, 1]$ to the data miner.

Mining the anonymized database. The data miner computes the collection
$\mathcal{F}(\sigma, f(D))$ of frequent itemsets and sends $\mathcal{F}(\sigma, f(D))$ with their frequencies
to the data owner.

Deanonymization of the σ-frequent itemsets. The data owner computes
$\mathcal{F}(\sigma, D) = f^{-1}(\mathcal{F}(\sigma, f(D)))$ and publishes $\mathcal{F}(\sigma, D)$.

If the data miner is honest then this procedure does not cause any privacy
problems and in fact the anonymization would be unnecessary. In this case the
only information revealed to possibly malicious parties is the collection of σ-
frequent itemsets in D and their frequencies. We assume that the data miner
computes the frequent itemsets correctly. However, there is no reason (excluding
possible ethical and legal reasons) why the data miner should not try to find out
more about the data. The frequent itemsets can leak some information about
the relationship between the anonymized and the original databases. Thus, the
anonymization approach is not secure in a strict cryptographic sense. The data
owner, however, can hope that not too much is leaked. As an extreme case,
the leakage could reveal the whole original database, i.e., it might be possible
to estimate the correct names of the items from the non-anonymized frequent
itemsets and their frequencies.

Problem 2 (deanonymization of transaction databases from σ-frequent itemsets).
Given an anonymized transaction database $f(D)$, the collections $\mathcal{F}(\sigma, f(D))$ and
$\mathcal{F}(\sigma, D)$ of σ-frequent itemsets, find a bijection $g : f(\mathcal{I}) \to \mathcal{I}$ such that $X \in
\mathcal{F}(\sigma, f(D)) \iff g(X) \in \mathcal{F}(\sigma, D)$.

One possible solution mapping g for the above problem is the inverse of
the original anonymization function f. Clearly, only the frequent items can be
deanonymized based on frequent itemsets since infrequent items do not occur
in the collection of frequent itemsets. Fortunately, the frequent items are often
considered the most important ones among all items.

Problem 2 is essentially about finding a matching between the items in $\mathcal{I}$
and the integers in $\{1, \ldots, |\mathcal{I}|\}$ that fulfills some additional constraints. Thus,
the problem can be formulated as solving integer inequalities as follows. For
each pair of itemsets $X \in \mathcal{F}(\sigma, D)$ and $Y \in \mathcal{F}(\sigma, f(D))$ with $|X| = |Y|$ and
$fr(X, D) = fr(Y, f(D))$ there is an indicator variable $u_{X,Y} \in \{0, 1\}$ indicating
whether all items in X are matched with the integers in Y.

The variables $u_{X,Y}$ are constrained as follows. It is required that each itemset
X in $\mathcal{F}(\sigma, D)$ is matched exactly to one itemset of same cardinality in $\mathcal{F}(\sigma, f(D))$
and the vice versa. That is,

$$\sum_{X \in \mathcal{F}(\sigma, D)} u_{X,Y} = 1 \quad \text{and}$$

$$\sum_{Y \in \mathcal{F}(\sigma, f(D))} u_{X,Y} = 1.$$

Furthermore, it is required that if $X \in \mathcal{F}(\sigma, D)$ and $Y \in \mathcal{F}(\sigma, f(D))$ are matched then all their items are matched with each other. This constraint can be expressed by requiring that the sum of indicator variables between the items of X and Y sum up at least to value $|X| u_{X,Y} = |Y| u_{X,Y}$, i.e.,

$$\sum_{A \in X, B \in Y} u_{A,B} - |X| u_{X,Y} \geq 0.$$

A bit vector u fulfilling these constraints is a solution to Problem 2: each itemset in $\mathcal{F}(\sigma, D)$ (and thus also each item) is matched with exactly one itemset in $\mathcal{F}(\sigma, f(D))$ with $|X| = |Y|$ and $fr(X, D) = fr(Y, f(D))$, and if itemsets $X \in \mathcal{F}(\sigma, D)$ and $Y \in \mathcal{F}(\sigma, f(D))$ are matched then the items in X are Y are matched with each other.

If the frequent itemsets are restrictive enough, it is possible to find out the correct names for the items, i.e., the data miner could construct the original data from the anonymized data and frequent itemsets.

The data owner is probably willing to use only quite simple techniques to obscure the frequent itemsets if she is not interested to mine the data by herself.

First, the data owner could perturb the frequencies of the frequent itemsets a small amount independently, e.g., by adding to each frequency a Gaussian zero-mean noise with small variance. With high noise levels this can reduce the chance to find a naming of the items that is close to the correct one but it also reduces the usefulness of the itemsets. Let $c_{X,Y}$ be the cost of matching $X \in \mathcal{F}(\sigma, D)$ with $Y \in \mathcal{F}(\sigma, f(D))$ with $|X| = |Y|$. Then the problem of solving the integer inequalities becomes an integer programming problem:

$$\min_{u} \sum_{X \in \mathcal{F}(\sigma, D), Y \in \mathcal{F}(\sigma, f(D))} c_{X,Y} u_{X,Y}$$

subject to

$$\sum_{X \in \mathcal{F}(\sigma, D)} u_{X,Y} = 1$$

$$\sum_{Y \in \mathcal{F}(\sigma, f(D))} u_{X,Y} = 1$$

$$\sum_{A \in X, B \in Y} u_{A,B} - |X| u_{X,Y} \geq 0$$

$$u_{X,Y} \in \{0, 1\}$$

If the additive noise $n_{X,Y}$ in the frequencies is Gaussian and loss of matching $X \in \mathcal{F}(\sigma, D)$ with $Y \in \mathcal{F}(\sigma, f(D))$ with $|X| = |Y|$ is defined to be $c_{X,Y} =$

$|fr(X, D) - fr(Y, f(D)) + n_{X,Y}|^2$ then the optimal solution of the integer program gives the maximum likelihood estimate of the correct item naming.

Second, the data owner could omit some frequent itemsets and add some false frequent itemsets. That situation can be modeled by adding one new indicator variable u_Z with cost c_Z for each $Z \in \mathcal{F}(\sigma, D) \cup \mathcal{F}(\sigma, f(D))$ indicating that itemset Z is not matched with any itemset and minimizing the cost. That is,

$$\min_u \sum_{Z \in \mathcal{F}(\sigma, D) \cup \mathcal{F}(\sigma, f(D))} c_Z u_Z$$

subject to

$$u_Y + \sum_{X \in \mathcal{F}(\sigma, D)} u_{X,Y} = 1$$

$$u_X + \sum_{Y \in \mathcal{F}(\sigma, f(D))} u_{X,Y} = 1$$

$$\sum_{A \in X, B \in Y} u_{A,B} - |X| u_{X,Y} \geq 0$$

$$u_{X,Y} \in \{0, 1\}$$

Combining the above integer programs it is possible to model the situation where both the frequencies and the structure of the collection are modified.

3   On Feasibility of Deanonymization

The integer programming formulations of Problem 2 and its generalizations consist of polynomial number of variables, equalities and inequalities in $|\mathcal{F}(\sigma, D)|$ and in $|\mathcal{F}(\sigma, f(D))|$. These numbers are clearly bounded above by the product $\mathcal{O}(|\mathcal{F}(\sigma, D)| |\mathcal{F}(\sigma, f(D))|)$ when both collections are non-empty. Thus, as integer programming is in NP, also Problem 2 is.

The complexity of the problem can be lower bounded by the complexity of the graph isomorphism problem:

Problem 3 (graph isomorphism). Given a graphs $G = \langle V, E \rangle$ and $G' = \langle V', E' \rangle$, find a bijection $f : V \to V'$ such that $\{u, v\} \in E \iff \{f(u), f(v)\} \in E'$.

On one hand the graph isomorphism is in $NP \cap co - NP$ thus it is not likely that it would be NP-complete, but on the other hand there are no polynomial time algorithms for the problem [17, 18].

Theorem 1. *Problem 2 is (polynomially) at least as hard as Problem 3.*

Proof. To show the hardness we construct from a graph $G = \langle V, E \rangle$ a transaction database D_G with polynomial number of transaction in the size of G. The transaction database consists of a transaction $\{A, B\}$ for each edge $\{A, B\} \in E$ and $\max_{C \in V} |B \in V : \{C, B\} \in E| - |B \in V : \{A, B\} \in E|$ copies of transactions $\{A\}$ for each vertex $A \in V$. Then $fr(A, D_G) = fr(B, D_G)$ for all $A, B \in V$,

$fr(X, D_G) = fr(Y, D_G)$ for each $X, Y \in E$ and $fr(X, D_G) = 0$ for all other X. Thus, choosing a positive minimum frequency threshold that is small enough $\mathcal{F}(\sigma, D) = V \cup E$. The transaction database D_G consists of at most $|V||E|$ transactions which is clearly polynomial in $|V|$ and $|E|$ and the size of each transaction is at most two.

Thus, both graphs G and G' can be represented by polynomial-size transaction databases in such a way that the frequent itemsets in D_G and $D_{G'}$ can be matched if and only if G and G' are isomorphic.   □

The practical vulnerability of the anonymized transaction databases depends on our ability to solve the computational task of finding the mapping in reasonable time. As we have the integer programming formulation of the problem, a natural solution is to solve the integer program. However, this is often computationally too demanding. A standard approach is to relax the integer program to a linear program. For linear programming there exist algorithms efficient in theory and practice [19]. In our case the relaxation of the integer program to a liner program means that the constraints $u_{X,Y} \in \{0,1\}$ are replaced by constraints $0 \leq u_{X,Y} \leq 1$. A major difficulty in this relaxation is that instead of matching the item names, the solution of the linear program gives for each itemset $Y \in \mathcal{F}(\sigma, f(D))$ values of matching with each $X \in \mathcal{F}(\sigma, D)$ that sum up to one. These values can be translated into integers by means of the following procedure:

1. If $\mathcal{F}(\sigma, f(D)) = \emptyset$ then halt. Otherwise choose Y randomly from $\mathcal{F}(\sigma, f(D))$.
2. Choose a pair $X \in \mathcal{F}(\sigma, D)$ for Y w.r.t. the distribution determined by $u_{Z,Y}$ for all Z.
3. Remove Y from $\mathcal{F}(\sigma, f(D))$ and X from $\mathcal{F}(\sigma, D)$ and set $u_{X,Z} = 0$ and $u_{Z,Y}$ for all Z.

Another possibility to find the matching between $\mathcal{I}$ and $\{1, \ldots, |\mathcal{I}|\}$ is to require that $u_{A,B} \in \{0,1\}$ for all $A \in \mathcal{I}, B \in \{1, \ldots, |\mathcal{I}|\}$. Usually the number of frequent items is much smaller than the number all frequent itemsets. There exist sophisticated techniques to solve this kind of mixed integer programs [20].

The frequent itemset collection often gives natural constraints that some items cannot be matched with some numbers. Obviously, this information can be used to prune the constraints and setting some variables to be zero in advance.

4   Experiments

To examine the practical feasibility of the integer programming solutions and its relaxations we conducted preliminary experiments with the course enrollment database of Department of Computer Science, University of Helsinki. More specifically, we wanted to study the effect of noise in frequencies to different estimation methods and different minimum frequency thresholds. The item namings were estimated using integer programming where variables were constrained to be either one or zero, mixed integer programming where only the variables corresponding to the singleton itemsets were required to be integer valued, and linear

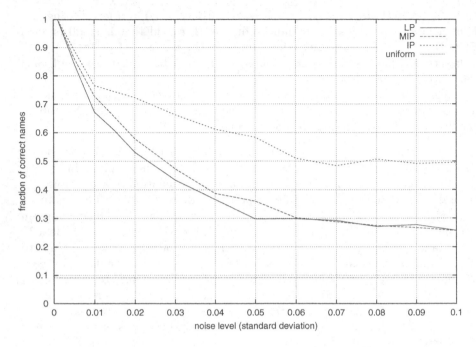

Fig. 1. Deanonymization based on 0.1-frequent itemsets.

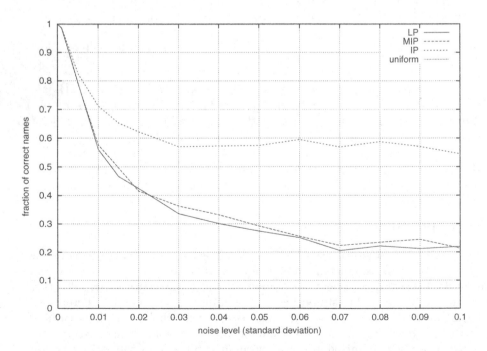

Fig. 2. Deanonymization based on 0.09-frequent itemsets.

programming where all values were constrained to be between zero and one. The quality of the solution was measured by computing the weight given to the correct item naming of the items divided by the number of frequent items. Thus in the case of integer and mixed integer programming the solutions were readily matchings whereas the linear programming solutions were able to vote several matchings. As the noise were added randomly, all results shown are averages of one hundred experiments.

In all our experiments all methods were always able to find the correct item namings when there were no noise in the frequencies. When the noise level were increased, then the differences between the methods became visible. The integer programming were able to find the most correct item namings. Mixed integer programming and linear programming were approximately equally good. However, the mixed integer programming solution was readily a proper item naming of the items instead of a probability distribution over the possible item names.

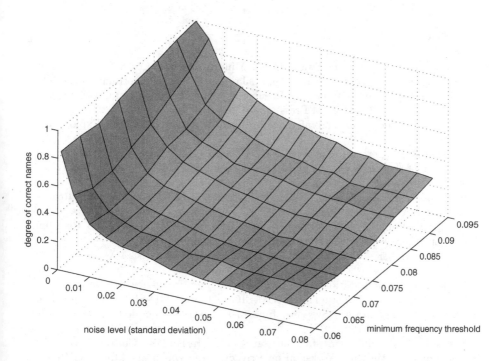

Fig. 3. Deanonymization using linear programming.

In Figure 1 and Figure 2 the results are shown for minimum frequency thresholds 0.1 and 0.09. The uniform voting for all matching pairs is shown as a baseline in both figures. Clearly, even linear programming were able to increase our knowledge about the correct item namings.

In Figure 3 experiments with linear programming for larger variety of different minimum frequency thresholds are shown. The behavior of the method were similar with different thresholds and the linear programming result were always better than the uniform assumption. Overall, the results show that the approach can be used to find namings of the items that are relatively close to the correct ones.

5   Conclusions

In this paper we have studied the privacy problems of renaming items in transaction databases. In particular, we considered the problem of finding the original names of the items using the anonymized transaction database and non-anonymized frequent itemsets. We examined the complexity of the problem and described practical solutions for the problems that can approximate the correct item namings reasonably well. The preliminary experiments on deanonymizing transaction databases shown promising results.

References

1. Agrawal, R., Imielinski, T., Swami, A.N.: Mining association rules between sets of items in large databases. In Buneman, P., Jajodia, S., eds.: Proceedings of the 1993 ACM SIGMOD International Conference on Management of Data, Washington, D.C., May 26-28, 1993. ACM Press (1993) 207–216
2. Farkas, C., Jajodia, S.: The inference problem: A survey. SIGKDD Explorations **4** (2002) 6–11
3. Verykios, V.S., Bertino, E., Fovino, I.N., Provenza, L.P., Saygin, Y., Theodoridis, Y.: State-of-the-art in privacy preserving data mining. SIGMOD Record **33** (2004) 50–57
4. Breiman, L., Friedman, J.H., Olshen, R.A., Stone, C.J.: Classification and Regression Trees. Wadsworth (1984)
5. Pearl, J.: Probabilistic Reasoning in Intelligent Systems: Networks of Plausible Inference. Revised second printing edn. Morgan Kaufmann (1988)
6. Han, J., Kamber, M.: Data Mining: Concepts and Techniques. Academic Press (2001)
7. Calders, T.: Computational complexity of itemset frequency satisfiability. In: Proceedings of the Twenty-Third ACM SIGACT-SIGMOD-SIGART Symposium on Principles of Database Systems, June 13-18, 2004, Maison de la Chimie, Paris, France. ACM (2004)
8. Mielikäinen, T.: On inverse frequent set mining. In Du, W., Clifton, C.W., eds.: Proceedings of the 2nd Workshop on Privacy Preserving Data Mining (PPDM), November 19, 2003, Melbourne, Florida, USA. IEEE Computer Society (2003) 18–23
9. Calders, T., Goethals, B.: Mining all non-derivable frequent itemsets. In Elomaa, T., Mannila, H., Toivonen, H., eds.: Principles of Data Mining and Knowledge Discovery, 6th European Conference, PKDD 2002, Helsinki, Finland, August 19-23, 2002, Proceedings. Volume 2431 of Lecture Notes in Artificial Intelligence. Springer (2002) 74–865
10. Saygin, Y., Verykios, V.S., Clifton, C.: Using unknowns to prevent discovery of association rules. SIGMOD Record **30** (2001) 45–54

11. Oliveira, S.R.M., Zaïane, O.R.: Privacy preserving frequent itemset mining. In Clifton, C., Estivill-Castro, V., eds.: IEEE Workshop on Privacy, Security, and Data Mining. Volume 14 of Conferences in Research and Practice in Information Technology. (2002)

12. Oliveira, S.R.M., Zaïane, O.R.: Protecting confidential knowledge by data sanitation. In Wu, X., Tuzhilin, A., Shavlik, J., eds.: Proceedings of the 3rd IEEE International Conference on Data Mining (ICDM 2003), 19-22 December 2003, Melbourne, Florida, USA. IEEE Computer Society (2003) 613–616

13. Verykios, V.S., Elmagarmid, A.K., Elisa Bertino, F., Saygin, Y., Dasseni, E.: Association rule hiding. IEEE Transactions on Knowledge and Data Engineering **16** (2004) 434–447

14. Atallah, M.J., Bertino, E., Elmagarmid, A.K., Ibrahim, M., Verykios, V.S.: Disclosure limitation of sensitive rules. In: Proceedings of 1999 Workshop on Knowledge and Data Engineering Exchange (KDEX '99). IEEE Computer Society (1999) 45–52

15. Oliveira, S.R.M., Zaïane, O.R., Saygin, Y.: Secure association rule sharing. In Dai, H., Srikant, R., Zhang, C., eds.: Advances in Knowledge Discovery and Data Mining, 8th Pacific-Asia Conference, PAKDD 2004, Sydney, Australia, May 26-28, 2004, Proceedings. Volume 3056 of Lecture Notes in Artificial Intelligence. Springer (2004) 74–85

16. Goethals, B., Zaki, M.J., eds.: Proceedings of the Workshop on Frequent Itemset Mining Implementations (FIMI-03), Melbourne Florida, USA, November 19, 2003. Volume 90 of CEUR Workshop Proceedings. (2003) http://CEUR-WS.org/Vol-90/.

17. Kreher, D.L., Stinson, D.R.: Combinatorial Algorithms: Generation, Enumeration and Search. CRC Press (1999)

18. Torán, J.: On the hardness of graph isomorphism. In: 41st Annual Symposium on Foundations of Computer Science, FOCS 2000, 12-14 November 2000, Redondo Beach, California, USA. IEEE Computer Society (2000) 180–186

19. Padberg, M.: Linear Optimization and Extensions. 2nd edn. Volume 12 of Algorithms and Combinatorics. Springer-Verlag (1999)

20. Martin, A.: General mixed integer programming: Computational issues for branch-and-cut algorithms. In Jünger, M., Naddef, D., eds.: Computational Combinatorial Optimization: Optimal and Provably Near-Optimal Solutions. Volume 2241 of Lecture Notes in Computer Science. Springer (2001) 1–25

A Methodology for Biologically Relevant Pattern Discovery from Gene Expression Data

Ruggero G. Pensa[1], Jérémy Besson[1,2], and Jean-François Boulicaut[1]

[1] INSA Lyon, LIRIS CNRS FRE 2672, F-69621 Villeurbanne cedex, France
{Ruggero.Pensa,Jeremy.Besson,Jean-Francois.Boulicaut}@insa-lyon.fr
[2] UMR INRA/INSERM 1235, F-69372 Lyon cedex 08, France

Abstract. One of the most exciting scientific challenges in functional genomics concerns the discovery of biologically relevant patterns from gene expression data. For instance, it is extremely useful to provide putative synexpression groups or transcription modules to molecular biologists. We propose a methodology that has been proved useful in real cases. It is described as a prototypical KDD scenario which starts from raw expression data selection until useful patterns are delivered. Our conceptual contribution is (a) to emphasize how to take the most from recent progress in constraint-based mining of set patterns, and (b) to propose a generic approach for gene expression data enrichment. The methodology has been validated on real data sets.

1 Introduction

Thanks to a huge research effort and technological breakthroughs, one of the challenges for molecular biologists is to discover knowledge from data generated at very high throughput. This is true not only for genomic data but also for gene expression data. Indeed, different techniques (e.g., microarray [1]) enable to study the simultaneous expression of (tens of) thousands of genes in various biological situations. Such data can be seen as expression matrices in which the expression level of genes (the attributes or columns) are recorded in various biological situations (the objects or lines). A toy example of a gene expression matrix is in Fig. 1a. Exploratory data mining techniques are needed that can, roughly speaking, be considered as the search for interesting *bi-sets*, i.e., sets of biological situations and sets of genes that are associated in some way. Indeed, it is interesting to look for groups of co-regulated genes, also known as *synexpression groups* [2], which, based on the guilt by association approach, are assumed to participate in a common function, or module, within the cell. Such an association between a set of co-regulated genes and the set of biological situations that gives rise to this co-regulation is called a *transcription module* and their discovery is a major goal in functional genomics. Various techniques can be used to identify a priori interesting bi-sets. Biologists often use clustering techniques to identify sets of genes that have similar expression profiles (see, e.g., [3]). Statistical methods can be used as well (see, e.g., [4,5]). Interesting pattern discovery techniques can be applied on boolean matrices that encode

E. Suzuki and S. Arikawa (Eds.): DS 2004, LNAI 3245, pp. 230–241, 2004.

expression properties of genes. Let $\mathcal{O}$ denote a set of biological situations and $\mathcal{P}$ denotes a set of genes. Expression properties, e.g., over-expression, can be encoded into $\mathbf{r} \subseteq \mathcal{O} \times \mathcal{P}$. $(o_i, g_j) \in \mathbf{r}$ denotes that gene j has the encoded expression property in situation i. For deriving a boolean context from raw gene expression data, we generally apply discretization operators that, depending on the chosen expression property, compute thresholds from which it is possible to decide between whether the true or the false value must be assigned. On our toy example, a value "1" for a biological situation and a gene means that the gene is up (greater than $|t|$) or down (lower than $-|t|$) regulated in this situation. Using threshold $t = 0.3$ leads to the boolean matrix in Fig. 1b. It is then possible to look for putative synexpression groups by computing the so-called frequent itemsets from the derived boolean contexts [6]. In our boolean toy example, the genes g_3 and g_5 have the same encoded expression property in situations o_1 and o_4. This observation might lead us to derive the bi-set $(\{o_1, o_4\}, \{g_3, g_5\})$ as being potentially interesting. Notice that sets of genes that are frequently co-regulated can be post-processed into association rules [7,8]. Stronger relationships between the components of a bi-set can increase their relevancy. For instance, $(\{o_1, o_4\}, \{g_2, g_3, g_5\})$ is one of the formal concepts (see, e.g., [9]) in the data from Fig. 1b. Informally, it means that $\{g_2, g_3, g_5\}$ is a maximal set of genes that have the recorded expression property in every situation from $\{o_1, o_4\}$ and that $\{o_1, o_4\}$ is a maximal set of situations which share the true value for every gene from $\{g_2, g_3, g_5\}$. Clearly, discovered concepts in this kind of boolean data provide putative transcription modules [10,11].

Sit.	\multicolumn{5}{c}{Genes}				
	g_1	g_2	g_3	g_4	g_5
o_1	0.36	0.42	0.56	0.124	0.35
o_2	-0.24	0.01	0.28	0.02	-0.32
o_3	0.25	0.35	0.55	0.012	-0.21
o_4	0.27	0.89	-1.02	0.71	0.52
o_5	0.53	0.24	0.64	-0.6	-0.01

(a)

Sit.	g_1	g_2	g_3	g_4	g_5
o_1	1	1	1	0	1
o_2	0	0	0	0	1
o_3	0	1	1	0	0
o_4	0	1	1	1	1
o_5	1	0	1	1	0

(b)

Fig. 1. A gene expression matrix (a) and a derived boolean context (b).

This paper is a methodological paper. It abstracts our practice in several real-life gene expression data analysis projects in order to disseminate a promising practice within the scientific community. Our methodology covers the whole KDD process and not just the mining phase. Starting from raw gene expression data, it supports the analysis and the discovery of relevant biological information via a constraint-based bi-set mining approach from computed boolean data sets. The generic process is described within the framework of inductive databases, i.e., each step of the process can be formalized as a query on data and/or patterns that satisfy some constraints [12,13]. It leads us to a formalization of *boolean gene expression data enrichment*. We already experimented a couple of practical instances of this approach and it has turned to be crucial for increasing the biological relevancy of the extracted patterns. An original validation of the methodology on a real data set w.r.t. a non trivial biological problem is provided.

In Section 2, the methodology is described by means of the definition of a prototypical KDD scenario. Each critical step is specified and difficulties for its execution are emphasized. In Section 3, we consider our recent contributions for supporting some of these steps. In other terms, we explain how we can execute specific instances of the given prototypical scenario by using our own data pre-processing tools (e.g., [14]), mining algorithms (e.g., [15]), and post-processing software (e.g., [16]). In Section 4, we illustrate an original application of the methodology for a real gene expression data analysis task. Section 5 concludes.

2   A Prototypical KDD Scenario

We assume that raw expression data, i.e., a function that assigns a real expression value to each couple $(o, g) \in \mathcal{O} \times \mathcal{P}$ is available and that some open problems have been selected by the molecular biologists. A typical example concerns the discovery of putative transcription modules that involve at least a given set of genes that are already known to be co-regulated in some class of biological situations, e.g., cancerous ones.

Due to the lack of space, we do not consider the typical data manipulation statements that are needed, e.g., for data normalization, data cleaning, gene and/or biological situation selection according to some background knowledge (e.g., removing the so-called housekeeping genes from consideration).

Discretization. The discretization step concerns gene expression property encoding and is obviously crucial. The simplest case concerns the computation of a boolean matrix $\mathbf{r} \subset \mathcal{O} \times \mathcal{P}$ which encode a simple expression property for each gene in each situation, e.g., over-expression. Different algorithms can be applied and parameters like thresholds have to be be chosen. For instance, [7] introduces three techniques for encoding gene over-expression:

- "Mid-Ranged". The highest and lowest expression values in a biological situation are identified for each gene and the mid-range value is defined. Then, for a given gene, all expression values that are strictly above the mid-range value give rise to value 1, 0 otherwise.
- "Max - X% Max". The cut off is fixed w.r.t. the maximal expression value observed for each gene. From this value, we deduce a percentage X of this value. All expression values that are greater than the (100 - X)% of the Max value give rise to value 1, 0 otherwise.
- "X% Max". For each gene, we consider the biological situations in which its level of expression is in X% of the highest values. These genes are assigned to value 1, 0 for the others.

The impact of the chosen algorithm and the used parameters on both the quantity and the relevancy of the extracted patterns is crucial. For instance, the density of the discretized data depends on the discretization parameters and the cardinalities of the resulting sets (collections of itemsets, association rules or formal concepts) can be very different. Therefore, we need to evaluate the goodness of a discretization process. Our thesis is that a good discretization might preserve some properties that can be already observed from raw data (see Section 3).

Boolean Gene Expression Data Enrichment. We can mine boolean gene expression matrices for frequent sets of genes and/or situations, association rules between genes and/or situations, formal concepts, etc. In the following, we focus on mining phases that compute concepts. When the extractions are feasible, many patterns are discovered (up to several millions) while only a few of them are interesting. It is however extremely hard to decide of the interestingness characteristics a priori. We now propose an extremely powerful approach for improving the relevancy of the extracted concepts by boolean data enrichment. It can be done a priori with some complementary information related to genes and/or situations. For instance, we can add information about the known functions of genes as it is recorded in various sources like Gene Ontology [17]. Other information can be considered like the associated transcription factors. A simple way to encode this kind of knowledge consists in adding a row to $\mathbf{r}$ for each gene property. Dually, we can add some properties to the situations vectors. For instance, if we know the class of a group of situations (e.g. cancerous vs. non cancerous cells) we can add a column to $\mathbf{r}$. We can also add boolean properties about, e.g., cell type or environmental features. Enrichment of boolean data can be performed by more or less trivial data manipulation queries from various bioinformatics databases. $\mathbf{r}' \subset \mathcal{O}' \times \mathcal{P}'$ will denote the relation of the enriched boolean context.

In Fig. 2a, we add two gene properties p_1 and p_2. A value "1" assigned to a property for some gene means that this gene has the property. For instance, p_1 could mean that the gene has a given function or is regulated by a given transcription factor. Dually, we consider two classes of situations c_1 and c_2.

Sit.	g_1	g_2	g_3	g_4	g_5	c_1	c_2
o_1	1	1	1	0	1	1	0
o_2	0	0	0	0	1	1	0
o_3	0	1	1	0	0	0	1
o_4	0	1	1	1	1	0	1
o_5	1	0	1	1	0	0	1
p_1	1	1	0	1	0	1	1
p_2	1	0	0	1	1	1	1

(a)

Sit.	g_1	g_2	g_3	g_4	g_5	c_1	c_2
o_1	1	1	1	0	1	1	0
o_2	0	0	0	0	1	1	0
o_3	0	1	1	0	0	0	1
o_4	0	1	1	1	1	0	1
o_5	1	0	1	1	0	0	1
p_1	1	1	0	1	0	1	1
p_2	1	0	0	1	1	1	1
p_3	1	0	1	1	0	1	1
p_4	1	1	1	0	1	1	1

(b)

Fig. 2. Two examples of enriched boolean microarray contexts.

A value "1" for a situation and a class means that this situation belongs to the class but this could be interpreted in terms of situation properties as well. For instance, c_1 could mean whether biological situations are cancerous ones or not. In the data in Fig. 2a, a formal concept like $(\{o_4, o_5\}, \{g_3, g_4, c_2\})$ informs us about a maximal rectangle that involves two genes in two situations that are of class c_2. This could reveal sets of genes that are co-regulated in non cancerous situations but not in cancerous ones. We discuss later how iterative enrichment is a key technique for improving the relevancy of the extracted patterns.

Constraint-Based Extraction of Formal Concepts. We consider here only formal concept discovery from eventually enriched boolean contexts. A formal concept is a maximal rectangle of "1" (1-rectangle) in the boolean matrix, and it can be represented as a bi-set of genes (eventually with situation properties) and situations (eventually with gene properties).

Definition 1. *(Concept and $\mathcal{C}_{Concept}$ constraint) A bi-set $(T, G) \in \mathcal{O} \times \mathcal{P}$ is a concept in $\mathbf{r}$ when it satisfies constraint $\mathcal{C}_{Concept}$ in $\mathbf{r}$ and $\mathcal{C}_{Concept}(T, G, \mathbf{r}) \equiv (T = \psi(G, \mathbf{r})) \wedge (G = \phi(T, \mathbf{r}))$ where ψ and ϕ are the Galois operators [9]. Let us recall that we have $\phi(T, \mathbf{r}) = \{g \in \mathcal{P} \mid \forall o \in O, (o, g) \in \mathbf{r}\}$ and $\psi(G, \mathbf{r}) = \{o \in \mathcal{O} \mid \forall g \in G, (o, g) \in \mathbf{r}\}$. (ϕ, ψ) is the Galois connection between $\mathcal{O}$ and $\mathcal{P}$.*

It is now possible to apply an algorithm for concept extraction to obtain the whole set of concepts and thus putative transcription modules. Notice that by construction, concepts are built on closed sets. It means that every algorithm that compute closed sets can be used to compute the concepts (see, e.g., [11] for the use of frequent closed set computation algorithms). Given Fig. 2a, $(\{o_1, o_4\}, \{g_2, g_3, g_5, \})$ and $(\{o_1, o_2, p_2\}, \{g_5, c_1\})$ are among the 29 concepts.

Mining every concept is not always tractable. If it is tractable, it provides potentially huge collections of patterns that have to be materialized for further post-processing guided by the molecular biologists. When the computation of every concept is not tractable, it is possible that pushing other user-defined constraints leads to tractable computations. For instance, we can extract concepts that contains a minimal or a maximal number of situations and/or genes, or that contains some particular situation and/or genes and/or their associated properties in the case of enriched contexts. Let us formalize such constraints:

Definition 2. *(Constraints on concept) A concept (T, G) is called frequent when it satisfies constraint $\mathcal{C}_t(\mathbf{r}, \sigma_1, T) \equiv |T| \geq \sigma_1$ (resp. $\mathcal{C}_g(\mathbf{r}, \sigma_2, G) \equiv |G| \geq \sigma_2$). A concept (T, G) satisfies a syntactical constraint of inclusion $\mathcal{C}_{Inclusion}(\mathbf{r}, X, G) \equiv X \subseteq G$ (resp. exclusion $\mathcal{C}_{Exclusion}(\mathbf{r}, X, G) \equiv X \not\subseteq G$). Dually, we can use $\mathcal{C}_{Inclusion}(\mathbf{r}, Y, T) \equiv Y \subseteq T$ (resp. $\mathcal{C}_{Exclusion}(\mathbf{r}, Y, T) \equiv Y \not\subseteq T$).*

It is quite useful to use these constraints in enriched contexts. For instance, we can specify that we want concepts whose situations belong to Class c_2 (say non cancerous cells) and such that the gene set contain some genes that are already known to participate to the studied regulatory way (e.g., g_1). It can be specified as the following inductive query:

$$q_1 : \mathcal{C}_{Concept}(T, G, \mathbf{r}) \wedge \mathcal{C}_{Inclusion}(\mathbf{r}, c_2, G) \wedge \mathcal{C}_{Inclusion}(\mathbf{r}, g_1, G).$$

Then, we can ask for a second collection with all the concepts (T, G) such that the class attribute c_1 is included in T:

$$q_2 : \mathcal{C}_{Concept}(T, G, \mathbf{r}) \wedge \mathcal{C}_{Inclusion}(\mathbf{r}, c_1, G).$$

Post-processing and Iteration. Concept extraction, even constraint-based mining, can produce large numbers of patterns, especially in the first iterations of the KDD process, i.e., when very few information can be used to further constrain the bi-sets to be delivered. Notice also that from a practical perspective, not all the specified constraints can be pushed into concept mining algorithms, in which case some of these constraints have to be checked in a post-processing phase.

KDD processes are clearly complex iterative processes for which every result can give rise to new ideas for more relevant constraint-based mining phases (inductive queries) or data manipulations. When a collection of patterns has been computed, it can be used for deriving new boolean properties. In particular, we have obtained two sets of patterns that can characterize two classes of genes and, dually, two classes of situations. Therefore, we can define two new class properties related to genes and their dual class properties related to situations. The boolean context $\mathbf{r}'$ can then be extended towards $\mathbf{r}'' \subset \mathcal{O}'' \times \mathcal{P}''$. Considering our running example, we can associate a new property p_3 (resp. p_4) for the genes belonging to the concepts which are returned by q_1 (resp. q_2). It leads to a new enriched boolean context given in Fig. 2b. New constraints on the classes can be used for the next mining phase. New set size constraints can be defined as well to avoid results due to noise. A new iteration will provide a new set of concepts. Each time a collection of concepts is available, we can decide either to analyze it by hand, e.g., studying each gene separately, or looking for new boolean data enrichment and new constraints for a new iteration.

In any case, at the end of the process, we have a set of putative interesting genes and a set of putative interesting situations. When considering our running example of putative transcription module discovery from an initial gene set (here $\{g_1\}$, called hereafter the seed set), it is interesting to stop iterations when the sets of genes include (almost) all the genes from the seed set and when the total number of genes which are not in the seed gene set can be studied in a reasonable time by means of biological experiments.

In our toy example (Fig. 2b), let us enforce the absence in T of p_4, i.e., those genes that are contained in the concepts concerning the situation belonging to Class c_1. The result is a single concept $(\{o_4, o_5, p_1, p_2, p_3\}, \{g_4, c_2\})$. The gene g_4 is co-regulated in two situations associated to Class c_2 but it is not in the seed set of genes known to be involved in the studied transcription process. It means that g_4 is a putative interesting gene that can be studied further to verify if its function is related to the studied biological problem. Notice also that genes to which we can associate new functions appear as interesting candidates for performing new iterations and take advantage of larger seed gene sets.

3   About Scenario Practical Executions

The prototypical scenario we have presented in the previous section can be executed in different ways, depending on available algorithms and tools. In this section, we explain how we can execute it on practical cases by taking the most from some recent advances on constraint-based set mining and gene expression data analysis. We do not provide here new results but evidence that such a prototypical scenario can be used by practitioners.

We have explained that discretization of raw gene expression data is a crucial phase. We clearly need a method to evaluate different boolean encoding (different techniques and/or various parameters) of the same raw data and thus a framework to support user decision about the discretization from which the mining process can start. Let E denote a gene expression matrix. Let $\{Bin_i, i = 1..b\}$ denote a set of different discretization operators and $\{r_i, i = 1..b\}$ a set of boolean contexts obtained by applying these operators, i.e. $\forall i = 1..b$, $r_i = Bin_i(E)$. Let $S : \mathbb{R}^{n,m} \longmapsto \mathbb{R}$ denote an evaluation function that measure the quality of the discretization of a gene expression matrix. We say that a boolean context r_i is more valid than another context r_j w.r.t the S measure if $S(r_i) > S(r_j)$. In [14], we recently studied an original method for such an evaluation. We suggest to compare the similarity between the dendrogram generated by a hierarchical clustering algorithm (e.g., [3]) applied to the raw expression data and the dendrograms generated by the same algorithm applied to each derived boolean matrix. Given a gene expression matrix E and two derived boolean contexts r_i and r_j for two distinct discretizations, we can choose the discretization that leads to the dendrogram which is the most similar to the one built on E. The idea is that a discretization that preserves the expression profile similarities is considered more relevant. In [14], a simple measure of similarity between dendrograms has been studied and experimentally validated on various gene expression data sets. It is used in Section 4 for our original application to the drosophila data set.

A second major problem concerns constraint-based mining of concepts. In our applications to gene expression data, we can get rather dense boolean contexts that are hard to process if further user-defined constraints are not only specified but also pushed deeply into the extraction algorithms. Using user-defined constraints enables to produce putative interesting concepts before any post-processing phase. Indeed, concept discovery techniques can provide huge collection of patterns and supporting post-processing on such collections is hard or even impossible. It motivates the a priori use of constraints on both $2^{\mathcal{O}}$ and $2^{\mathcal{P}}$. We saw typical examples of constraints on the size of T and G. The recent algorithm D-MINER introduced in [15] computes concepts under monotonic constraints and can be used in dense boolean data sets when the previous algorithms (concept lattice discovery algorithms or frequent closed set computation algorithms) generally fail.

Another important problem concerns the postprocessing of concept collections. As the number of concepts to analyze starts to be huge, we need efficient exploration techniques to support the subjective search of interesting concepts. In [16], we propose an "Eisen-like" visualization technique, that allows to group similar concepts by means of a hierarchial clustering algorithm using an original definition of distance. Thanks to this approach we can reduce the effects of concept multiplication due to noise in data and support the post-processing of thousands of concepts with a graphical approach.

4   An Application

We have used our methodology on a real gene expression data analysis problem for which it was possible to evaluate the relevancy of the results thanks to the available documentation [18]. It concerns the gene expression of the Drosophila melanogaster during its life cycle. The paper considers the expression level of 4 028 genes for 66 sequential time periods from the embryonic state till the adulthood. The related data set is available on line[1]. The total number of samples is 81 since the gene expression during the adult state is measured for male and female individuals and the expression level of more genes is available. For our experiment we have selected a set of 4 137 genes and 20 time periods concerning adult individuals. This set is composed of 8 male adult samples, 8 female adult samples, 2 male and 2 female tudor samples. We selected 4 of the 4 137 genes which are known to be "male somatic genes", i.e., a class of genes that characterize the male individuals (Genes CG2858, CG2267, CG17843, and CG2082). Let us denote this set by $KG = \{kg_1, kg_2, kg_3, kg_4\}$. We want to discover new knowledge about this group, i.e., adding other genes to the seed set KG by applying our methodology. Notice that the genes from KG have been selected randomly among the known male somatic genes. In this experiment, our goal is to demonstrate that, given a small gene set, we are able to increase our knowledge with two simple iterations of the method. In other terms, we do not claim that we want to find all "male somatic genes" but we want to rediscover part of this knowledge thanks to the available biological results from [18].

Preprocessing. We marked a group of 351 genes as being always underexpressed (in all the 20 situations). Another group of 353 genes has been marked since it is over-expressed in more than 10 biological situations. We performed a projection on non-marked genes and we obtain at the end a $20 \times 3\,433$ expression matrix denoted E. To discretize E, we choose the "Max - X% Max" method:

$$\bar{b}_i = Bin(\bar{e}_i)$$

where, for each gene vector $\bar{b}_i$,

$$b_{ij} = \begin{cases} 1, & if\ (1 - X) \max_j (e_{ij}) < e_{ij} < \max_j (e_{ij}) \\ 0, otherwise \end{cases} \tag{1}$$

where $j = 1..20$ and $\bar{e}_i \in E$.

---

[1] http://genome-www5.stanford.edu/

Different values can be chosen for X and we applied the method described in [14] when considering X values between 0.01 and 0.9. The result of this analysis for gene dendrograms are summarized in Fig. 3. The best value for our similarity score is when $X = 0.54$. Consequently this is the threshold we used in order to derive the boolean gene expression data.

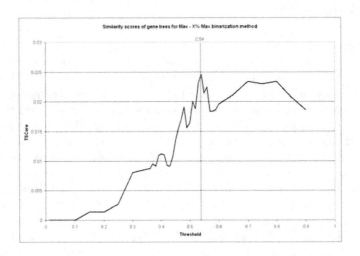

Fig. 3. Gene similarity scores for "Max - X%Max" on E when X varies.

Then, we associated two sex class properties to situations by adding two columns to the boolean matrix. The first property c_M is set to "1" for all male individuals while the second one c_F gets the "1" value for all female individuals. This enriched boolean context $\mathbf{r}$ has been the starting point for our concept mining process.

Extraction. We performed the sequence of operations described in Section 2. First we tried to get the whole collection of concepts:

$$GT = \left\{ (T,G) \in 2^{\mathcal{O}} \times 2^{\mathcal{P}} \mid \mathcal{C}_{Concept}(T,G,\mathbf{r}) \right\}.$$

It has been feasible in this context and $|GT| = 14\,884$ (excluding those containing only situation and gene properties).

The following step has been to further constrain the solution set. We decided to focus on the collection of concepts that concern male individuals and that contains at least one gene from KG. The associated constraint $\mathcal{C}_M$ is:

$$\mathcal{C}_M(T,G,\mathbf{r}) \equiv \mathcal{C}_{Concept}(T,G,\mathbf{r}) \wedge \mathcal{C}_{Inclusion}(\mathbf{r},c_M,G) \wedge \mathcal{C}_a(\mathbf{r},KG,G)$$

where $\mathcal{C}_a(\mathbf{r},KG,G)$ is a "at-least-one" constraint, and it is satisfied if $\exists kg \in KG \mid \mathcal{C}_{Inclusion}(\mathbf{r},kg,G)$.

Let GT_M denote this set, D-MINER can compute it and $|GT_M| = 440$.

Then, we have been looking for concepts that concern only female individuals. Furthermore, to tackle noisy data in the boolean context, we specified also a constraint of minimal size for gene sets ($\sigma_g = 20$) and situation sets $\sigma_t = 5$:

$$\mathcal{C}_F(T, G, \mathbf{r}) \equiv \mathcal{C}_{Concept}(T, G, \mathbf{r}) \wedge \mathcal{C}_{Inclusion}(\mathbf{r}, c_F, G) \wedge \mathcal{C}_g(\mathbf{r}, \sigma_g, G) \wedge \mathcal{C}_t(\mathbf{r}, \sigma_t, T)$$

where $\mathcal{C}_t$ and $\mathcal{C}_g$ are constraints on minimal size that are efficiently pushed into the computation by the D-MINER algorithm.

The result denoted by GT_F is a collection of $|GT_F| = 515$ concepts.

Boolean Context Enrichment. To reduce the size of concepts and thus the number of genes to analyze, we have defined a new class for genes that are not in the GT_F set. The genes contained in such a set do not characterize male individuals and can be excluded from consideration at the next extraction task. We added a new row called r_M that is true (1) for all genes not contained in $\bigcup G \mid (T, G) \in GT_F$. 713 genes were contained in such a set and thus the r_M property is true for 2 720 genes. Let $\mathbf{r}'$ denote this new boolean context.

Second Extraction. We then processed the new boolean context using a new constraint $\mathcal{C}_{MG}$ using the r_M property:

$$\mathcal{C}_{MG}(T, G; \mathbf{r}') \equiv \mathcal{C}_{Concept}(T, G, \mathbf{r}') \wedge \mathcal{C}_{Inclusion}(\mathbf{r}', r_M, T) \wedge \mathcal{C}_a(\mathbf{r}', KG, G)$$

where $\mathcal{C}_a$ is defined as before. We obtained a reduced set GT'_M of 295 concepts.

We decided to further reduce the size of the collection of concepts by means of a minimal size constraint on situations. We wanted to keep only concepts that contains at least 6 situations, i.e., one more than $1/2$ of the total number of male individuals:

$$\mathcal{C}'_{MG}(T, G, \mathbf{r}') \equiv \mathcal{C}_{MG}(T, G; \mathbf{r}') \wedge \mathcal{C}_t(\mathbf{r}', 6, T)$$

It has given a set GT''_{MG} of 83 concepts. This has been considered as a relatively small set for subjective exploration.

Final Post-processing. We finally performed some post-processing on GT''_{MG}. We selected the genes contained in the concepts of GT''_{MG} when they were appearing in at least $0.5 \cdot |GT''_{MG}|$ concepts, i.e. genes that were fairly represented. As result, we got a quite small collection of 11 genes. None of the genes from our seed set KG occurs in this collection. On the other hand, three of these genes are already described in [18] as belonging to the "male somatic gene" class. This result has been obtained by analyzing in detail only 11 genes among the 3 433 genes of the expression matrix. Another important result is the presence of a very interesting concept in the last set of concepts we built (Tab. 1). It concerns 8 male individuals and 14 genes, 5 of them being presented in [18] as "male somatic genes". Among these, only one was present in our seed set KG.

Table 1. A concept concerning 14 genes (5 somatic) and 8 male individuals. Each cell in the table contains the original expression value. Only somatic genes are represented.

Situations		Genes					
	...	CG17843	CG6761	CG10096	CG18284	CG7157	...
A03M	...	1.789	2.199	2.659	4.159	3.749	...
A05M	...	2.628	2.728	4.168	4.788	3.858	...
A10M	...	2.29	1.83	2.89	3.53	3.86	...
A15M	...	2.048	1.588	4.728	4.998	4.628	...
A20M	...	2.587	2.127	3.377	3.597	4.967	...
A25M	...	2.336	1.886	3.636	4.516	3.716	...
A30M	...	2.568	1.958	3.048	3.858	3.808	...
AT05M	...	3.505	1.925	5.125	5.535	5.385	...

5   Conclusion

We have designed a new data mining methodology to analyze gene expression data thanks to constraint-based mining of formal concepts. We have described a prototypical KDD scenario that has been proved useful in several real-life gene expression data analysis problems. Boolean data enrichment is a very powerful technique for supporting the iterative search of relevant patterns w.r.t. a given analysis task. It is indeed related to the many contribution to feature construction techniques. We are currently applying the whole method on the data from [19] to improve our understanding of insulino-regulation.

Acknowledgements

The authors want to thank Dr. Sophie Rome for stimulating discussions and her help for preparing the drosophila data set. Jérémy Besson is funded by INRA.

References

1. DeRisi, J., Iyer, V., Brown, P.: Exploring the metabolic and genetic control of gene expression on a genomic scale. Science **278** (1997) 680–686
2. Niehrs, C., Pollet, N.: Synexpression groups in eukaryotes. Nature **402** (1999) 483–487
3. Eisen, M., Spellman, P., Brown, P., Botstein, D.: Cluster analysis and display of genome-wide expression patterns. Proc. Natl. Acad. Sci. USA **95** (1998) 14863–14868
4. Ihmels, J., Friedlander, G., Bergmann, S., Sarig, O., Ziv, Y., Barkai, N.: Revealing modular organization in the yeast transcriptional network. Nature Genetics **31** (2002) 370–377
5. Bergmann, S., Ihmels, J., Barkai, N.: Iterative signature algorithm for the analysis of large-scale gene expression data. Physical Review **67** (2003)
6. Agrawal, R., Mannila, H., Srikant, R., Toivonen, H., Verkamo, A.: Fast discovery of association rules. In: Advances in Knowledge Discovery and Data Mining, AAAI Press (1996) 307–328

7. Becquet, C., Blachon, S., Jeudy, B., Boulicaut, J.F., Gandrillon, O.: Strong-association-rule mining for large-scale gene-expression data analysis: a case study on human sage data. Genome Biology **12** (2002)
8. Creighton, C., Hanash, S.: Mining gene expression databases for association rules. Bioinformatics **19** (2003) 79 – 86
9. Wille, R.: Restructuring lattice theory: an approach based on hierarchies of concepts. In Rival, I., ed.: Ordered sets. Reidel (1982) 445–470
10. Rioult, F., Boulicaut, J.F., Crémilleux, B., Besson, J.: Using transposition for pattern discovery from microarray data. In: Proceedings ACM SIGMOD Workshop DMKD'03, San Diego (USA) (2003) 73–79
11. Rioult, F., Robardet, C., Blachon, S., Crémilleux, B., Gandrillon, O., Boulicaut, J.F.: Mining concepts from large sage gene expression matrices. In: Proceedings KDID'03 co-located with ECML-PKDD 2003, Catvat-Dubrovnik (Croatia) (2003) 107–118
12. Boulicaut, J.F., Klemettinen, M., Mannila, H.: Modeling KDD processes within the inductive database framework. In: Proceedings DaWaK'99. Volume 1676 of LNCS., Florence, I, Springer-Verlag (1999) 293–302
13. De Raedt, L.: A perspective on inductive databases. SIGKDD Explorations **4** (2003) 69–77
14. Pensa, R.G., Leschi, C., Besson, J., Boulicaut, J.F.: Assessment of discretization techniques for relevant pattern discovery from gene expression data. In: Proceedings BIOKDD'04 co-located with SIGKDD'04, Seattle, USA (2004) To appear.
15. Besson, J., Robardet, C., Boulicaut, J.F.: Constraint-based mining of formal concepts in transactional data. In: Proceedings PAKDD'04. Volume 3056 of LNAI., Sydney (Australia), Springer-Verlag (2004) 615–624
16. Robardet, C., Pensa, R., Besson, J., Boulicaut, J.F.: Using classification and visualization on pattern databases for gene expression data analysis. In: Proceedings PaRMa'04 co-located with EDBT 2004. Volume 96 of CEUR Proceedings., Heraclion-Crete, Greece (2004) 107–118
17. Ashburnerand, M., Ball, C., Blake, J., Botstein, D., et al.: Gene ontology: tool for the unification of biology. the gene ontology consortium. Nature Genetics **25** (2000) 25–29
18. Arbeitman, M., Furlong, E., Imam, F., Johnson, E., Null, B., Baker, B., Krasnow, M., Scott, M., Davis, R., White, K.: Gene expression during the life cycle of drosophila melanogaster. Science **297** (2002) 2270–2275
19. Rome, S., Clément, K., Rabasa-Lhoret, R., Loizon, E., Poitou, C., Barsh, G.S., Riou, J.P., Laville, M., Vidal, H.: Microarray profiling of human skeletal muscle reveals that insulin regulates 800 genes during an hyperinsulinemic clamp. Journal of Biological Chemistry (2003) 278(20):18063-8.

Using the Computer to Study
the Dynamics of the Handwriting Processes

Gilles Caporossi[1], Denis Alamargot[2], and David Chesnet[2]

[1] GERAD and HEC Montréal, Canada
Gilles.Caporossi@hec.ca
[2] LACO CNRS and Université de Poitiers, France
{Denis.Alamargot,David.Chesnet}@univ-poitiers.fr

Abstract. In this paper, we present tools to help understanding the dynamics of cognitive processes involved in handwriting and in text composition. Three computer systems or programs used for this analysis are explained, the results obtained by their mean are exposed and their potential meaning discussed.

1 Introduction

For cognitive sciences, there is a great interest in understanding the writing processes. Indeed, text composition is an increasingly present human activity in our society. The historical development of computer has strengthened the use of keyboard but the forthcoming Tablet-PC and the well-diffused PDA give back a privileged role to handwriting for text composition and for data capture. Identifying the mental mechanisms underlying the handwriting activity brings up several outcomes.

1. At a fundamental level: to study the strategies of the writer in order to improve our knowledge of human rules to process the information.
2. At an educational level: to teach children and students to use expert strategies for writing and for composing.
3. At an ergonomic level: to design writing tools (systems for data capture, text processing, etc.) adapted to human strategies.
4. At an artificial intelligence level: to generate written code making the computer able to describe a database and to communicate with natural language.

According to cognitive psychology, writing is a complex activity implying transformation of a multidimensional knowledge structure (domain knowledge) into a linear sequence of words (the text). This transforming must respect linguistic conventions (linguistic knowledge: spelling) and communicative conventions (pragmatic knowledge: legibility, relevance). Several processes are involved in this transformation.

1. The planning process generates and organizes text content by retrieving domain knowledge from long-term memory or by encoding domain information from the environment (documentary sources, for instance).

E. Suzuki and S. Arikawa (Eds.): DS 2004, LNAI 3245, pp. 242–254, 2004.

2. The formulation process translates semantical representation into linguistic structures.
3. The revising process allows the writer to evaluate and to modify conceptual and linguistic characteristics of the text produced so far.
4. Finally, the execution process performs the written message by fulfilling graphomotor plans.

These processes have been well described in theoretical models of writing [1] and the functioning of each one has been studied in numerous experiments, involving various participants [16]. Nevertheless the question of the dynamics of the writing processes still remains difficult to answer. At least, writing is a dual processing activity. The writer encodes visual information as input (text produced so far; potential documentary sources) and executes graphomotor plans as output (to form letters on the screen or on the sheet of paper). The synchronization between input and output depends on the nature and on the number of writing processes engaged to assume the transformation of information. Those 'intermediary' writing processes can be fired simultaneously but due to the limited capacity of human cognitive system, some of them (mainly the most demanding in terms of cognitive resources) need to be postponed and engaged sequentially. Several mechanisms like metacognitive control [12] or processing capacity [10] have been evoked to explain why a process is fired (in terms of decisive criteria) and how it is fired (in parallel or sequentially). Indeed, the actual difficulty to understand such rules of writing dynamics is to precisely identify the moment when a process is engaged and the nature of this process. This scientific discovery may significantly be improved by the use of the machine.

2   Aim of the Research

The aim of this paper is to present a computer environment which allows the scientist to understand the engagement of writing processes and to discover the rules underlying this dynamics. If complete automation is a goal for research in artificial intelligence, Langley [14] and Valdés-Perès [19] also consider the computer as a collaborator for the researcher.

   To describe the dynamics of writing processes, we ask a participant to compose a text from sources and we use the machine to analyze the temporal relationships between the activities of the eye (taking visual information from source, in the task environment) and of the pen (writing the text, making pauses, etc.). We argue that relationships between the eye and the pen constitute pertinent indicators of the writer's mental activity and notably of the writing processes engaged.

   The use of two systems as well as data mining programs lead us to make the evidence of some previously suspected relations but also to discover some new knowledge about the writer behavior. As a first step, the *Eye and Pen* system records data while a participant compose an instructional text (directions required to build a turbine) from documentary sources (pictures and names of the different parts of the turbine). These data are extracted from a larger experimental study conducted with sixteen participants [2]. The second step consists in the use of some clustering programs in order to find different processing sequences associated to the handwriting processes. As a third step, the DyDA program is used for visualization and analysis of the results of the clustering programs as well as the visualization of the data.

3　Description of the Eye and Pen System

The Eye and Pen paradigm [8] was designed to study the dynamics of writing processes. Based on the synchronized measurements of ocular (the input) and graphomotor (the output) activities of the writer, the system provides a very fine-grained description of the temporal characteristics of written production. Eye and Pen relies on two main devices: a digitizing graphic tablet (to record spatial coordinates and pressure of the pen on the tablet surface) and an eye-tracker (to record eye movements). The whole is controlled by two PC type micro-computers. The first one, devoted to the eye movements acquisition sends the gaze position coordinates to the second PC which also simultaneously records the informations given by the graphic tablet. All these observations are stamped with a common base millisecond timing.

Eye and Pen data can be analyzed by a module which is an evolution of G-Studio software [9]. From the text written by the subject, and digitalized by the tablet, one may rebuild forward and backward on the screen, the trace leaved by the pen and the gaze position at the same time (synchronized events). Each event (movement or stop of the pen, saccades or fixations of the gaze) is numbered and may have a code assigned to select, sort and classify data.

4　Using Clustering to Study Writing Processes

In order to highlight the structure underlying the data, three clustering methods were used. The first one is the classical centroid based clustering model using sum of squared distance between each observation and the center of its cluster. As this model did not help finding any useful structure in the data, two other clustering models were tried. These two methods, based upon linear models are hyperplane clustering and clusterwise regression. All the three models are described below.

4.1　The Centroid Based Clustering Model

The centroid based model may be defined as follows:

$$Min \sum_{i=1}^{m} \sum_{k=1}^{K} z_{ik}(x_{ij} - c_{kj})^2 \tag{1}$$

$$Subject\,To$$

$$\sum_{k=1}^{K} z_{ik} = 1 \; \forall i = 1 \dots m \tag{2}$$

$$z_{ik} \in \{0, 1\} \; \forall i = 1 \dots m \; \forall k = 1 \dots K. \tag{3}$$

Where $z_{ik} = 1$ if the observation i belongs to cluster k and equals 0 otherwise and x_{ij} represents the jth coordinate of observation i. The centroid of cluster k's coordinates are computed using the following equations

$$c_{kj} = \frac{\sum_{i=1}^{m} z_{ik} x_{ij}}{\sum_{i=1}^{m} z_{ik}} \; \forall k = 1 \dots K \; \forall j = 1 \dots n. \tag{4}$$

The objective is to cluster the data in such a way that the sum of squares of distance between each observation and the centroid of its cluster is minimized. Equation (2) means that each observation belongs to exactly one cluster. The centroid based model is usually referred to as k-means. Although, k-means makes reference to an algorithm to obtain the clusters instead of just the model. We make this distinction here because if we consider the same problem, the way we search for the clusters is improved compared to the standard k-means algorithm.

4.2 The Hyperplane or k-Plane Clustering Model

As explained by Caporossi and Hansen [5] [6], a way to extract information from a database is by finding an hyperplane (relation) fitting the data. This method finds relations using the mathematics of the principal component analysis but instead of considering the largest eigenvalues of the variance-covariance matrix, one concentrates to the smallest. Should this smallest eigenvalue be close to 0, a relation is found. If this method was successful in graph theory, it also reveals a great potential in other fields when applied within a clustering scheme. Indeed it is possible that one single hyperplane does not fit the data but if two or more equations are considered, any observation is close enough to at least one of them. In this case, we have a more complex relation of the kind "either the relation A holds, or the relation B does". The k-plane algorithm proposed by Bradley and Mangasarian [4] uses the principle of k-means to handle this clustering problem. The *hyperplane* clustering problem may be formulated in the same way as the previous model, except that the objective function is:

$$Min \sum_{i=1}^{m} \sum_{k=1}^{K} z_{ik} d_{ik}. \tag{5}$$

Where dik is the distance between the hyperplane corresponding to cluster k and the observation i. The hyperplane corresponding to cluster k is deduced from the last eigenvector (corresponding to the smallest eigenvalue) of the variance-covariance matrix associated to cluster k.

4.3 The Clusterwise Regression Model

The main difference between k-plane clustering and the clusterwise regression is that a dependent variable that one needs to predict is used in the last case.

Another difference is that the error corresponding to a given observation is not computed in the same way for both models although this last difference is not of great importance in practice.

The clusterwise regression constraints the model to involve a chosen variable in its model. As first exposed by Späth [18] in 1979, in the clusterwise regression model, each cluster is represented by a regression model fitting the corresponding data and each observation is assigned to the cluster that best fits it. The clusterwise regression model may also be written in the same way as the previous ones except that the objective function is [15]:

$$Min \sum_{i=1}^{m} \sum_{k=1}^{K} z_{ik}(y_i - \sum_{j=1}^{n} b_{jk}x_{ij} + b_{0k})^2 \qquad (6)$$

where the variables b_{jk} are regression coefficients for the cluster k.

4.4   Optimization Algorithm Used for Clustering the Data

For clustering data using the centroid model, the *k-means* algorithm is usually used. In the case of hyperplane clustering, the *k-plane* algorithm proposed by Bradley and Mangasarian uses the same scheme. In the case of the clusterwise regression, Späth [18] proposes a different algorithm taking advantage of the possible update of the models. All these algorithms thus lead to a local optimum and none explicitly consider routines to get out of it (except multistart).

The *alternate descent* algorithm upon which are based *k-means* and *k-plane* is the following:

1. Randomly choose the cluster corresponding to each data.
2. For each cluster, compute the best model *i.e.*, the centroid, the parameters of the hyperplane that best fits the data or the regression parameters, depending on the model used.
3. If an observation best fits another cluster *i.e.*, it is closer to another centroid or to another model, move it to that cluster. If at least one observation was moved, go back to Step 2 otherwise, the descent is complete.

This *alternate descent* algorithm converges quickly but the solution so found highly depends on the initial solution used, which makes it a bad optimization technique as it often misses the global optimum. Alone or within a *multistart* scheme, it leads to rather poor results on which one cannot rely enough to identify the models underlying the data. On the other hand, is is much faster than other descent methods we know for this problem.

As the main reason for the use of a clustering algorithm in discovery science is to find a partition of the data that leads to interpretation and allows the construction of hypothesis, the optimization algorithm must be fast and reliable. For this reason, in the context of a project with the Bell Laboratories, an efficient framework for optimizing clustering problems was developed using a heuristic based upon the Variable Neighborhood Search (VNS) [17] [13].

The Variable Neighborhood Search Algorithm

Repeat until the stopping condition is met:

1. Set $k = 1$;
2. Until $k = k_{max}$, repeat the following steps
    (a) (*shaking*) generate a partition x' at random from the kth neighborhood of x (i.e., $x' \in N_k X(x)$):
    (b) (*descent*) Apply alternate local search with x' as initial solution: denote by x'' the local optimum obtained;
    (c) (*improvement or continuation*) If the solution x'' so obtained is better than the best known one x, move there ($x \leftarrow x''$) and continue the search within $N_1(x)(k = 1, l = 1)$; otherwise set $l \leftarrow l + 1$. If $l = l_{max}$ then set $k \leftarrow k + 1$ and $l \leftarrow 1$.

In the case of clustering, shaking in the kth neighborhood is done by choosing k times 2 clusters at random and randomly reassign their observations to one or the other cluster. This shaking is a local perturbation that only affects few clusters. By the rules of VNS, if the local search does not succeed after l_{max} shakings and local searches, the shaking magnitude (k) is increased by 1. The stopping criterion may either be $k = k_{max}$ or the total cpu time used.

5   Description of the DyDA Program

In order to easily understand the Eye and Pen data, the first task achieved was to build a specific program that allows a visualization of the data in high dimension with an accurate representation of the time component. The main purpose of DyDA is to provide the researcher a graphical interface that helps understanding the whole process underlying the data. To achieve this goal, the program needs

- to take the dynamical aspect into account,
- to show many features at the same time,
- to show correspondence between different variables.

5.1   Using DyDA to Have a Dynamic Overview of the Process

As the first interest for the researcher is a spatial view of the data, one may, as illustrated on Figure 1, plot position of the pen and position of the eye on the same window.

An animation showing the evolution of the eye and pen position during time improves the reading of the data but is still difficult to follow. The alternative we propose here is to continuously show a "slice" of data corresponding to one or few seconds so that a reasonable amount of information is displayed at the same time. Wholes seconds of the process may then be understood by a single look as we can see on a screen capture represented on Figure 2.

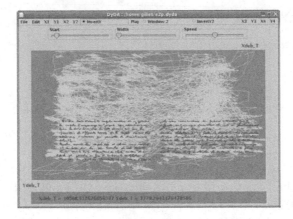

Fig. 1. View of the main window of DyDA showing the movements of the eye and pen.

5.2   Increasing the Dimension of the View and Exhibiting Correspondence Between Variables

In order to have a good understanding of the dynamics of writing, at each time, the researcher needs to see the position and pressure of the pen, the position of the eye and eventually some combination of them. As it would be difficult to show pressure or any different information on the same window as the spatial representation, a second window was added to DyDA. On that second window may be drawn speed and the pressure of the pen. The right window on Figure 2, shows a slice of the writing and the corresponding trajectory of the eye while the left window displays pressure against the speed of the pen. This figure clearly

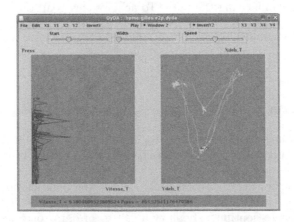

Fig. 2. A "slice" with pressure against speed of the pen curve on the first window and the corresponding written sequence as well as the position of the eye on the second one.

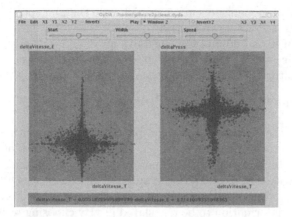

Fig. 3. variation of the pen speed against variation of the eye speed or variation of pressure.

shows that the pressure is rather high but varies while the pen is slow but still writing. On the right window we notice that the writer does not look at his writing (this is a parallel sequence).

To exhibit the relation between the different points representing the same data, the user may select and highlight some data on the main curve of the first window and see the corresponding points highlighted on the other curves. as illustrated on Figure 4.

6   Application Examples and New Results

The analysis of the data was achieved in two steps; the DyDA program alone was first used to have low level (almost mechanical) overview of data. Clustering was then applied to have a deeper understanding of the various activities involved at a higher level (more related to cognitive activities).

6.1   First Overview of the Data with DyDA

Showing Dependencies Between Eye and Pen Variations

Some intuition about the process came from drawing variation of eye or pen speed as well as pressure of the pen one against the other. Indeed, Figure 3 shows the variation of the eye speed (left window) and variation of pressure (on the right window) against the variation of the pen speed. This figure suggests that variation of the pen speed very rarely occurs simultaneously with the variation of eye speed or pressure.

Our attention was thus driven to the study of the variations of speed and pressure. The analysis of the dynamic process indicates that the eye acceleration usually occurs in three cases:

- when a parallel sequence occurs, in which case the pen speed does not vary much, even if it slows down a little,
- when the pen is paused and the eye looks away,
- and when the pen is about to be repositioned. In this case, the eye moves first to the future position of the pen and this last one only moves after.

Exhibiting Sequences of Parallel Processing

In most studies, the presence of parallel processes during writing is more discussed than objectively demonstrated [7]. One of the interest of the Eye and Pen paradigm is to bring out specific periods during which ones a new visual information is process in parallel with graphomotor execution. During these periods, the transcription of a visual information is accomplished while the eye is looking for a new information to be translated [2]. By choosing to highlight data with large distance between eye and pen (*i.e.* the pen is out of the parafoveal vision) and with pen speed greater than 0 (the pen is writing), it is possible to exhibit so called parallel sequences. The Figure 4 shows how DyDA may be used to attest those parallel sequences. Few seconds are needed to exhibit them precisely. Once these sequences are selected, the user may decide to activate the animation and see in real time the process of writing with the selected sequences highlighted. As long as the selection is not erased, it is possible to switch to the single window view or to display other variables, eventually while the animation is still running.

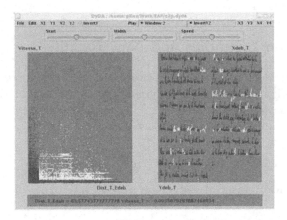

Fig. 4. DyDA with parallel sequences highlighted.

6.2   Classifying Different Phases of Visual Activities During Writing

To identify cognitive activities related to the handwriting task, we first needed to "clean" the data. Indeed, most variance of the raw data refers to rather mechanical components. For example, nystagmus, the constant eye movement around the watched object is not of great interest at a cognitive level. Indeed all eye movements with velocity smaller than 30 degrees per second were aggregated into a single point: the barycenter of all the corresponding eye positions. This threshold of 30 degrees per second is commonly used by researchers working on that topic to define the so called fixations reflecting the position the participant is willing to look at (for example, see [11]). A now important dimension in the data is the duration of each fixation as well as the distance the eye has achieved between fixations. As each observation is associated to a given clock time but not to a period of time, or sequence, the euclidian metric is appropriate. Indeed this would not be the case if each observation was associated to a sequence, as it is the case in hand writing recognition [20][3].

The first attempt to find information in the data with clustering methods was achieved using the centroid based model. Unfortunately, it did not provide any understandable clusters. The second step was to use *hyperplane* or *k-plane* clustering. If the results were more encouraging, the only information extracted by this mean was of the kind: either the pen (or the eye) is stopped, or its speed is almost constant. As a third attempt we used the clusterwise regression model. To use such a method, one needs a dependent variable. The distance achieved by the eye being related to the kind of composing activity conducted by the writer we decided to try clusterwise regression with eye distance as dependent variable. After running the program for 2 to 7 clusters, we noticed that each additional cluster, up to 4, yields an important error reduction. However, this is no more the case when increasing further this number of clusters, suggesting that the structure of the data consists in 4 clusters.

Conducted for each cluster, the analysis with DyDA of the temporal characteristics of visual and of graphomotor movements indicates four kinds of eye behavior.

1. Moving away from the pen with high speed, the eye can next stay relatively static, making few and long fixations on specific part of the sources before finally quickly joining the pen. This behavior should attest the writer's necessity to encode and to generate a new piece of content which could be immediately translated.
2. Still far from the pen, the eye can also explore, in a very dynamic way, the different parts of the source, by doing numerous fixations, moving with a very high speed and browsing on a large distance. This second kind of behavior could underly an important planning activity occurring during a long writing pause (the writer explores the source to plan a large part of text to be produced). It could also be due to a revision activity that consists in verifying agreement between the text produced so far and the source's information. In this case, the eye quickly browses the source as well as the text (to read). Two periods corresponding to such a revision may be described in the protocol.
3. During graphomotor execution, the eye is generally close to the pen (except during a period of parallel processing). In this case, the writer accomplishes long and static fixations, moving slowly to rejoin regularly the pen. This behavior can be interpreted as an ocular control of graphomotoric execution.

4. In some cases, an unexpected phenomenon occurs. When the eye comes back from an intensive source exploration, or when the pen moves quickly from the end to the beginning of a line, the eye can make a long fixation, relatively far from the pen executing the trace, but also far from source's information. This behavior could correspond to a peripheral (or parafoveal) monitoring of the trace. Here it is interesting to note that data-mining could be helpful to bring out two kinds of relationships between eye and graphomotor execution: a local control and a peripheral monitoring.

7   Conclusion and Future Work

The Eye and Pen device and DyDA software represent key elements to study the dynamics of writing from source. They constitute a powerful computer environment able to catch very fine grained phenomena related to ocular and graphomotor activities. They also provide to the researcher an easy-to-use tool helping to understand the temporal and the spatial relationships between the behavioral indicators. Used in association with Eye and Pen and DyDA, data-mining and discovery methods offer interesting perspective regarding the nature of recorded data. In our study, data-mining is used to infer the upstream cognitive processes. By now, it has been used to categorize different temporal patterns in the graphomotoric execution of written words and one main result can already be stressed: clusterwise regression appears to be very relevant to segment and categorize the flow of writing from source. By isolating four different relationships between the visual input and the graphomotoric output of writing, the clusterwise regression allows us to interpret four different patterns in the writing activity. One of them, the fourth cluster, appears to be original and its cognitive signification to be precised. Mostly descriptive, these first results must be reinforced and validated by systematical investigations.

The future work will aim, first, at analyzing the dynamics of the four writing processes configurations, that is to say the order in which these configurations occur all along the text composition; second at comparing these sequences in various contexts either involving different participants or different kinds of writing tasks. Indeed, further data mining techniques such as dynamic time warping will then need to be used to handle correctly the temporal dimension and its distortion from one writer to another.

In every cases, these issues give a great insight of the possibilities given by the machines to explore new dimensions in human cognitive activities.

Acknowledgments

The authors thank C. Ros - from laboratory LaCo-CNRS - Université de Poitiers and C. Dansac from laboratory LTC-CNRS Université de Toulouse for their contribution on the data analysis. This work was supported by NSERC - Canada and ACI-MSHS - France.

References

1. D. Alamargot and L. Chanquoy. *Through the models of writing*. Dordrecht-Boston-London: Kluwer Academic Publishers, 2001.
2. D. Alamargot, C. Dansac, and D. Chesnet. Parallel processing around pauses: A conjunct analysis of graphomotor and eye movements during writing instructions. In M. Torrance, D. Galbraith, and L. v. Waes, editors, *Recents developpements in writing-process research (Vol. 2)*. Dordrecht-Boston-London: Kluwer Academic Press, In press.
3. C. Bahlmann and H. Burkhardt. The writer independent online handwriting recognition system frog on hand and cluster generative statistical dynamic time warping. *IEEE Trans. on Pattern Analysis and Machine Intelligence*, 26(5), 2004.
4. P.S. Bradley and O.L. Mangasarian. k-plane clustering. *Journal of Global Optimization*, 16:23–32, 2000.
5. G. Caporossi and P. Hansen. Finding Relations in Polynomial Time. In *Proceedings of the XVI International Joint Conference on Artificial Intelligence*, pages 780–785, 1999.
6. G. Caporossi and P. Hansen. Variable Neighborhood Search for Extremal Graphs: 5. Three Ways to Automate Finding Conjectures. *Discrete Mathematics*, 276:81–94, 2004.
7. L. Chanquoy, J.N. Foulin, and M. Fayol. Temporal management of short text writing by children and adults. *Cahiers de Psychologie Cognitive*, 10(5):513–540, 1990.
8. D. Chesnet and D. Alamargot. Analyses en temps réel des activités oculaires et graphomotrices du scripteur: intérêt du dispositif 'eye and pen'. *L'Année Psychologique*, In press.
9. D. Chesnet, F. Guillabert, and E. Espéret. G-studio: un logiciel pour l'étude en temps réel des paramètres temporels de la production écrite. *L'Année Psychologique*, 94:115–125, 1994.
10. J. Grabowski. Writing and speaking: Common grounds and differences. toward a regulation theory of written language production. In *The Science of Writing: Theory, methods, individual differences and applications*, pages 73–91. Mahwah (NJ): Laurence Erlbaum Associated, 1996.
11. Z.M. Griffin and K. Bock. What the eyes say about speaking. *Psychological Science*, 11(4):274–279, 2000.
12. D.J. Hacker. Comprehension monitoring of written discourse across early-to-middle adolescence. *Reading and Writing*, 9(3):207–240, 1997.
13. P. Hansen and N. Mladenović. Variable neighborhood search: Principles and applications. *European Journal of Operations Research*, 130:449–467, 2001.
14. P. Langley. The computer-aided discovery of scientific knowledge. *Proceeding of the First International Conference on Discovery Science*, 1998.
15. Kin-Nam Lau, Pui-Iam Leung, and Ka-Kit Tse. A mathematical programming approach to clusterwise regression. *European Journal of Operations Research*, 116:640–652, 1999.
16. C.M. Levy and S. Ransdell. *The science of writing: Theories, methods, individual differences, and applications*. Hillsdale (NJ): Laurence Erlbaum Associated, 1996.
17. N. Mladenović and P. Hansen. Variable neighborhood search. *Computers and Operations Research*, 24:1097–1100, 1997.
18. H. Spaeth. Clusterwise linear regression. *Computing*, 22:367–373, 1979.

19. R.E. Valdés-Peréz. Principles of Human Computer Collaboration for Knowledge Discovery in Science. *Artificial Intelligence*, 107:335–346, 1999.
20. M. Vlachos, M. Hadjieleftheriou, D. Gunopulos, and E. Keogh. Indexing multidimensional time-series with support for multiple distance measures. In Lise Getoor, Ted E. Senator, Pedro Domingos, and Christos Faloutsos, editors, *Proceedings of the Ninth ACM SIGKDD International Conference on Knowledge Discovery and Data Mining*. ACM, 2003.

Product Recommendation in e-Commerce Using Direct and Indirect Confidence for Historical User Sessions

Przemysław Kazienko

Wrocław University of Technology, Department of Information Systems,
Wybrzeże S. Wyspiańskiego 27, 50-370 Wrocław, Poland
kazienko@pwr.wroc.pl
http://www.pwr.wroc.pl/~kazienko

Abstract. Product recommendation is an important part of current electronic commerce. Useful, direct and indirect relationships between pages, especially product home pages in an e-commerce site, can be extracted from web usage i.e. from historical user sessions. The proposed method introduces indirect association rules complementing typical, direct rules, which, in the web environment, usually only confirm existing hyperlinks. The direct confidence, the basic measure of direct association rules, reflects pages' co-occurrence in common user sessions, while the indirect confidence exploits an additional, transitive page and relationships existing between, not within sessions. The complex confidence, combining both direct and indirect relationships, is engaged in the personalized process of product recommendation in e-commerce. Carried out experiments have confirmed that indirect association rules can deliver the useful knowledge for recommender systems.

1 Introduction

The effective usage of accessible information about customer behaviour is one of the greatest challenges for current electronic commerce. The promotion and recommendation of particular products, performed on e-commerce web pages, are good indicators of increasing sales. Data mining methods are very useful in the recommendation process, especially association rules which determine with what probability the given product appears to be sold together with another one based on historical data about customer transactions [26]. The efficiency of this approach was investigated and proven in [6]. Demographic filtering [24], collaborative filtering [3, 17] and content based filtering [21, 25] are other typical methods of recommendation that have been studied for the last 10 years.

The main data source of user activities in the web environment is user sessions, to which data mining methods are used for recommendation [4, 14] or personalization purposes [5, 20]. According to [8] we can say that the recommendation based on web usage mining is positively effective. Typically, the recommendation is used to support navigation across the web site [12, 15, 20]. However, it may also become the tool for motivating a visitor to buy a product in e-commerce [14, 27].

Implementing general concept of association rules to user sessions we can explore sets of pages (itemsets) frequently visited together during one session [1, 19, 23, 31]. However, this standard approach reflects only direct associations between web pages

E. Suzuki and S. Arikawa (Eds.): DS 2004, LNAI 3245, pp. 255–269, 2004.
© Springer-Verlag Berlin Heidelberg 2004

derived from single sessions. The majority of discovered rules only confirm "hard" connections resulting from hyperlinks, excepting relationships between pages, which do not occur frequently in the same user sessions. This oversight especially concerns pages not connected directly with hyperlinks. Thus, typical association rules (called in this paper *direct*) correspond to relationships existing "within" user sessions. Due to the hypertext nature of the web, standard parameters of direct association rules (support and confidence) have usually the greatest values for pages "hard" connected with hyperlinks.

Following the idea of citations in the scientific literature [7, 16] and hyperlinks in hypermedia systems like WWW [9, 30], direct associations can be extended with indirect (transitive) ones. In an indirect association, if two documents (pages) are independently similar to the third, transitive document, they both can be expected to be similar to each other. In other words, two pages that both separately, relatively frequently co-occur in sessions with another, third page can be considered as "indirect associated". The purpose of this document is to describe a method of product pages recommendation in an e-commerce web site that exploits both direct and indirect relationships between pages. The method also includes the time factor that reduces the influence of old sessions on the recommendation process. Direct and indirect association measures (confidence functions) are estimated offline using historical sessions. Next, they are combined into one function and they can be confronted against the current user session delivering the personalized recommendation of products.

Previous research work in mining indirect associations was carried out by Tan and Kumar [28, 29]. However, their indirect patterns differ from those presented in this paper.

The proposed method is a part of the ROSA (Remote Open Site Agents) project (http://www.zsi.pwr.wroc.pl/rosa) based on the multi-agent architecture [11, 13] and developed in cooperation with Fujitsu España.

2 e-Commerce Environment

The method presented in this paper operates on a single e-commerce web site which is treated as the set D of independent web pages (documents) $d_i \in D$. A special subset D^P of *product pages* can be extracted from the set D, $D^P \subset D$. Each product page $d_i^P \in D^P$ is the home page for a single product coming from the e-commerce offer and stored in the product database (the product set P). We assume that there exists exactly one (the symbol $\exists!$) product page $d_i^P \in D^P$ for each product p_k from the database and each product $p_k \in P$ has only one related product page. There may exist some products from P that do not possess corresponding home pages at all:

$$\left(\forall d_i^P \in D^P\right)\left(\exists! \, p_k \in P\right)\left(d_i^P \text{ is the home page of the product } p_k\right),$$

$$\left(\forall d_i^P, d_j^P \in D^P\right)\left(d_i^P, d_j^P \text{ are product pages for the product } p_k \in P \Rightarrow d_i^P = d_j^P\right) \tag{1}$$

We have a one to one relationship between products and their web pages (Fig. 1). It means that a product page with a unique, separate URL corresponds to only one product record in the database. However, every e-commerce site also contains other documents (denominated here *normal* or *non-product pages*) that possess more general content: the latest company news, product reviews, product group description, manuals, technologies overviews, practical advise, etc. Such pages are independent and they are poorly or completely un-related to specific products [14].

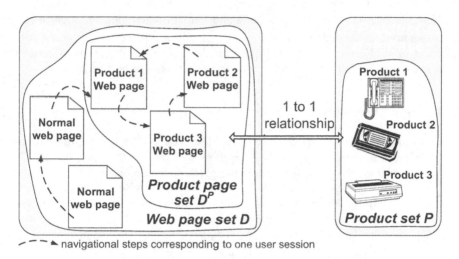

- - - ➤ navigational steps corresponding to one user session

Fig. 1. The e-commerce web site and related product database.

The recommendation relies on suggestion of the list of product home pages (i.e. products) separately on each page from the e-commerce web site. Thus, the goal of the method is to determine the appropriate product list for each page from the web site based on historical users' behaviors gathered by the systems (web usage mining).

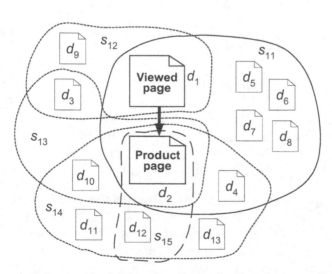

Fig. 2. Web sessions (page sets) related to viewed page d_1 and product page d_2, which is considered as a potential candidate for recommendation.

A user navigating throughout the web site visits both product and non-product (normal) pages and the e-commerce system stores these user behaviors in the form of sessions. Each session s_i is a set of pages viewed during one user's visit in the web site. There are five sessions on the Fig. 2: $s_{11}=\{d_1, d_2, d_4, d_5, d_6, d_7, d_8\}$, s_{12}, s_{13}, s_{14}, s_{15}. Note that the session set is unordered and without repetitions although in many papers sessions are treated as a sequence of pages following one another according to the order of HTTP requests. In our approach the order does not seem to be useful because association rules, unlike sequential patterns, do not respect sequences.

3 Method Overview

The general concept of the method could be described as follows: the more often a product page was visited together with the given web page in the past, the more this product page should be recommended on the given page at present. This goal can be achieved directly, using user sessions. All product pages visited together with the given page in any sessions may be potentially recommended. There is only one such session with direct influence on recommendation on the Fig. 2 - s_{11}. Using *the direct confidence function* described below we can estimate the belief level that the specific product page should be recommended on the given page. The e-commerce environment – like other sales channels – has one interesting feature: customers change their behaviors over the course of time. For that reason a time factor is included into the confidence function giving *the time weighted direct confidence*, and in this way, the older sessions have less influence onto a confidence value than the latest ones.

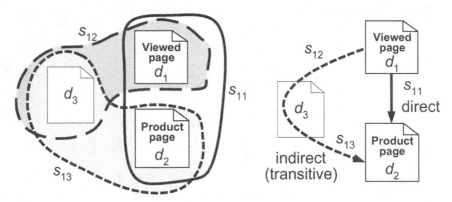

Fig. 3. Direct and indirect relationships.

However, pages may be related not only directly but also indirectly through a "relay document" (Fig. 3). The session s_{12} contains the considered page d_1 and the page d_3 ("relay document"). On the other hand, d_3 belongs also to the session s_{12} together with the product page d_2. Thus, the page d_1 indirectly (transitive, through the page d_3) co-occurs with the product page d_2. This general concept of transitiveness was widely used in citation analysis for scientific literature [7, 16] and in hypertext systems [9, 30]. It can be briefly expressed as follows: two documents cite or link to another, third document, so they seem to be similar. This analogue case occurs when two documents are cited or linked by the third one.

There are two "transitive links" (indirect associations) between d_1 and d_2 on the Fig. 2: $s_{12} \rightarrow s_{13}$ (through d_3) and $s_{11} \rightarrow s_{14}$ (through d_4). Note that the session s_{11} is the source for both direct and indirect relations. The measure of the strength of indirect association is *the indirect confidence function*. It treats all web site pages as potential "relay documents" and exploits prior estimated time weighted direct confidence function to and from such documents.

The final recommendation list (list of suggested product pages) depends on both direct and indirect confidence. These two functions are combined into one *complex confidence function* (Fig. 4). The influence on a direct and indirect confidence value can be adjusted within the complex confidence function so that direct co-occurrences may be more emphasized.

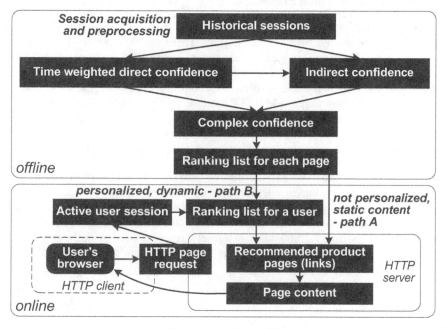

Fig. 4. Method overview.

Complex confidence function value is evaluated, for each pair: any page – product page. Having these values the separate product page list for each web page (ranking list) can be created using descending order of function values. Product pages with the highest complex confidence value are on the top of the ranking list for the given page and they are candidates for recommendation. Note that ranking lists are fixed and independent of active users. Owing to this, time consuming ranking calculations can be performed offline.

Hyperlinks to product pages from the top of ranking lists may be statically incorporated into all web site pages (offline). In this way, we obtain a non-personalized web site with pages of fixed content (links) – path A on the Fig. 4. Another approach (path

B) takes into account the active session of a user. The ranking list is verified towards the products the active user was recommended on previous pages visited during the current session. Such products have less chance being recommended again. This is the personalized way of recommendation and it requires web pages to be generated dynamically (appropriate links have to be inserted online into the page's HTML content).

Please note that all offline tasks should be periodically repeated having data inconstancy in view (for example a new product appears and another one disappears). The update problem was solved in the ROSA system by the introduction of multi-agent architecture and the update method presented in [13]

4   Recommendation Process

4.1   Session Acquisition and Preprocessing

The first step of the method is the acquisition of user sessions – the extraction of sets of pages that were visited together in the past. There are two major approaches to achieve this goal. The former (used among others in the ROSA system [12, 15]) is based on the online capture of all pages visited by a user and their storage in the database. Although HTTP is a stateless protocol cookies and rewriting URL encoding, permit web developers to join single HTTP requests into sessions. If for any reasons it would not be possible to recognize web sessions online, they must be retrieved offline from web log files. HTTP fields (IP address, user agent, referrer, etc.) and data-time stamps allow the HTTP requests to be joined with a certain probability.

Definition 1. Let the tuple $s=(V,t^s)$ be the user session. V is the set of single page visits performed by a user during one site session; $V \subset D$. t^s is the session start time.

It happens relatively often that some users visit the web site by accident and quickly leave it. Sessions for such visits are very short and they should be omitted from further processing. Additionally, extremely long sessions usually come from web crawlers and they do not contain human being behaviour. Thus, only sessions s_i containing more than certain number of pages min_p and shorter than max_p, $min_p \leq card(V_i) \leq max_p$ are included; typically $min_p=3$ and $max_p=200$. The set of all such reliable sessions is denoted by S.

4.2   Direct Confidence

Definition 2. A *direct association rule* is the implication $d_i \rightarrow d^P_j$, where $d_i \in D$, $d^P_j \in D^P$ and $d_i \neq d^P_j$. A direct association rule is described by two measures: *support* and *confidence*. The direct association rule $d_i \rightarrow d^P_j$ has the support $sup(d_i \rightarrow d^P_j)=sup(d_i,d^P_j)/card(S)$; where $sup(d_i,d^P_j)$ is the number of sessions $s_k \in S$ containing both d_i and d^P_j; $d_i, d^P_j \in V_k$. d_i is called the *body* and d^P_j is the *head* of the rule $d_i \rightarrow d^P_j$.

Direct association rule $d_i \rightarrow d^P_j$ reflects the direct relationship from d_i to d^P_j.

The direct confidence function – $con(d_i{\rightarrow}d^P{}_j)$ denotes with what belief the product page $d^P{}_j$ may be recommended to a user while watching the page d_i. In other words, the direct confidence factor is the conditional probability $P(d^P{}_j|d_i)$ that a session containing the page d_i also contains the product page $d^P{}_j$:

$$con\!\left(d_i \rightarrow d^P_j\right)=\begin{cases} P\!\left(d^P_j\big|d_i\right)\approx\dfrac{n_{ij}}{n_i}, & \text{if } n_{ij}>\upsilon \\[2mm] 0, & \text{if } n_{ij}\le\upsilon \end{cases} \qquad (2)$$

where n_{ji} – the number of sessions with both d_i and $d^P{}_j$ page; n_i – the number of sessions containing d_i. The threshold $\bar{o}$ is used for removing pages $d^P{}_j$ occurring with the page d_i only occasionally, by accident. It prevents such $d^P{}_j$ product to be recommended on the page d_i what may happen e.g. if the page d_i is new – with the small number of sessions (small n_i). Typically, $\bar{o}=2$. Intuitive, the threshold $\bar{o}$ corresponds in a sense to the minimum confidence typically used in association rules methods. The main task of $\bar{o}$ is to exclude rare direct associations between pages.

It was assumed that all pages are statistically independent of each other. But this is not the case. Some pages are connected by links (but most pairs are not), some were recommended by the system while other ones were not, and some are placed deeper in the web site structure. Thus, from the statistical point of view the probability value (n_{ji}/n_i) is only an approximation.

4.3   Time Factor

Products' and pages' fads, which have gone a long time ago, are a significant problem with the equation (2). Users often change their behavior, so we should not rely on older sessions with the same confidence as on newer ones. If the given product $d^P{}_j$ was visited together with the page d_i many times but only in the past, then such product should not be recommended so much at present. For all these reasons, the introduction of the time factor is proposed. Numbers of sessions n_{ji} and n_i in (2) are replaced with *the time weighted numbers of sessions*: n'_{ji} and n'_i, respectively; as follows:

$$con^t\!\left(d_i \rightarrow d^P_j\right)=\frac{n'_{ij}}{n'_i}=\frac{\sum_{k:\,s_k\in S;\,d_i,d^P_j\in s_k}(\tau)^{tp_k}}{\sum_{k:\,s_k\in S;\,d_i\in s_k}(\tau)^{tp_k}}, \quad \text{if } n'_i>0, n_{ij}>\upsilon \qquad (3)$$

where: $con^t(d_i{\rightarrow}d^P{}_j)$ – the time weighted direct confidence; τ – the constant time coefficient from the interval $[0,1]$; tp_k – the number of time periods since beginning of the session s_k until the processing time. In other words, while calculating n'_{ji} and n'_i each session s_k is counted not as 1 (like in n_{ji} and n_i) but as $(\tau)^{tp_k}$. Time period length (a unit of measure for tp_k) depends on how often users enter the web site. The time coefficient τ denotes changeability of the site content and behaviour of users. The more often the site changes, the smaller should be the τ value. In this way, older sessions have less influence on recommendation results.

4.4 Indirect Confidence

The similarity of pages is expressed not only with direct associations derived from user sessions but also in the indirect way (Fig. 3).

Definition 3. *Partial indirect association rule* $d_i \rightarrow^\circ d^P_j, d_k$ is the *indirect* implication from d_i to d^P_j with respect to d_k, for which two direct association rules exist: $d_i \rightarrow d_k$ and $d_k \rightarrow d^P_j$, where $d_i, d_k \in D$, $d^P_j \in D^P$; $d_i \neq d^P_j \neq d_k$. The page d_k, in the partial indirect association rule $d_i \rightarrow^\circ d^P_j, d_k$, is called *the transitive page*. The set of all possible transitive pages d_k, for which the partial indirect association rule from d_i to d^P_j exists, is called T_{ij}.

Another function – *the partial indirect time weighted confidence function* $con^\circ(d_i \rightarrow^\circ d^P_j, d_k)$ describes the quality of the partial indirect association. It denotes with what confidence the product page d^P_j can be recommended on the page d_i indirectly (transitive) with respect to the single page d_k.

$$con^\circ(d_i \rightarrow^\circ d^P_j, d_k) = P(d_k | d_i) * P(d^P_j | d_k) \qquad (4)$$

The function $con^\circ(d_i \rightarrow^\circ d^P_j, d_k)$ can be expressed applying the time weighted direct confidence (3):

$$con^\circ(d_i \rightarrow^\circ d^P_j, d_k) = con^t(d_i \rightarrow d_k) * con^t(d_k \rightarrow d^P_j) \qquad (5)$$

$$con^\circ\left(d_i \rightarrow^\circ d^P_j, d_k\right) = \begin{cases} \dfrac{n'_{ki}}{n'_i} * \dfrac{n'_{jk}}{n'_k}, & \text{if } n'_i > 0, n'_k > 0 \\ 0, & \text{otherwise} \end{cases} \qquad (6)$$

The function $con^\circ(d_i \rightarrow^\circ d^P_j, d_k)$ takes into consideration only one transitive document – d_k. Complete indirect association rules are introduced to accumulate values of partial indirect time weighted confidence function for all transmitters.

Definition 4. *Complete indirect association rule* $d_i \rightarrow^\# d^P_j$ aggregates all partial indirect association rules from d_i to d^P_j with respect to all possible transitive pages $d_k \in T_{ij}$; where $d_i \in D$, $d^P_j \in D^P$; $d_i \neq d^P_j$.

The complete indirect confidence – $con^\#(d_i \rightarrow^\# d^P_j)$ is introduced as the measure for the usefulness of complete indirect association. Its value is the average of the values of all partial indirect confidence functions:

$$con^\#(d_i \rightarrow^\# d_j^P) = \frac{\sum_{k=1}^{card(T_{ij})} con^\circ\left(d_i \rightarrow^\circ d_j^P, d_k\right)}{max_T}, \ d_k \in T_{ij} \qquad (7)$$

$$con^{\#}\left(d_i \to^{\#} d_j^P\right) = \frac{1}{n'_i \, max_T}\left(\sum_{k=1}^{card(T_{ij})} n'_{ki} \, \frac{n'_{jk}}{n'_k}\right), \; d_k \in T_{ij} \qquad (8)$$

where $max_T = \max\limits_{d_i \in D; d_j^P \in D^P}\left(card(T_{ij})\right)$. Typical value of max_T is about 3–10% of all web

pages. It was examined for sets of 1,000–4,000 web pages.

Please note that complete indirect rules differ from those proposed by Tan *et al.* in [28, 29]. We have not introduced any assumption that pages d_i and d_j^P are not directly correlated like Tan *et al.* did. Moreover, direct associations are an important part of the complex confidence described below. Another difference is the dissimilar meaning of the term "indirect association rule". Tan *et al.* proposed rules in a sense similar to complete indirect rules (def. 4). However, their rules need to have the assigned cardinality of the set of transitive pages T_{ij} (called a mediator set) and this set is treated as one whole. In such approach d_i and d_j^P have to co-occur with a complete set of other pages instead of with a single transitive page. There are also no partial rules – components of complete rules in that approach. Additionally, one pair d_i, d_j^P may possess many indirect rules with many mediator sets, which often overlap. However, in recommendation systems, we need one measure that helps us to find out whether the considered page d_j^P should or should not be suggested to a user on the page d_i. An appropriate method for integration of rules should be only worked out for Tan's *et al.* rules.

4.5 Complex Confidence

Having direct and complete indirect confidence for rules from d_i to d_j^P, $d_i{\neq}d_j^P$, we can combine them into one final function – *the complex confidence function* $r(d_i \to d_j^P)$, which is also *the ranking function* used further for recommendation:

$$r(d_i \to d_j^P) = \alpha * con^t(d_i \to d_j^P) + (1-\alpha) * con^{\#}(d_i \to^{\#} d_j^P), \quad d_i{\neq}d_j^P \qquad (9)$$

$$r\left(d_i \to d_j^P\right) = \alpha \frac{n'_{ij}}{n'_i} + \frac{(1-\alpha)}{n'_i (N-2)}\left(\sum_{k=1,k\neq i,k\neq j}^{N} n'_{ik} \, \frac{n'_{jk}}{n'_k}\right), \quad d_i{\neq}d_j^P \qquad (10)$$

where: α – direct confidence reinforcing factor, $\alpha \in [0,1]$. Setting α we can emphasize or damp the direct confidence at the expense of the indirect one. Taking into account normalization performed in (7) and (8) factor α should be closer to 0 rather than to 1. According to performed experiments, the proper balance between direct and indirect confidence is reached for $\alpha{=}0.2$.

Sessions including d_i, d_j^P and transitive documents d_k (e.g. s_{11} on the Fig. 1) are used within equations (9) and (11) at least twice. At first, as direct factor increasing n'_{ij}. Next, in $con^{\#}(d_k \to^{\#} d_j^P)$ enlarging the value of n'_{jk}. Longer sessions containing more pages (e.g. m pages) may be exploited many times – up to m-1 times.

The complex confidence function is estimated for every pair d_i, d_j^P, where $d_i \in D$ and $d_j^P \in D^P$. In this way, we obtain the matrix of similarities between every page and every product page in the web site. Remember that D includes product pages ($D^P \subset D$)

as well as all normal pages. Note that "the direction" of all direct and indirect rules is always "from" any page "to" a product page and only product pages are considered to be recommended.

4.6 Confidence Calculation

Direct confidence can be calculated using any typical association rules algorithm [2, 22]. However, these algorithms use minimal confidence and minimal support as main parameters instead of the threshold δ from (2). Another slight modification of original algorithms is needed to include time factor (3).

Partial indirect rules may be obtained using prior extracted direct rules. Note that according to (5), we need only two direct confidence values to calculate one partial indirect confidence value. No access to source user sessions is necessary in the calculation. Taking advantage of this property, the IDARM (In-Direct Association Rules Miner) algorithm will be introduced in [10] to extract complete indirect association rules with assigned confidence values from direct rules.

Since Tan *et al.* extract their indirect patterns from source user sessions [28, 29], the way of calculation is another fundamental difference distinguishing presented indirect association rules from those by Tan *et al.*

4.7 Recommendation

Values of complex confidence function for all product pages $d^P_j \in D^P$ are calculated for every page d_i in the web site and in the next step these values are placed in a descending order. In this way, we obtain the ranking list of product pages r^i_i for each page d_i in the web site. We select the highest ranked product pages from the list r^i_i as candidates for recommendation.

However, the recommendation may be performed in a static way, without personalization (Fig. 4, path A) or as the personalized, dynamic process (Fig. 4, path B). In the former approach, the first M product pages from the top of the list r^i_i are incorporated in the form of static hyperlinks into the HTML content of the page d_i. This task is executed offline, just after prior calculations. In such case, every user requesting a page from the web site is recommended the same list of products. The number M is usually about 3-5 and it depends on project assumptions for the user interface. More links appearing in the recommendation window may result in information overload.

The latter solution (personalized recommendation) requires the active user session to be monitored. The system stores not only the whole sequence of current user requests but also product pages recommended on each, previous page. To limit the amount of necessary data, only K pages lately visited by the user are kept in the extended form i.e. with the list of recommendation for each page. In this way, we retain the separate $M \times K$ matrix of URLs for each active user.

Let L_k be the set (list) of pages recommended by the system on the k-th page of the active user, $k=1, 2, \ldots, K$. The last visited page, from which the user came to the just being generated page, has the index 1, the previous one - 2, etc.

Next, the ranking list for requested page r^l_i is recalculated using *personalized rank-ing function* $r'(d_i \rightarrow d^P_j)$ to damp pages that belong to the list L_k:

$$r'\!\left(d_i \rightarrow d^P_j\right) = \begin{cases} r\!\left(d_i \rightarrow d^P_j\right) * \prod_{k:\, d^P_j \in L_k}\!\left(\dfrac{k-1}{K}\right), & \text{if } \exists k : d^P_j \in L_k \\[2mm] r\!\left(d_i \rightarrow d^P_j\right), & \text{otherwise} \end{cases}, \quad d_i \neq d^P_j \qquad (11)$$

The ranking function $r(d_i \rightarrow d^P_j)$ is reduced here with the factor $(k-1)/K$. For the previous page (the one exposed just before the actual one) this factor is equal to 0 ($k=1$) and product pages, which have been suggested on such a page, are excluded from current recommendation. Note that a product page d^P_j may be recommended on many pages within the last K ones. For such a page d^P_j its ranking function value is decreased several times by $(k-1)/K$ factor although this does not prevent such a page being recommended. The recommendation is possible while its ranking function value is much greater than for other product pages. We are only sure that the page d^P_j will not be suggested on the next page visited by the user. The index k of actual page will then be equal to 1, so all now recommended products will be excluded from the next recommendation.

Owing to (1) and (11) a user is recommended with the products from the e-commerce offer related with product web pages.

5 Experiments

Some experiments were carried out to discover association rules and ranking lists of all kinds. They were performed for the HTTP server log files coming from the certain Polish hardware e-commerce portal, which included 4,242 web pages. The original set consisted of 100,368 user sessions derived from 336,057 HTTP requests (only for HTML pages). However, only 16,127 sessions left after cleaning – too short and too long sessions were excluded.

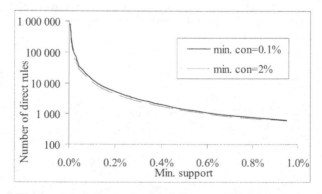

Fig. 5. The number of direct rules in relation to minimal support threshold.

The number of direct rules extracted from user sessions, first and foremost depends on minimal support threshold (Fig. 5). Typical minimal confidence and minimal sup-

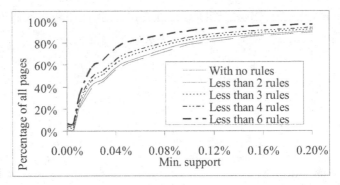

Fig. 6. The number of pages (as percentage of all 4,242 web pages) with zero, at most 1, 2, 3, 5 direct rules; min. confidence = 1%.

port coefficients were used in research instead of the threshold $\bar{o}$ (2) because of available software libraries. The most suitable value of minimal support for the test collection seems to be from the range [0.01%;0.1%] and for minimal direct confidence about 1-2%. Actually, such values probably exclude most unreliable direct associations. Greater values of thresholds strongly reduce number of rules and in consequence the system would have too short ranking lists. Anyway, 24-83% of all portal pages possess less than 3 direct rules (for min.sup∈ [0.01%;0.1%]) – Fig. 6. For these pages direct associations deliver too few rules and the system receives the insufficient number of items for recommendation.

The number of discovered partial indirect rules strictly depends on the number of available direct ones, so it is directly related to minimal support and confidence (Fig. 7 left). The number of complex rules was obviously only a little greater than the number of complete indirect ones (the difference only amounted several hundreds) but it was always about 20 times greater than the number of direct rules (Fig. 7 right).

Up to 18% (818) pages may have a very short ranking list (less than 5 positions) derived from direct rules (Table 1) while this rate reaches only max. 0.2% (9 pages) for complex rules. This data justifies visibly the usage of indirect rules in web recommendation systems. They considerably enrich complex rules and as a result they significantly lengthen ranking lists.

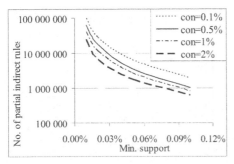
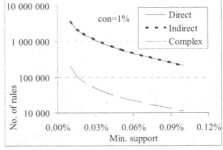

Fig. 7. The number of partial indirect rules (left); the comparison of the number of direct, complete indirect and complex rules (right) discovered for different values of minimal support.

Table 1. The number of pages (bodies of rules) with the certain length of rankings based on direct and complex rules; minimal confidence equals 1%.

Min. sup [%]	No. of pages with only 1 rule		No. of pages with 1 to 2 rules		No. of pages with 1 to 3 rules		No. of pages with 1 to 4 rules		No. of pages with 1 to 5 rules	
	direct	complex	direct	complex	direct	complex	direct	complex	direct	complex
0.01	69	2	167	3	257	3	390	3	502	9
0.02	94	2	253	2	386	2	621	2	775	2
0.05	71	0	229	0	372	1	674	2	818	2
0.10	60	2	167	2	267	2	457	2	526	2

Note that the number of pages with no rankings at all is the same for rankings based on both direct and complex rules (see Fig. 6, the lowest curve for approximate values). It comes from the def. 3. Partial and consequently also complete indirect rules may exist for the given page d_i if and only if there exists at least one direct rule for this page. Thus, indirect rules can only extend the non-empty set of rules starting from the page d_i.

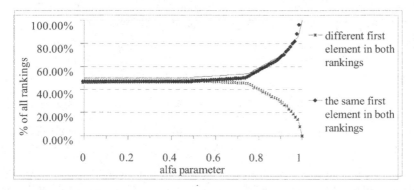

Fig. 8. The diversity of first positions in two kinds of rankings: based on only direct and complex rules, in relation to *á*, for first 1,000 pages (rankings); min.sup=0.02%, min.con=2%.

The integration of direct and indirect rules in (9) and (10) may result not only in the length of ranking lists but it also influences on the order in rankings. For example, first position in rankings may be different depending on which direct or indirect rules have had greater influence on values of complex confidence (Fig. 8). This influence may be adjusted by setting appropriate values of α in (9) and (10). 49,4% of all ranking lists (i.e. pages) possessed the distinct first element for α=0. In such case, complex rules consist of only indirect associations. The last tests were carried out based on [18].

6 Conclusions and Future Work

The presented method uses information about historical users' behaviour (web usage mining) to recommend products in the e-commerce web site. Direct and indirect associations coming out of user sessions are used to estimate helpfulness of individual product pages to be suggested on the just visited page. This also may be verified (personalized) at the last stage of the method in view of current user behaviour.

The ranking list of products (11) is created for both product and non-product pages, that means for almost every page in the web site.

Indirect confidence function includes transitive similarity of pages. Note, that in (7) $d_k \in D$, the product page may be recommended with respect to both product and non-product pages. Since the threshold $\tilde{o}$ is not used in the partial indirect confidence (6), the complete indirect confidence may promote products popular in the site. This is valid for new source pages, which have not yet been visited in many sessions.

Performed experiments have shown that indirect association rules may deliver the meaningful knowledge for recommender systems. They significantly extend and reorder ranking lists, what is essential for pages not having many direct rules.

There is also other data, besides web usage information, available in many e-commerce systems, like data about products placed into the basket, purchased products [4] and web pages' content [12, 14, 15]. They may be combined with the proposed method in the future. Another important extension is to reduce the complex confidence of product pages, to which static HTML links from the current page exist.

The relationship between products and product pages is "one to one" (1). The method could be expanded to include also "one to many" and "many to many" relationships.

References

1. Adomavicius G., Tuzhilin A.: Using Data Mining Methods to Build Customer Profiles. IEEE Computer, 34 (2) (2001) 74–82.
2. Agrawal R., Imieliński T., Swami A.: Mining association rules between sets of items in large databases. ACM SIGMOD International Conference on Management of Data, Washington D.C., ACM Press (1993) 207–216.
3. Buono, P., Costabile, M.F., Guida, S., and Piccinno, A.: Integrating User Data and Collaborative Filtering in a Web Recommendation System. OHS-7, SC-3, and AH-3 (2001), LNCS 2266, Springer Verlag, (2002) 315–321.
4. Cho Y.H., Kim J.K., Kim S.H.: A personalized recommender system based on web usage mining and decision tree induction. Expert Systems with Applications, 23 (3) (2002) 329–342.
5. Datta A, Dutta K., VanderMeer D., Ramamritham K., Navathe S.B.: An architecture to support scalable online personalization on the Web. The VLDB Journal The International Journal on Very Large Data Bases. 11 (1) (2001) 114–117.
6. Geyer-Schulz A., Hahsler M.: Comparing Two Recommender Algorithms with the Help of Recommendations by Peers. WebKDD 2002. LNCS 2703. Springer Verlag (2003) 137–158.
7. Goodrum A., McCain K.W., Lawrence S., Giles C.L.: Scholarly publishing in the Internet age: a citation analysis of computer science literature. Information Processing and Management 37 (5) (2001) 661–675.
8. Ishikawa H., Ohta M., Yokoyama S., Nakayama J., Katayama K.: On the Effectiveness of Web Usage Mining for Page Recommendation and Restructuring. NODe 2002, LNCS 2593, Springer Verlag (2003) 253–267.
9. Kazienko P.: Hypertekst Clustering based on Flow Equivalent Trees. Wrocław University of Technology, Department of Information Systems, Ph.D. Thesis, in Polish (2000) http://www.zsi.pwr.wroc.pl/~kazienko/pub/Ph.D.Thesis2000/PhD.zip.

10. Kazienko P.: Mining Indirect Association Rules for the Web. (2004) to appear.
11. Kazienko P.: Multi-agent Web Recommendation Method Based on Indirect Association Rules. KES'2004, 8th International Conference on Knowledge-Based Intelligent Information & Engineering Systems, LNAI, Springer Verlag (2004).
12. Kazienko P., Kiewra M.: Link Recommendation Method Based on Web Content and Usage Mining. IIPWM'03, Advances in Soft Computing, Springer Verlag (2003) 529-534, http://www.zsi.pwr.wroc.pl/~kazienko/pub/IIS03/pkmk.pdf.
13. Kazienko P., Kiewra M.: ROSA - Multi-agent System for Web Services Personalization. AWIC 2003, LNAI 2663, Springer Verlag (2003) 297–306.
14. Kazienko P., Kiewra M.: Integration of Relational Databases and Web Site Content for Product and Page Recommendation. 8th International Database Engineering & Applications Symposium. IDEAS '04, IEEE Computer Society (2004) 111–116.
15. Kazienko P., Kiewra M.: Personalized Recommendation of Web Pages. Chapter 10 in: Nguyen T. (ed.) Intelligent Technologies for Inconsistent Knowledge Processing. Advanced Knowledge International, Adelaide, South Australia (2004) 163–183.
16. Lawrence S., Giles, C.L., Bollacker K.: Digital Libraries and Autonomous Citation Indexing. IEEE Computer 32 (6) (1999) 67–71.
17. Lee D., Choi H.: Collaborative Filtering System of Information on the Internet. Computational Science - ICCS 2002, Part III, LNCS 2331, Springer Verlag (2002) 1090–1099.
18. Matrejek M.: Knowledge Discovery from Data about Behavior of Web Users based on Indirect Association Rules. Master Thesis. Wrocław University of Technology, Department of Information Systems, 2004, in Polish.
19. Mobasher B., Cooley R., Srivastava J.: Automatic Personalization Based on Web Usage Mining. Communications of the ACM, 43 (8) (2000) 142–151.
20. Mobasher B., Dai H., Luo T., Nakagawa M.: Effective Personalization Based on Association Rule Discovery from Web Usage Data. WIDM01, ACM (2001) 9–15.
21. Mooney, R.J., Roy, L.: Content-based book recommending using learning for text categorization. 5th ACM Conference on Digital Libraries (2000) 195–204.
22. Morzy T., Zakrzewicz M.: Data mining. Chapter 11 in Błażewicz J., Kubiak W., Morzy T., Rubinkiewicz M (eds): Handbook on Data Management in Information Systems. Springer Verlag, Berlin Heidelberg New York (2003) 487–565.
23. Nakagawa M., Mobasher B.: Impact of Site Characteristics on Recommendation Models Based On Association Rules and Sequential Patterns. IJCAI'03 Workshop on Intelligent Techniques for Web Personalization, Acapulco, Mexico (2003).
24. Pazzani M.: A Framework for Collaborative, Content-Based and Demographic Filtering. Artificial Intelligence Rev. 13 (5-6) (1999) 393–408.
25. Pazzani, M., Billsus, D.: Learning and revising user profiles: The identification of interesting web sites. Machine Learning, 27 (1997) 313–331.
26. Sarwar B.M., Karypis G., Konstan J.A., Riedl J.: Analysis of Recommendation Algorithms for E-Commerce. ACM Conference on Electronic Commerce (2000) 158–167.
27. Schafer J.B., Konstan J.A., Riedl J.: E-Commerce Recommendation Applications. Data Mining and Knowledge Discovery 5 (1/2) (2001) 115–153.
28. Tan P.-N., Kumar V.: Mining Indirect Associations in Web Data. WEBKDD 2001. LNCS 2356 Springer Verlag (2002) 145–166.
29. Tan P.-N., Kumar V., Srivastava J.: Indirect Association: Mining Higher Order Dependencies in Data. PKDD 2000, LNCS 1910 Springer Verlag (2000) 632–637.
30. Weiss R., Velez B., Sheldon M.A., Namprempre C., Szilagyi P., Duda A., Gifford D.K.: HyPursuit: A Hierarchical Network Search Engine that Exploits Content-Link Hypertext Clustering. 7th ACM Conference on Hypertext. ACM Press (1996) 180–193.
31. Yang H., Parthasarathy S.: On the Use of Constrained Associations for Web Log Mining. WEBKDD 2002, LNCS 2703, Springer Verlag (2003) 100–118.

Optimal Discovery
of Subword Associations in Strings
(Extended Abstract)

Alberto Apostolico[1,*], Cinzia Pizzi[2], and Giorgio Satta[2]

[1] Dipartimento di Ingegneria dell' Informazione, Università di Padova, Padova, Italy
and Department of Computer Sciences, Purdue University,
Computer Sciences Building, West Lafayette, IN 47907, USA
axa@dei.unipd.it
[2] Dipartimento di Ingegneria dell' Informazione, Università di Padova,
Via Gradenigo 6/A, 35131 Padova, Italy
{cinzia.pizzi,satta}@dei.unipd.it

Abstract. Given a textstring x of n symbols and an integer constant d, we consider the problem of finding, for any pair (y, z) of subwords of x the number of times that y and z occur in tandem (i.e., with no intermediate occurrence of either one of them) within a distance of d symbols of x. Although in principle there might be n^4 distinct subword pairs in x, we show that it suffices to consider a family of only n^2 such pairs, with the property that for any neglected pair (y', z'), there is a corresponding pair (y, z) contained in our family and such that: (*i*) y' is a prefix of y and z' is a prefix of z, and (*ii*) the tandem index of (y', z') equals that of (y, z). We show that an algorithm for the construction of the table of all such tandem indices can be built to run in optimal $O(n^2)$ time and space.

Keywords: Pattern Matching, String Searching, Subword Tree, Substring Statistics, Tandem Index, Association Rule.

1 Introduction

The problem of characterizing and detecting unusual events such as recurrent subsequences and other streams or over/under-represented words in sequences arises ubiquitously in diverse applications and is the subject of much study and interest in fields ranging from Computer and Network Security to Data Mining, from Speech and Natural Language Processing to Computational Molecular Biology. It also gives rise to interesting modeling and algorithmic questions, some of which have displayed independent interest.

* Work Supported in part by an IBM Faculty Partnership Award, by the Italian Ministry of University and Research under the National Projects FIRB RBNE01KNFP, and PRIN "Combinatorial and Algorithmic Methods for Pattern Discovery in Biosequences", and by the Research Program of the University of Padova, by NSF Grant CCR-9700276 and by NATO Grant CRG 900293.

E. Suzuki and S. Arikawa (Eds.): DS 2004, LNAI 3245, pp. 270–277, 2004.
© Springer-Verlag Berlin Heidelberg 2004

Among the problems in this class we find, for instance, the detection of all squares or palindromes in a string, for which optimal $O(n \log n)$ algorithms have long been known (refer, e.g., [3, 7] and references therein). It is not difficult to extend those treatments to germane problems such as the discovery of pairs of occurrences, within a given distance, of a same string, or of a string and its reverse, and so on.

In this paper, we concentrate on the problem of detecting repetitive phenomena that consist of unusually frequent *tandem* occurrences, within a pre-assigned distance in a string, of two distinct but otherwise unspecified substrings. By the two strings occurring in tandem, we mean that there is no intermediate occurrence of either one in between.

Specifically, we consider the following problem. Let x be a string of n symbols over some alphabet Σ and d some fixed non-negative integer. For any pair (y, z) of subwords of x, their *tandem index* $I(y, z)$ relative to x is the number of times that z has a closest occurrence in x within a distance of d from a corresponding, closest occurrence of y to its left. We are interested in finding pairs of subwords with surprisingly high tandem index.

The problem can be cast in the emerging contexts of *data mining* and *information extraction*. As is well known, while traditional data base queries aim at retrieving records based on their isolated contents, these contexts focus on the identification of patterns occurring across records, and aim at the retrieval of information based on the discovery of interesting rules present in large collection of data. Central to these developments is the notion of *association rule*, which is an expression of the form $S_1 \rightarrow S_2$ where S_1 and S_2 are sets of data attributes endowed with sufficient *confidence* and *support*. Sufficient support for a rule is achieved if the number of records whose attributes include $S_1 \cup S_2$ is at least equal to some pre-set minimum value. Confidence is measured instead in terms of the ratio of records having $S_1 \cup S_2$ over those having S_1, and is considered sufficient if this ratio meets or exceeds a pre-set minimum. Clearly, a statistic of the number of records endowed with the given attributes must be computed as a preliminary step, and this is often a bottleneck for the process of information extraction. We refer to [1], [8] and [14] for a broader discussion of these concepts.

Back to our problem, we observe that, in principle, there might be n^4 distinct pairings of subwords of in x. We show, however, that it suffices to restrict attention to a family containing only only n^2 pairs, after which for any neglected pair (y', z'), there is a pair (y, z) in the family such that: (*i*) y' is a prefix of y and z' is a prefix of z, and (*ii*) the tandem index of (y', z') equals that of (y, z). We show that an algorithm for the construction of the table of all such tandem indices can be built to run in optimal $O(n^2)$ time and space for a string x of n symbols.

A generalization of the notion of tandem, called *proximity word-association pattern*, has been proposed in [4] with similar motivations. A proximity word-association pattern is defined as a tuple of k string components, each matching the input text at a distance smaller or equal than a given d from the matching of the previous component[1]. Among several results, [4] implicitly defines an algorithm for counting matches of such patterns in an input string of length n, reporting a worst case time complexity of $O(d^k n^{k+1} \log n)$. A related algorithm was also proposed in [18], running in time $O(n^{k+1})$ but requiring $O(n^k)$ scans of the input string. In the present proposal we are therefore concerned with the case $k = 2$, and offer an asymptotical improvement over the algorithm in [4] with a worst case running time of $O(n^2)$ and a constant number of passes over the input string.

As mentioned, a tandem statistics of the kind collected by our algorithm finds application in many domains. In Section 4, we will briefly describe its uses in the discovery of dyadic structures in genome analysis and Part-of-Speech Tagging in Natural Language Processing.

2   Preliminaries

We begin by recalling an important "left-context" property from [5]. Given two words x and y, the *end-set* of y in x is the set of ending positions of *occurrences* of y in x, i.e., $endpos_x(y) = \{i : y = x_{i-|y|+1} \dots x_i\}$ for some i, $|y| \le i \le n$. Two strings y and z are equivalent on x if $endpos_x(y) = endpos_x(z)$. The equivalence relation instituted in this way is denoted by $\equiv^x$ and partitions the set of all strings over Σ into equivalence classes. Thus, $[y]$ is the set of all strings that have occurrences in x ending at the same set of positions as y. In the example of the string $abaababaabaabababaababa$, for instance, $\{ba, aba\}$ forms one such equivalence class and so does $\{aa, baa, abaa\}$. A symmetric equivalence relation can be defined in terms of *start-set* and denoted by $\equiv_x$, with obvious meaning. Recall that the *index* of an equivalence relation is the number of equivalence classes in it.

Fact 1 *The index k of the equivalence relation $\equiv^x$ (alternatively, $\equiv_x$) obeys $k < 2n$.*

Fact 1 suggests that we might only need to look among $O(n^2)$ substring pairs of a string of n symbols in order to find unusually frequent pairs. The following considerations show that this statement can be made precise giving an indirect proof of it. We recall the notion of the *suffix tree* T_x associated with x. This is essentially a compact trie with $n + 1$ leaves and at most n internal nodes that collects all suffixes of $x\$$, where \$ a symbol not in Σ. We assume familiarity of the reader with the structure and its clever $O(n \log |\Sigma|)$ time and linear space constructions such as in [12, 16, 19]. The word ending precisely at vertex α of T_x is denoted by $w(\alpha)$. The vertex α is called the *proper locus* of $w(\alpha)$. The *locus* of w is the unique vertex of T_x such that w is a prefix of $w(\alpha)$ and $w(\text{FATHER}(\alpha))$ is a proper prefix of w.

Having built the tree, some simple additional manipulations make it possible to count and locate the distinct (possibly overlapping) instances of any pattern w

---

[1] In [4] a different notation is used, with d representing the number of string components and k representing the maximum distance.

in x in $O(|w|)$ steps. For instance, listing all occurrences of w in x is done in time proportional to $|w|$ plus the total number of such occurrences, by reaching the locus of w and then visiting the subtree of T_x rooted at that locus. Alternatively, a trivial bottom-up computation on T_x can weigh each node of T_x with the number of leaves in the subtree rooted at that node. This weighted version serves then as a statistical index for x, in the sense that, for any w, we can find the frequency of w in x in $O(|w|)$ time. Note that the counter associated with the locus of a string reports its correct frequency even when the string terminates in the middle of an arc. This is, indeed, nothing but a re-statement and a proof of Fact 1, where the equivalence classes are represented by the nodes of the tree. From now on, we assume that a tree with weighted nodes has been produced and is available for our constructions.

3   Algorithms

For simplicity of exposition we assume that Σ is the binary alphabet $\Sigma = \{a, b\}$, but it should be clear that this assumption can be removed without penalty on our constructions.

In view of Fact 1 we may restrict attention to the $O(n)$ equivalence classes of $\equiv_x$ represented by strings that end precisely at the internal nodes and leaves of T_x. Even these latter, each being formed by some consecutive prefixes of a singleton string, may be neglected as uninteresting.

As a warmup for the discussion, consider the problem of computing the following relaxed notion of a tandem index, which we denote by $\bar{I}(y, z)$ and consists of the number of instances of z that fall within d positions of a closest occurrence of y to its left. The difference with respect to $I(y, z)$ is that we still forbid intervening occurrences of y, but now allow possibly intervening occurrences of z.

It is easy to compute $\bar{I}(y, z)$ for a fixed word y and all z's with a proper locus in T_x in overall linear time. This is done as follows. We can assume that, as a trivial by-product of the construction of T_x, there is access from each leaf of T_x to the corresponding position of x and *vice versa*. Now, let ν be the node of T_x such that $w(\nu) = y$. Visit the subtree of T_x rooted at ν and place a special mark on the positions of x that correspond to leaves in this subtree. This marks all the starting positions of occurrences of y in x. Next, scan the positions of x in ascending order: tag those positions that fall within d positions of an occurrence of y, and assign a weight of 1 to every leaf that corresponds to such a position. Visit now T_x bottom up, and assign to each node a weight equal to the sum of the weights of its children. Clearly, this accomplishes computation of $\bar{I}(y, z)$ for all z's. Iterating the process for all y's fills the table of $\bar{I}(y, z)$ values for all pairs of words with a proper locus in T_x, and this takes (optimal) $O(n^2)$ time and space.

Consider now computing the $I(y, z)$ values for a given y and all strings z. If, in the bottom up computation above, we wanted to compute $I(y, z)$ instead of $\bar{I}(y, z)$, then we should prevent the weighting process from counting more than one occurrence of z within the d-range of each closest occurrence of y.

Let $\mathcal{L}$ be sorted list of positions of x, with occurrences of y suitably marked and the positions falling within a distance of d from such occurrences tagged as like above. Define a *clump* in $\mathcal{L}$ as a maximal run of tagged positions falling between two consecutive occurrences of y (or following the last occurrence of y). Let us say further that a clump is *represented* at a node μ of T_x if at least one element of the clump is found in the subtree of T_x rooted at μ. Clearly, we want each clump to contribute precisely one unit weight at all nodes where the clump itself is represented.

It is possible to achieve this by the following preprocessing of T_x. With k the total number of clumps, initialize k empty *clump lists*. Visit the (leaves of the) tree from left to right. For each leaf encountered, check on its corresponding entry in $\mathcal{L}$ whether this leaf belongs to a clump. In this case, assign to the leaf a *rank* equal to the ordinal number of arrival of the leaf in its clump according to the visit. At the end, the concatenation of the clump lists constitutes a sorted list of clumps such that leaves within each clump are met in order of ascending rank. This means that scanning each clump list now meets the leaves in the clump in the same order as they would be encountered when visiting T_x from left to right. At this point we invoke as a subroutine the following well known Lowest Common Ancestor (l.c.a.) Algorithm (see, e.g., [15]): given a tree T with n leaves, it is possible to preprocess T in time linear in the number of nodes and in such a way that, after preprocessing, for any two leaves i and j it is possible to give in constant time the lowest common ancestor of i and j.

For a given y, we pre-process all clumps in succession as follows. Singleton clumps are not touched. With reference now to the generic non-singleton clump, consider its leaves in order of ascending rank. For each leaf, find its l.c.a.'s relative to the leaf and its successor and give the weight of this node a -1 handicap. At the end of the process, weigh the tree by the bottom up weighting procedure as before.

We claim that the weights thus assigned to any node μ represent $I(y, w(\mu))$. This is seen by induction on the sparsification $\bar{T}$ of T_x intercepted by the leaves of some arbitrary clump and the associated l.c.a. nodes. The assertion is true on any deepest internal node of $\bar{T}$. In fact, since T_x is binary then so is $\bar{T}$. Considering any deepest internal node of $\bar{T}$, the leftmost one of its two leaves will have caused the algorithm to give the node a handicap of -1. This would combine with the weight of 2 resulting from the bottom up weighting to yield a weight of 1. Assume now the assertion true for all descendants of a node μ of $\bar{T}$, we have that the two children α and β of μ will be assigned a weight of 1 in the bottom up process. Considering the rightmost leaf of the subtree rooted at α and the leftmost one in the subtree rooted at β, we have that μ, their l.c.a., has been given a handicap of -1 during pre-processing of the clump. Therefore, the sum of weights at μ will again be 1. The following theorem summarizes our discussion.

Theorem 2. *Given a string x of n symbols, the tandem indices for all pairs of subwords of x can be computed in $O(n^2)$ time and space.*

3.1 Implementation

In the tasks at hand, briefly discussed in Section 4, it is reasonable to assume that $|y|, |z|$, and d are much smaller than $|x|$. Therefore the algorithm was implemented on a *truncated* [13] rather than standard suffix tree data structure. This choice allows us to save a considerable amount of the working space at each run of the experiments. Truncated suffix trees are suffix trees in which the maximum length of a label from the root to a leaf is bounded [13] by some parameter N. The algorithm described in the previous sections needs some adjustments in order to work still correctly on a truncated suffix tree. The main problem is due to the fact that in a truncated suffix tree we loose the suffix tree property for which each leaf corresponds to one and only one string suffix. Therefore it can happen that the same leaf maps to several marked positions in the input string. To deal with this, we modify the algorithm as follows. Initially, each leaf is weighted with the sum of the corresponding number of marked positions. If these positions belong to different clumps, then no problem will arise in the total weight calculation, since each position will be treated separately at each clump analysis. On the other hand, if two (or more) positions belong to the same clump, then their associated node appears two or more consecutive times in the ordered clump list. In this case the LCA procedure is not executed. Instead, we decrement the value of the weight of that leaf. Then we proceed by comparing the next pair of nodes in the clump, and follow the above procedure if the nodes are again the same, or the standard algorithm procedure otherwise.

4 Applications

Here we describe two example applications of our algorithm, the first one to Bioinformatics, the second to Natural Language Processing.

4.1 Dyad Analysis

Dyad analysis for discovering cis-acting regulatory patterns in gene promoter regions are based on statistics on co-occurrences of short solid patterns. The method is based on the observation that many regulatory sites can be modeled by a structure $w_1 n_s w_2$, where w_1 and w_2 are solid words of length 3, s the distance between the two, and n_s a sequence of s unspecified nucleotides. We refer to [17] for a detailed description and validation of the method. Although experimental results proved the method to be efficient for the detection of sites bound by $C_6 Zn_2$ binuclear cluster proteins and other transcription factors, the exhaustive search through all the possible trinucleotides and all the possible pairs among them is time consuming and prevents a more efficient use of the tool. Moreover, especially for longer patterns, reasoning in terms of the equivalence classes of $\equiv_x$ instead of fixed length words is more meaningful in so far as the dyads detected in this way are *maximal* in an obvious sense.

4.2   Part-of-Speech Tagging

Another target applications of the algorithms developed in this paper is computational learning for natural language processing. More specifically, we focus on the task of part-of-speech tagging [10], where words in the text must be classified for their lexical category (Noun, Verb, Adjective, etc.).

A common solution in the design of part-of-speech taggers is to exploit sets or cascades of rules that are automatically induced from examples of correct classifications. Usually, these rules classify words on the basis of finite size contextual windows, centered around the word of interest, as described for instance in [6]. However, such systems do not achieve 100% accuracy in classification, as compared with classifications provided by human experts. One of the reasons for such a gap is due to the limited size of the adjacency windows that are used by the rule schemata described above. It has been shown in [9] that certain contextual patterns called *barriers*, that are highly effective in classification, can appear at variable distances from the word of interest, demanding for more powerful rule patterns than those described above.

The tandem patterns that are discovered by the algorithm of Section 3 can be applied to the task of automatic extraction of contextual rules for part-of-speech tagging, by simply modifying the algorithm to count the end-to-start distance instead than the previously described start-to-start distance between tandem components. Preliminary results on the CommonNoun/Adjective confusion class showed that this method could be helpful when a set of fixed single window rules fail to give a clear classification.

5   Conclusions

We have investigated the problem of collecting counts on the number of times any pair of strings occur in tandem (i.e., with no intermediate occurrence of either one of them) in a given input text and within a given distance. We have provided an algorithm for the construction of the table of all such tandem counts, running in optimal $O(n^2)$ time and space. This improves on previous results that are found in the literature. We have also discussed two possible application of the algorithm: dyad analysis in bioinformatics, and part-of-speech tagging in natural language processing.

References

1. Agrawal, R., Imielinski, T., Swami, A.: Mining Association Rules between Sets of Items in Large Databases. In *Proc. ACM SIGMOD*, 207–216, Washington DC May 1993.
2. Aho, A.V., Hopcroft, J.E., Ullman, J.D.: *The Design and Analysis of Computer Algorithms*, Addison-Wesley, Reading, Mass (1974)
3. Apostolico, A., Galil,Z. (Eds.): *Pattern Matching Algorithms*, Oxford University Press, New York (1997)

4. Arimura, H., Arikawa, S.: Efficient Discovery of Optimal Word-Association Patterns in Large Text Databases. *New Generation Computing*, 18: 49–60 (2000)
5. Blumer, A., Blumer, J., Ehrenfeucht, A., Haussler, D., Chen, M.T. and Seiferas, J.: The Smallest Automaton Recognizing the Subwords of a Text. *Theoretical Computer Science* , 40: 31–55 (1985).
6. Brill, E.: Transformation-Based Error-Driven Learning and Natural Language Processing: A Case Study in Part of Speech Tagging, *Computational Linguistics* (1995)
7. Crochemore, M., Rytter, W.: *Text Algorithms*, Oxford University Press, New York (1994)
8. Han, J., and Kamber, M.: Data Mining: Concepts and Techniques, Morgan Kaufmann Publishers, 2000.
9. Karlsson, F., Voutilainen, A., Heikkilä, F., and Anttila, A.: Constraint Grammar. A Language Independent System for Parsing Unrestricted Text. Mouton de Gruyter (1995)
10. Manning C.D. , Schütze, H.: Foundations of Statistical Natural Language Processing, MIT Press (1999)
11. Marcus, M. P., Santorini, B., and Marcinkiewicz, M.A.: Building a Large Annotated Corpus of English: The Penn Treebank, *Computational Linguistics*, 19(2):313–330 (1993)
12. McCreight, E.M.: A Space-Economical Suffix Tree Construction Algorithm. *Journal of the ACM*, 23(2): 262–272, April 1976.
13. Na, J.C., Apostolico A., Iliopoulos C.S., Park K., Truncated Suffix Trees and their Application to Data Compression *Theoretical Computer Science*, Vol. 304 , Issue 1-3, pp.87–101 (2003)
14. Piatesky-Shapiro, G., Frawley W.J. (eds.): *Knowledge Discovery in Databases.* AAAI Press/MIT Press (1991).
15. Schieber, B., Vishkin, U.: On Finding Lowest Common Ancestors: Simplifications and Parallelizations, *SIAM Journal on Computing*, 17:1253–1262 (1988)
16. Ukkonen, E.: On-line Construction of Suffix Trees. *Algorithmica*, 14(3): 249–260 (1995)
17. van Helden, J., Rios, A.F., Collado-Vides, J.: Discovering Regulatory Elements in Non-coding Sequences by Analysis of Spaced Dyads. *Nucleic Acid Research*, Vol.28, No.8: 1808-1818, 2000.
18. Wang, J. T.-L., Chirn, G.-W., Marr, T.G., Shapiro, B., Shasha, D., Zhang, K.: Combinatorial Pattern Discovery for Scientific Data: Some Preliminary Results. In *Proceedings of 1994 SIGMOD*, 115–125 (1994)
19. Weiner, P.:, Linear Pattern Matching algorithm. In *Proceedings of the 14th Annual IEEE Symposium on Switching and Automata Theory*, pages 1–11, Washington, DC (1973)

Tiling Databases

Floris Geerts[1], Bart Goethals[2], and Taneli Mielikäinen[2]

[1] Laboratory for Foundations of Computer Science
School of Informatics, University of Edinburgh
fgeerts@inf.ed.ac.uk
[2] HIIT Basic Research Unit
Department of Computer Science
University of Helsinki, Finland
{goethals,tmielika}@cs.Helsinki.FI

Abstract. In this paper, we consider 0/1 databases and provide an alternative way of extracting knowledge from such databases using tiles. A tile is a region in the database consisting solely of ones. The interestingness of a tile is measured by the number of ones it consists of, i.e., its area. We present an efficient method for extracting all tiles with area at least a given threshold.
A collection of tiles constitutes a tiling. We regard tilings that have a large area and consist of a small number of tiles as appealing summaries of the large database. We analyze the computational complexity of several algorithmic tasks related to finding such tilings. We develop an approximation algorithm for finding tilings which approximates the optimal solution within reasonable factors. We present a preliminary experimental evaluation on real data sets.

1 Introduction

Frequent itemset mining has become a fundamental problem in data mining research and it has been studied extensively. Many efficient algorithms such as Apriori [1], Eclat [18] and FP-growth [7] have been developed to solve this problem.

More recently, a new setting, finding the top-k (closed) most frequent itemsets of a given minimum length, has been proposed [8]. For small k this provides a small and comprehensive representation of the data, but at the same time it raises the question whether frequency is the right interestingness measure. The use of frequency has been criticized in the context of association rules as well and many alternatives have been offered [16]. Unfortunately, finding good and objective interestingness measures for itemsets seems to be a hard problem. Additionally, allowing such measures to prune the huge search space of all itemsets might be even harder. Currently, most techniques rely on the monotonicity property of the frequency of itemsets. Indeed, supersets of infrequent itemsets cannot be frequent, and can therefore be pruned away from the huge space of all itemsets.

In this paper we introduce a new and objective interestingness measure for itemsets. It is based on the concept of a tile and its area. Informally, a tile consists of a block of ones in a 0/1 database as shown in Figure 1.

E. Suzuki and S. Arikawa (Eds.): DS 2004, LNAI 3245, pp. 278–289, 2004.

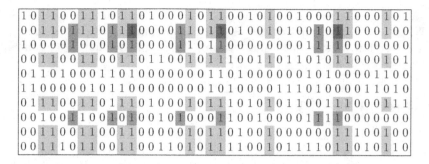

Fig. 1. Example of a 0/1 database a tiling consisting of two overlapping tiles (darkest shaded area correspond to intersection of the two tiles).

We call the number of ones in a tile the area of the tile in the database. A collection of (possibly overlapping) tiles constitutes a tiling (see Figure 1). The area of a tiling in the database is the total number of ones which are part of the tiles in the tiling. Obviously, the larger the tile or tiling, the more knowledge they represent about the database. Moreover, a large tiling consisting of only a small number of tiles can be extremely informative when one is curious about the content of a 0/1 database. Indeed, the tiling implicitly determines an upper bound on the number of different databases that are consistent with a database, and hence, it presents a small characterization of the database at hand.

In this paper we consider the following problems: the maximum k-tiling problem which asks for a tiling consisting of at most k tiles having the largest possible area; the minimum tiling problem which asks for a tiling of which the area equals to the total number of ones in the database and consists of the minimum number of tiles; the large tile mining problem which asks for all tiles in the database each having at least some minimum area; and the top-k tiles problem which asks for the k tiles that have the largest area. Furthermore, we present hardness results on the complexity of the problems listed above; we propose algorithms that solve these problems or give good approximations when optimal solutions are intractable; and we present promising experimental results on several large real world sparse and dense databases.

As we will show, the itemset constraints considered here belong to the class of length decreasing support constraints. The algorithms we propose use an adaptation and improvement of the known pruning strategies for length decreasing support constraints [15, 17].

The concept of a tile is not entirely new, but the problems we consider have not yet been investigated. The notion of tiles is similar to combinatorial rectangles (communication complexity [10]), co-clusters (clustering [5]), bi-sets (formal concepts [3]), or conjunctive clusters (learning [14]). The difference with our setting is that we look for exact tiles, while most research instantly relaxes the condition of exactness and studies approximations of the tiles.

This paper is organized as follows. We formally state the problems in Section 2 and give some complexity results. An algorithm for each problem is presented in Section 3. Experimental results are reported in Section 4. We conclude the paper in Section 5.

2   Problem Statement

Let $\mathcal{I}$ be a set of n items. We can assume that $\mathcal{I}$ is a finite set $\{1, \ldots, n\}$. An itemset I is a non-empty subset of $\mathcal{I}$. A transaction is a tuple $\langle tid, T \rangle$ where tid is a transaction identifier and T is an itemset. A (transactional) database D is a set of m transactions. The size of a transaction is the number of items in it; the size of a database is the sum of the sizes of the transactions in the database. A transaction database D can also be regarded as a binary matrix of size $m \times n$ where m is the number of transactions in D and n is the number of items in $\mathcal{I}$. The (i,j)th element in the binary matrix corresponding to D is equal to 1 if the ith transaction contains item j, and it is equal to 0 otherwise. Hence, the size of D is equal to the number of ones in this matrix. We will often mix these two interpretations of a transaction database in the remainder of the paper.

Given an itemset I, a transaction $\langle tid, T \rangle$ covers I if $I \subseteq T$. The cover of an itemset I, denoted by $cover(I, D)$, consists of the transaction identifiers of transactions in D which cover I. The support of an itemset I with respect to a database D is the cardinality of $cover(I, D)$. We denote this number by $supp(I, D)$, or simply $supp(I)$ when D is clear from the context.

Definition 1 (tiles, tilings and their area). *Let I be an itemset and D be a database. The tile corresponding to I is defined as*

$$\tau(I, D) = \{(tid, i) \mid tid \in cover(I, D), i \in I\}.$$

When D is clear from the context, we write $\tau(I)$ instead. The area of $\tau(I)$ is equal to its cardinality. Moreover,

$$area(\tau(I), D) = |\tau(I)| = |I| \cdot |cover(I, D)|.$$

A tiling $\mathcal{T} = \{\tau(I_1), \ldots, \tau(I_k)\}$ consists of a finite number of tiles; its area is $area(\mathcal{T}, D) = |\tau(I_1) \cup \cdots \cup \tau(I_k)|$.

We call a tile *maximal* if the corresponding itemset is closed. Recall that an itemset is closed if it is not contained in a larger itemset having the same support. All tiles considered in this paper will be maximal.

The area of a tiling $\mathcal{T}$ reflects how many possible databases there exist which are consistent with the tiling. We actually can compute an upper bound on the number of such databases easily based on the area of the tiling. Let m be the number of rows and n the number of single items. Then there are maximum 2^{nm} possible databases. Since a tiling fixes $2^{area(\mathcal{T},D)}$ entries in a database, there are only $2^{nm - area(\mathcal{T},D)}$ possible databases left which are consistent with the tiling $\mathcal{T}$. Based on this relationship between the number of databases consistent with

$\mathcal{T}$ and the area of D which the tiling $\mathcal{T}$ covers, the most interesting tilings are the ones with the maximum area. Indeed, these tilings reduce the number of databases consistent with the tiling the most.

The main goal of this paper is to find an algorithm which computes the maximum k-tiling.

Problem 1 (MAXIMUM k-TILING). Given a database D and a positive integer k, find a tiling $\mathcal{T}$ consisting of at most k tiles with maximum area $area(\mathcal{T}, D)$.

Related to Problem 1 is the problem of finding the minimum number of tiles covering the database. This number is a measure for the complexity of the database. When the complexity of the database is small, then the maximum k-tiling for small values of k can be regarded as an excellent representation of the database.

Problem 2 (MINIMUM TILING). Given a database D, find a tiling of D with area equal to the size of the database and consisting of the least possible number of tiles.

In order to solve Problem 1 and Problem 2, the main algorithmic task is to find all maximal tiles with area at least a given minimum area threshold.

Problem 3 (LARGE TILE MINING). Given a database D and a minimum area threshold σ, find all tiles in D with an area at least σ.

Since the collection of large tiles can be very large, it is only natural to look at a few largest tiles only:

Problem 4 (TOP-k TILES). Given a database D and a positive integer k, find k tiles with largest areas.

To analyze the computational complexity of finding tiles and tilings, let us first consider the simple subproblem of finding the largest tile in a given database D:

Problem 5 (MAXIMUM TILE). Given a database D, find the tile with the largest area in D.

A solution for this problem is important for the approximations of good tilings.

Theorem 1. MAXIMUM k-TILING, MINIMUM TILING, LARGE TILE MINING, TOP-k TILES *and* MAXIMUM TILE *are* NP-*hard*.

Proof. The MAXIMUM TILE problem can be seen also by viewing the database as an adjacency matrix of a bipartite graph $G = (V_1, V_2, E)$ and looking for the largest complete bipartite subgraph of G, i.e., the largest subset E' of edges such that $E' = V_1' \times V_2'$ for some $V_1' \subseteq V_1$ and $V_2' \subseteq V_2$. This problem is known as the MAXIMUM EDGE BICLIQUE problem which is shown to be NP-hard [13]. Based on this correspondence, also the MINIMUM TILING problem is NP-hard [12].

The other complexity results now follow directly since MAXIMUM TILE is a special case of the TOP-k TILES problem, with $k = 1$. It is also a special case of the LARGE TILE MINING problem when σ is assigned to be the maximum area of a tile in D. Finally, the MAXIMUM TILE problem is a special case of the MAXIMUM k-TILING problem with $k = 1$. $\qquad\square$

3   Algorithms

In this section, we first present the LTM algorithm for finding large tiles. LTM uses a branch and bound search strategy combined with several pruning techniques. Based on the LTM algorithm, we give an algorithm, k-LTM which finds the top-k tiles. Finally, we use k-LTM to find an approximation of a maximum k-tiling of a database. The approximation algorithm k-Tiling is a greedy algorithm and ensures an approximation within a constant factor from the optimal solution.

3.1   Large Tile Mining

Given a database D and a minimum area threshold σ, we want to find all maximal tiles which have an area at least σ.

In our algorithm, we will refine pruning techniques used in the context of length-decreasing support constraints on itemsets [15,17]. Let $f(x)$ be a monotone decreasing function, i.e., $f(x) \geq f(x+1) \geq 1$; f is called a length-decreasing support constraint and an itemset I is called frequent w.r.t. f if $supp(I, D) \geq f(|I|)$. An itemset I which correspond to a tile $\tau(I)$ of area at least σ can be seen as an itemset frequent w.r.t the length- decreasing support constraint $f(|I|) = \sigma/|I|$. Indeed, $area(I, D) = |I| \cdot supp(I, D) \geq \sigma$ iff $supp(I, D) \geq \sigma/|I|$.

First, our algorithms will follow a similar depth-first search strategy and counting mechanism as used in the Eclat algorithm and its variants [18,19]. The same search strategy was later also successfully used in the FP-growth algorithm [7], and is based on a divide and conquer mechanism.

Denote the set of all tiles in D with area at least σ, corresponding to itemsets with the same prefix $I \subseteq \mathcal{I}$ by $\mathcal{T}[I](D, \sigma)$.

The main idea of the search strategy is that all large tiles containing item $i \in \mathcal{I}$, but not containing any item smaller than i, can be found in the so called i-conditional database [7], denoted by D^i. That is, D^i consists of those transactions from D that contain i, and from which all items before i, and i itself are removed. In general, for an itemset I, we can create the I-conditional database, D^I, consisting of all transactions that contain I, but from which all items before the last item in I and that item itself have been removed. Whenever we compute the area of a tile in D^I, we simply need to add $|I|$ to the width of the tile and multiply this with the support of the corresponding itemset. Then, after adding I to the itemset, we found exactly all large tiles containing I, but not any item before the last item in I which is not in I, in the original database, D.

The large tile mining algorithm LTM as shown in Figure 2, recursively generates for every item $i \in \mathcal{I}$ the set $\mathcal{T}[\{i\}](D^i, \sigma)$. (Note that $\mathcal{T}[\{\}](D, \sigma) = \bigcup_{i \in \mathcal{I}} \mathcal{T}[\{i\}](D^i, \sigma)$ contains all large tiles.)

In order to compute the area of a tile, we need to know the support of the itemset representing that tile. This value is computed exactly as in the Eclat algorithm. That is, the algorithm stores the database in its vertical layout which means that each item is stored together with its cover instead of listing explicitly all transactions. In this way, the support of an itemset I can be easily

Input: D, σ, I (initially called with $I = \{\}$)
Output: $T[I](D, \sigma)$
 1: $T[I] := \{\}$;
 2: $Prune(D, \sigma, I)$.
 3: **for all** i occurring in D **do**
 4:     **if** $|cover(\{i\})|(|I| + 1) \geq \sigma$ **then**
 5:        Add $\tau(I \cup \{i\})$ to $T[I]$;
 6:     **end if**
 7:     $D^i := \{\}$;
 8:     **for all** j occurring in D such that $j > i$ **do**
 9:        $C := cover(\{i\}) \cap cover(\{j\})$;
10:        Add (j, C) to D^i;
11:     **end for**
12:     Compute $T[I \cup \{i\}](D^i, \sigma)$ recursively;
13:     Add $T[I \cup \{i\}]$ to $T[I]$;
14: **end for**

Fig. 2. The LTM algorithm.

Input: D, σ, I
 1: **repeat**
 2:    **for all** i occurring in D **do**
 3:       **if** $UB_{I \cup \{i\}} < \sigma$ **then**
 4:          Remove i from D.
 5:       **end if**
 6:       **for all** $tid \in cover(\{i\})$ **do**
 7:          **if** $size(tid) < ML_{I \cup \{i\}}$ **then**
 8:             Remove tid from $cover(\{i\})$.
 9:          **end if**
10:       **end for**
11:    **end for**
12: **until** nothing changed

Fig. 3. The Prune procedure.

computed by simply intersecting the covers of any two subsets $J, K \subseteq I$, such that $J \cup K = I$. This counting mechanism is perfectly suited for our purposes since it immediately gives us the list of transaction identifiers that, apart from the itemset itself, constitutes a tile.

We now describe the LTM algorithm in more detail. First, the algorithm is initialized (line 1). Then, on line 2, the main pruning mechanism is executed, as will be explained later. This mechanism will remove certain items from the candidate set that can no longer occur in a large tile and remove transactions that can no longer contribute to the area of a large tile. On line 3, the main loop of the algorithm starts by considering each item separately. On lines 4–6, each large tile is added in the output set. After that, on lines 7–11, for every item i, the i-projected database $\mathcal{D}^i$ is created. This is done by combining every item j with i, such that $j > i$ and computing its cover by intersecting the covers of both items (line 10). On line 12, the algorithm is called recursively to find all large tiles in the new database $\mathcal{D}^i$. However, in every such conditional database, each item must be treated as the tile represented by the itemset $I \cup \{i\}$.

For reasons of presentation, we did not add the details of restricting the search space to closed itemsets such that only maximal tiles are considered. Nevertheless, the algorithm can be easily extended with the techniques used in the CHARM algorithm [19] such that only closed itemsets are generated.

Unfortunately, if the area of a tile does not meet the minimum area threshold, we cannot simply prune away all tiles in its branch, because its supersets considered in that branch might still represent tiles with large areas. Nevertheless, it is possible to compute an upper bound on the area of any tile that can still be generated in the current branch. Based on this upper bound, the *Prune* procedure, as shown in Figure 3, is able to prune some items from the search space and at the same time it reduces the size of the database.

The *Prune* procedure consist of a repeated application of the node pruning methods used in LPMiner [15] and BAMBOO [17]. More specifically, for each item i in the database D (line 2) we compute an upper bound $UB_{I\cup\{i\}}$ of the largest possible tile containing $I\cup\{i\}$. To obtain this upper bound for a given i, we count how many transactions in the current I-conditional database, containing i, have size at least ℓ, for all occurring ℓ. Denote this number by $supp_{\geq\ell}(i, D^I)$. Then, the upper bound on the size of the largest possible area of a tile containing $I \cup \{i\}$ is given by $UB_{I\cup\{i\}} = \max\{(|I| + \ell) \cdot supp_{\geq\ell}(i, D^I) \mid \ell \in \{1, 2, \ldots\}\}$. Obviously, to compute this number, we need the size of each transaction in the current conditional database. This size is perfectly derivable from the vertical representation of the conditional database. In practice, however, it is more efficient to store a separate array in which the size of each transaction is stored for each conditional database. Also, note that the size for a given transaction in the current conditional database can be much smaller than the size of that transaction in the original database.

If the upper bound for a given item i in the database is smaller than the minimum area threshold (line 3), then this means that i will never be part of an itemset corresponding to a tile of area larger than σ, and therefore, it can be removed from D (line 4).

This also implicitly means that the size of all transactions is decreased, which may affect the upper bounds of the other items, and hence, they can be recomputed. This process can be repeated until no more items can be removed.

Even if an item cannot be completely removed from the database, it is sometimes possible to remove it from several transactions. More specifically, consider the following number $ML_{I\cup\{i\}} = \min\{\ell \mid \ell \cdot supp_{\geq\ell}(i, D^I) \geq \sigma\}$. This number gives the minimum size of a transaction containing i, that can still generate a tile with area at least σ. Hence, from each transaction containing i that is shorter, i can be removed (lines 7–8). In the Prune procedure in Figure 3, we find the size of a transaction *tid* using *size(tid)*. Again, this removal can have an effect on the upper bound of the other items, such that their upper bounds can be recomputed and maybe some of them can still be removed.

3.2 Top-k Tiles

In order to find the top-k largest tiles, we adapt the LTM algorithm as follows. Initially the minimum area threshold is zero. Then, after the algorithm has generated the first k large tiles, it increases the minimum area threshold to the size of the smallest of these k tiles. From here on, the minimum area threshold can be increased every time a large tile is generated w.r.t. the current threshold. All generated tiles that do not have an area larger than the increased minimum area threshold can of course be removed.

3.3 Finding the Tiling

Even if finding tilings close to the best ones with reasonable guarantees is not feasible in theory, we would like to find good tilings anyway.

For the MINIMUM TILING problem it is clear what the best tiling is: it is a complete tiling of the database with the smallest number of tiles. The best k-tiling can be similarly considered to be the k-tiling with the largest area.

However, in data mining this might not be enough: due to the exploratory nature of data mining the data mining tool should support interactive use. For example, determining the value k in advance might be an unreasonable requirement for the data analyst. Instead, the tool should make it as convenient as possible to explore different values of k. This is not the case if the suggested k-tiling and $k + 1$-tiling differ a lot from each other. In the best case (for the data analyst) the suggested k-tiling and $k + 1$-tiling would differ only by one tile.

Such k-tilings for all values of k can be determined simply by fixing an ordering for the tiles and considering the k first tiles in the ordering as the best k-tiling [11]. Clearly, this kind of ordering does not provide k-tilings with maximum area for all values of k but some approximation ratio might be guaranteed for all values of k simultaneously if the ordering of the tiles would be determined in a good way. For example, if the ordering of tiles is constructed greedily by adding the tile that covers the largest area of the uncovered parts of the database, we get decent upper bounds for the approximation ratios of the MINIMUM TILING and the MAXIMUM k-TILING problems.

Theorem 2. *The* MINIMUM TILING *problem can be approximated within the factor* $\mathcal{O}(\log nm)$ *and* MAXIMUM k-TILING *can be approximated within the factor* $e/(e-1)$ *for all values of k simultaneously, given an oracle that finds for any database D and tiling $\mathcal{T}$ the tile $\tau(I)$ such that* $area(\mathcal{T} \cup \{\tau(I)\}, D) = \max_{I' \subseteq I} area(\mathcal{T} \cup \{\tau(I')\}, D)$.

Proof. These problems can be interpreted as instances of the MINIMUM SET COVER problem and the MAXIMUM k-COVERAGE problem, respectively: the set that is to be covered corresponds to the ones in the database and the sets that can be used in the cover are the maximal tiles in D. The only problem in the straightforward reduction to set cover is that the number of maximal tiles can be exponential in the size of the database. Fortunately, the greedy algorithm for set cover that gives the desired approximation bounds for both variants of the problem depends only on the ability of finding the tile $\tau(I, D)$ that maximizes the area $area(\mathcal{T} \cup \{\tau(I)\}, D)$ for a given collection $\mathcal{T}$ of tiles [6].

If we have such an algorithm, then the approximation bounds follow from the bounds on the MINIMUM SET COVER problem and the MAXIMUM k-COVERAGE problem [2]. □

The oracle that gives the tile covering the most extra area outside a given tiling, can be implemented reasonably efficiently by adapting the LTM algorithm. Whenever we compute the area of a tile (line 4) in the LTM algorithm in Figure 2, we subtract the part of the area that was already covered earlier. Additionally, we can also improve the Prune procedure by computing a second upper bound which takes the already covered area into account. Indeed, the current upper bound needs to take the size of the original transactions into account and we can not simply remove what is already covered from them, since the remaining part might not even be a tile anymore. We can, however, store a second array containing only the uncovered size of each transaction. Using this array, the uncovered area of any candidate tile in the current conditional database, which contains item i, is at most the sum of of these sizes of the transactions that contain i. Finally, we take the minimum of these two bounds: $UB^*_{I \cup \{i\}} = \min\{UB_{I \cup \{i\}}, \sum_{tid \in cover(\{i\})} size^*(tid)\}$, which can replace the old upper bound (line 4) in the Prune procedure in Figure 3. Note that $size^*$ now stores the sizes of the uncovered parts of the transactions, containing item i.

Thus, in practice we can solve the problems MINIMUM TILING and MAXIMUM k-TILING with very good approximation bounds.

4 Experimental Evaluation

We implemented our algorithms in C++ and experimented on a 2GHz Pentium-4 PC with 1GB of memory, running Linux.

We present the evaluation of the algorithms on two substantially different datasets. The mushroom data set is a dense dataset, containing characteristics of various species of mushrooms and was originally obtained from the UCI repository of machine learning databases [4]. It consists of 119 single items, 8 124 transactions, resulting in 186 852 ones. The BMS-Webview-1 dataset is a sparse dataset, containing several months of click-stream data from an e-commerce website, and is made publicly available by Blue Martini Software [9]. It consists of 497 items, 59 602 transactions, resulting in 149 639 ones.

In the first series of experiments we ran the LTM algorithm for various minimum area thresholds, which is shown in Figure 4. As can be seen, the execution times show the feasibility of our approach, taking into account the extremely low minimum area thresholds. Indeed, a minimum area of 250 for the BMS-Webview-1 dataset, corresponds to $250/149\,639 = 0.3\%$ of the ones in the database. Similarly, for the mushroom dataset, a minimum area of 250 corresponds to $250/186\,852 = 0.1\%$ of the ones in the database.

The number of tiles found and the number of generated candidate tiles by the LTM algorithm for various minimum area thresholds is shown in Figure 5. Here, it shows that the dense mushroom dataset indeed contains a huge number of large tiles, while the sparse BMS-Webview-1 dataset contains a modest number of large tiles, but a lot smaller ones. For the same reason, the number of candidate tiles is much closer to the actual number of large tiles in the mushroom dataset as compared to the BMS-Webview-dataset. Note, since the mushroom database contains only 8 124 transactions, the minimum area threshold of 9 000 forces that no discovered tile consists of only a single item. Similarly, a minimum area threshold which is larger than the size of all transactions, guarantees that no single transaction can be a large tile.

In the second series of experiments we ran the k-LTM algorithm for various minimum area thresholds, which is shown in Figure 4. As can be seen, the execution times are larger than the execution times of the standard LTM algorithm, but it still shows the feasibility of our approach.

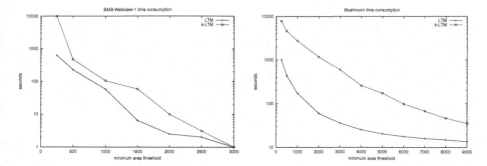

Fig. 4. Large tile mining on the BMS-Webview-1 (left) and Mushroom (right) data set. The plots show the execution times of the LTM and k-LTM algorithms for various minimum area thresholds.

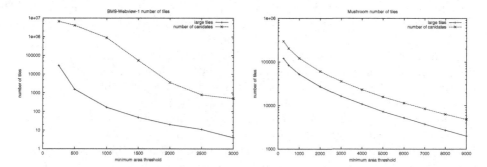

Fig. 5. Large tile mining on the BMS-Webview-1 (left) and Mushroom (right) data set. The plots show the number of tiles found and the number of tiles checked by LTM algorithm for various minimum area thresholds.

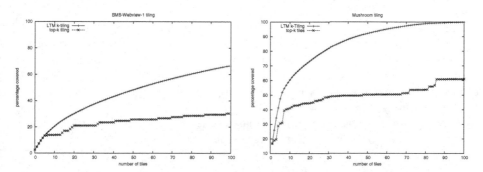

Fig. 6. Comparison of the covered area of tilings for the BMS-Webview-1 dataset (left) and Mushroom dataset (right) obtained by running the k-LTM algorithm and the k-Tiling algorithm on these data sets for different values of k.

Finally, to show the effectiveness of the k-Tiling algorithm, we tested the algorithm for a varying number k of tiles. To compare, we plot the percentage of the number of ones covered by the k-Tiling algorithm, together with the number of ones covered by taking the top-k tiles generated by the k-LTM algorithm. The results, shown in Figure 6, are striking for the mushroom dataset. Indeed, already 60% of the ones in the database are covered by only 10 tiles! To obtain a tiling that covers 90% of the database, only 45 tiles are needed. For the BMS-Webview-dataset, slightly more tiles are needed to obtain a similar tiling, which is of course expected, given the sparsity of this database.

5   Conclusions and Future Work

We introduced the concepts of tiles and tilings in the context of transaction or 0/1 databases and argued that the area of a tile and tilings is a good interestingness measure. Indeed, a tiling implicitly determines an upper bound on the number of different databases that are consistent with a database, and hence, it presents a small characterization of the database at hand.

We presented some computational challenges to computing different types of tilings. A theoretical analysis of these challenges shows that they are all NP-hard. In practice, however, computing these tilings might be still feasible.

Therefore, based on a powerful iterative upper bounding mechanism, we developed an elegant branch and bound algorithm that discovers all large tiles with respect to a minimum area threshold. This algorithm prunes away a lot of tiles from the huge search space of all possible tiles. Based on this algorithm, we derive two other algorithms using small adaptations for the problem of finding the k largest tiles and finding the largest tiling consisting of k tiles only.

Our experiments verify the efficacy and efficiency of the algorithms. From a knowledge discovery point of view, we show that using only a marginal number of tiles, we are already able to present a very good picture of the databases at hand.

This work is preliminary in the sense that the the knowledge extraction aspect is barely investigated. The main goal of this paper was to introduce the problem and show the feasibility of our approach. In future work, we will investigate the quality of tilings as knowledge representations. A more thorough experimental evaluation and comparison with existing approaches (frequent itemsets, top-k frequent itemsets) is postponed to a follow-up paper as well. Finally, we want to improve upon the upper bound of databases consistent with a given tiling, since that would result in more accurate interestingness measures.

Acknowledgments

We wish to thank Blue Martini Software for contributing the KDD Cup 2000 data [9].

References

1. R. Agrawal, H. Mannila, R. Srikant, H. Toivonen, and A.I. Verkamo. Fast discovery of association rules. In *Advances in Knowledge Discovery and Data Mining*, chapter 12, pages 307–328. AAAI/MIT Press, 1996.
2. G. Ausiello, P. Crescenzi, V. Kann, A. Marchetti-Spaccamela, and M. Protasi. *Complexity and Approximation: Combinatorial Optimization Problems and Their Approximability Properties*. Springer-Verlag, 1999.
3. J. Besson, C. Robardet, and J.-F. Boulicaut. Constraint-based mining of formal concepts in transactional data. *Proceedings of PAKDD'04*, pages 615–624, 2004.
4. C.L. Blake and C.J. Merz. UCI repository of machine learning databases. http://www.ics.uci.edu/~mlearn/MLRepository.html, 1998.
5. I.S. Dhillon, S. Mallela, and D.S. Modha. Information-theoretic co-clustering. *Proceedings of KDD'03*, pages 89–98, 2003.
6. U. Feige. A threshold of ln n for approximating set cover. *Journal of the Association for Computing Machinery*, 45(4):634 – 652, 1998.
7. J. Han, J. Pei, Y. Yin, and R. Mao. Mining frequent patterns without candidate generation: A frequent-pattern tree approach. *Data Mining and Knowledge Discovery*, 8(1):53–87, 2004.
8. J. Han, J. Wang, Y. Lu, and P. Tzvetkov. Mining top-k frequent closed patterns without minimum support. In *Proceedings of ICDM'02*, pages 211–218, 2002.
9. R. Kohavi, C. Brodley, B. Frasca, L. Mason, and Z. Zheng. KDD-Cup 2000 organizers' report: Peeling the onion. *SIGKDD Explorations*, 2(2):86–98, 2000. http://www.ecn.purdue.edu/KDDCUP.
10. E. Kushilevitz and N. Nisan. *Communication Complexity*. Cambridge, 1996.
11. T. Miclikäinen and H. Mannila. The pattern ordering problem. In *Proceedings of PKDD'03*, volume 2838 of *Lecture Notes in Artificial Intelligence*, pages 327–338. Springer-Verlag, 2003.
12. J. Orlin. Containment in graph theory: covering graphs with cliques. *Indigationes Mathematicae*, 39:211–128, 1977.
13. R. Peeters. The maximum edge biclique is NP-complete. *Discrete Applied Mathematics*, 131:651–654, 2003.
14. D. Ron, N. Mishra, and R. Swaminathan. On conjunctive clustering. *Proceedings of COLT'03*, pages 448–462, 2003.
15. M. Seno and G. Karypis. LPMiner: An algorithm for finding frequent itemsets using length-decreasing support constraint. *Proceedings of ICDM'01*, pages 505–512, 2001.
16. P.-N. Tan, V. Kumar, and J. Srivastava. Selecting the right interestingness measure for association patterns. In *Proceedings of KDD'02*, pages 32–41, 2002.
17. J. Wang and G. Karypis. BAMBOO: Accelerating closed itemset mining by deeply pushing the length-decreasing support constraint. *Proceedings of SIAM DM'04*, 2004.
18. M.J. Zaki. Scalable algorithms for association mining. *IEEE TKDE*, 12(3):372–390, 2000.
19. M.J. Zaki and C.-J. Hsiao. CHARM: An efficient algorithms for closed itemset mining. In R. Grossman, J. Han, V. Kumar, H. Mannila, and R. Motwani, editors, *Proceedings of SIAM DM'02*, 2002.

A Clustering of Interestingness Measures

Benoît Vaillant[1], Philippe Lenca[1], and Stéphane Lallich[2]

[1] GET ENST Bretagne / Dpt. LUSSI – CNRS TAMCIC, France
[2] Laboratoire ERIC - Univ. Lumière - Lyon 2, France

Abstract. It is a common issue that KDD processes may generate a large number of patterns depending on the algorithm used, and its parameters. It is hence impossible for an expert to sustain these patterns. This may be the case with the well-known APRIORI algorithm. One of the methods used to cope with such an amount of output depends on the use of interestingness measures. Stating that selecting interesting rules also means using an adapted measure, we present an experimental study of the behaviour of 20 measures on 10 datasets. This study is compared to a previous analysis of formal and meaningful properties of the measures, by means of two clusterings. One of the goals of this study is to enhance our previous approach. Both approaches seem to be complementary and could be profitable for the problem of a user's choice of a measure.

Keywords: Rule interestingness measure, clustering, experimental study.

1 Introduction

One of the main objectives of Knowledge Discovery in Databases (KDD) is the production of rules, interesting from the point of view of a user. A rule can be labelled as interesting providing that it is *valid, novel, potentially useful, and ultimately understandable* [1]. These generic terms cover a wide range of aspects, and the quality of a rule is information which is quite hard to grasp. Moreover, the (potentially) large number of rules generated by the algorithms commonly used makes it impossible for an expert to take all the rules into consideration without some automated assistance. A common way of reducing the number of rules, and hence enabling the expert to focus on what should be of interest to him, is to pre-filter the output of KDD algorithms according to interestingness measures. By enabling the selection of a restricted subset of "best rules" out of a larger set of potentially valuable ones, interestingness measures play a major role within a KDD process. We show in [2] that the selection of the best rules implies the use of an adapted interestingness measure, as the measures may generate different rankings. Different properties may give hints, and a multi-criteria decision approach can be applied in order to recommend a restricted number of measures. However, the choice of a measure responding to a user's needs is not easy: some properties defined on the measures are incompatible ([3], [4]). Therefore there is no *optimal* measure, and a way of solving this problem is to try to find good compromises. In [5], [6], we focused on helping the user select an adapted interestingness measure with regards to his goals and preferences.

E. Suzuki and S. Arikawa (Eds.): DS 2004, LNAI 3245, pp. 290–297, 2004.

The aim of the experiments developed in this paper is to complete our formal approach with an analysis of the behaviour of measures on concrete data. This two-fold characterisation is useful since some properties of the measures are hard to evaluate in a formal way, such as the robustness to noise [7].

In section 2 we present an experimentation scheme using our tool, HERBS. Experimental results are summarised in section 3. The experimental clustering of the measures is compared to a hierarchical ascendant clustering of the measures in section 4. We conclude in section 5.

2   A Short Description of the Analysis Tool, HERBS

The aim of HERBS [8] is to analyse and compare rule datasets or interestingness measures. It has been designed to be an interactive *post-analysis* tool, and hence case datasets, rule datasets and interestingness measures are seen as inputs.

2.1   Analysis Background, Notations

We use in this article the following notations. $\mathcal{C}$ is a case dataset containing n cases and $\mathcal{R}$ a rule dataset containing η rules. Each case is described by a fixed number of attributes. The dataset may have been cleaned and we assume that there are no missing values. We consider association rules of the form $A \rightarrow B$ where A and B are conjunctions of tests on the values taken by the attributes. In the AIS algorithm initially proposed by [9], the attributes are all binary, the rules focus only on co-occurrences, and the conclusion is restricted to a single test. This kind of analysis is a particular case of the analysis of a contingency table, introduced by [10] within the GUHA method and developed later on by [11] in the 4FT-MINER tool. We note n_a (resp. n_b, n_{ab}, $n_{a\bar{b}}$) the number of cases of $\mathcal{C}$ matching A (resp. B, A and B, A but not B); hence $n_{a\bar{b}} = n_a - n_{ab}$. It is important to keep in mind the fact that we do not bind the rules with the dataset used to create them. Thus, n_a, n_b, n_{ab} and $n_{a\bar{b}}$ can potentially take any integer value from 0 to n.

2.2   Experimentation Scheme

The analysis and the experimental comparison of the measures we propose rely on their application to a couple $(\mathcal{C}, \mathcal{R})$. For a synthetic comparison of the rankings of the rules by two given measures (each measure is defined in order to have a ranking of the rules from the best to the worst, this interestingness relying highly on the users' aims), we compute a preorder agreement coefficient, τ_1 which is derived from Kendall's τ (see [12]).

We intentionally limited our study to measures related to the interestingness of association rules, such as defined in [9]. These measures are functions of n, n_a, n_b and n_{ab}. They do not take into account several others quality factors such as attribute costs, misclassification costs [13], cognitive constraints [14]... Such issues are hence not considered in this paper. We detail the selection criterion of the measures in [5], [6]. We remind their definitions and abbreviations in table 1.

Table 1. Studied measures.

Measure	Abbreviation	Definition
support	SUP [9]	$\frac{n_a - n_{a\bar{b}}}{n}$
confidence	CONF [9]	$1 - \frac{n_{a\bar{b}}}{n_a}$
linear correlation coefficient	R [15]	$\frac{n n_{ab} - n_a n_b}{\sqrt{n n_a n_b n_{\bar{a}} \cdot n_{\bar{b}}}}$
centred confidence	CENCONF	$\frac{n n_{ab} - n_a n_b}{n n_a}$
conviction	CONV [16]	$\frac{n_a n_{\bar{b}}}{n n_{a\bar{b}}}$
Piatetsky-Shapiro	PS [17]	$\frac{1}{n}(\frac{n_a n_{\bar{b}}}{n} - n_{a\bar{b}})$
Loevinger	LOE [18]	$1 - \frac{n n_{a\bar{b}}}{n_a n_{\bar{b}}}$
information gain	IG [19]	$\log(\frac{n n_{ab}}{n_a n_b})$
Sebag-Schoenauer	SEB [20]	$\frac{n_a - n_{a\bar{b}}}{n_{a\bar{b}}}$
lift	LIFT [21]	$\frac{n n_{ab}}{n_a n_b}$
Laplace	LAP [22]	$\frac{n_{ab}+1}{n_a+2}$
least contradiction	LC [23]	$\frac{n_{ab} - n_{a\bar{b}}}{n_b}$
odd multiplier	OM [4]	$\frac{(n_a - n_{a\bar{b}})n_{\bar{b}}}{n_b n_{a\bar{b}}}$
example and counter example rate	ECR	$\frac{n_a - 2n_{a\bar{b}}}{n_a - n_{a\bar{b}}}$
Kappa	KAPPA [24]	$2\frac{n n_a - n n_{a\bar{b}} - n_a n_b}{n n_a + n n_b - 2n_a n_b}$
Zhang	ZHANG [25]	$\frac{n n_{ab} - n_a n_b}{\max\{n_{ab} n_{\bar{b}}, n_b n_{a\bar{b}}\}}$
implication index	-IMPIND [26]	$\frac{n n_{a\bar{b}} - n_a n_{\bar{b}}}{\sqrt{n n_a n_{\bar{b}}}}$
intensity of implication	INTIMP [27]	$P[poisson(\frac{n_a n_{\bar{b}}}{n}) \geq n_{a\bar{b}}]$
entropic intensity of implication	EII [28]	$\{[(1 - h_1(\frac{n_{a\bar{b}}}{n})^2) \times$ $(1 - h_2(\frac{n_{a\bar{b}}}{n})^2)]^{1/4}{}_{\text{INTIMP}}\}^{1/2}$
probabilistic discriminant index	PDI [29]	$P[\mathcal{N}(0,1) >_{\text{IMPIND}} {}^{CR/\mathcal{B}}]$

- $h_1(t) = -(1 - \frac{n \cdot t}{n_a})\log_2(1 - \frac{n \cdot t}{n_a}) - \frac{n \cdot t}{n_a}\log_2(\frac{n \cdot t}{n_a})$ if $t \in [0, n_a/2\,n[$; else $h_1(t) = 1$
- $h_2(t) = -(1 - \frac{n \cdot t}{n_{\bar{b}}})\log_2(1 - \frac{n \cdot t}{n_{\bar{b}}}) - \frac{n \cdot t}{n_{\bar{b}}}\log_2(\frac{n \cdot t}{n_{\bar{b}}})$ if $t \in [0, n_{\bar{b}}/2\,n[$; else $h_2(t) = 1$
- *poisson* stands for the Poisson distribution
- $\mathcal{N}(0,1)$ stands for the centred and reduced normal repartition function
- IMPIND $^{CR/\mathcal{B}}$ corresponds to IMPIND, centred reduced (CR) for a rule set $\mathcal{B}$

3   Experimental Results on 10 Couples $(\mathcal{C}, \mathcal{R})$

Experiments have been carried out on databases retrieved from the UCI Repository (ftp.ics.uci.edu/). When there is no ambiguity, we refer indifferently to couples, case and rule sets by their names in the Repository. We note BCW the *breast-cancer-wisconsin* database. The parameters of the APRIORI algorithm [30] have been fixed experimentally in order to obtain rule sets of an acceptable size in terms of computational cost. The strong differences in sizes of the rule sets is related to the number of modalities of the different attributes of the case databases. A particular option was used in order to compute Cmc: APRIORI which usually explores a restricted number of nodes of the lattice formed by the different modalities of the attributes as been forced to explore the entire lattice. $Cmc2$ has been obtained by filtering Cmc, with a minimum lift of 1.2. The *Solarflare* database is divided into two case sets, $\mathcal{SF}_1$ and $\mathcal{SF}_2$, described by the same attributes. $\mathcal{R}_1$ (resp. $\mathcal{R}_2$) is the rule set issued from $\mathcal{SF}_1$ (resp. $\mathcal{SF}_2$). We

Table 2. Summary of the different sets used, and APRIORI parameters.

name	n	sup_{min}	$conf_{min}$	η
Autompg	392	5	50	49
BCW	683	10	70	3095
Car	1728	5	60	145
Cmc	1473	5	60	2878
Cmc2	n/a	n/a	n/a	825

name	n	sup_{min}	$conf_{min}$	η
$(\mathcal{SF}_1, \mathcal{R}_1)$	323	20	85	5402
$(\mathcal{SF}_2, \mathcal{R}_2)$	1066	20	85	6312
$\mathcal{R}_1^1$	n/a	n/a	n/a	4130
$\mathcal{R}_1^2$	n/a	n/a	n/a	2994

filtered $\mathcal{R}_1$, with the method exposed in [31] following the results of [32] to keep only rules that are significant from a statistical point of view. Using $\mathcal{SF}_1$ (resp. $\mathcal{SF}_2$), we obtained the rule set $\mathcal{R}_1^1$ (resp. $\mathcal{R}_1^2$). The characteristics of the sets are summarised in table 2.

We generated 10 preorder comparison matrices, summarised in table 3 (the value of τ_1 is proportional to the radius of the corresponding portion of disc). The AMADO method [33] has been applied to the average matrix of the results (the rows and the columns of the matrix are reorganised in order to highlight the block structures). The results are quite in agreement, and we can make out 4 groups of measures, although some differences appear depending on which database is considered. This justifies once more the need for empirical studies of the properties and behaviour of the measures.

4   Experimental *vs.* Formal Approach

In order to have a better understanding of the behaviour of the measures, we compared the typology in 4 classes coming from our experiments with the formal typology developed in [5], [6]. The formal approach can be synthesised with a 20×8 decision matrix, containing the evaluation of the 20 measures on 8 criteria. We kept only 6 of the criteria for the comparison, as two of them – namely g_7 (easiness to fix a threshold) and g_8 (intelligibility) – do not influence the experimental results at all:

g_1 (**asymmetric processing of A and B [13]**): Since A and B may have a very different signification, it is desirable to distinguish measures that give different evaluations of rules A $\rightarrow$ B and B $\rightarrow$ A from those which do not.

g_2 (**decrease with n_b [17]**): Given n_{ab}, $n_{a\bar{b}}$ and $n_{\bar{a}b}$, it is of interest to relate the interestingness of a rule to n_b. If the number of records verifying B but not A increases, the interestingness of the rule should decrease.

g_3 (**reference situations: independence [17]**): To avoid keeping rules that contain no information, it is necessary to eliminate the A $\rightarrow$ B rule when A and B are independent, meaning when the probability of obtaining B is independent of the fact that A is true or not. A comfortable way of dealing with this is to require that a measure's value at independence should be constant.

g_4 (**reference situations: logical rule**): In the same way, the second reference situation we consider is related to the value of the measure when there is no counter example. It is desirable that the value should be constant or infinite.

Table 3. Preorder comparisons for 20 measures for 10 couples ($\mathcal{C}$, $\mathcal{R}$).

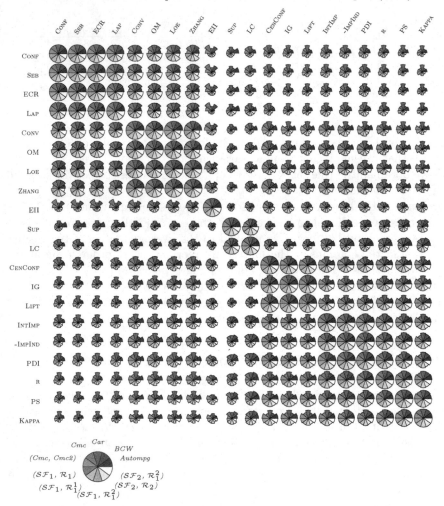

g_5 (**linearity with** $n_{a\bar{b}}$ **around** 0^+): Some authors [34] express the desire to have a weak decrease in the neighbourhood of a logic rule rather than a fast or even linear decrease (as with confidence or its linear transformations). This reflects the fact that the user may tolerate a few counter examples without significant loss of interest, but will definitely not tolerate too many. However, the opposite choice may be preferred, as a convex decrease with $n_{a\bar{b}}$ around the logic rule increases the sensitivity to a false positive.

g_6 (**sensitivity to** n): Intuitively, if the rates of presence of A, A → B, B are constant, it may be interesting to see how the measure reacts to a global extension of the database (with no evolution of rates). The preference of the user might be indifferent to having a measure which is invariant or not with the dilation of data. If the measure increases with n and has a maximum value, then

Table 4. Cross-classification of the measures.

Formal \ Experimental	Class 1	Class 2	Class 3	Class 4
Class 1	PS, KAPPA, IG CENCONF, LIFT, R -IMPIND, PDI			
Class 2	INTIMP	EII	LOE, ZHANG, OM, CONV	
Class 3			CONF, SEB, ECR	
Class 4			LAP	LC, SUP

there is a risk that all the evaluations might come close to this maximum. The measure would then lose its discrimination power.

All criteria are bivaluate except g_5 which is trivaluate. The 20×6 matrix hence obtained was re-encoded in a 20×13 normal disjunctive matrix composed of boolean values. These values do not represent any judgement on the measures, but only list the properties shared by the different measures. The typology in 4 classes (see table 4) coming from this matrix is obtained with a hierarchical ascendent clustering, following the WARD criterion, applied to the square of the euclidean distance (that is, in our case, twice the number of differing properties).

Table 4 shows that both approaches globally lead to similar clusterings, but some shifts are interesting. The first class, {PS, KAPPA, IG, CENCONF, LIFT, R, -IMPIND, PDI} is the same in both cases. What is more, three groups of measures are stable: {LOE, ZHANG, OM, CONV}, {CONF, SEB, ECR} and {SUP, LC}. Within these groups the measures are similar both from the formal and experimental point of view.

The main difference is the presence of a third class in the experimental approach that spans over classes 2, 3 and 4 of the formal clustering. The behaviour of the measures LOE, ZHANG, OM and CONV (formal class 2) is close to that of CONF, SEB and ECR (formal class 3). However, INTIMP and EII that also belong to formal class 2 behave differently. INTIMP shifts to the experimental class 1 (with LIFT and CENCONF), and EII has an original behaviour. These differences strengthen our formal analysis since EII and INTIMP cannot be distinguished with our formal criteria. EII is defined as the product of INTIMP with an inclusion index whose role is to make EII more discriminant. This explains the experimental differences.

LAP shifts to LC and SUP (class 2), LOE, ZHANG, OM and CONV shift to INTIMP and EII (class 4), whereas the core of class 3 consists of CONF, SEB and ECR. The reasons are that LOE, ZHANG, OM and CONV have many properties in common with INTIMP and EII (g_1, g_2, g_3, g_4), which is not the case for CONF, SEB and ECR for g_2 and g_3. However, these measures verify g_1 and g_4.

Property g_4 has an important impact on experimental results. When it is verified, all the logical rules are evaluated with a maximal value, no matter what the conclusion is made up of. Another reason for these shifts is that LAP, really close to SUP in our formal study, can differ from CONF experimentally only for values of n_a close to 0 (nuggets). The minimum thresholds of the APRIORI algorithms make this impossible, and this can be seen as an algorithmic bias.

5   Conclusion

Association rule quality measures play a major role within a KDD process, but they have a large diversity of properties, that have to be studied on real data in order to use a measure adapted to the experimental context. We have presented results coming from a tool we developed, HERBS, and compared the behaviour of 20 measures on 10 datasets. We were then able to identify 4 main groups of measures. This clustering was then compared to a clustering coming from a formal analysis we had done previously. The experimental approach seems to be an important addition to the formal approach. Indeed, it has first confirmed the validity of the list of formal properties we thought were worth studying. What is more, it has also led to a new reflection on the importance of these properties. For example, requiring that a rule quality measure should have a fixed value for a logical rule has the bias of favouring logical rules with a large conclusion.

References

1. Fayyad, U., Piatetsky-Shapiro, G., Smyth, P., Uthurusamy, R., eds.: Advances in KDD. AAAI/MIT Press (1996)
2. Lenca, P., Meyer, P., Vaillant, B., Picouet, P.: Aide multicritère à la décision pour évaluer les indices de qualité des connaissances – modélisation des préférences de l'utilisateur. EGC 2003 **1** (2003) 271–282
3. Tan, P.N., Kumar, V., Srivastava, J.: Selecting the right interestingness measure for association patterns. In: ACM SIGKDD Int. Conf. on KDD. (2002) 32–41
4. Lallich, S., Teytaud, O.:   Évaluation et validation de l'intérêt des règles d'association. RNTI-E-1 (2004) 193–217
5. Lenca, P., Meyer, P., Picouet, P., Vaillant, B., Lallich, S.: Critères d'évaluation des mesures de qualité en ECD. RNTI (2003) 123–134
6. Lenca, P., Meyer, P., Vaillant, B., Lallich, S.:   A multicriteria decision aid for interestingness measure selection. Technical Report LUSSI-TR-2004-01-EN, Dpt. LUSSI, ENST Bretagne (2004)
7. Azé, J., Kodratoff, Y.: A study of the effect of noisy data in rule extraction systems. In: EMCSR. (2002) 781–786
8. Vaillant, B., Picouet, P., Lenca, P.: An extensible platform for rule quality measure benchmarking. In Bisdorff, R., ed.: HCP'2003. (2003) 187–191
9. Agrawal, R., Imielinski, T., Swami, A.: Mining association rules between sets of items in large databases. In Buneman, P., Jajodia, S., eds.: ACM SIGMOD Int. Conf. on Management of Data. (1993) 207–216
10. Hajek, P., Havel, I., Chytil, M.: The GUHA method of automatic hypotheses determination. Computing 1 (1966) 293–308
11. Rauch, J., Simunek, M.: Mining for 4ft association rules by 4ft-miner. In: Int. Conf. on Applications of Prolog. (2001) 285–294
12. Giakoumakis, V., Monjardet, B.: Coefficients d'accord entre deux préordres totaux. Statistique et Analyse des Données **12** (1987) 46–99
13. Freitas, A.: On rule interestingness measures. KBSJ (1999) 309–315
14. Le Saux, E., Lenca, P., Picouet, P.: Dynamic adaptation of rules bases under cognitive constraints. EJOR **136** (2002) 299–309

15. Pearson, K.: Mathematical contributions to the theory of evolution. regression, heredity and panmixia. Philosophical Trans. of the Royal Society **A** (1896)
16. Brin, S., Motwani, R., Ullman, J.D., Tsur, S.: Dynamic itemset counting and implication rules for market basket data. In Peckham, J., ed.: ACM SIGMOD 1997 Int. Conf. on Management of Data. (1997) 255–264
17. Piatetsky-Shapiro, G.: Discovery, analysis and presentation of strong rules. In Piatetsky-Shapiro, G., Frawley, W., eds.: KDD, AAAI/MIT Press (1991) 229–248
18. Loevinger, J.: A systemic approach to the construction and evaluation of tests of ability. Psychological monographs **61** (1947)
19. Church, K.W., Hanks, P.: Word association norms, mutual information an lexicography. Computational Linguistics **16** (1990) 22–29
20. Sebag, M., Schoenauer, M.: Generation of rules with certainty and confidence factors from incomplete and incoherent learning bases. In Boose, J., Gaines, B., Linster, M., eds.: EKAW'88. (1988) 28–1 – 28–20
21. Brin, S., Motwani, R., Silverstein, C.: Beyond market baskets: generalizing association rules to correlations. In: ACM SIGMOD/PODS'97. (1997) 265–276
22. Good, I.J.: The estimation of probabilities: An essay on modern bayesian methods. The MIT Press, Cambridge, MA (1965)
23. Azé, J., Kodratoff, Y.: Evaluation de la résistance au bruit de quelques mesures d'extraction de règles d'assocation. EGC 2002 **1** (2002) 143–154
24. Cohen, J.: A coefficient of agreement for nominal scale. Educational and Psychological Measurement **20** (1960) 37–46
25. Terano, T., Liu, H., Chen, A.L.P., eds.: Association Rules. In Terano, T., Liu, H., Chen, A.L.P., eds.: PAKDD 2000. Volume 1805 of LNCS., Springer (2000)
26. Lerman, I., Gras, R., Rostam, H.: Elaboration d'un indice d'implication pour les données binaires, i et ii. Mathématiques et Sciences Humaines (1981) 5–35, 5–47
27. Gras, R., Ag. Almouloud, S., Bailleuil, M., Larher, A., Polo, M., Ratsimba-Rajohn, H., Totohasina, A.: L'implication Statistique, Nouvelle Méthode Exploratoire de Données. Application à la Didactique, Travaux et Thèses. La Pensée Sauvage (1996)
28. Gras, R., Kuntz, P., Couturier, R., Guillet, F.: Une version entropique de l'intensité d'implication pour les corpus volumineux. EGC 2001 **1** (2001) 69–80
29. Lerman, I., Azé, J.: Une mesure probabiliste contextuelle discriminante de qualité des règles d'association. EGC 2003 **1** (2003) 247–262
30. Borgelt, C., Kruse, R.: Induction of association rules: APRIORI implementation. In: 15th Conf. on Computational Statistics. (2002)
31. Vaillant, B., Lenca, P., Lallich, S.: Association rule interestingness measures: an experimental study. Technical Report LUSSI-TR-2004-02-EN, Dpt. LUSSI, ENST Bretagne (2004)
32. Lallich, S.: Mesure et validation en extraction des connaissances à partir des données. Habilitation à Diriger des Recherches – Université Lyon 2 (2002)
33. Chauchat, J.H., Risson, A.: 3. In: Visualization of Categorical Data. Blasius J. & Greenacre M. ed. (1998) 37–45 New York : Academic Press.
34. Gras, R., Couturier, R., Blanchard, J., Briand, H., Kuntz, P., Peter, P.: Quelques critères pour une mesure de qualité de règles d'association. RNTI-E-1 (2004) 3–31

Extracting Minimal
and Closed Monotone DNF Formulas*

Yoshikazu Shima[1], Shinji Mitsuishi[2],[**], Kouichi Hirata[2], and Masateru Harao[2]

[1] Graduate School of Computer Science and Systems Engineering
shima@dumbo.ai.kyutech.ac.jp
[2] Department of Artificial Intelligence
Kyushu Institute of Technology
Kawazu 680-4, Iizuka 820-8502, Japan
mituishi@dumbo.ai.kyutech.ac.jp
{hirata,harao}@ai.kyutech.ac.jp

Abstract. In this paper, first we introduce *minimal* and *closed* monotone DNF formulas as extensions of *maximal* and *closed* itemsets. Next, by incorporating the algorithm *dnf_cover* designed by Hirata *et al.* (2003) with the algorithm CHARM designed by Zaki and Hsiao (2002), we design the algorithm *cdnf_cover* to extract closed monotone DNF formulas with a *pruning* as same as *dnf_cover*. Finally, we implement *cdnf_cover* and apply it to bacterial culture data.

1 Introduction

The purpose of *data mining* is to extract hypotheses that explain a database. An *association rule* is one of the most famous forms of hypotheses in data mining or association rule mining [1, 2]. In order to extract association rules from a transaction database, the algorithm APRIORI, introduced by Agrawal *et al.* [1, 2], first extracts *frequent itemsets* as sets of items satisfying the *minimum support*. Then, by combining items in each frequent itemset, we can extract association rules satisfying both the minimum support and the *minimum confidence*.

However, the frequent itemset is inappropriate when we extract hypotheses that explain a database *nearly overall*, because it just reflects the items with very high frequency, which are not interesting in general. As an appropriate form of hypotheses in such a case, Hirata *et al.* [4] have regarded an itemset as a *monotone monomial* and extended it to a *monotone DNF formula* as a disjunction of monotone monomials. Then, they have designed the algorithm *dnf_cover*, which is an extension of APRIORI, to extract monotone DNF formulas.

Roughly speaking, *dnf_cover* first not only extracts monotone monomials satisfying the minimum support δ but also stores monotone monomials not satisfying δ but satisfying the *minimum monomial support* σ to a *seed*. Next, it

---

* This work is partially supported by Grand-in-Aid for Scientific Research 15700137 and 16016275 from the Ministry of Education, Culture, Sports, Science and Technology, Japan, and 13558036 from the Japan Society for the Promotion of Science.
** Current address: Obic Co., Ltd.

E. Suzuki and S. Arikawa (Eds.): DS 2004, LNAI 3245, pp. 298–305, 2004.

constructs monotone DNF formulas by connecting each element of a seed to a disjunction $\vee$ while satisfying δ and the *maximum overlap* η. However, in the previous work [4], it has remained the problems that the extracted monotone DNF formulas by *dnf_cover* always contain redundant formulas and the number of them is too large to analyze.

On the other hand, in order to reduce the number of frequent itemsets, many researchers have studied *maximal frequent itemsets* [3] and *frequent closed itemsets* [5, 9, 10]. A frequent itemset is *maximal* if it is not a subset of any other frequent itemset, and *closed* if it has no proper superset with the same frequency. It is known that $MC = M \subseteq C \subseteq F$ [5], where M, C, MC and F denote the set of all maximal frequent itemsets, all frequent closed itemsets, all maximal frequent closed itemsets and all frequent itemsets, respectively.

In this paper, in order to solve the above problems for *dnf_cover*, first we introduce *minimal* and *closed* monotone DNF formulas as extensions of maximal and closed itemsets as follows. A monotone DNF formula is *minimal* if it is *minimal* as a set of monotone monomials and its monotone monomial is maximal. Also a monotone DNF formula is *closed* if it has no proper *subset* with the same frequency as a set of monotone monomials and its monotone monomial is closed. Then, we show the similar relationship $MCD = MD \subseteq CD \subseteq FD$ as $MC = M \subseteq C \subseteq F$, where MD, CD, MCD and FD denote the set of all minimal monotone DNF formulas, all closed monotone DNF formulas, all minimal closed monotone DNF formulas and all frequent monotone DNF formulas, respectively.

Next, we design the algorithm *cdnf_cover*, by incorporating the algorithm *dnf_cover* with the algorithm CHARM designed by Zaki and Hsiao [10] to extract frequent closed itemsets. Note that the algorithm *dnf_cover* does not store all monotone monomials not satisfying δ but satisfying σ to a seed; it stores just monotone monomials satisfying σ when not satisfying δ by APRIORI. This method is called a *pruning* in this paper. Then, we also adopt such a pruning in the algorithm *cdnf_cover*.

Finally, we implement the algorithm *cdnf_cover* and apply it to bacterial culture data, which are full version of data in [6–8]. We use 6 kinds of data, MRSA, Anaerobes, Bacteroides, Fusobacterium, Prevotella and Streptococcus data. The number of records in them is 118, 1046, 498, 154, 157 and 155, respectively. Then, we evaluate the extracted monotone DNF formulas by *cdnf_cover*.

2   Minimal and Closed Monotone DNF Formulas

Let $\mathcal{X}$ and $\mathcal{I}$ be finite sets. We call an element of $\mathcal{X}$ an *item* and $\mathcal{I}$ an *id*. Also we call $X \subseteq \mathcal{X}$ an *itemset*. Then, $\mathcal{D} \subseteq \mathcal{I} \times \mathcal{X}$ is a *transaction database*.

For an itemset $X \subseteq \mathcal{X}$, the *cover set* $cvs_{\mathcal{D}}(X)$ of X for $\mathcal{D}$ is defined as $cvs_{\mathcal{D}}(X) = \{i \in \mathcal{I} \mid \forall x \in X, (i, x) \in \mathcal{D}\}$. The *frequency* of X in $\mathcal{D}$ is defined as $|cvs_{\mathcal{D}}(X)|$ and denoted by $freq_{\mathcal{D}}(X)$.

For the *minimum support* δ $(0 < \delta \leq 1)$, we say that an itemset X is *frequent* (*for* $\mathcal{D}$ *w.r.t.* δ) if $freq_{\mathcal{D}}(X) \geq \delta|\mathcal{D}|$. Then, F_{δ} denotes the set of all frequent itemsets for $\mathcal{D}$ w.r.t. δ, that is, $F_{\delta} = \{X \subseteq \mathcal{X} \mid freq_{\mathcal{D}}(X) \geq \delta|\mathcal{D}|\}$.

Definition 1. We say that a frequent itemset $X \in F_\delta$ is *maximal* (*for $\mathcal{D}$ w.r.t.* δ) if there exists no $Y \in F_\delta$ such that $Y \supset X$.

Definition 2. We say that a frequent itemset $X \in F_\delta$ is *closed* (*for $\mathcal{D}$ w.r.t.* δ) if there exists no $Y \in F_\delta$ such that $Y \supset X$ and $freq_\mathcal{D}(Y) = freq_\mathcal{D}(X)$.

Let M_δ, C_δ and MC_δ denote the set of all maximal frequent itemsets, all frequent closed itemsets and all maximal frequent closed itemsets, respectively. Then, it is known the following theorem.

Theorem 1 (Pasquier *et al.* [5]). $MC_\delta = M_\delta \subseteq C_\delta \subseteq F_\delta$.

In this paper, we regard an item $x \in \mathcal{X}$ as a *variable*, and an itemset $X \subseteq \mathcal{X}$ as a *monotone monomial* over $\mathcal{X}$, that is, a conjunction of variables. Then, we extend a monotone monomial over $\mathcal{X}$ to a *monotone DNF formula* $X_1 \vee \cdots \vee X_m$ over $\mathcal{X}$ as a disjunction of monotone monomials $X_1, \ldots, X_m$. We sometimes identify it with a set $\{X_1, \ldots, X_m\}$.

Let f be a monotone DNF formula $X_1 \vee \cdots \vee X_m$. Then, the *cover set* $cvs_\mathcal{D}(f)$ of f for $\mathcal{D}$ is defined as $cvs_\mathcal{D}(f) = cvs_\mathcal{D}(X_1) \cup \cdots \cup cvs_\mathcal{D}(X_m)$. The *frequency* of f in $\mathcal{D}$ is defined as $|cvs_\mathcal{D}(f)|$ and denoted by $freq_\mathcal{D}(f)$.

Note that a monotone DNF formula is a *disjunction* of monotone monomials, while a monotone monomial is a *conjunction* of variables. By paying our attention to the duality between a conjunction and a disjunction, we introduce *minimal* and *closed* monotone DNF formulas corresponding to maximal and closed itemsets as follows.

Let σ be the minimum support for monotone monomials, called the *minimum monomial support*. Then, let $FD_{\delta,\sigma}$ be the following set of all *frequent monotone DNF formulas* (*for $\mathcal{D}$ w.r.t.* δ *and* σ).

$$FD_{\delta,\sigma} = \{f = X_1 \vee \cdots \vee X_m \mid X_i \in F_\sigma (1 \le i \le m) \wedge freq_\mathcal{D}(f) \ge \delta|\mathcal{D}|\}.$$

Definition 3. We say that $f = X_1 \vee \cdots \vee X_m \in FD_{\delta,\sigma}$ is *minimal* (*for $\mathcal{D}$ w.r.t.* δ *and* σ) if the following statements hold.

1. $X_i \in M_\sigma$ for each i $(1 \le i \le m)$.
2. There exists no $g \in FD_{\delta,\sigma}$ such that $g \subset f$.

Definition 4. We say that $f = X_1 \vee \cdots \vee X_m \in FD_{\delta,\sigma}$ is *closed* (*for $\mathcal{D}$ w.r.t.* δ *and* σ) if the following statements hold.

1. $X_i \in C_\sigma$ for each i $(1 \le i \le m)$.
2. There exists no $g \in FD_{\delta,\sigma}$ such that $g \subset f$ and $freq_\mathcal{D}(g) = freq_\mathcal{D}(f)$.

Let $MD_{\delta,\sigma}$, $CD_{\delta,\sigma}$ and $MCD_{\delta,\sigma}$ denote the set of all minimal frequent, all closed frequent and all minimal frequent closed DNF formulas, respectively. As similar as Theorem 1, the following theorem also holds.

Theorem 2. $MCD_{\delta,\sigma} = MD_{\delta,\sigma} \subseteq CD_{\delta,\sigma} \subseteq FD_{\delta,\sigma}$.

Proof. By Definition 3, 4 and Theorem 1, it is obvious that $MCD_{\delta,\sigma} \subseteq MD_{\delta,\sigma} \subseteq CD_{\delta,\sigma} \subseteq FD_{\delta,\sigma}$.

Let f be a minimal monotone DNF formula $f = X_1 \vee \cdots \vee X_m \in MD_{\delta,\sigma}$. Since $M_\sigma \subseteq C_\sigma$, it holds that $X_i \in C_\sigma$. Suppose that f is not closed. Then, there exists $g \in FD_{\delta,\sigma}$ such that $g \subset f$ and $freq_{\mathcal{D}}(g) = freq_{\mathcal{D}}(f)$. This implies that f is not minimal by Definition 3, which is a contradiction. Hence, f is closed, so $MCD_{\delta,\sigma} = MD_{\delta,\sigma}$. $\square$

3   Extraction Algorithm

In this section, we design the algorithm *cdnf_cover*, by incorporating the algorithm *dnf_cover* designed by Hirata *et al.* [4] with the algorithm CHARM designed by Zaki and Hsiao [10]. In the reminder of this paper, $\mathcal{D}$, δ and σ denote a transaction database, the minimum support and the minimum monomial support, respectively.

In our algorithm, we use the following modification of CHARM. Here, we denote the original algorithm for $\mathcal{D}$ under δ by CHARM$(\mathcal{D}, \delta)$, which returns all frequent closed itemsets.

1. CHARMLEAVES$(\mathcal{D}, \delta)$:
   It is a procedure to output the frequent closed itemsets on leaves in a search tree for CHARM$(\mathcal{D}, \delta)$. Note that the leaves are corresponding to the maximal frequent itemsets, because $MC_\delta = M_\delta$.
2. CHARMPRUNING$(\mathcal{D}, \delta, \sigma)$:
   It is a procedure to output (C_1, C_2), where C_1 is an output of CHARM$(\mathcal{D}, \delta)$. On the other hand, let N be the set of nodes removed by CHARM$(\mathcal{D}, \delta)$ in a search tree. Then, C_2 is defined as follows.

$$C_2 = \{X \in N \mid cl(X) = X \wedge freq_{\mathcal{D}}(X) \geq \sigma|\mathcal{D}|\}.$$

   Here, *cl* is a *closure operator* [5]. This process is based on searching for a search tree constructed by CHARM$(\mathcal{D}, \delta)$. Note that C_2 in CHARMPRUNING is corresponding to a pruning for *cdnf_cover*.

We design the algorithm *cdnf_cover* to extract closed monotone DNF formulas with a pruning as Figure 1. Note that we can also design the algorithms *mdnf_cover_all* to extract *all* minimal monotone DNF formulas *without a pruning* and *mdnf_cover* to extract minimal monotone DNF formulas *with a pruning*, by adopting CHARMLEAVES instead of CHARMPRUNING and by modifying the while-loop in Figure 1. The empirical result for *mdnf_cover_all* will be discussed in Section 5.

Lemma 1. For a monotone DNF formula f, if there exist closed monotone monomials X and Y in f such that $X \subseteq Y$, then f is not closed.

Proof. Let f be of the form $g \vee X \vee Y$. By the definition of $cvs_{\mathcal{D}}$, if $X \subseteq Y$, then $cvs_{\mathcal{D}}(X) = cvs_{\mathcal{D}}(X \vee Y)$. Hence, it holds that $cvs_{\mathcal{D}}(g \vee X) = cvs_{\mathcal{D}}(g \vee X \vee Y)$, that is, $freq_{\mathcal{D}}(g \vee X) = freq_{\mathcal{D}}(g \vee X \vee Y)$. $\square$

```
procedure cdnf_cover(D, δ, σ)
    (DNF₁, SEED) ←CHARMPRUNING(D, δ, σ);
    l ← 0; S₀ ← ∅; S₁ ← SEED;
    while Sₗ₊₁ ≠ ∅ do begin
        l ← l + 1; DNFₗ₊₁ ← ∅; Sₗ₊₁ ← ∅;
        forall f ∈ Sₗ do begin
            forall lexicographically larger elements X ∈ SEED
                   than all monomials in f do
(*)             if freqᴅ(f ∨ X) > freqᴅ(f) then
                    Sₗ₊₁ ← Sₗ₊₁ ∪ {f ∨ X};
                    if freqᴅ(f ∨ X) ≥ δ|D| then
                        DNFₗ₊₁ ← DNFₗ₊₁ ∪ {f ∨ X};
        end /* forall */
    end /* while */
                l
    return     ⋃ DNFᵢ;
               i=1
```

Fig. 1. The algorithm *cdnf_cover*.

Theorem 3. The algorithm *cdnf_cover* in Figure 1 extracts the closed monotone DNF formulas satisfying δ and σ.

Proof. Since *SEED* from CHARMPRUNING is a set of closed monotone monomials, it is sufficient to show the correctness of the statement (*), that is, for monotone monomials X and Y such that $X \subseteq Y$, if $f \vee X$ is closed but $f \vee X \vee Y$ is not closed, then $f \vee X \vee Y \vee Z$ is not closed for each Z. By Lemma 1, it holds that $freq_D(f \vee X) = freq_D(f \vee X \vee Y)$. Since $X \subseteq Y$, it holds that $freq_D(f \vee X \vee Z) = freq_D(f \vee X \vee Y \vee Z)$, so $f \vee X \vee Y \vee Z$ is not closed. □

4   Empirical Results from Bacterial Culture Data

In this section, we give the empirical results obtained by applying the algorithm *cdnf_cover* to bacterial culture data. We use 6 kinds of data, MRSA, Anaerobes, Bacteroides (Bact.), Fusobacterium (Fuso.), Prevotella (Prev.) and Streptococcus (Stre.) data, The number of records in them is 118, 1046, 498, 154, 157 and 155, respectively. The last four data are a part of Anaerobes data corresponding to 4 species of Anaerobes. All of them consist of data between four years (from 1995 to 1998) with 93 attributes, containing 16 antibiotics for benzilpenicillin (PcB), synthetic penicillins (PcS), augmentin (Aug), anti-pseudomonas penicillins (PcAP), 1st generation cephems (Cep1), 2nd generation cephems (Cep2), 3rd generation cephems (Cep3), 4th generation cephems (Cep4), anti-pseudomonas cephems (CepAP), aminoglycosides (AG), macrolides (ML), tetracyclines (TC), lincomycins (LCM), chloramphenicols (CP), carbapenems (CBP) and vancomycin (VCM). The computer environment is that CPU and RAM are Pentium 4 2.8 GHz and 2 GB, respectively.

Figure 2 describes the results obtained by applying *cdnf_cover* to such data. The number of extracted closed monotone DNF formula by *cdnf_cover* is identical between $\sigma = 5, 10, 15, 20, 25$ and 30%. On the other hand, the number of extracted monotone DNF formulas by *dnf_cover* under $\delta = 75\%$ and $\sigma = 10\%$ from MRSA data is 161 ($\eta = 5\%$), 1503 (10%), 8569 (15%), 68946 (20%) and 605391 (25%). By comparing them with Figure 2, we can conclude that the number of extracted formulas by *cdnf_cover* is almost moderate.

data	δ	#SEED	#DNF	sec.	data	δ	#SEED	#DNF	sec.
MRSA	90%	1	0	0.00 − 0.01	Fuso.	90%	3	8	0.00 − 0.01
	85%	1	2	0.01 − 0.02		85%	5	35	0.01 − 0.02
	80%	5	23	0.01 − 0.02		80%	8	103	0.01 − 0.02
	75%	16	2258	0.07 − 0.08		75%	7	130	0.01 − 0.02
Anaerobes	85%	0	7	0.03 − 0.05	Prev.	70%	0	5	0.00 − 0.02
	80%	0	7	0.03 − 0.05		65%	0	6	0.00 − 0.02
	75%	0	7	0.04 − 0.05		60%	2	9	0.01 − 0.02
	70%	2	18	0.04 − 0.05		55%	6	68	0.01 − 0.02
Bact.	80%	0	7	0.01 − 0.03	Stre.	80%	0	4	0.00 − 0.02
	75%	0	15	0.01 − 0.03		75%	4	24	0.00 − 0.02
	70%	2	19	0.01 − 0.03		70%	4	29	0.00 − 0.02
	65%	22	276240	10.78 − 10.96		65%	3	36	0.00 − 0.02

Fig. 2. The cardinality of a seed, the number of extracted closed monotone DNF formulas and the running time for *cdnf_cover*, where the range of σ is from 5% to 30%.

Furthermore, the extracted closed monotone DNF formulas by *cdnf_cover* have the following characterization.

1. The extracted formulas from all data always consist of closed monotone monomials with information in 16 antibiotics.
2. The extracted formulas are always *non-redundant*, that is, they contain no formulas such as (**year** = **95**) ∨ (**year** = **96**) ∨ (**year** = **97**) ∨ (**year** = **98**) or (**gender** = **male**) ∨ (**gender** = **female**).

Figure 3 describes the sensitivity of antibiotics occurring in the extracted closed monotone DNF formulas by *cdnf_cover*.

data	PcB	PcS	Aug	PcAP	Cep1	Cep2	Cep3	Cep4	CepAP	AG	ML	TC	LCM	CP	CBP	VCM
MRSA	R	R			R					R	R	R	R			S
Anaerobes						S	S					S		S	S	
Bact.	R			S	R	S	S					S		S	S	
Fuso.	S			S	S	S	S			S	S	S	S	S		
Prev.				S	S	S	S			S	S	S	S	S		
Stre.				S	S		S			S	S	S		S		

Fig. 3. The sensitivity of antibiotics containing extracted closed monotone DNF formulas. Here, R and S denote resistant and susceptibility, respectively.

From MRSA data, we can extract the resistant for not only PcB, PcS, Cep1 and AG which are antibiotics to determine MRSA, but also ML, TC and LCM.

From Bact. data, we can extract the resistant for PcB and Cep1, and the susceptibility for PcAP, Cep2, Cep3, TC, CP and CBP. On the other hand, from Anaerobes, Fuso., Prev. and Stre. data, the extracted susceptibility is very similar and we can extract no resistant for antibiotics. Furthermore, Figure 4 describes the examples of extracted closed monotone DNF formulas by *cdnf_cover*.

MRSA	(AG = R ∧ ML = R ∧ PcS = R) ∨ (Cep1 = R ∧ LCM = R ∧ ML = R ∧ PcB = R ∧ beta = none) ∨(Cep1 = R ∧ LCM = R ∧ PcB = R ∧ PcS = R ∧ TC = R ∧ VCM = S ∧ beta = none) ∨(LCM = R ∧ PcB = R ∧ gender = male) ∨ (ML = R ∧ gender = male) ∨(beta = none ∧ gender = male)	89.0%
	(AG = R ∧ TC = R) ∨ (Cep1 = R ∧ gender = male) ∨(Cep1 = R ∧ LCM = R ∧ PcB = R ∧ PcS = R ∧ TC = R ∧ VCM = S ∧ beta = none) ∨(LCM = R ∧ ML = R ∧ PcB = R ∧ PcS = R ∧ beta = none) ∨ (PcS = R ∧ gender = male)	88.1%
Anae-	(CBP = S ∧ Cep3 = S) ∨ (Cep2 = S ∧ Cep3 = S)	70.6%
robes	CBP = S ∧ CP = S ∧ Cep2 = S ∧ TC = S	71.8%
Bact.	(CBP = S ∧ CP = S ∧ Cep1 = R ∧ Cep3 = S) ∨ (CBP = S ∧ Cep2 = S ∧ PcB = R)	70.1%
	(CBP = S ∧ CP = S ∧ Cep1 = R ∧ Cep2 = S) ∨ (CBP = S ∧ CP = S ∧ Cep1 = R ∧ Cep3 = S) ∨(CBP = S ∧ CP = S ∧ Cep2 = S ∧ PcB = R)	69.7%
Fuso.	(CBP = S ∧ CP = S ∧ Cep1 = S ∧ PcAP = S ∧ gender = male) ∨(CBP = S ∧ CP = S ∧ LCM = S ∧ TC = S ∧ gender = male)	52.6%
	(CBP = S ∧ CP = S ∧ Cep1 = S ∧ Cep2 = S ∧ TC = S ∧ gender = male) ∨(CBP = S ∧ CP = S ∧ Cep1 = S ∧ PcAP = S ∧ gender = male) ∨(TC = S ∧ beta = none ∧ sample = catheter)	72.1%
Prev.	(CBP = S ∧ CP = S ∧ Cep2 = S ∧ ML = S ∧ beta = none) ∨(CBP = S ∧ CP = S ∧ Cep2 = S ∧ PcAP = S ∧ TC = S ∧ beta = none)	56.7%
	(CBP = S ∧ CP = S ∧ Cep2 = S ∧ ML = S ∧ beta = none) ∨(CBP = S ∧ CP = S ∧ LCM = S ∧ sample = catheter)	70.1%
Stre.	(CBP = S ∧ LCM = S) ∨ (CBP = S ∧ ML = S) ∨ (Cep1 = S ∧ LCM = S) ∨(Cep3 = S ∧ TC = S ∧ beta = none)	72.3%
	(CBP = S ∧ LCM = S) ∨ (CBP = S ∧ ML = S) ∨ (Cep1 = S ∧ LCM = S) ∨ (LCM = S ∧ ML = S)	58.1%

Fig. 4. The examples of extracted closed monotone DNF formulas by *cdnf_cover*.

5   Discussion

In this paper, we have formulated *minimal* and *closed* monotone DNF formulas, as extensions of maximal and closed itemsets. Then, we have designed the algorithm *cdnf_cover* to extract closed monotone DNF formulas, by incorporating *dnf_cover* with CHARM. Finally, we have applied *cdnf_cover* to bacterial culture data, and succeeded to extract just non-redundant formulas consisting of monomials with information in antibiotics. With a pruning as similar as *dnf_cover*, the number of extracted monotone DNF formulas by *cdnf_cover* is almost moderate.

We have already implemented the algorithm *mdnf_cover_all* to extract *all* minimal monotone DNF formulas under δ, σ and the maximum overlap η [4]. Figure 5 describes the empirical results for *mdnf_cover_all* applying to MRSA data. Then, without a pruning, the cardinality of a seed is too large to extract minimal monotone DNF formulas of which number is moderate. Hence, by Theorem 1 and 2, even if we implement the algorithm to extract all (closed) monotone DNF formulas under δ, σ and η instead of *mdnf_cover_all*, the number of extracted (closed) monotone DNF formulas is excessive.

Other algorithms to extract frequent closed itemsets have developed as A-CLOSE [5] and LCM [9]. It is a future work to design the algorithm to extract

δ	σ	dnf_cover	cdnf_cover	mdnf_cover_all	σ	#SEED	η	#DNF	sec.
80%	5%	106	5	56575	20%	1600	5%	34717	709.00
	10%	68	5	12904			10%	1082829	13815.30
	15%	50	5	5128	25%	830	5%	1449	9.85
75%	5%	196	16	56575			10%	8764	104.20
	10%	158	16	12904	30%	477	5%	336	0.88
	15%	140	16	5128			10%	765	1.16

Fig. 5. The cardinality of a seed (left) and the cardinality of a seed, the number of extracted minimal monotone DNF formulas and the running time for *mdnf_cover_all* under δ = 80% (right) from MRSA data.

closed monotone DNF formulas with a pruning based on them, in order to solve that the cardinality of a seed is not monotonic as Figure 2. Furthermore, it is a future work to apply our algorithms to other data except bacterial culture data and to evaluate the results.

References

1. R. Agrawal, H. Mannila, R. Srikant, H. Toivonen, A. I. Verkamo: *Fast discovery of association rules*, in U. M. Fayyed, G. Piatetsky-Shapiro, P. Smyth, R. Uthurusamy (eds.): *Advances in Knowledge Discovery and Data Mining*, AAAI/MIT Press, 307–328, 1996.
2. R. Agrawal, R. Srikant: *Fast algorithms for mining association rules in large databases*, Proc. of 20th VLDB, 487–499, 1994.
3. D. Burdick, M. Calimlim, J. Gehrke, MAFIA: *A maximal frequent itemset algorithm for transaction databases*, Proc. 17th International Conference on Data Engineering, 443–452, 2001.
4. K. Hirata, R. Nagazumi, M. Harao: *Extraction of Coverings as Monotone DNF Formulas*, Proc. 6th International Conference on Discovery Science, LNCS **2843**, 165–178, 2003.
5. N. Pasquier, Y. Bastide, R. Taouil, L. Lakhal: *Discovering frequent closed itemsets for association rules*, Proc. 7th International Conference on Database Theory, LNCS **1540**, 398–416, 1999.
6. E. Suzuki: *Mining bacterial test data with scheduled discovery of exception rules*, in [7], 34–40.
7. E. Suzuki (ed.): Proc. International Workshop of KDD Challenge on Real-World Data (KDD Challenge 2000), 2000.
8. S. Tsumoto: *Guide to the bacteriological examination data set*, in [7], 8–12.
9. T. Uno, T. Asai, Y. Uchida, H. Arimura: *LCM: An efficient algorithms for enumerating frequent closed item sets*, Proc. Workshop on Frequent Itemset Mining Implementation, 2003.
10. M. J. Zaki, C.-J. Hsiao: CHARM: *An efficient algorithm for closed itemset mining*, Proc. 2nd SIAM International Symposium on Data Mining, 457–478, 2002.

Characterizations of Multivalued Dependencies and Related Expressions
(Extended Abstract)

José Luis Balcázar[1] and Jaume Baixeries[1]

Dept. Llenguatges i Sistemes Informàtics
Universitat Politècnica de Catalunya
c/ Jordi Girona, 1-3
08034 Barcelona
{balqui,jbaixer}@lsi.upc.es

Abstract. We study multivalued dependencies, as well as the propositional formulas whose deduction calculus parallels that of multivalued dependencies, and the variant known as degenerated multivalued dependencies. For each of these sorts of expressions, we provide intrinsic characterizations in purely semantic terms. They naturally generalize similar properties of functional dependencies or Horn clauses.

1 Introduction

Multivalued dependencies (MVD) are a natural generalization of functional dependencies, and an important notion in the design of relational databases. The presence of functional dependencies that do not result from keys indicates the possibility of decomposing a relation with no information loss. Multivalued dependencies precisely characterize the relations in which a lossless-join decomposition can be performed.

Early works on multivalued dependencies focused on providing a consistent and complete calculus for entailment between dependencies ([6], [7], [8], [19], [20]). In fact, the very same set of deduction rules is consistent and complete both for logical entailment between Horn clauses and for semantic entailment between functional dependencies. One way of explaining the connection is the "comparison-based binarization" of a given relation r: a relation derived from r, formed by binary tuples, each obtained from a pair of original tuples from r by attribute-wise comparison. Then it is easy to see that a functional dependency holds in r if and only if the comparison-based binarization, seen as a theory (i.e. a set of propositional models), satisfies the corresponding Horn clause.

Similarly, in [16] a family of propositional formulas is identified, for which a consistent and complete calculus not only exists but, additionally, corresponds to a syntactically identical consistent and complete calculus for multivalued dependencies. However, the connection is much less clear than simply considering the comparison-based binarization; and, in particular, the database expressions that

E. Suzuki and S. Arikawa (Eds.): DS 2004, LNAI 3245, pp. 306–313, 2004.

most naturally correspond to the propositional formulas under a comparison-based binarization are so-called "degenerate multivalued dependencies".

We prove here semantic, intrinsic characterizations of multivalued dependencies, degenerate multivalued dependencies, and multivalued dependency formulas. Our statements present alternative properties that hold for a relation r, or for a propositional theory T, exactly when a given multivalued dependency, respectively degenerated multivalued dependency, holds for r, or when a given multivalued dependency formula holds in T.

2 Multivalued Dependencies

Our definitions and notations from relational database theory are fully standard. We denote $R = \{A_1, \ldots, A_n\}$ the set of attributes, each with a domain $Dom(A_i)$; then a tuple t is a mapping from R into the union of the domains, such that $t[A_i] \in Dom(A_i)$ for all i. Alternatively, tuples can be seen as well as elements of $Dom(A_1) \times \cdots \times Dom(A_n)$. A relation r over R is a set of tuples. We will use capital letters from the end of the alphabet for sets of attributes, and do not distinguish single attributes from singleton sets of attributes. We denote by XY the union of the sets of attributes X and Y. Our binarization process is standard:

Definition 1. *For tuples t, t' of a relation r, $ag(t, t')$ (read: "agree") is the set X of attributes on which t and t' have coinciding values: $A \in ag(t, t') \Leftrightarrow t[A] = t'[A]$.*

The following is the standard definition of multivalued dependency. Let X, Y, and Z be disjoint sets of attributes whose union is R. For tuples t and t', denote them as xyz and $xy'z'$ meaning that $t[X] = t'[X] = x$, $t[Y] = y$, etc.

Definition 2. *A **multivalued dependency** $X \twoheadrightarrow Y|Z$ holds in r if and only if for each two tuples xyz and $xy'z'$ in r, also $xy'z$ appears.*

Our characterization mainly rests on the following definitions:

Definition 3. *For a set of attributes X of R, $\tau_r(X)$ (but when r is clear from the context we drop the subscript) is the set of all pairs of tuples from r whose agree set is X:*

$$\tau_r(X) = \{\langle t, t' \rangle \,|\, ag(t, t') = X\}$$

Definition 4. *For sets T, T' of pairs of tuples, we denote $T \bowtie T'$ the set*

$$T \bowtie T' = \{\langle t, t' \rangle \,|\, \exists t''(\langle t, t'' \rangle \in T \wedge \langle t'', t' \rangle \in T') \wedge \exists t'''(\langle t', t''' \rangle \in T \wedge \langle t, t''' \rangle \in T')\}$$

Our main result about mutivalued dependencies is as follows:

Theorem 1. *Let X, Y, Z be pairwise disjoint sets of attributes of R, such that their union XYZ includes all the attributes. Then the multivalued dependency $X \twoheadrightarrow Y|Z$ holds in r if and only if, for each $X' \supseteq X$, $\tau(X') = \tau(X'Y') \bowtie \tau(X'Z')$, where $Y' = Y \setminus X'$ and likewise $Z' = Z \setminus X'$.*

Here we provide, as a sequence of lemmas, the major steps in the proof.

Lemma 1. $\tau(XY) \bowtie \tau(XZ) \subseteq \tau(X)$.

Lemma 2. *If* $X \twoheadrightarrow Y|Z$ *holds in* R, *then* $\tau(X) = \tau(XY) \bowtie \tau(XZ)$.

Lemma 3. *If* $X \twoheadrightarrow Y|Z$ *holds in* R, *and* $X' \supseteq X$, *then* $\tau(X') = \tau(X'Y') \bowtie \tau(X'Z')$, *where* $Y' = Y \setminus X'$ *and likewise* $Z' = Z \setminus X'$.

Lemma 4. *If, for each* $X' \supseteq X$, $\tau(X') = \tau(X'Y') \bowtie \tau(X'Z')$, *where* $Y' = Y \setminus X'$ *and likewise* $Z' = Z \setminus X'$, *then* $X \twoheadrightarrow Y|Z$ *holds in* R.

3   Multivalued Dependency Clauses

In this section we will work only with binary tuples, so that the attributes now play the role of propositional variables, and each binary tuple can be seen as a propositional model. The ordering between models is the boolean-cube bitwise partial order, denoted $x \leq y$, or $x < y$ for the proper order. Operations $\wedge$ and $\vee$ on models apply bitwise. We denote $\top$ the model consisting of all trues. Bitwise unions and intersections are extended to theories (that is, sets of models) in the usual way; we also agree to the standard convention that the union of an empty theory is the all-false model $\bot$, whereas the intersection of an empty theory is the top model $\top$. The Hamming weight of a model is the number of variables it assigns to true; the Hamming weight of a set of models is the sum of the Hamming weights of its elements.

The following characterization is known since the earliest works on Horn logic. The proof can be found in a number of references (e.g. in [13]).

Theorem 2. *A propositional theory is a Horn theory, that is, can be axiomatized by a conjunction of Horn clauses, if and only if it is closed under intersection.*

As mentioned in the introduction, the calculus for entailment in functional dependencies mirrors a calculus for Horn clauses, in the sense that it is easy to associate a Horn clause to each functional dependency in such a way that this mapping commutes with the logical consequence relation; moreover the rules of the calculus are syntactically equal in both sides. Similarly, it turns out that there is a calculus for multivalued dependencies that mirrors, in the same sense of commuting with the logical consequence relation, a calculus for a specific family of propositional formulas: multivalued dependency formulas. They are defined as conjunctions of clauses of the form $X \to Y \vee Z$, for disjoint terms X, Y, and Z that satisfy the additional condition that their union is R, the set of all the variables. These implications are naturally called multivalued dependency clauses. See [16] for details on all these issues.

Our main result in this section is a characterization, in the spirit of the closure under intersection of Horn theories, for theories defined by multivalued dependency formulas. Our main technical ingredient is as follows.

Definition 5. *Consider a set of binary tuples T. We say that $x \in T$ is a focus of T if, for every $y \in T$, $x \vee y$ is not the top model $\top$.*

Note that, for $y = x$, this implies that the focus x itself is not $\top$.

Definition 6. *Consider a propositional theory T. A focused intersection of T is a model that can be obtained as the intersection of all the members of a subtheory $T' \subseteq T$ that has at least one focus (of T').*

Theorem 3. *A propositional theory can be axiomatized by a conjunction of Horn or multivalued dependency clauses if and only if it is closed under focused intersection.*

Again we only provide a sequence of lemmas highlighting the major steps in the proof; we only will prove the lemma that departs more strongly from the previously known facts.

Lemma 5. *Consider a multivalued dependency clause, and a propositional theory T that satisfies it. Assume that T has a focus. Then the intersection of all the members of T satisfies the clause.*

Lemma 6. *Assume that T is axiomatized by a conjunction of Horn or multivalued dependency clauses. Then T is closed under focused intersection.*

Let us prove the converse now. Assuming that T is closed under focused intersection, we simply consider all the clauses that are true for all of T. We prove that their conjunction axiomatizes T. Since these clauses are all true for all of T, it remains to see that a model x that is not in T violates some such clause that is true of T. For a theory T and a model x, the subset $T_x \subseteq T$ is

$$T_x = \{y \in T \mid x \leq y\}$$

Lemma 7. *Let $x \notin T$. Consider the intersection z of all the models in T_x. If $x \neq z$, then x violates a Horn clause that is true of T.*

The final case, in whose proof we depart significantly from previous related results, corresponds to x being indeed the intersection of all of T_x. The special case $x = \top$, or $T_x = \emptyset$, is handled separately in a trivial way. Thus from now on we assume that $x < \top$. We prove that this case is covered by the multivalued dependency clauses.

Lemma 8. *Let T be a theory closed under focused intersection, and let $x \notin T$, with $x < \top$. Assume that the intersection of all the models in T_x is precisely x. Then x violates a multivalued dependency clause that is true of T.*

Proof. Consider a subset $T' \subseteq T_x \subseteq T$ such that x is still the intersection of all the models in T', but T' has, under this condition, minimal Hamming weight. Note that, in particular, this implies that each pair y, z in T' reaches $y \vee z = \top$; otherwise, $y \wedge z$ would belong to T by closure under focused intersection, thus to T_x as well, and replacing both in T' by this intersection would reduce the Hamming weight.

As a consequence, fixed any $y \in T'$, all the other elements of T', and their intersection as well, have value true for all those variables that y sets to false. Note also that x is not all true and thus T' is nonempty.

Pick any arbitrary $y \in T'$; since $x \notin T$ but $y \in T_x$, they differ, and $x < y$. Let z_y be the intersection of $T' - \{y\}$, so that $x = y \wedge z_y$. As just argued in the previous paragraph, $y \vee z_y = \top$. Also, $T' - \{y\} \neq \emptyset$ since otherwise $x = y$ (but note that z_y may not be in T).

Let X be the variables satisfied by x, and likewise Y and Z for y and z_y respectively. Consider the clause $\phi = (X \rightarrow Y \vee Z)$, which is then a multivalued dependency clause (technically, the ones of x should be removed from both disjuncts of the right hand side but this is in fact irrelevant). The minimality of T' (and the fact that $x < y$, so y is not all zeros) implies that $x < z_y$ since otherwise we could cross y off from T' and reduce Hamming weight.

Therefore, $x < y$ and $x < z_y$, which jointly imply that x falsifies ϕ. We prove now that in fact T satisfies it, so that it belongs to the axiomatization we constructed in the first place, and this completes the proof that each model not in T falsifies at least one of the axioms, which is our current claim.

Assume, therefore, that some model $w \in T$ falsifies this clause; that is, it satisfies its left hand side but falsifies both disjuncts of the right hand side. Satisfying the term X means $x \leq w$, so that $w \in T_x$. Falsifying Y implies that $w \wedge y < y$. If we can prove that $w \wedge y \in T$ then we are done, since both are above x, thus both are in T_x and $w \wedge y \in T_x$ as well: it could have been used instead of y in T', contradicting again the minimality of T'.

Here is where closure under focused intersection plays its role: we simply prove that $w \vee y < \top$, and since both w and y are in T, their focused intersection must be as well. Thus it only remains to prove that w and y have a common zero, and for this we use the single remaining property of w, that of not satisfying the second disjunct of the right hand side of ϕ. Namely, $w \not\models Z$ means that Z intersects the zeros of w. Now, recalling $x \leq w$, the zeros of w, say $0(w)$, must be also zeros of x, say $0(x)$ with $0(w) \subseteq 0(x)$; hence $0(w) = 0(w) \cap 0(x)$, and $x = y \wedge z_y$ so that $0(x) = 0(y) \cup 0(z_y)$ for likewise defined $0(y)$ (the complement of Y) and $0(z_y)$ (the complement of Z):

$$Z \cap 0(w) = Z \cap 0(w) \cap 0(x) = Z \cap 0(w) \cap (0(y) \cup 0(z_y)) = Z \cap 0(w) \cap 0(y)$$

Given that $Z \cap 0(w) \neq \emptyset$, the set $Z \cap 0(w) \cap 0(y)$ is equally nonempty and the larger set $0(w) \cap 0(y)$ is nonempty too, as was to be shown.   ∎

Taken together, the lemmas prove the main theorem in this section. We can use the same techniques to characterize the case of using only multivalued dependency clauses, without the company of Horn clauses.

Theorem 4. *A propositional theory T can be axiomatized by a conjunction of multivalued dependency clauses if and only if it is closed under focused intersection, contains $\top$ and, for each model $x \notin T$, $x = \bigwedge T_x$.*

4   Degenerated Multivalued Dependencies

Motivated by the clauses that we have studied in the previous section, we now consider expressions on a relation r that correspond exactly to imposing a multivalued dependency clause on the comparison-based binarization of r.

Indeed, this means that we consider pairs of tuples from r, of the form $\langle t, t' \rangle$, and, on the basis of a multivalued dependency clause $X \to Y \vee Z$, we require that, if $t[X] = t'[X]$, then either $t[Y] = t'[Y]$ or $t[Z] = t'[Z]$. Note the connection with the comparison-based binarization.

Definition 7. *A degenerated multivalued dependency (DMVD) $X \Rrightarrow Y|Z$ holds in a relation if for each pair of tuples t, t' such that $t[X] = t'[X]$ then $t[Y] = t'[Y]$ or $t[Z] = t'[Z]$.*

One simple way of characterizing them is as follows. Consider the following more relaxed form of comparing two tuples on some attributes: $\rho(X) = \{\langle t_1, t_2 \rangle | t_1[X] = t_2[X]\}$. Its difference with τ is, clearly, that it is not necessary that those two tuples disagree in the rest of the attributes. The following relationship trivially holds:

Proposition 1. $\rho(X) = \bigcup_{X \subseteq X'} \tau(X')$

With that notation, we can easily see the following characterization:

Proposition 2. *A degenerated multivalued dependency $X \Rrightarrow Y|Z$ holds if and only if $\rho(XY) \cup \rho(XZ) = \rho(X)$.*

A bit less trivial is the fact that we can also use τ to characterize these formulas, using essentially the same argumentation from the perspective of the larger sets of attributes X':

Proposition 3. *$X \Rrightarrow Y|Z$ holds if and only if, for each $X' \supseteq X$ with $\tau(X') \neq \emptyset$, either $XY \subseteq X'$ or $XZ \subseteq X'$.*

That is, for any proper subsets $Y' \subset Y$ and $Z' \subset Z$, $\tau(XY'Z')$ must be empty.

5   Discussion

We have described a semantic characterization of multivalued dependencies in relational databases, as well as a semantic characterization of the propositional theories axiomatized by conjunctions of Horn clauses and multivalued-dependency clauses, that are their counterpart in the realm of propositional logic. Specifically, we have identified a form of closure under intersection that holds exactly for these theories. This can be seen as analogous to the characterization of Horn theories, that are the parallel in the propositional realm to functional dependencies, as exactly the theories that are closed under unrestricted intersection.

Our interest in these properties stems from recent studies of related data mining problems. Whereas dependencies and other integrity constraints are expected to be identified at the time of designing a database schema, it may actually happen that some such correlation went undetected in the design of the database, and, therefore, it may be possible that actually the relation can be decomposed, i.e. a certain multivalued dependency actually holds on the relation, but is not explicitly documented in the intensional database. One may wish to explore the possibility that some implicit dependency actually holds on the extensional database, that is, the tuples themselves, in order to improve the design, efficiency, and understanding of the phenomena that the database is intended to reflect. In fact, then, as argued in [9], the problem becomes one of inductive analysis, in the standard machine-learning setting of learning from examples.

Yet another process that falls in the same analogy is the search for deterministic association rules [1], [15]. Indeed, under the name of Discrete Deterministic Data Mining, the proposal has been put forth of computing, from relational data, so-called deterministic association rules, which are association rules with no condition on the support but 100% confidence; and it has been shown that, particularly in scientific domains amenable to automated scientific discovery processes, where correlations between observations are ubiquitous since they are due to underlying natural laws, this sort of data mining process is highly effective [15]. We contributed to that study [4] by proving that, from a point of view that can be seen as Knowledge Compilation [18], the process of discovery of deterministic association rules is actually constructing (an axiomatization of) the empirical Horn approximation: the smallest Horn theory that contains the given tuples. The translation of these facts into functional dependencies through the comparison-based binarization is quite simple, see [11].

Several algorithms have been proposed in the literature to find functional dependencies from the extensional database; we should mention here [5], [14], [12], [9] and [17]. In particular, TANE [12] is based on partitions of the set of tuples. A recent work [3] has characterized this approach as well in terms of Formal Concept Analysis, and, by combining it with the so-called dependency basis, this partition-based connection with Concept Lattices has been extended to incorporate multivalued dependencies into the framework [2]. With respect to [9], their algorithm fdep is argued there to be more efficient than TANE in many empirical evaluations of the computation of functional dependencies. Besides, an important property of the approach of [9] is that, by encapsulating into some subroutines the test of whether a dependency holds for a database, the same algorithmic schemas can be applied to the discovery of multivalued dependencies [10]. We believe that further advances may be possible through our present fundamental study, based on formally proving intrinsic combinatorial characterizations of multivalued dependencies and related expressions. These may suggest either alternative algorithmic avenues, or improvements on existing algorithms by means of, e.g., more aggressive pruning of the search spaces. Further research along these lines is under development.

References

1. Agrawal R., Mannila H., Srikant R., Toivonen H., Verkamo I. *Fast Discovery of Association Rules.* Advances in Knowledge Discovery and Data Mining, p. 307-328. AAAI Press, 1996.
2. Balcázar, J.L., Baixeries J. *Using Concept Lattices to Model Multivalued Dependencies.* Submitted: http://www.lsi.upc.es/~jbaixer/recerca/index_recerca.html.
3. Baixeries J. *A Formal Concept Analysis Framework to Model Functional Dependencies.* Mathematical Methods for Learning (2004).
4. Balcázar, J.L., Baixeries J. *Discrete Deterministic Data Mining as Knowledge Compilation.* Workshop on Discrete Mathematics and Data Mining in SIAM International Conference on Data Mining (2003).
5. Castellanos M., Saltor F. *Extraction of Data Dependencies.* Information Modelling and Knowledge Bases V.IOS Press, Amsterdam, 1994, pp. 400-420.
6. Fagin R. *Multivalued dependencies and a new normal form for relational databases.* ACM TODS 2, 3, Sept. 1977, pp. 262-278.
7. Fagin R., Beeri C., Howard J. H. *A complete axiomatization for functional and multivalued dependencies in database relations.* Jr. Proc. 1977 ACM SIGMOD Symposium, ed. D. C. P. Smith, Toronto, pp. 47-61.
8. Fagin R., Vardi Y. V. *The theory of data dependencies: a survey.* Mathematics of Information Processing, Proceedings of Symposia in Applied Mathematics, AMS, 1986, vol. 34, pp. 19-72.
9. Flach P., Savnik I. *Database dependency discovery: a machine learning approach.* AI Communications,volume 12 (3): 139–160, November 1999.
10. Flach P., Savnik I. *Discovery of multivalued dependencies from relations.* Intelligent Data Analysis,volume 4 (3,4): 195–211, November 2000.
11. Ganter, B., Wille R. *Formal Concept Analysis. Mathematical Foundations.* Springer, 1999.
12. Huhtala Y., Kärkkäinen J., Porkka P., Toivonen H. *TANE: An Efficient Algorithm for Discovering Functional and Approximate Dependencies.* The Computer Journal 42(2): 100 - 111, 1999.
13. Khardon R., Roth D. *Reasoning with Models* Artificial Intelligence 87, November 1996, pages 187-213.
14. Kivinen J., Mannila H. *Approximate inference of functional dependencies from relations.* Theoretical Computer Science 149(1) (1995), 129-149.
15. Pfaltz, J.L., Taylor, C.M. *Scientific Discovery through Iterative Transformations of Concept Lattices.* Workshop on Discrete Mathematics and Data Mining at 2nd SIAM Conference on Data Mining, Arlington. Pages 65-74. April 2002.
16. Sagiv Y., Delobel D., Scott Parker D., Fagin R. *An equivalence between relational database dependencies and a fragment of propositional logic.* Jr. J. ACM 28, 3, July 1981, pp. 435-453. Corrigendum: J. ACM 34, 4, Oct. 1987, pp. 1016-1018.
17. Savnik I., Flach P. *Bottom-up Induction of Functional Dependencies from Relations.* Proc. of AAAI-93 Workshop: Knowledge Discovery in Databases. 1993.
18. Selman B, Kautz H. *Knowledge compilation and theory approximation* Journal of the ACM Volume 43, Issue 2 (March 1996) Pages: 193 - 224, 1996
19. Ullman J.D. *Principles of Database and Knowledge-Base Systems.* Computer Science Press, Inc. 1988.
20. Zaniolo C., Melkanoff M. A. *On the Design of Relational Database Schemata.* ACM TODS 6(1): 1-47 (1981).

Outlier Handling in the Neighbourhood-Based Learning of a Continuous Class

Fabrice Muhlenbach[1], Stéphane Lallich[2], and Djamel A. Zighed[2]

[1] EURISE – Faculté des Sciences et Techniques, Université Jean Monnet,
23 rue du Docteur Paul Michelon, 42023 Saint-Etienne Cedex 2, France
`fabrice.muhlenbach@univ-st-etienne.fr`
[2] ERIC Lyon, Université Lumière – Lyon 2,
5 avenue Pierre Mendès-France, 69676 Bron Cedex, France
`{lallich,zighed}@univ-lyon2.fr`

Abstract. This paper is concerned with the neighbourhood-based supervised learning of a continuous class. It deals with identifying and handling outliers. We first explain why and how to use the neighbourhood graph issued from predictors in the prediction of a continuous class. Global quality of the representation is evaluated by a neighbourhood autocorrelation coefficient derived from the Moran spatial autocorrelation coefficient. Extending the analogy with spatial analysis, we suggest to identify outliers by using the scattering diagram associated to the Moran coefficient. Several experiments realized on classical benchmarks show the interest of removing the outliers with this new method.

1 Introduction

The goal of a supervised learning method is to predict an example value of a class variable Y knowing the p attributes description X_1, X_2, ..., X_p (the predictive variables) of this example. The learning algorithms try to establish how to predict the class value of unknown data by using the present information in a set of known data – the learning set (LS) – during a previous stage. The class variable Y or the p attributes can be categorical or numerical. If Y is continuous, the learning method is referred to as "regression learning" [14].

The regression [20] – and particulary the linear multivalued regression – is one of the most widely used method in the regression learning algorithm family. The value to predict Y is calculated through a weighted sum of all predictive attributes and the learning method consists in finding the weights minimizing the sum of squared distances between real values and predicted values. Some decision trees such as *CART* [6] or *M5* [13], neural networks [15] or instance-based learning methods – based on the k-nearest neighbours – can also be used to obtain results when Y is numerical.

In the case of model-based methods (e.g., decision tree, multiple regression, neural network, bayesian network), the prediction rule is a model built during the learning step. The distance-based methods are limited to the data storage and the prediction rule consists in assigning to an unknown example the value

E. Suzuki and S. Arikawa (Eds.): DS 2004, LNAI 3245, pp. 314–321, 2004.
© Springer-Verlag Berlin Heidelberg 2004

that looks the most like its nearest neighbours in the representation space. Such a method is characterized by three key-points:

1. the choice of a dissimilarity or similarity function adapted to the predictive attributes: Euclidian distance for a continuous attribute, Hamming distance for a boolean attribute;
2. the definition of the "nearest neighbour" (e.g., the first or k^{th});
3. the computation of a prediction rule with these neighbours: mean or local regression in the case of a numerical value [10].

The method we have adopted belongs to the distance-based learning algorithm family with the particularity of using a neighbourhood graph instead of the nearest neighbours in the 2nd key-point [21]. This choice has been motivated by the following reasons: a neighbourhood graph is a symmetrical and connected graph that syntheses the information supplied by the predictive attributes; contrary to the k-NN, there is no arbitrary parameter such the k value to provide for constructing a neighbourhood graph.

The neighbourhood graphs constitute a structure that can be adapted to a representation space from which it is possible to define quality representation indicators. Actually, on the other side of the prediction, the neighbourhood graph constitutes a representation space which facilitates the navigation in an example database and the visualization of the contextual information.

It is then important to define global and local representation quality indicators. These indicators are very useful in some data mining applications such as the feature selection (i.e., to discover which predictive attributes are the more relevant) or the outlier detection.

An outlier is an example which has a value appearing to be in contradiction with the rest of the data set [3]. Outliers can have various origins [12]. The abnormal characteristic of outliers disturbs the generalization process, particulary if the learning method used the mean or standard deviation of the variables. This is why the outlier search is an important stage in knowledge data discovery [4].

2   Global Evaluation of the Representation Quality

The information provided by the predictive attributes about the class variable Y is synthesized on the neighbourhood graph. We choose the relative neighbourhood graph of Toussaint (RNG) [19] because the complexity computation is reduced, and it can be built from a simple dissimilarity matrix. In a RNG, two examples α and β are neighbours if there is not another example γ closer from α than β and closer from β than α: $d(\alpha, \beta) \leq Max\left(d\left(\alpha, \gamma\right), d\left(\beta, \gamma\right)\right) \forall \gamma, \gamma \neq \alpha, \beta,$ where $d(u, v)$ is a distance measure between the examples u and v (figure 1).

The RNG can be built from a simple dissimilarity matrix. The n vertices of the neighbourhood graph obtained from the p predictive attributes $X_j, j = 1, 2, ..., p$ correspond to the examples, and there are a edges connecting examples which are relative neighbours in the graph.

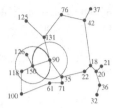

Fig. 1. *RNG* built with two numerical predictive attributes. Examples with the values 90 and 150 are connected because the lune (in gray) between these examples is empty.

Let $\mathbf{V}$ be the boolean connection matrix of the graph and $\mathbf{W}$ the weight matrix associated to the edges (the weights can be the distances, the ranks, etc.).

The statistical signification of this information is exploited by testing the independence of the neighbourhood graph and the distribution of the values of Y on the graph vertices. By analogy with the spatial contiguity graph, we have used the neighbourhood graph with the methodological framework and the tools of spatial analysis, especially spatial autocorrelation (SAC) analysis tools [7]. In spatial analysis, the spots of a variable Y are described in $\mathbb{R}^2$ or $\mathbb{R}^3$, and the SAC is positive when the neighbour spots have similar Y values and negative when the Y values are different.

To test the lack of SAC of a continuous variable (null hypothesis H_0), the autocorrelation coefficients of Moran, Geary and Getis are classically used. The cross-product statistic [9] allows a unified view of all these coefficients. The general form of the cross-product statistic for the spatial analysis is $\Gamma = \sum_{i=1}^{n} \sum_{j=1}^{n} w_{ij} U_{ij}$, where W is a matrix evaluating the spatial proximity between the vertices i et j and U_{ij} is a proximity measure between i and j in another dimension. The cross-product statistic can be split in the sum of local components: the local component i is $\Gamma_i = \sum_{j=1}^{n} w_{ij} U_{ij}$. This property will be useful in the outlier search.

The distribution structure of the example values of Y on the neighbourhood graph vertices will provide information on the capacities of the p attributes to predict the value of Y: we will test the hypothesis of lack of structure (H_0) or no neighbourhood autocorrelation between the values of Y. If H_0 is not rejected, it will be illusive to try to learn Y using a neighbourhood-based learning method.

To test the no structure hypothesis, we propose to build the neighbourhood autocorrelation coefficients in a similar way of the spatial autocorrelation coefficient developed by Moran, Geary and Getis (cf. Cliff and Ord [7]). These coefficients are expressed as a particular cross-product statistic except the normalization coefficient.

We have retained the Moran I coefficient which is unanimously considered as the best choice because it presents local properties useful for outlier detection [2]. The Moran I coefficient is the quotient of a local covariance by the total variance. The Moran I equals to 0 under the lack of SAC hypothesis. The I sign indicates the positive or negative nature of the spatial autocorrelation. It can exceptionally be greater than 1 when the weights are very unbalanced and when the extreme values of Y are affected by the stronger weights.

Whichever the cross-product form retained, two probabilistic schema exist: the Gaussian scheme N where all observations are the result of n independent drawings in a Gaussian population, and the randomized scheme R – less restrictive – where the n observed values of Y are the result of a pure random drawing on the $n!$ permutations of all possible locations.

The computation of the moments related to the law that follows the variable I under H_0 according to the N or R schema is described in [7]. With few restricting conditions the law of I under H_0 is asymptotically standard as long as $\sum_{j=1}^{n}(w_{ij}+w_{ji})y_j$ quantities are not much different. The p-value of I_{obs} (observed) can be calculated by standard approximation. On the R scheme the exact statistics can be used if n is not important, and the Moran coefficient moments under H_0 hypothesis are :

$$E(I) = \frac{-1}{n-1} \text{ and } E(I^2) = \frac{n\left[(n^2-3n+3)S_1 - nS_2 + 3S_0^2\right] - b_2\left[(n^2-n)S_1 - 2nS_2 + 6S_0^2\right]}{(n-1)^{(3)}S_0^2}$$

with $\sum_2 w_{ij} = \sum_{i=1}^{n}\sum_{j=1,i\neq j}^{n} w_{ij}$, $S_0 = \sum_2 w_{ij}$, $S_1 = \frac{1}{2}\sum_2 (w_{ij}+w_{ji})^2$ and $S_2 = \sum_{i=1}^{n}(w_{i+}+w_{+i})^2$ where w_{i+} is the row sum.

3   Outlier Detection

In a work dedicated to spatial data mining, Shekhar, Lu and Zhang [18] propose a general definition of outliers associated with the spatial distribution of a variable on a spatial graph that can be usefully adapted to the neighbourhood graphs. Given a learning set LS and a weight matrix W associated to the edges of a neighbourhood graph built from the predictive attributes, let $NG = \{LS, W\}$ the corresponding neighbourhood graph with $N(i) = \{j \in LS : w_{ij} \neq 0\}$.

- Let U be an attribute with $U : LS \to \mathbb{R}$,
- let V be a neighbourhood aggregation function of U, $V : \mathbb{R}^N \to \mathbb{R}$,
- let D be a difference function with $D : \mathbb{R}^2 \to \mathbb{R}$,
- let T be a decision function with $T : \mathbb{R} \to \{True, False\}$.

Definition. An example i is a $NG\text{-}outlier(U, V, D, T)$ iff the boolean function $T\{D(U_i; V_i)\}$ is true.

Given a class variable Y, let Y^* the corresponding standardized variable and Y' the corresponding weighted neighbourhood sum. Then $y_i^* = \frac{y_i - \overline{y}}{s_y}$ and $y_i^{*'} = \sum_{j=1}^{n} w_{ij}(y_j - \overline{y})$ where $s_y^2 = \frac{1}{n}\sum_{i=1}^{n}(y_i - \overline{y})^2$ is the variance of Y. Let $y_i^{*'}$ the weighted sum of Y standardized in the neighbourhood of i and $y_i^{*'*}$ will express the standardized value of $y_i^{*'}$.

Scattering Diagram Associated to the Moran Coefficient. Outliers can be defined from the Moran scattering diagram that shows local tendencies [2]. The example scatterplot is represented in a $(y_i^*, y_i^{*'})$ plan which associates the standardized values of Y to their corresponding neighbourhood means. The Moran line is the line that well-fits this scatterplot defined by $y_i^{*\prime} = a_{y^{*\prime}/y^*} y_i^* + b_{y^{*\prime}/y^*}$. Insofar as y_i^* is standardized, $a_{y^{*\prime}/y^*} = \frac{1}{n} \sum_{i=1}^n y_i^* y_i^{*\prime} = \frac{S_0}{n} I$ and $b_{y^{*\prime}/y^*} = \overline{y^{*\prime}}$. In the 1-normalized case, $a_{y^{*\prime}/y^*} = I$. Nevertheless it is more interesting to draw the scatterplot by using y_i^* for the X-axis and $y_i^{*\prime*}$ for the Y-axis: the Moran line will cross the zero point and its slope will be given in the 1-normalized case by $a_{y^{*\prime*}/y^*} = r(y^*, y^{*\prime}) = a_{y^{*\prime}/y^*} \frac{s_{y^*}}{s_{y^{*\prime}}} = \frac{I}{s_{y^{*\prime}}}$.

A Moran scattering diagram outlier (SD-Moran-outlier) will be a vertex i for which the $y_i^{*\prime*}$ value will not appear in its own confidence interval computed from the Moran line with a α risk. If Y is Gaussian, $\frac{y_i^{*\prime*} - a_{y^{*\prime*}/y^*} y_i^{*\prime*}}{s_{y^{*\prime*}/y^*}}$ is roughly Gaussian for $n > 30$, where $s_{y^{*\prime*}/y^*}$ is the standard deviation of the prediction:

$$s^2_{y^{*\prime*}/y^*} = \frac{n}{n-2}\left(1 - r^2(y^*, y^{*\prime})\right)\left(1 + \frac{1}{n} + \frac{y_i^{*2}}{n}\right) = \frac{n+1+y_i^{*2}}{n-2}\left(1 - \frac{a^2_{y^{*\prime}/y^*}}{s^2_{y^{*\prime}}}\right)$$

The vertex i will be a SD-Moran-outlier if $\left|\frac{y_i^{*\prime*} - a_{y^{*\prime*}/y^*} y_i^{*\prime*}}{s_{y^{*\prime*}/y^*}}\right| > u_{1-\alpha/2}$.

4   Experimentations

Organization of the Experiments. The method of detecting and removing the outliers presented in the previous section has been applied on 7 domains of the UCI Machine Learning Repository [5] and the CMU StatLib site. All the chosen domains have numerical class attributes. For each domain, we have removed the missing data examples, the categorical predictive attributes have been reencoded on a complete disjunctive form $(0/1)$, and the numerical predictive attributes have been standardized.

The databases have been randomly split in two parts with the same number of examples, one part for the learning set and the other part for the test set. We introduce from 0 to 20% of noise in the learning set of each domain: we change some Y values by replacing them from the values given by a uniform law simulated between the extreme values of Y (it produces more importance to the highest and lowest values of the Y distribution).

A neighbourhood graph is built with these noisy dataset on which we detect the Moran scattering diagram outliers (SD-Moran-outliers). When all outliers are detected, we can remove them from the learning set or relabel them with a neighbourhood weighted mean or a weighted local regression. To compare the performance of the new learning set to the original one when they are used to predict the values of the class variable Y of the test set, some specific indicators of the prediction quality are needed.

The most widely used indicators are the root mean square error ($RMSE = \sqrt{\frac{1}{n} \sum_{i=1}^n (y_i - \widehat{y}_i)^2}$), the mean absolute deviation ($MAD = \frac{1}{n} \sum_{i=1}^n |y_i - \widehat{y}_i|$) and the mean absolute percent error ($MAPE = \frac{1}{n} \sum_{i=1}^n \frac{|y_i - \widehat{y}_i|}{y_i}$). The comparison of these indicators depending on whether the learning set is filtered or not allows to demonstrate the efficiency of the proposed filtering process.

Table 1. Results on 7 domains when we introduce 10% of noise.

	RMSE		MAPE		MAD		noisy data correctly detected	true data (correctly untouched)	non detected noisy data	true data falsely detected
	all	filtered	all	filtered	all	filtered				
autos	4.16	3.63	0.14	0.11	3.00	2.61	6.28	172.0	13.72	3.96
auto-MPG	4.09	3.58	0.13	0.11	2.97	2.57	7.04	172.4	12.96	3.60
CPU	140.9	136.9	1.94	1.70	78.21	73.70	1.08	89.16	8.92	4.84
housing	5.78	5.32	0.19	0.16	3.88	3.47	7.80	221.7	17.20	6.32
plasma ret.	252.6	252.8	0.37	0.37	192.8	193.4	1.04	134.4	14.96	6.64
pw-linear	2.87	2.86	—	—	2.29	2.28	1.08	86.36	8.92	3.64
servo	1.67	1.64	1.57	1.47	1.22	1.19	1.60	72.32	6.40	2.68

We use the noisy learning set with all data and the same noisy learning set after having removed the outliers to proceed at the prediction of the test set with a mean of the 3-nearest neighbours. This process is 25 times repeated. We compare the average performances of the prediction with these two learning sets by using the three quality indicators aforementioned. On the table 1, we present the average results obtained with the 7 domains with 10% of noise.

Analysis of the Results. At the first attemps, three of the 7 domains have given satisfactory results: *autos*, *auto-MPG* and *housing*. The behaviours of the three quality indicators *RMSE*, *MAPE* and *MAD* are relatively similar. As an illustrative example, we present in figure 2 the results obtained with the *autos* database.

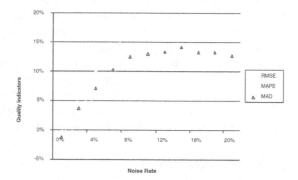

Fig. 2. Relative quality gain with *autos* when there is from 0 to 20% of noise.

For these domains, as we introduce noise, the number of non noisy data falsely detected tends to decrease when the number of outliers tends to grow. The removing method of the outliers can decrease from 10 to 20% of the values of the *RMSE*, *MAPE* and *MAD* indicators.

Data Symmetrization. The excessive dissymmetry of the class attribute values perturbs the outlier detection method. By calculating the Fisher γ_1 dissymmetry coefficient, we can rapidly detect this particular kind of situation.

Let σ^2 be the variance of Y defined by $n\sigma^2 = \sum_{i=1}^{n} (y_i - \bar{y})^2$.

The Fisher coefficient is $\gamma_1 = \frac{\sum_{i=1}^{n}(y_i - \bar{y})^3}{n\sigma^3}$.

Usually a dissymmetric variable falls flat on the right, this produces a highly positive value of γ_1. In this case, a $\ln(1 + Y)$ transformation can symmetrise the distribution of Y. For example, this is the case of the *CPU* database: $\gamma_1 = 3.89$. The Napierian logarithm transformation of the data produces a γ_1 of .57 and gives better results for the filtering method as we can see in figure 3.

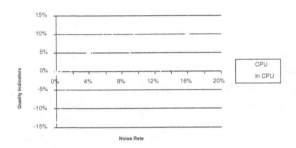

Fig. 3. Relative *RMSE* reduction for the *CPU* database as a function of the noise rate with a logarithmic transformation (ln CPU) – or not (CPU) – of the class attribute Y.

Limits. The remaining three databases (*plasma retinol, servo* and *pw-linear*) do not present significative improvement of the prediction performance.

For the *plasma retinol* database, this result was expected because the global Moran coefficient (presented in section 2) is not significant and indicates that a distance-based learning algorithm can not be applied.

The result of the *servo* database is just weak (only 5% of improvement).

The sole disappointing result concerns the *pw-linear* database, an artificial dataset with a significant global Moran coefficient, a label distribution not far from the symmetry, and where some values can be negative. The specificity of the results obtained with this database needs some further studies.

5   Conclusion and Future Work

By using the neighbourhood graphs in the data mining domain, we propose an original approach of the supervised learning which can (1) test the pertinence of the representation space, (2) detect the outliers or (3) select the relevant attributes in the categorical case [12] as well as in the continuous case.

Moreover, when a new example is examined, the incremental use of the neighbourhood graph obtained with the learning algorithm can produce an intelligent navigation of the labelled example base and a visualization of the contextual information with a reduced complexity [16].

Lastly, the neighbourhood correlation concept that we have used in the supervised learning domain has several applications. In the case of a categorical class Y, it has been used to reduce the size of the learning set in a boosting optimization problem [17]. When Y is continuous, it can be used to perform image segmentation in the pattern recognition domain [11]: it allows to control locally and globally the image segmentation process through an iterative decimation of the pixel contiguity graph.

References

1. Aha, D.W., Kibler, D., and Albert, M.K. Instance-based learning algorithms, ML, **6**, 37–66, 1991.
2. Anselin, L. Local indicators of spatial association, LISA, Geographical Analysis, **27**, 93–115, 1995.
3. Barnett, V. and Lewis, T. Outliers in statistical data, Wiley, 2nd edition, 1984.
4. Beckman, R.J. and Cooks, R.D. Outlier...s, Technometrics, **25**, 119–149, 1983.
5. Blake, C.L. and Merz, C.J. UCI Repository of machine learning databases, Irvine, CA: University of California, Department of Information and Computer Science [http://www.ics.uci.edu/~mlearn/MLRepository.html], 1998.
6. Breiman, L. Friedman, J.H. Olshen R.A., and Stone C.J. Classification and regression trees, Wadsworth International Group, Belmont, CA, 1984.
7. Cliff, A.D. and Ord, J.K. Spatial processes, models and applications, Pion Ltd, 1981.
8. Getis, A. and Ord, J.K. Local spatial statistics: an overview, Spatial analysis: Modelling in a GIS environment, Longley, P. & Batty M. (eds.), Wiley, 261–277, 1996.
9. Hubert, L. and Schultz, L. Quadratic assignment as a general data analysis strategy, British Journal of Math. & Stat. Psychology, **29**, 190–241, 1976.
10. Kibler, D., Aha, D.W., and Albert, M.K. Instance-based prediction of real-valued attributes, Computational Intelligence, **5**, 51–57, 1989.
11. Lallich, S., Muhlenbach, F., and Jolion, J.M. A test to control a region growing process within a hierarchical graph, Pattern Recog., **36**, 2201–2211, 2003.
12. Muhlenbach, F., Lallich, S., and Zighed, D.A. Identifying and handling mislabelled examples, JIIS, **22**(1), 89–109, 2004.
13. Quinlan, J.R. Learning with Continuous Classes, Proc. of the 5th Australian Joint Conference on AI, Hobart, Australia, 343–348, 1992.
14. Quinlan, J.R. Combining Instance-Based and Model-based Learning, Proc. of the 10th ICML, Amherst, MA, USA, 236–243, 1993.
15. Rumelhart, D.E. and McClelland, J.L. Parallel Distributed Processing, MIT Press, 1986.
16. Scuturici, V.M., Clech, J., and Zighed, D.A. Topological query in image databases, Proc. of CIARP'03, Havana, Cuba, 144–151, 2003.
17. Sebban, M., Nock, R., and Lallich, S. Stopping criterion for boosting-based data reduction techniques, JMLR, **3**, 863–885, 2002.
18. Shekhar, S., Lu C.T., and Zhang, P. Unified approach to detecting spatial outliers, GeoInformatica, **7**(2), 139–166, 2003.
19. Toussaint, G.T. Proximity graphs for Nearest Neighbor decision rules: Recent progress, Proc. Interface 2002, 34th Symp. on Comput. & Statistics, 2002.
20. Uysal, I. and Guvenir, H.A. An overview of regression techniques for knowledge discovery, Knowledge Engineering Review, **14**, 319–340, 1999.
21. Zighed, D.A., Lallich, S., and Muhlenbach, F. Separability index in supervised learning, Proc. of PKDD'02, Helsinki, Finland, 475–487, 2002.

A New Clustering Algorithm
Based on Cluster Validity Indices

Minho Kim and R.S. Ramakrishna

Department of Information and Communications, GIST
1 Oryong-dong, Buk-gu, Gwangju 500-712, Republic of Korea
{mhkim,rsr}@gist.ac.kr

Abstract. This paper addresses two most important issues in cluster analysis. The first issue pertains to the problem of deciding if two objects can be included in the same cluster. We propose a new similarity decision methodology which involves the idea of cluster validity index. The proposed methodology replaces a qualitative cluster recognition process with a quantitative comparison-based decision process. It obviates the need for complex parameters, a primary requirement in most clustering algorithms. It plays a key role in our new validation-based clustering algorithm, which includes a random clustering part and a complete clustering part. The second issue refers to the problem of determining the optimal number of clusters. The algorithm addresses this question through complete clustering which also utilizes the proposed similarity decision methodology. Experimental results are also provided to demonstrate the effectiveness and efficiency of the proposed algorithm.

1 Introduction

Clustering operation attempts to partition a set of objects into several subsets. The idea is that the objects in each subset are indistinguishable under some criterion of similarity [1], [5], [6], [9].

The core problem in clustering is *similarity decision*. We look for similarity while deciding whether or not two objects may be included in the same group, i.e., cluster. For the purpose of measuring similarity (or dissimilarity) between two objects, the concept of distance between them is most widely used. The most common way to arrive at a similarity decision employs thresholding. The easiest way to decide similarity of two objects is to compare the distance between them (subject to the user-specified threshold value). That is, if the distance is less than the threshold, they can be included in the same cluster; otherwise, each of them should be placed in two different clusters [9]. Agglomerative/divisive hierarchical clustering algorithms [6] also employ a similar method (as outlined above). In the hierarchical structure resulting from agglomeration/division (for example, dendrogram), each independent subgraph below a user-specified distance is a cluster and thus objects below that subgraph are in the same cluster and can be said to be similar. In partitional clustering algorithms [6], the similarity decision may be affected by the predetermined number

E. Suzuki and S. Arikawa (Eds.): DS 2004, LNAI 3245, pp. 322–329, 2004.

k of clusters. It will also be influenced by the threshold. The threshold and threshold-related parameters play a very important role in cluster analysis. They are usually determined by trial and error, or through very complex processes. These methodologies are thought to be computationally quite involved.

The focus in cluster analysis has shifted to cluster validity indices in recent times. The main objective is to determine the optimal number of clusters [1], [3], [4], [7], [8], [10]. However, the indices are not specifically targeted at similarity decision, per se.

In this paper, we address the relationship between similarity decision and cluster validity index, and also propose *a validation-based clustering algorithm* (*VALCLU*) centered around the (extracted) relationship. VALCLU consists of two parts: *random clustering* and *complete clustering*. The former builds a *clustering pool* in a random fashion and, then iteratively decides the similarity of two objects. The optimal number of clusters is found by complete clustering. We also present a new cluster validity index, *VI*, which can be used in complete clustering. VALCLU finds the optimal number of clusters without requiring painstakingly determined complex parameters for (similarity) decision making.

The rest of the paper is organized as follows. Section 2 discusses similarity decision. In section 3, the proposed VALCLU algorithm is described. Experimental results and conclusions are given, in sections 4 and 5, respectively.

2 Similarity Decision

How can we decide similarity of objects? Let us discuss the problem through an example. (Please see Fig. 1)

Fig. 1. Recognition of similarity between two white objects.

Under what condition(s) can we say that the two objects within the oval are similar and hence, can be included in the same cluster? To begin with, assume that there are only two white objects within the oval and a gray object and a black object without. It might be difficult to conclude that the two white objects are dissimilar to the gray object, and that only the two white objects can be grouped together and they are similar. Let us consider another situation wherein there are two white objects and a black object (instead of the gray object). It is now possible to decide that the two white objects belong to the same cluster while the black object does not.

The examples above indicate that an important factor that affects the recognition of a cluster is *relativity* (relative similarity/dissimilarity). That is, if two objects (two white objects in Fig. 1) are relatively similar (are located closely) compared with the other object (the black one in Fig. 1), one can easily separate the objects.

What kind of adjustment in our recognition affects the decision (of similarity)? The intra-cluster distance of a new cluster generated by merging two objects appears to be greater than those of two independent objects. In other words, we have to sacrifice compactness when merging two objects. However, merging two objects makes it easier to discriminate objects from one another, since they are thought of as a single (abstract) object (cluster). That is, we gain 'separability' of objects (clusters). This discussion is summarized in below.

Observation 1. Two objects may be comprehended as being similar (and hence, may be included in the same group of objects) if we sacrifice (relatively small) compactness. In the process, we gain separability while merging two objects.

Obs. 1 provides a qualitative methodology for testing similarity. However, we need to note that the rationale is the same as that of a cluster validity index. A cluster validity index is a quantitative parameter. Therefore, adopting a cluster validity index in Obs. 1 leads to a new quantitative methodology for similarity decision as outlined below.

Definition 1. Two objects are similar (and hence, can be included in the same group of objects) if the corresponding value of a cluster validity index is smaller after merging (them) than before merging.

(Here we are assuming that an optimal value of validity index is its minimum).

By virtue of the above definition, we can make similarity decisions by comparing values of the cluster validity index before and after merging. This is the key idea in the VALCLU algorithm we propose in the next section.

3   Validation-Based Clustering Algorithm

For purposes of efficiency, we divide the proposed algorithm into two parts according to the way the clusters to be tested are chosen. The *random clustering* part uses only a small number of clusters while the *complete clustering* part uses all the clusters. We note here that an object is a unit of similarity test in section 2, but that a cluster, which is a group of objects, is a unit of the test in the validation-based clustering algorithm.

3.1   Random Clustering

All the existing cluster validity index based clustering algorithms make use of all the clusters at each level. Thus, the larger the number of clusters at a level, the higher is the complexity of computing the cluster validity index. Moreover, we encounter a large number of singleton clusters if a level is close to the initial state (of cluster analysis).

However, in order to make similarity decisions by means of cluster validity index proposed in section 2, we need two clusters for merging and at least one cluster for comparison.

As for choosing two clusters to be merged, we may choose a pair of clusters with minimum distance from the whole set of clusters (i.e., two clusters with global minimum distance). But this is computationally very expensive. Calculating the cluster validity index turns out to be expensive as well for the same reason.

In order to resolve these problems, we propose *random clustering*, which randomly forms a *clustering pool* (CP) with size |CP|, from the set of all the clusters. Similarity decision as proposed in section 2 follows thereafter. Ideally, $|CP| \geq 3$ from the above considerations. The algorithm is outlined in below. The details about $EXIT_1$ and $EXIT_2$ are described later in section 3.3.

```
Randomly compose a cluster pool (CP);
Calculate index(|CP|);
Virtually merge 2 clusters with minimum distance in CP;
Calculate index(|CP| - 1);
IF index(|CP| - 1) is optimal THEN
   Merge the 2 clusters;
   IF EXIT1 condition THEN GOTO complete clustering;
ELSE
   IF EXIT2 condition THEN GOTO complete clustering;
GOTO step 1
```

3.2 Cluster Validity Indices

Cluster validity indices can be classified into two categories. *Ratio type* indices are characterized by the ratio of intra-cluster distance to inter-cluster distance. The *summation type* index is defined to be the sum of intra-cluster distance and inter-cluster distance with appropriate weighting factors. However, the ratio type index cannot be used in the initial state of random clustering, which is mainly comprised of singleton clusters. This is so because the intra-cluster distance of a cluster with one member is 0 and the index value is 0 or ∞, and hence, similarity comparisons are not meaningful. Therefore, only summation type indices can be used for random clustering. Recently proposed summation type indices include: SD [4], v_{sv} [7]. (Due to space limitations, details have been omitted.)

Since complete clustering can use indices of ratio type (unlike the random clustering), we propose a new cluster validity index in this category. This validity index, *VI* is defined in eqn. (1):

$$VI(nc) = \frac{1}{nc} \sum_{i=1}^{nc} \left(\frac{\max\limits_{k=1...nc, k \neq i} \{S_i + S_k\}}{\min\limits_{l=1...nc, l \neq i} \{d_{i,l}\}} \right)$$

$$S_i = \frac{1}{n_i} \sum_{x \in X_i} \|c_i - x\|, \quad d_{i,J} = \|c_i - c_l\|$$

(1)

In the equations above, nc stands for the number of clusters, c_i for a representative of cluster i. The optimal value of the index VI is its minimum value.

The proposed index VI can be explained by focusing on three features. First, the term $\min_{l=1...nc, l \neq i} \{d_{i,l}\}$ in eqn. (1) points to the clusters which need to be merged by virtue of their exhibiting a very small value, implying very high VI-value. Second, a large $\max_{k=1...nc, k \neq i} \{S_i + S_k\}$ implies that unnecessary merging has taken place. Finally, the averaging in eqn. (1) combines the total information contained in the current state of the cluster structure and thereby imparts robustness.

3.3 Complete Clustering

Random clustering using the clustering pool has two weaknesses. First, it does not use the full set of clusters when arriving at similarity decisions by computing the cluster validity index. This may lead to wrong decisions and may fail to find the optimal nc. In order to address this problem, we need to make use of the whole set of clusters as a clustering pool. That is, switching to an algorithm with $|CP| = nc$ is imperative. The switching time could be $nc = \sqrt{N}$ as per a well known rule of thumb. Here N is the total number of data objects. This is an $EXIT_1$ condition in section 3.2, i.e., *Exit if $nc < \sqrt{N}$*.

Another weakness of random clustering is that it can be trapped in a local minimum. This can be avoided by merging repeatedly until $nc = 1$ irrespective of the similarity decision and looking for just the cluster structure with the optimal value of the cluster validity index. The intuitive way to decide the entrapment in a local minimum is to see if merging has failed repeatedly over a certain number (e.g., $_{nc}C_{|CP|}$) of tests. This is the $EXIT_2$ condition in section 3.2.

Now, we propose the second part of the validation-based clustering algorithm, called *complete clustering*, by taking the above points into account. The complete clustering algorithm is given in below.

```
[Initialize]
  nc = # clusters;
  index_optimal = MAX_VALUE;

WHILE ( nc >= 2 ) {
  Calculate index(nc);  //(|CP| == nc)
    IF index(nc) is optimal than index_optimal THEN
      index_optimal = index(nc);
      Store the current configuration of clusters;
  Merge two clusters with minimum distance;
  nc--;
}
```

VALCLU is similar to agglomerative hierarchical clustering algorithms. However, one of the major drawbacks of the latter is the absence of refinement. VALCLU also suffers from this drawback. In order to address this problem, we refine the results of random clustering and complete clustering through the well known k-means algorithm. As is well known, k-means algorithm guarantees acceptable results, given proper initial representatives and number of clusters, k. Random clustering and complete clustering satisfy both these requirements.

4 Experimental Results

For the purpose of evaluating the effectiveness of the proposed VALCLU algorithm, five synthetic datasets and one real world dataset were used. The synthetic datasets are shown in Fig. 2. We used iris dataset for the real world dataset test [2].

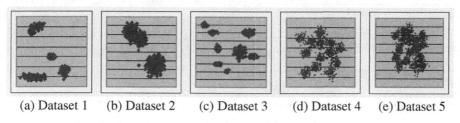

(a) Dataset 1   (b) Dataset 2   (c) Dataset 3   (d) Dataset 4   (e) Dataset 5

Fig. 2. Synthetic datasets.

To begin with, for the purpose of evaluating the performance of random clustering, we measured the average number of similarity decision tests for each nc until it succeeded in merging two clusters ($n1$). If we look for two clusters with global minimum distance (d_{min}) as in agglomerative hierarchical clustering, the number of distance computations for each nc is $nc \cdot (nc - 1)/2$. Thus, the total number of evaluations over the range $\sqrt{N} \leq nc \leq N$ (the same range as that of the random clustering), is $n2 = (N + 1) \cdot N \cdot (N - 1)/6 - (\sqrt{N} + 1) \cdot \sqrt{N} \cdot (\sqrt{N} - 1)/6$. On the other hand, for a clustering pool in random clustering, we need to perform $n3 = |CP| \cdot (|CP| - 1)/2 + (|CP| - 1) \cdot (|CP| - 2)/2 = (|CP| - 1)^2$ evaluations.

With a view to fairly compare random clustering with clustering using global d_{min}, we look at $n1$ and $n2/(n3 \cdot (N - \sqrt{N})) = n4$. Table 1 shows the results. Since any value greater than or equal to 3 can be selected for $|CP|$, we arbitrarily selected $|CP| = 15$.

In Table 1, it can be seen that the value $n1$ of (random) clustering pool is much smaller than the value $n4$ of clustering using global d_{min}. That is, random clustering is much more efficient than clustering using entire clusters as in agglomerative hierarchical clustering. While comparing v_{sv} with SD, the latter requires fewer tests than the former with random clustering. In addition, v_{sv} and SD have almost the same clustering error rate (in average 0.026 and 0.024, respectively), where clustering error rate refers to the rate at which member data differs from the majority class in the same cluster. Therefore, SD is seen to be a more efficient index for random clustering.

We will now examine the performance of complete clustering in computing the optimal nc. Here, we adopt the recently proposed index I [8] and the index VI proposed in section 3.2, as well as the indices used in random clustering. Fig. 3 shows the findings. Note that in Fig. 3, the optimal nc is not provided for real data, i.e., the

Table 1. Comparisons of $n1$, $n4$, and N for v_{sv} and SD with respect to various datasets in random clustering.

Data	index	$n1$	N	$n4$
Dataset 1	v_{sv}	28.215	500	222.34
	SD	1.033		
Dataset 2	v_{sv}	69.673	800	563.93
	SD	1.073		
Dataset 3	v_{sv}	34.663	550	268.43
	SD	1.000		
Dataset 4	v_{sv}	33.047	130 0	1,477.97
	SD	1.004		
Dataset 5	v_{sv}	30.255	130 0	1,477.97
	SD	1.014		
Real data	v_{sv}	13.225	150	20.78
	SD	1.040		

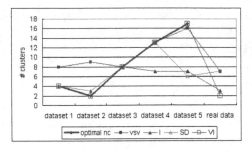

Fig. 3. Found number of clusters through complete clustering by using various cluster validity indices.

Iris dataset. The reason why we did not show the optimal nc for this dataset is that the optimal nc for the this dataset is debatable [1]. Therefore, in this paper we work with $nc = 2$ as well as $nc = 3$. From Fig. 3, we see that the index VI and the optimal nc almost perfectly match in performance while the others have some mismatches. In other words, the index VI gives the best result in complete clustering.

Table 2. Error rate comparisons between VALCLU and K-means.

	Dataset 1	Dataset 2	Dataset 3	Dataset 4	Dataset 5	Iris
VALCLU	0.0000	0.0000	0.0000	0.0062	0.0135	0.0000
K-means	0.1400	0.0000	0.1233	0.1084	0.1364	0.0200

In Table 2, error rates of clustering results are provided to demonstrate the labeling performance of VALCLU. Also, the results for K-means algorithm are included for performance comparison purposes. It is well known that the clustering result of K-means algorithm depends on the initialization of its seeds and the number of clusters. In our evaluation, seeds are randomly initialized. Even though we provide the exact number of clusters in the evaluation, the algorithm yields different error rates for each run. Thus, for error rates of K-means algorithm (the second row) in Table 2, we show the average error rates obtained from 30 runs. From Table 2, we can see that the performance on the clustering quality of VALCLU surpasses that of K-means algorithm and it guarantees identical results for each run.

5 Conclusions

In this paper we have proposed a validation-based clustering algorithm, called VALCLU, that utilizes cluster validity indices; and evaluated its effectiveness and efficiency. The methodology proposed for deciding the similarity between two clusters (or objects) are based on cluster validity indices. It plays a key role in the two main parts of the validation-based clustering algorithm, viz., random clustering and complete clustering. It can effectively determine if two clusters can be merged into one cluster through a (quantitative) change of values of the cluster validity index. Also, it determines the optimal number of clusters in complete clustering. Experimental results show that random clustering requires much less computations than agglomerative hierarchical clustering. As for similarity decision, several cluster validity indices were evaluated. Experimental results indicate that the index SD is the most efficient for random clustering and that the index VI proposed in this paper shows the best results among various indices for complete clustering. Also, VALCLU performs better than the well known K-means algorithm. Further work on various aspects of cluster validity indices is in progress.

Acknowledgement

This work was supported by the Ministry of Education (MOE) through the Brain Korea 21 (BK21) project.

References

1. Bezdek, J.C., Pal, N.R.: Some new indexes of cluster validity. IEEE Trans. Sys., Man, and Cyber. PART B: Cyber. 28(3) (1998) 301-315
2. Blake, C.L., Merz, C.J.: UCI Repository of machine learning databases (http://www.ics.uci.edu/~mlearn/MLRepository.html). Univ. of California, Irvine, Dept. of Info. & Comp. Sci. (1998)
3. Davies, D.L., Bouldin, D.W.: A cluster separation measure. IEEE Trans. Pattern Analysis and Machine Intelligence (PAMI) 1(2) (1979) 224-227
4. Halkidi, M., Vazirgiannis, M.: Quality scheme assessment in the clustering process. European Conf. Principles and Practice of Knowledge Discovery in Databases (PKDD). Lecture Notes in Artificial Intelligence Vol. 1910 (2000) 265-276
5. Han, J., Kamber, M.: Data mining: concepts and techniques. Morgan Kaufmann (2001)
6. Jain, A.K., Murty, M.N., Flynn, P.J.: Data clustering: a review. ACM Computing Surveys 31(3) (1999) 264-323
7. Kim, D.-J., Park, Y.-W., Park, D.-J.: A novel validity index for determination of the optimal number of clusters. IEICE Trans. Inf. & Syst. E84-D(2) (2001) 281-285
8. Maulik, U., Bandyopadhyay, S.: Performance evaluation of some clustering algorithms and validity indices. IEEE Trans. Pattern Analysis and Machine Intelligence (PAMI) 24(12) (2002) 1650-1654
9. Monmarché, N., Slimane, M., Venturini, G.: On improving clustering in numerical databases with artificial ants. European Conf. Advances in Artificial Life (ECAL). Lecture Notes in Artificial Intelligence Vol. 1974 (1999) 626-635
10. Schwarz, G.: Estimating the dimension of a model. Annals of Statistics 6(2) (1978) 461-464

An Efficient Rules Induction Algorithm
for Rough Set Classification

Songbo Tan[1] and Jun Gu[2]

[1] Software Department, ICT, CAS, P.O. Box 2704, Beijing, 100080, P.R. China
Graduate School of the Chinese Academy of Sciences, P.R. China
tansongbo@software.ict.ac.cn
[2] Department of Computer Science, Hong Kong University of Science and Technology
eecs@263.net

Abstract. The theory of rough set provides a formal tool for knowledge discovery from imprecise and incomplete data. Inducing rules from datasets is one of the main tasks in rough set based data mining. According to Occam Principle, the most ideal decision rules should be the simplest ones. Unfortunately, induction of minimal decision rules turns out to be a NP-hard problem. In this paper, we propose an heuristic minimal decision rules induction algorithm *RuleIndu* whose time complexity is $O(|A|*|U|^2)$ and space requirement is $O(|A|*|U|)$. In order to investigate the efficiency of proposed algorithm, we provide the comparison between our algorithm *RuleIndu* and some other rules induction algorithms on some problems from UCI repository. In most cases, our algorithm *RuleIndu* outmatches some other rules induction algorithms not only in classification time but also in classification accuracy.

1 Introduction

Inducing rules from datasets, representing sets of learning examples described by attributes, is one of the main tasks in rough set based data mining. Pawlak [2] presents a method to simplify decision tables that consists of following three steps: computation of condition attributes, elimination of duplicate rows and deletion of superfluous attribute values.

According to Occam Principle, the most ideal decision rules should be the simplest ones. A simplified rule (a minimal rule) is the one in which the conditions in its antecedent is minimal, i.e., the rule with minimal complexity. Unfortunately, induction of minimal decision rules turns out to be a NP-hard problem [4].

Therefore, methods to solve this NP-hard problem play an important role in the development of rough set-based data mining. In this paper, we propose an efficient rules induction algorithm *RuleIndu* whose time complexity is $O(|A||U|^2)$ and space requirement is $O(|A||U|)$. ($|A|$ denotes the number of attributes and $|U|$ denotes the number of objects of information system). In the experiment, our algorithm achieves significant performance compared with some other rules induction algorithms in ROSETTA system [5].

The rest of this paper is organized as follows: In the next section, we briefly overview the previous work. Our algorithm *RuleIndu* is introduced in Section III. Experimental results of *RuleIndu* and some other rules induction algorithms are given in Section IV. Finally Section V concludes this paper.

E. Suzuki and S. Arikawa (Eds.): DS 2004, LNAI 3245, pp. 330–337, 2004.

2 Prior Work

Several rough set-based rules induction algorithms have been developed. In this section, we briefly review some rules induction algorithms for rough set classification [3][7]. The conventional algorithms are mainly grouped under two categories: the rules induction algorithms based on finding of minimal rules set [7] and the rules induction algorithms based on generation of complete set of all rules in the given syntax [3].

Hassanien [7] proposes a simplification rules generation algorithm via dropping some condition attributes. For every rule r, a condition is dropped from rule r, and the rule r is checked for decision consistency. If the rule r is inconsistent, then restore the dropped condition. The step is repeated until every condition of the rule r has been dropped once. Before the acceptance of the rule r as a decision rule, its redundancy must be checked. Note that its time and space requirements are $O(|A|^2|U|^2)$ and $O(|A||U|)$ respectively.

After introducing one rules induction algorithms based on finding of minimal rules set, we continue to review a rule induction algorithm based on generation of a complete set of all rules. Bjorvand [3] first generates all possible and impossible rules based on a complete Cartesian product of attributes value sets. After the generation process, some criteria can be set to remove some obviously unnecessary rules. This methodology makes an attempt to find excellent rules among all possible rules, but with the increase of attribute numbers, the search will become quickly computationally intractable.

In this paper, a new efficient rules induction algorithm based on finding of minimal rules set is developed.

3 Proposed Algorithm

3.1 The Outline of Our Rule Induction Algorithm

First we obtain the attribute reduct based on total discernibility matrix of given decision table via any attribute reduction algorithm [1][5]. We delete the columns of the decision table involved with attributes not contained in the obtained reduct and remove the duplicate rows. We look each row of decision table after attribute reduction as one rule. Then our rules induction algorithm *RuleIndu* eliminates superfluous attribute values of rules one by one via calling the attribute value reduction algorithm *HeuriAttriValRed* (See Fig.1). Obviously, our algorithm *RuleIndu* mainly deals with the third stage of the three-step method for decision rules proposed by Pawlak [2].

Rather than construct one total dicernibility matrix for all pairs of objects with different decision values, for each row i we construct one partial dicernibility matrix between the row i and other rows whose decision values are different from row i. Then we design a heuristic attribute value reduction algorithm *HeuriAttriValRed* (See Fig.1) for each partial dicernibility matrix, and after each execution of the algorithm *HeuriAttriValRed* one *partial* reduct (attribute value reduct) with respect to the row i is found. According to the *partial* reduct, an approximate minimal rule is generated by dropping redundant attribute values. This procedure is repeated until each rule (row) is operated once. Finally we delete the duplicate rules.

For simplicity, if an attribute is involved with an entry in partial discernibility matrix, we say the attribute covers the entry, or the entry is covered by the attribute. Naturally we can regard the attribute value reduction problem as a special set-covering problem.

The attribute value reduction algorithm *HeuriAttriValRed* attempts to choose the attribute covering the maximum non-empty entries of the partial dicernibility matrix, which are so far not covered any selected attribute. This procedure is repeated until all nonempty entries in partial dicernibility matrix are covered by at least one attribute, which is equivalent to finding one *partial* reduct (attribute value reduct) for the rule (row).

Given a decision table with ten condition attributes $(a_1,...,a_{10})$ and one decision attribute d. After attribute reduction based on total discernibility matrix, one reduct is found which consists of four conditional attributes a_1, a_2, a_3, a_4. We only consider the object (row) *1* and other four objects with different decision values from object *1*. Then a partial decision table can be illustrated as Table 1.

Table 1. The partial decision table with respect to object 1.

U	a_1	a_2	a_3	a_4	d
1	TRUE	TRUE	FALSE	FALSE	0
2	FALSE	FALSE	TRUE	TRUE	1
3	FALSE	TRUE	TRUE	FALSE	1
4	FALSE	TRUE	FALSE	TRUE	1
5	FALSE	FALSE	FALSE	FALSE	1

Then we can obtain the partial discernibility matrix with respect to object *1* as follows:

$$\begin{pmatrix} \phi & a_1a_2a_3a_4 & a_1a_3 & a_1a_4 & a_1a_2 \\ \phi & \phi & \phi & \phi & \phi \\ \phi & \phi & \phi & \phi & \phi \\ \phi & \phi & \phi & \phi & \phi \\ \phi & \phi & \phi & \phi & \phi \end{pmatrix}. \tag{1}$$

In order to compute attribute value reduct for each rule based on each partial discernibility matrix more efficiently, we introduce four matrixes, i.e., *NormMat*, *CovMat*, *IsBecovered* and *CovUncovItem*, with respect to row *i*. For the sake of simplicity, in the rest of this paper, we substitute "partial discernibility matrix" with "discernibility matrix".

3.2 Necessary Denotations

1. Be-Covered (or NormMat) Matrix

   If we take the non-empty entries of discernibility matrix as rows, then we can construct a so-called *Be-Covered* (or *NormMat*) matrix, in which each row records the attributes covering one non-empty entry. The first column of *NormMat* denotes the number of attributes covering one non-empty entry of discernibility matrix. In order to illustrate *NormMat*, we take a simple partial decision table, e.g., Table 1, as an example. Then we obtain our *NormMat* as following formula (2) and (3).

$$\begin{pmatrix} 4 & a_1 & a_2 & a_3 & a_4 \\ 2 & a_1 & a_3 & \phi & \phi \\ 2 & a_1 & a_4 & \phi & \phi \\ 2 & a_1 & a_2 & \phi & \phi \end{pmatrix} . \tag{2}$$

$$\begin{pmatrix} 4 & 1 & 2 & 3 & 4 \\ 2 & 1 & 3 & \phi & \phi \\ 2 & 1 & 4 & \phi & \phi \\ 2 & 1 & 2 & \phi & \phi \end{pmatrix} . \tag{3}$$

In above matrix (3), 1, 2, 3 and 4, excluding in the first column, denote the attributes of the partial decision table.

2. Covering Matrix (*CovMat*)

On the other hand, if we take attributes of the partial decision table as rows, then we can construct a so-called *CovMat* matrix, in which each row records the non-empty entries of discernibility matrix covered by one attribute. The first column of *CovMat* is the number of entries covered by one attribute. Then we obtain *CovMat* of above simple partial decision table as matrix (4). *CovMat* facilitates to quickly find all non-empty entries covered by one attribute without scanning all entries of discernibility matrix.

$$\begin{pmatrix} 4 & 1 & 2 & 3 & 4 \\ 2 & 1 & 4 & \phi & \phi \\ 2 & 1 & 2 & \phi & \phi \\ 2 & 1 & 3 & \phi & \phi \end{pmatrix} . \tag{4}$$

In matrix (4), 1, 2, 3 and 4, excluding the first column, denote non-empty entries of discenibility matrix $a_1a_2a_3a_4$, a_1a_3, a_1a_4, and a_1a_2 respectively.

3. IsBeCovered Vector

IsBeCovered vector has the same rows as *NormMat* and has only one column, which indicates whether the corresponding row of *NormMat* is covered by at least one attribute. Each entry of *IsBeCovered* vector has value 0 or other numbers larger than 0, i.e., the NO. of attribute, which denotes the corresponding row of *NormMat* is not covered or covered. For example, if the *IsBeCovered* vector is (0 0 0 1) indicates the forth row of *NormMat*, i.e., the entry "a_1a_2" of discernibility matrix, is covered by the attribute *1*, i.e., a_1. First we initialize the *IsBeCovered* vector as (0 0 0 0). Then if we first select the attribute a_2, we change the *IsBeCovered* vector to (2 0 0 2).

4. CovUncovItem Vector

CovUncovItem vector has the same rows as *CovMat* and has only one column, which records the number of entries of discernibility matrix covered by one attribute and not covered by any selected attributes. The *CovUncovItem* vector facilitates to pick out the attribute covering the maximum non-empty entries of dicernibility matrix, which are so far not covered any selected attribute.

First we initialize the *CovUncovItem* as the first column of *CovMat*. For example, we obtain initial *CovUncovItem* of above the partial decision table as (4 2 2 2). If we select an attribute, we set the corresponding entry of *CovUncovItem* to –1. Then we reduce the entry of *CovUncovItem* corresponding to each unselected attribute by the number of entries covered by this unselected attribute among all entries covered by justly selected attribute but not covered by previously selected attributes.

For example, if we first select a_2, then *CovUncovItem*(2)=-1; *CovUncovItem*(1)=4-2=2; *CovUncovItem*(3)=2-1=1; *CovUncovItem*(4)= 2-1=1.

Therefore, when a reduct is found, the entries of *CovUncovItem* corresponding to selected attributes equal –1 and other entries of *CovUncovItem* equal 0.

In our algorithm, discernibility matrix is unnecessary, because we can directly construct all four matrixes from partial decision table.

3.3 Necessary Variable

NormRows: the number of rows of *NormMat*; *NormColumns*: the number of columns of *NormMat*; *CovRows*: the number of rows of *CovMat*; *CovColumns*: the numbers of columns of *CovMat* *UnCoverRows*: the number of rows of *NormMat*, which are not covered by selected attributes; *CovRowNum*: the number of those rows in *NormMat* covered by one attribute; *CovRow*: the current row of *NormMat* covered by one attribute; *NormColNum*: the number of attributes covering one row of *NormMat*; *NormCol*: the current attribute covering one row of *NormMat*;

3.4 The RuleIndu Algorithm

Initialization: to start, procedure *LoadData*() loads the decision table. By calling *AttriRed*(), we obtain the attribute reduct. The attribute reduction algorithm may be any efficient attribute reduction algorithm [1][5]. Then according to the obtained reduct, the function *GetRedData*() deletes redundant columns and removes the duplicate rows of given decision table.

Heuristic Rules Induction: in one iteration of *for-loop* (See Fig.2), according to the reduct decision table, we obtain the *NormMat*, *CovMat*, *IsBecovered* and *CovUncovItem* matrixes by calling function *Initilization*(i) with respect to one object *i*. Then *HeuriAttriValRed*() and *GetOneRule*() are employed to compute approximately minimal rule for the rule (object) *i*.

In this function *HeuriAttriValRed*(), we employ the heuristic information: the number of entries of discernibility matrix covered by each unselected attribute, which are not covered by selected attributes. Then we choose the attribute *max_j* that covers the maximum entries of discernibility matrix that are not covered by selected attributes. And then we scan the row of *CovMat* corresponding to justly selected attribute *max_j*. If the row of *NormMat* corresponding to scanned entry of *CovMat* is not covered by selected attributes excluding justly selected attribute *max_j*, then we scan the row of *NormMat*. For each scanned attribute *j*, if it is not selected, we execute *CovUncovItem*(j)--.

Running Time: Given a decision table with $|U|$ objects and $|A|$ attributes.

In one iteration of *for-loop* of algorithm *RuleIndu*, the non-empty entries of discernibility matrix with respect to the object *i* is smaller than $|U|$. Accordingly, the running time of *Initialization*() is $O(|A||U|+|A||U|+|U|+|A|)$, i.e., $O(|A||U|)$.

```
void HeuriAttriValRed (){
    UnCoverRows=NormRows;
    while(UnCoverRows!=0){
        // choose max_j that maximize{ CovUncovItem(i)}
        max_j=SelectMaxCovAttri();
        CovUncovItem[max_j]=-1;
        CovRowNum=CovMat[max_j *CovColumns+0];
        for(j=0;j< CovRowNum;j++){
            CovRow=CovMat[max_j *CovColumns+(j+1)];
            //Only scan rows of NormMat that are not covered by selected attributes
            if(IsBeCovered[CovRow -1]==0){
                IsBeCovered[CovRow -1]= max_j;
                UnCoverRows--;
                NormColNum= NormMat[(CovRow -1)*NormColumns+0];
                for(k=0;k< NormColNum;k++){
                    NormCol=NormMat[(CovRow -1)*NormColumns+(k+1)];
                    if(CovUncovItem[NormCol -1]!=-1)
                        CovUncovItem[NormCol -1]--;
                }
            }
        }
    }//End while
}
```

Fig. 1. The Function of *HeuriAttriValRed*.

```
void RuleIndu(){
    LoadData();              //load the decision table data
    AttriRed();             //Calculate the reduct by calling reduction algorithm
    //the rules induction part
    GetRedData();           //delete redundant columns, and remove the duplicate rows
    For each object i in decision table Do
        Initilization(i);       //Obtain the NormMat, CovMat, IsBecovered and CovUncovItem
        HeuriAttriValRed();     //the heuristic attribute values reduction procedure
        GetOneRule();           //According to CovUncovItem, obtain a new rule with respect to the object I
    End For
    DelDupRule();           //delete the duplicate rules
    OutPutResult();         //print the result
}
```

Fig. 2. The Algorithm of *RuleIndu*.

$SelectMaxCovAttri()$ can be done in $O(|A|)$. In each execution of *while-loop* (See Fig.1) only one row of *CovMat* and some rows of *NormMat* covered by justly selected attribute *max_j*, which are not covered by selected attributes excluding justly selected attribute *max_j*. Therefore the computation with respect to *CovMat* is at most one time scan of *CovMat* and the computation with respect to *NormMat* is just one time scan of *NormMat*. As a result, the executing time of *HeuriAttriValRed()* is $O(|A||U|+|A||U|+|A|)$, i.e., $O(|A||U|)$.

$GetOneRule(i)$ can be done in $O(|A|)$.

Therefore, the total executing time of *for-loop* of algorithm *RuleIndu* is:

$$O\left(\sum_{i=1}^{|U|}(|A||U|+|A||U|+|A|)\right)=O\left(|A|\sum_{i=1}^{|U|}(|U|+|U|+1)\right)=O\left(|A||U|^2\right). \tag{5}$$

$DelDupRule()$ will take at most $O(|A||U|^2)$.

Therefore, the total running time of rules induction part of algorithm *RuleIndu*, which does not consist of the attribute reduction (i.e. *AttriRed()*), is $O(|A||U| + |A||U|^2 + |A||U|^2)$, i.e., $O(|A||U|^2)$.

Space Requirement: the main space requirements of rules induction part of algorithm *RuleIndu* are the four matrixes: *NormMat*, *CovMat*, *IsBecovered* and *CovUncovItem* which requires $O(|A||U|)$, $O(|A||U|)$, $O(|U|)$ and $O(|A|)$ respectively. Consequently the space complexity is $O(|A||U|)$.

4   Experiment Results

In this section, we give experimental results of proposed rules induction algorithm *RuleIndu* on some problems from UCI repository [6]. Our algorithm was coded in C and tested under a personal computer with single 700 MHZ processor and 128M memory.

Our algorithm *RuleIndu* obtains the attribute reducts of datasets by *Johnson reduct* procedure in ROSETTA system [5]. The *Johnson induction algorithm* applies *Johnson reduct* procedure for attribute reduction and *rules generation* procedure for rules induction. The *genetic induction algorithm* adopts *genetic reduct* procedure for attribute reduction and *rules generation* procedure for inductive rules.

Eight datasets are randomly selected from UCI repository. The leftmost column of TABLE 2 is the name of the dataset from UCI repository. The 2nd, 3rd, 4th columns are train set numbers, test set numbers and original attribute numbers respectively. The 5th, 8th, 11th columns are test correct rate by using rules generated by three algorithms respectively. The 6th, 9th, 12th columns are the classifying time by using rules generated by three algorithms respectively. The 7th, 10th, 13th columns are decision rule numbers generated by three algorithms respectively.

First, in terms of classifying time, our algorithm *RuleIndu* has a much faster speed than *Johnson induction algorithm* and *genetic induction algorithm* on all datasets. For most datasets, our algorithm *RuleIndu* is about two orders magnitude faster than *genetic induction algorithm*.

Table 2. Comparision between *RuleIndu* and rules induction algorithms in ROSETTA.

Instances	Train Set	Test Set	Ori. Attri.	*RuleIndu*			ROSETTA *Johnson Induction*			ROSETTA *Genetic Induction*		
				Test Cor.	Time	Rules	Test Cor.	Time	Rules	Test Cor.	Time	Rules
Voting	300	135	16	89.6	0.01	36	69.6	<1.0	99	86.7	3.0	10742
Zoo	67	34	16	85	0.0	12	79.4	<1.0	16	85	1.0	917
Mofn-3-7-10	300	1024	10	100	0.0	56	96.88	1.0	124	96.88	1.0	124
Rand20	100	3000	20	47.8	0.01	75	4.8	3.0	99	42.5	101.0	19749
Monk2	169	432	6	77	0.12	105	39.12	<1.0	169	39.12	<1.0	169
Chess	1500	1696	36	66.8	0.02	189	21.76	4.0	743	23.11	13.0	3050
SoybeanSmall	31	16	55	25	0.00	5	25	<1.0	6	25	<1.0	672
Lymphography	98	50	18	80.0	0.0	47	18.0	<1.0	88	70.0	1.0	8471

Then in terms of classification quality, our algorithm *RuleIndu* achieves much higher classification accuracy than *Johnson induction algorithm* and *genetic induction algorithm* with exception of two datasets, i.e., **Zoo and** SoybeanSmall.

Finally with regard to decision rule numbers, the *genetic induction algorithm* produces quite much more rules than our algorithm *RuleIndu*. For every dataset our algorithm induces much less rules than *Johnson induction algorithm*.

Accordingly, we can say, our algorithm is an effective rules induction algorithm.

5 Conclusion

In this paper we develop an efficient rules induction algorithm *RuleIndu* for minimal decision rules problem whose time complexity is $O(|A||U|^2)$ and space requirement is $O(|A||U|)$. The comparison between our algorithm *RuleIndu* and two rules induction algorithms in ROSETTA system is conducted. The experimental results indicate the effectiveness of our algorithm.

References

1. Nguyen Sinh Hoa, Nguyen Hung Son.: Some Efficient Algorithms For Rough Set Methods.
2. Pawlak, Z.: Rough sets: theoretical aspects of reasoning about data. Kluwer Academic Publishers. Dordrecht (1991)
3. Bjorvand, A. T.: Rough Enough-A System Supporting the Rough Sets Approach.
4. Hong, J. R.: AE1: Extension Matrix Approximate Method for the General Set Covering Problem. International Journal of Computer & Information Science (1985)
5. Øhrn, A., Komorowski, J.: ROSETTA-A rough set toolkit for analysis of data. In: Proceeding of Third International Joint Conference on Information Sciences (JCIS'97). Durham, NC, USA, March 1-5, 3(1997), pp. 403-407
6. Merz, C.J., Murphy, P.: UCI repository of machine learning database.
7. Hassanien, A. E., Ali, J. M. H.: Rough Set Approach for Generation of Classification Rules of Classification Rules of Breast Cancer Data. Informatica (2004), Vol. 15, No. 1, pp. 23-38

Analysing the Trade-Off Between Comprehensibility and Accuracy in Mimetic Models*

Ricardo Blanco-Vega, José Hernández-Orallo, and María José Ramírez-Quintana

Departamento de Sistemas Informáticos y Computación
Universidad Politécnica de Valencia, C. de Vera s/n, 46022 Valencia, Spain
{rblanco,jorallo,mramirez}@dsic.upv.es

Abstract. One of the main drawbacks of many machine learning techniques, such as neural networks or ensemble methods, is the incomprehensibility of the model produced. One possible solution to this problem is to consider the learned model as an oracle and generate a new model that "mimics" the semantics of the oracle by expressing it in the form of rules. In this paper we analyse experimentally the influence of pruning, the size of the invented dataset and the confidence of the examples in order to obtain shorter sets of rules without reducing too much the accuracy of the model. The experiments show that the factors analysed affect the mimetic model in different ways. We also show that by combining these factors in a proper way the quality of the mimetic model improves significantly wrt. other previous reports on the mimetic method.

1 Introduction

In this paper we analyse and improve a general method for converting the output of any incomprehensible model into one simple and comprehensible representation: set of rules. The goal of converting any data mining model into a set of rules may seem a chimera, but there are two feasible ways of achieving it. First, using many specific "translators" to convert each kind of model into rules. Secondly, using a general "translator" to convert any model into sets of rules.

There have been many techniques developed for the first approach, especially to convert neural networks into rules (rule extraction techniques), and also for other representations, such as support-vector machines. The second approach, even though it would be more generally applicable, it has not been analysed in the same extent, probably because it was unclear in which way a set of rules could be extracted from any kind of model, independently of its representation.

The solution to this problem cannot be easier, but it was recently been presented by Domingos [2][3]: we can treat the learned model as an oracle and generate a new labelled dataset with it (invented dataset). Next, the labelled dataset is used for learning a decision tree, such as C4.5, which ultimately generates a model in the form of rules. The process is shown in Figure 1. The first learning stage (top) uses any data mining modelling technique to obtain an accurate model, called the *oracle*. With this model we label a random dataset R, and jointly with the training set T, we train a second model, using a comprehensible data mining technique. The second model is called the *mimetic* model.

---

* This work has been partially supported by CICYT under grant TIN 2004-7943-C04-02, Acción Integrada Hispano-Austríaca HU2003-003 and Generalitat Valenciana.

E. Suzuki and S. Arikawa (Eds.): DS 2004, LNAI 3245, pp. 338–346, 2004.

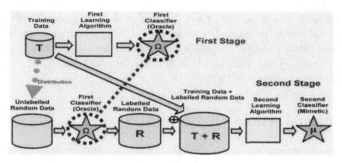

Fig. 1. Mimetic Technique.

The inventor of the technique, Domingos called it CMM (Combined Multiple Models) and used bagging [2] as oracle and C4.5rules as the final comprehensible model. He used a fixed number of randomly generated examples (1,000 for all the datasets). These were also joined to the original training set for learning the decision tree. In [4] we further analysed the method (which we called "mimetism") experimentally, for boosting, a different ensemble method. The results are quite consistent with Domingos. Additionally, in [5] we analysed the technique theoretically, proving that 100% fidelity is achievable with unpruned decision trees, as long as a sufficient large random sample is generated.

From all these results, we have learned some things about how "mimetic classifiers" work: A great number of random examples is necessary to achieve high fidelity, but the number of rules is also high. The use of the original training set (jointly with the random examples) is beneficial for the accuracy.

However, there are other issues where the behaviour of the method is not so clear: The method has not been applied to other oracles, especially oracles which are not ensembles, such as neural networks. The relationship between the comprehensibility of the resulting model and some factors such as: degree of pruning, number of random examples generated, etc., is also unknown.

In this work we analyse the method regarding these issues, concentrating especially on how short the sets of rules can be obtained without sacrificing too much the fidelity of the mimetic model with respect to the oracle.

In order to settle a precise reference metric, we define the following "quality metric", which represents a trade-off between comprehensibility (roughly represented here by the number of rules of the mimetic model) and accuracy:

$$Q = \frac{(\,Acc(Mim) - Acc(Ref)\,)\,/\,Acc(Ref)}{(Rules(Mim) - Rules(Ref))\,/\,Rules(Ref)} \tag{1}$$

The "reference model" (*Ref*) represents a comprehensible model learned directly with the original training set, such as C4.5 while Mim represents the mimetic model (possibly C4.5 as well). Obviously, if the results with the mimetic procedure are not better than with the reference model there would be no point in using the mimetic technique. As we will see, the factors that affect this quality metric Q are manifold and complex.

The paper is organised as follows. Section 2 introduces the experimental setting which has been used to perform the analysis of the mimetic technique. Section 3 studies the influence of pruning on the quality and fidelity of mimetic classifiers using neural networks and boosting as oracles. The relation between the invented dataset size and the quality of the mimetic model is analysed in Section 4. A function to estimate the optimal size of the random dataset for each problem is also included. Section 5 modifies the random dataset by taking the confidence of the oracle into account. Section 6 includes a joint analysis about the combination of factors. Section 7 presents the conclusions and future work.

2 Experimental Setting

In this section we present the experimental setting used for the analysis of the mimetic method described in this paper. For the experiments, we have employed 20 datasets (to see Table 1) from the UCI repository [1]. For the generation of the invented dataset we use the technique proposed in [4].

Table 1. Information about datasets used in the experiments.

No.	Dataset	Attr.	Num.Attr.	Nom.Attr.	Classes	Size	Missing
1	anneal	38	6	32	6	898	No
2	audiology	69	0	69	24	226	Yes
3	balance-scale	4	4	0	3	625	No
4	breast-cancer	9	0	9	2	286	Yes
5	cmc	9	2	7	3	1,473	No
6	colic	22	7	15	2	368	Yes
7	diabetes	8	8	0	2	768	No
8	hayes-roth	4	0	4	3	132	No
9	hepatitis	19	6	13	2	155	Yes
10	iris	4	4	0	3	150	No
11	letter	16	16	0	26	20,000	No
12	monks1	6	0	6	2	556	No
13	monks2	6	0	6	2	601	No
14	monks3	6	0	6	2	554	No
15	mushroom	22	0	22	2	8,124	Yes
16	sick	29	7	22	2	3,772	Yes
17	vote	16	0	16	2	435	Yes
18	vowel	13	10	3	11	990	No
19	waveform-5000	40	40	0	3	5,000	No
20	zoo	17	1	16	7	101	No

We have considered two kinds of oracles: Neural Networks and Boosting, using their implementations in the Weka data mining package (MultilayerPerceptron and AdaBoostM1, respectively). Also, the reference and the mimetic classifiers are constructed with the J48 algorithm included in Weka. The number of boost iterations is 10 in the AdaBoostM1 algorithm. In what follows, we denote the neural network oracle as NN, the Boosting oracle as Boost, the reference classifier as J48 and the mimetic classifier as Mim. Finally, when we show average results of many datasets, we will use the arithmetic mean of all datasets. For all the experiments, we use 10-fold cross-validation.

3 Analysis of Pruning

In this section we analyse how the quality metric and the mimetic classifier fidelity are affected by the use of different pruning degrees in the Mim algorithm. To do this, several experiments have been performed modifying the confidence threshold for pruning in the J48 algorithm; we have considered the following values: 0.0001, 0.001, 0.01, 0.02, 0.05, 0.10, 0.15, 0.20, 0.25, and 0.3. In all cases, we have fixed the size of the invented random dataset to 10,000.

Figure 2 shows the average results when the oracle is a neural network (Mim NN) and when the oracle is Boosting (Mim Boost). We have also included as reference the accuracy obtained by the oracles and by the J48 algorithm learned with the original training set.

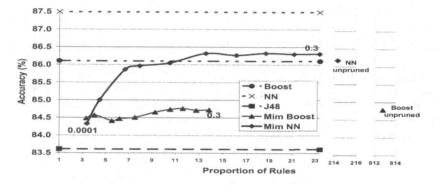

Fig. 2. Accuracy vs Proportion of rules of Mim for several degrees of pruning.

As we can see, Mim has a better behaviour when the oracle is the neural network. The reason is that the neural network is in average a better oracle than Boosting (in terms of accuracy). Also, both of them are better than J48. In Figure 2 we can see that the increase in accuracy practically reaches its maximum with a pruning degree of 0.01 at a proportion of rules around 7 times more than the reference J48 classifier. This "optimal" point is corroborated by the quality metric which is also maximum for this point.

From all the previous results, it seems that pruning, or at least the pruning method included in J48, gives poor manoeuvrability to get good accuracy results with fewer rules. For these datasets, the best qualities are obtained with an increase of almost 2 points in accuracy but with decision trees which are 7 times larger than the original ones. Hence, in the following section, we study the influence of other more interesting factors such as the invented dataset size.

4   Analysis of the Invented Dataset Size

Previous works on mimetic classifiers [2][3][4] have considered a fixed size for the invented dataset (usually between 1,000 or 10,000). However, it is clear that this value is relatively small for datasets such as "letter" and relatively large for datasets such as "hayes-roth". In this section we want to better analyse the relationship between the invented dataset size and the quality of the mimetic model (in terms of the quality metric defined in Section 1).

In order to study this factor, we have performed experiments with several sizes for the invented dataset, from $0.3n$ to $6n$, where n is the size of the training set. Each

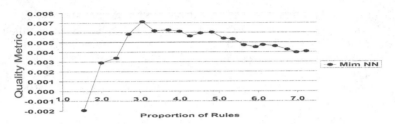

Fig. 3. Quality Metric vs Proportion of rules of Mim for several sizes.

increment is 0.3, making a total of 20 different sizes per dataset. We use the neural network as oracle and J48 as mimetic classifier.

Figure 3 shows how the quality metric evolves for increasing size of the invented dataset (the horizontal axis shows the proportion of rules). As we can see, there is a maximum point at (3.06, 0.0071), with an invented dataset of 1.5 times larger than the training dataset, which means that, as expected, very short datasets have very low accuracy but, on the other side, it is not beneficial to generate datasets that are too big.

Since accuracy does not grow linearly, it is clear that we have a saturation point for the quality metric as the one shown in the picture. However, this saturation point is not reached at the same point for each dataset. Hence, considering 1.5 as a good value for all datasets would not be a good choice.

Having this in mind, we try to estimate the optimal size for each dataset. In order to do this, first, with the previous experiments, we will determine the size of the invented dataset for which the quality metric gives the best value. With this, we will have a different best factor for each of the 20 datasets. What we will do next is to use this data for estimating a function that returns the size of the random dataset which would be optimal for a new dataset.

For this, we use the following variables for each of the 20 datasets: the number of nominal attributes (NomAttr), the number of numerical attributes (NumAttr), the number of classes (Classes) and the size of the training dataset (Size). The output of the function is the factor φ (size random/size train) with respect to the original training set. More formally, we want to estimate the following function:

$$\varphi = f(\text{NomAttr, NumAttr, Classes, Size})$$

Due to the small number of examples for this estimation (20 datasets) we have used a simple modelling technique: multiple linear regression (with and without independent coefficient). In this way, we can estimate the optimal invented dataset as follows:

$$n = \varphi \times Siz. \quad \varphi = 0.05097 \times NumAttr - 0.01436 \times NomAttr + 0.17077 \times Classes + 0.00003 \times Size$$

Table 2 shows the values estimates by the previous equation and the actual values for 5 fresh datasets.

As we can see, the estimation is not perfect, and the estimated values (Q n calc) are usually below the real values (Q max). However, if we look at the average values, the average quality obtained by using this estimation is significantly better (0.018) than that obtained by a fix invented dataset size of 150% which is 0.01. This corroborates the idea of considering an appropriate size for each dataset, depending on, at least, the previous factors (number of nominal and numeric attributes, number of classes, and, size of the training set).

Table 2. Estimated and real quality results for 5 fresh datasets. n max = Size when Q is maximum (actual). n calc = Size calculated with the formula (estimated). Q max = Actual quality metric. Q n calc = Estimated Q. Q 150% size = Q for fix invented dataset size of 150%.

No	Dataset	Attr	NumAttr	NomAttr	Classes	Size	n max	n calc	Q max	Q n calc	Q 150% size
1	Autos	25	15	10	7	205	1051	374	-0.01716	-0.03600	-0,02439
2	CarsW	6	0	6	4	1728	466	1124	0.14232	0.09200	0,05983
3	Credit-a	15	6	9	2	690	2235	372	-0.00042	-0.00335	-0,00084
4	Heart-c	13	6	7	5	303	81	324	0.07948	0.04000	0,01813
5	Hypothyroid	29	7	22	4	3772	1018	3172	-0.00132	-0.00230	-0,00219
	Average								0.04058	0.01807	0.010108

5 Use of Confidence in the Mimetic Method

When an invented dataset is used to learn a mimetic model, it is usually generated without taking into account the confidence of the oracle over this set of examples. We use the confidence of an example as the estimated probability of the predicted class. It seems reasonable to think that the quality of the mimetic model would improve if we use only those examples of the invented dataset for which the oracle gives a high confidence. In this section we analyse how the confidence of the invented dataset can be used to improve the mimetic technique.

In order to do this, once the invented dataset R has been constructed and added to the training set T, we process this set in the following way. First, we remove from $R+T$ all examples whose confidence value is below a confidence threshold t_c. Note that the examples of T are never removed because they have a confidence value of 1. Next, we remove the repeated examples. Finally, we duplicate the remaining examples a certain number of times depending on its confidence value. The resulting set, which we denote as D_{RT}, is used for training the mimetic model. The number of times that an example must occur in D_{RT} is defined as follows: let C_e be the confidence value of an example e and F a given repetition factor, then the number of occurrences of e in D_{RT} is $occ(e)= round(C_e \times F)$. Since the objective of the occ function is to determine whether the example must be duplicated or not, each example e for which $occ(e)=0$ remains in D_{RT}. Note that, as the repetition factor increases, examples with high confidence become more significant, and they may occur more than once in D_{RT}.

For the experiments the confidence thresholds t_c used were 0, 0.3, 0.9, 0.95, 0.98, 0.99 and 1.0, and for each one we used a repetition factor ranging from 1 to 4. The size of the invented dataset R was 10,000. Figure 4 shows these results.

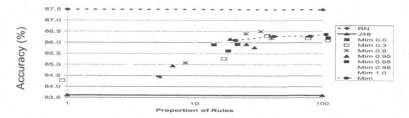

Fig. 4. Accuracy vs Proportion of Rules of Mim depending on a confidence and a repetition factor

We have observed that in the case of confidence threshold $t_c=1.0$ and repetition factor $F=1$, the process in general does not add invented examples to T (only adds a small number of invented examples for two datasets for which there were invented examples with confidence=1). In some problems, we even got smaller datasets D_{RT} than the original training sets. This is caused by the fact that some original datasets had repeated examples that were eliminated.

For the case of $t_c=0.99$ and $F=1$, we get a size of invented examples in D_{RT} around 3,000. If we contrast this value to a size of invented examples in D_{RT} around 7,000 when we use $t_c=0.0$ (3,000 invented examples approx. are removed because they are repeated), we see that an important percentage of examples are given a confidence $\geq$ 0.99 by the NN. Consequently, with $t_c=0.99$ we have an intermediate situation which is more on the left of Figure 4 than the original Mim. Additionally, the accuracy is almost totally reached with this case (86.1).

Regarding the repetition factor, the behaviour is quite similar for all cases, but has different interpretations. For instance, for $t_c=0.99$ and for $F=1$ all the remaining examples are included once and for $F=2$ all the remaining examples are included twice. The important accuracy increase between these two cases can be justified by the fact that J48 has a limitation on the minimum number of examples per node, and this duplication allows J48 to be more detailed.

To confirm these observations we show in Figure 5 the quality metric for these experiments. As we can see, the best quality metric is obtained using a threshold confidence of 0.99 (Mim 0.99) with a repetition factor of 2.

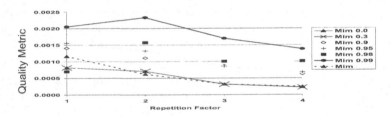

Fig. 5. Quality Metric vs Repetition Factor for different confidence threshold.

6   Combination of Factors

Finally, we made an experiment combining some of the results obtained in the previous experiments. We used pruning at 0.01 and 0.1, the size of the invented dataset was set to the value predicted by the estimated model in Section 4, and the repetition factor was set to 2 and the level of confidence to 0.99. Table 3 shows the results of these both scenarios.

The results with pruning level at 0.01 show that the three main factors considered (pruning, invented dataset size and relevance of the examples), if used together, can dramatically reduce the number of rules. In fact, the average results show that the number of rules is even below J48 with its default parameters. In this scenario, however, the increase in accuracy is mild (from 83.61 to 84.36). The picture changes when we see the results with pruning level at 0.1. In this case, accuracy increases to

84.68 with a size of the models which only rises to 2.52 times more rules than the original J48 model. The quality is 0.0084.

Table 3. Experimental results obtained by the combination of factors.

No. Dataset	NN Acc	J48 Acc	J48 Rules	Mim 0.01 Acc	Mim 0.01 Rules	Mim 0.01 Ratio	Mim 0.1 Acc	Mim 0.1 Rules	Mim 0.1 Ratio
1	98.89	98.56	39.50	98.18	48.60	1.23	98.39	53.05	1.34
2	83.21	77.33	30.20	85.05	53.30	1.77	85.76	53.10	1.76
3	90.84	78.40	39.60	77.88	32.83	0.83	79.60	52.45	1.32
4	67.96	74.08	7.50	70.39	3.17	0.42	72.17	19.15	2.55
5	50.86	51.57	155.70	54.51	47.67	0.31	51.69	228.55	1.47
6	81.94	85.13	5.50	84.40	5.20	0.95	85.44	6.25	1.14
7	74.42	74.19	19.20	74.53	27.20	1.42	72.53	63.50	3.31
8	81.20	68.58	19.00	74.74	22.63	1.19	77.25	24.65	1.30
9	80.06	79.43	9.40	79.35	2.27	0.24	79.63	5.55	0.59
10	96.81	94.96	4.70	95.33	4.77	1.01	94.67	4.65	0.99
11	82.08	87.98	1,158.10	87.65	1,037.20	0.90	85.69	16,655.10	14.38
12	100.00	97.12	30.10	100.00	28.00	0.93	100.00	28.00	0.93
13	100.00	63.29	24.50	65.72	1.00	0.04	65.72	1.00	0.04
14	98.49	98.92	14.00	96.95	9.97	0.71	98.92	13.70	0.98
15	100.00	100.00	25.00	99.90	90.10	3.60	99.98	161.70	6.47
16	96.84	98.68	28.60	98.32	10.65	0.37	98.32	10.65	0.37
17	94.49	96.55	5.80	95.49	2.33	0.40	96.22	5.70	0.98
18	93.15	79.75	128.00	79.50	155.75	1.22	82.93	494.9	3.87
19	95.02	92.39	8.30	92.82	12.03	1.45	92.68	14.15	1.70
20	83.54	75.36	290.70	76.54	177.10	0.61	76.02	1450.2	4.99
Avg.	87.49	83.61		84.36		0.98	84.68		2.52

7 Discussion and Conclusions

Summing up, from previous works and after the analysis on some of the separated factors (especially pruning), it seemed that it was almost impossible to improve the quality metric. Reducing the number of rules systematically entailed a reduction of accuracy and vice versa. However, the study of factors such as the size of the invented dataset and the modification of the distribution of examples are better tools to maintain significant improvements in accuracy while significantly reducing the number of rules. These final combined results suggest that there is still margin to pursue in this line, and that good compromises can be found, turning the mimetic technique originally introduced by Domingos, into a real useful and general technique for knowledge discovery.

Additionally, this work provides a further insight on how mimetic classifiers work. The use of the confidence of the oracle in order to modify the distribution of examples is one of the main new contributions of this work and suggests that the increase in number of rules can be partially due to overfitting to low-confidence examples generated by the oracle.

Finally, as future work, we would like to investigate several issues. For instance, instance selection methods could be useful for reducing the size of the invented dataset. The evaluation of mimetic models with other metrics, such as AUC (Area Under the ROC Curve), would also be interesting, since decision trees have better AUCs when the tree is not pruned [6]. Another issue to study would be to analyse the use of confidence without the training set, thus making the mimetic technique more generally applicable.

References

1. Black C. L.; Merz C. J.UCI repository of machine learning databases, 1998.
2. Domingos, P. Knowledge Discovery Via Multiple Models. *Intelligent Data Analysis*, 2(1-4): 187-202, 1998.
3. Domingos, P. Learning Multiple Models without Sacrificing Comprehen-sibility, *Proc. of the 14th National Conf. on AI*, pp:829, 1997.
4. Estruch, V.; Ferri, C.; Hernandez-Orallo, J.; Ramirez-Quintana, M.J. Simple Mimetic Classifiers, *Proc. of the Third Int. Conf. on Machine Learning and Data Mining in Pattern Recognition*, LNCS 2734, pp:156-171, 2003.
5. Estruch, V.; Hernández-Orallo, J.: Theoretical Issues of Mimetic Classifiers, TR DSIC, http://www.dsic.upv.es/~flip/papers/mim.ps.gz, 2003.
6. Provost, F.; Domingos, P. Tree induction for probability-based rankings, *Machine Learning*, 52 (3): 199-215, 2003.

Generating AVTs Using GA for Learning Decision Tree Classifiers with Missing Data*

Jinu Joo[1], Jun Zhang[2], Jihoon Yang[1], and Vasant Honavar[2]

[1] Department of Computer Science, Sogang University
1 Shinsoo-Dong, Mapo-Ku, Seoul 121-742, Korea
jujoo@mllab.sogang.ac.kr, jhyang@ccs.sogang.ac.kr
[2] Artificial Intelligence Research Laboratory, Department of Computer Science
Iowa State University, Ames, IA 50011 USA
{jzhang,honavar}@cs.iastate.edu

Abstract. Attribute value taxonomies (AVTs) have been used to perform AVT-guided decision tree learning on partially or totally missing data. In many cases, user-supplied AVTs are used. We propose an approach to automatically generate an AVT for a given dataset using a genetic algorithm. Experiments on real world datasets demonstrate the feasibility of our approach, generating AVTs which yield comparable performance (in terms of classification accuracy) to that with user supplied AVTs.

1 Introduction

In a typical inductive learning scenario, instances to be classified are represented as ordered tuples of attribute values. However, attribute values can be grouped together to reflect assumed or actual similarities among the values in a domain of interest or in the context of a specific application. Such a hierarchical grouping of attribute values yields an attribute value taxonomy (AVT) [1]. For example, Fig. 1 shows an AVT defined over color. Under *color* node there are more specific values defined which have more specific values defined under each sub node. In one instance the color attribute may be specified as sky blue while in another instance it can be described as just blue.

AVTs have been exploited in learning classifiers from data due to its various advantages [1]. In particular, the AVTs are used in AVT based decision tree learning algorithm (AVT-DTL) [2] which classifies partially missing and totally missing values more efficiently than traditional algorithms such as ID3 [3] and C4.5 [4].

An AVT needs to be specified a priori in a classification task. However, in many domains, AVTs specified by human experts are unavailable. Even when a human-supplied AVT is available, it is interesting to explore whether alternative groupings of attribute values into an AVT might yield more accurate or

---

* This research was supported in part by grants from the National Science Foundation (grant 021969) and the National Institutes of Health (GM066387) to Vasant Honavar.

E. Suzuki and S. Arikawa (Eds.): DS 2004, LNAI 3245, pp. 347–354, 2004.

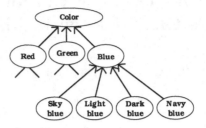

Fig. 1. An AVT on color.

more compact classifiers. Against this background, we propose a method that produces an AVT automatically for a given dataset. Due to the huge search space of possible AVTs, we need to consider methods (or algorithms) that finds *near-optimal* solutions. Here we resort to a genetic algorithm to automate the generation of AVTs for AVT-DTL algorithm on datasets with missing values.

2   AVT-Guided Decision Tree Learning

AVT-DTL algorithm [2] classifies partially specified data based on a user provided AVT. This algorithm performs a AVT-guided hill climbing search in a decision tree hypothesis space. AVT-DTL works top-down starting at the root of each AVT and builds a decision tree that uses the most abstract attribute values that are sufficiently informative for classifying the training set consisting of the partially specified instances. It is straightforward to extract classification rules from the set of pointing vectors associated with the leaf nodes of the decision tree constructed by AVT-DTL. For example, Fig. 2 shows pointing vectors that points to two high-level attribute values *blue, polygon* in the two taxonomies of *Color* and *Shape*. If this pointing vector appears in the leaf node of the decision tree with class label +, the corresponding rule will be: If (*Color=blue Shape=polygon* ...) then Class =+. See [2] for detailed descriptions on AVT-DTL.

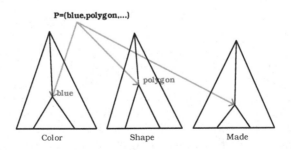

Fig. 2. Pointing vector.

3 GA Based AVT Generation

3.1 Genetic Algorithm

Genetic algorithm (GA) [5–7] is an efficient nonlinear heuristic search algorithm, only using simple operators like mutation, crossover, and selection. It is an evolving iterative process working with a solution set called *population*. A population is a set of solutions called *chromosomes*. In each generation different chromosomes form a population, and each population contains better chromosomes than the former generation. By using operators like selection, crossover, and mutation the population set evolves to converge to the optimal solution. During these operations, the fitness function is used to evaluate the *fitness* (or *appropriateness* or *propriety*) value of each individual. Our GA-based approach for generating AVTs proceeds as follows:

GA-AVT

1. Create the initial population of an AVT, where each chromosome is a bit string representation and each attribute is represented by a set of *blocks*. Blocks are binary bits which represent node bindings on their level in the tree. More details on blocks are described in Section 3.2.
2. Evaluate the fitness of each individual in the population set. The fitness function is the classification accuracy estimated by 10-fold cross validation.
3. Enter the population in the roulette wheel selection.
4. Execute the crossover and mutation operation on the roulette wheel selected population.
5. Repeat step 1 through 5 until a certain number of iterations are reached or when the solution does not improve.

3.2 AVT Representation

To modify and generate AVTs in GA, we need to form an appropriate representation for AVTs. We introduce a binary string format to represent an AVT. The difficulty of finding a good representation for an AVT is first, an AVT has a certain number of leaf nodes that never change. Second, the traditional representation of trees would cause a duplication of nodes after genetic operation. Therefore we devised a new type of chromosome in binary string that represents an AVT which satisfies above two constraints.

Fig. 3 is an abstract structure of how a chromosome is formed with *genes* and blocks. Inside a chromosome there are n genes where n is the number of attributes. Each gene represents its attribute's AVT respectively (i.e. in *nursery* dataset, the first gene represents the AVT of the first attribute, *parents occupation*). A gene consists of a set of bit strings called blocks. Again there are m blocks in one gene where m is the logarithm of leaves. In other words, If a_i is the i_{th} attribute and l_{a_i} represents the number of its leaf nodes, then the number of *blocks* b_{a_i} is $b_{a_i} = \lfloor \log_2 l_{a_i} \rfloor$. Finally, a block contains k binary bits where $k = l_{a_i}$.

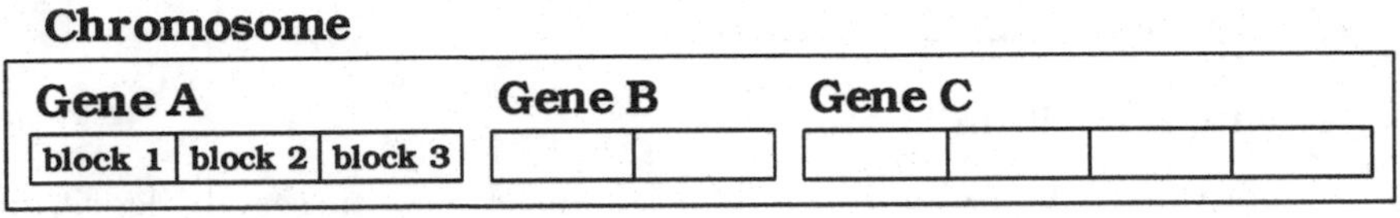

Chromosome

Gene A			Gene B		Gene C		
block 1	block 2	block 3					

Fig. 3. Structure of a Chromosome.

Taking this newly defined chromosome, we can build an AVT for a fixed number of leaf nodes. If h_j is the block number of the i_{th} attribute and has a range of $1 \leq h_j \leq b_{a_i}$, then each $block_{h_j}$ represents the information of how the nodes or subtrees should be grouped in step j. Building AVTs with blocks should start from the block with $j = 1$ and end when j reaches b_{a_i}. Bit strings in a block tells how to group nodes and subtrees by comparing a bit from its adjacent bits. For instance , the k_{th} bit will be compared with $(k-1)_{th}$ bit and $(k+1)_{th}$ bit. If the k_{th} bit is the same bit as the $(k+1)_{th}$ bit, the corresponding leaf nodes are considered to be grouped and have the same parent abstract node. If the k_{th} bit has a toggle difference compared to the $(k+1)_{th}$ bit, the two corresponding leaf nodes stay separate. When comparing subtrees, the bit representing the corresponding subtree is the first starting bit of the tree representation. If a subtree corresponds to bits starting from k to $k + p$ (where $1 \leq p \leq l_{a_i} - k$), then bit k will be the representative bit of that subtree to compare with other bits excluding bits k to $k + p$. By executing each step of blocks until h_j reaches b_{a_i} will accomplish an AVT structure.

The example below demonstrates how to group up an AVT with binary bit blocks 000110 - 0XX1X1 corresponding to a certain sorted leaf node set a, b, c, d, e, f. The full process is illustrated in the following sequence. (X represents don't care bits which means either 0 or 1 is available in this position, and hyphen distinguish one block from another.):

1. Consider the first block, 000110. The first three nodes a, b, c will group to a certain abstract node since the first three bits are all 0's. Hence d, e and f will group up in a similar way. Let us name the abstract node that groups $\{a, b, c\}$, A; $\{d, e\}$, B; and f stands alone. Therefore the nodes considered on the next step will be A, B, and f (Fig. 4a).

2. To create the next level of the tree we consider the second block, 0XX1X1. Since nodes a, b, c are child nodes to abstract node A, and d, e are to B, which forms a subtree, only first, fourth and sixth bits in $block$ 2 are needed to proceed the next step. Because the second block's first and fourth bits turn out to be different, nodes A and B don't group for now. In a similar manner the fourth and sixth bits are compared, and the two nodes B and f are combined since the two bits are equal (Fig. 4b).

3. Finally since there are no other $blocks$ following, the final remaining nodes A and C are grouped to the root node which is the attribute name (Fig. 4c).

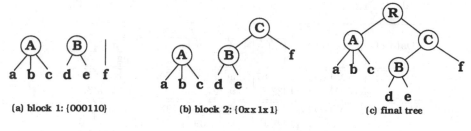

(a) block 1: {000110}   (b) block 2: {0xx1x1}   (c) final tree

Fig. 4. Decoding AVT.

3.3   Genetic Operators for GA

A one point crossover operation is applied to every block respectively. This type of crossover is necessary because each block contains the information of how nodes are to be constructed on the level of the tree. Selection operation is basically a roulette wheel selection. But since classification accuracy (or fitness value) might differ only slightly from individual to individual, adjusting these values are necessary. Here adjustment was performed by first searching the chromosome that has the minimum fitness value in the population. Second, multiply the fitness value difference between the original fitness value and the minimum fitness value by three. This fitness value adjustment is performed to all individuals. Mutation operation is performed with mutation rate 0.01. Once a chromosome is selected to perform a mutation, one bit per every block is randomly chosen to be mutated.

4   Experiment

4.1   Datasets and Experimental Setup

Several experiments were performed to explore the performance of the GA generated AVT (GA-AVT) compared to the predefined AVT given by Zhang and Honavar (*Nursery*) [2] and Taylor *et al.* (*Mushroom*) [8]. In each case, the error rate was estimated by 10-fold cross validation. The experiments compared error rates from the original AVT with GA generated AVT. The GA-AVT was generated with datasets containing totally missing values in this experiment. Experiments were conducted with different pre-specified percentage (0%, 5%, 10%, 20%, 30%, 40%, or 50%) of totally missing attribute values. The GA parameters were set as, population: 20; max generation: 50; mutation rate: 0.01; crossover rate: 0.6; number of elite chosen: 2.

4.2   Experimental Results

We obtained the results shown in (Table 1). In the case of *Nursery* data, the error rate of GA-AVT was somewhat lower than the error rate of the original AVT, especially when missing attribute percentage was high. In general, GA-AVT has

Table 1. Error rate estimation with AVT-DTL and GA-AVT-DTL.

Dataset	Method	Missing percentage						
		50%	40%	30%	20%	10%	5%	0%
		% Error rates estimated using 10-fold						
Nursery	AVT-DTL	39.87	33.50	27.82	21.20	12.94	8.37	2.90
	GA-AVT-DTL	37.72	32.98	27.29	21.07	12.52	14.73	3.02
Mushroom	AVT-DTL	6.04	3.87	2.33	1.60	0.64	0.26	0
	GA-AVT-DTL	6.01	3.63	2.39	1.50	0.60	0.19	0

better or similar performance compared with the original AVTs. Because AVT-DTL itself has a long computation time, average of 2 to 3 minutes, running GA AVT-DTL to search for AVTs will give 33 to 34 hours based on our experimental environment of 20 populations and 50 generations. However once the AVT is generated by GA, this AVT will have chances to be reused in different datasets of the same domain. Therefore in the long run using GA to generate AVTs will provide more accurate decision trees than using a user-supplied AVT. Fig. 5 is an example graph of the evolution of AVTs evolving by GA with datasets of missing value at 50%. As we can see in the graph, best solutions were attained within 5 generations while average fitness keeps increasing.

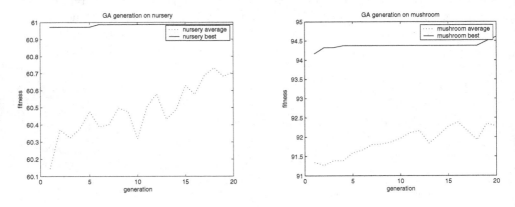

Fig. 5. AVT evolving by GA (datasets of missing value at 50%).

5   Related Work

Cimiano et al. [9, 10] used agglomerative clustering for learning taxonomies from text. Gibson and Kleinberg [11] introduced STIRR, an iterative algorithm based on non-linear dynamic systems for clustering categorical attributes. Ganti et al. [12] designed CACTUS, an algorithm that uses intra-attribute summaries to cluster attribute values. However both of them didn't make taxonomies and used the generated for improving classification tasks. Pereira et al. [13] describe distributional clustering for grouping words based on class distributions associated with the words in text classification. Yamazaki et al. [14] described an algorithm for extracting hierarchical groupings from rules learning by FOCL (an inductive learning algorithm) [15] and report improved performance on learning translation rules from examples in a natural language processing task. Slonim and Tishby [16, 17] described a technique (called the agglomerative information bottleneck method) which extends the distributional clustering approach described by Pereira et al. [13], using Jensen-Shannon divergence for measuring distance between document class distributions associated with words and applied it to a text classification task. Baker and McCallum [18] report improved performance on text classification using a technique similar to distributional clustering and a distance measure, which upon closer examination, can be shown to be equivalent to Jensen-Shannon divergence [16]. Zhang and Honavar recently proposed an algorithm (AVT-Learner) for automated construction of AVTs from data that uses Hierarchical Agglomerative Clustering (HAC) to cluster attribute values based on the distribution of classes that co-occur with the values [1].

6   Summary and Future Work

Through this paper we have presented the need for automatic generation of AVTs, and introduced a GA based approach. We have performed two experiments with *Nursery* and *Mushroom* datasets to generate optimal AVTs and compared them with the original AVTs. Experimental results demonstrate the feasibility of our approach in terms of the classification accuracy. In particular, our GA-based automatic induction of AVTs obviate the need for ad hoc construction of AVTs that are possibly inefficient and inaccurate.

Some directions for future work include: First, improve the representation of AVTs and corresponding genetic operators to make a more effective GA model in generating AVTs. Second, adjust the AVT-DTL algorithm to achieve better performance and running time. Third, conduct well-designed experiments (possibly with more real-world data), and compare the performance (e.g. classification accuracy, tree size, etc.) of GA-based approach with that of other methods (e.g. AVT-Learner). Fourth, develop GA-AVT combined with a variety of machine learning algorithms (e.g. Naive Bayes classifier).

References

1. Kang, D.K., Silvescu, A., Zhang, J., Honavar, V.: Generation of attribute value taxonomies from data and their use in data driven construction of accurate and compact naive bayes classification. In: Proceedings of the ECML/PKDD Workshop on Knowledge Discovery and Ontologies. (2004)
2. Zhang, J., Honavar, V.: Learning decision tree classifiers from attribute value taxonomies and partially specified data. In: Proceedings of the Twentieth International Conference on Machine Learning (ICML 2003). (2003)
3. Quinlan, R.: Induction of decision trees. Machine Learning 1 (1986) 81–106
4. Quinlan, R.: C4.5: Programs for Machine Learning. Morgan Kaufmann, San Mateo, CA (1992)

5. Mitchell, M.: An Introduction to Genetic algorithms. MIT Press, Cambridge, MA (1996)
6. Goldberg, D.: Genetic Algorithms in Search, Optimization, and Machine Learning. Addison-Wesley, New York (1989)
7. Yang, J., Honavar, V.: Feature subset selection using a genetic algorithm. IEEE Intelligent Systems **13** (1998) 44–49
8. Taylor, M., Stoffel, K., Hendler, J.: Ontology-based induction of high level classification rules. In: SIGMOD Workshop on Research Issues on Data Mining and Knowledge Discovery. (1997)
9. Cimiano, P., Staab, S., Tane, J.: Automated acquisition of taxonomies from text: Fca meets nlp. In: Proceedings of the ECML/PKDD Workshop on Adaptive Text Extraction and Mining, Cavtat-Dubrovnik, Croatia. (2003) 10–17
10. Cimiano, P., Hotho, A., Staab, S.: Comparing conceptual, partitional and agglomerative clustering for learning taxonomies from text. In: Proceedings of the European Conference on Artificial Intelligence(ECAI'04). (2004)
11. Gibson, D., Kleinberg, J., Raghavan, P.: Clustering categorical data: An approach based on dynamical systems. VLDB Journal:Very Large Data Bases **8** (2000) 222–236
12. Ganti, V., Gehrke, J., Ramakrishnan, R.: Cactus - clustering categorical data using summaries. In: Proceedings of the fifth ACM SIGKDD international conference on Knowledge discovery and data mining, ACM press (1999) 73–83
13. Pereira, F., Tishby, N., Lee, L.: Distributional clustering of english words. In: 31st Annual Meeting of the ACL. (1993) 183–190
14. Yamazaki, T., Pazzani, M., Merz, C.: Learning hierarchies from ambiguous natural language data. In: International Conference on Machine Learning. (1995) 575–583
15. Pazzani, M., Kibler, D.: The role of prior knowledge in inductive learning. Machine Learning **9** (1992) 54–97
16. Slonim, N., Tishby, N.: Agglomerative information bottleneck. In: NIPS-12. (1999)
17. Slonim, N., Tishby, N.: Document clustering using word clusters via the information bottleneck method. In: Proceedings of the 23rd annual international ACM SIGIR conference on Research and Development in Information Retrieval, ACM press (2000) 208–215
18. Baker, L.D., McCallum, A.K.: Distributional clustering of words for text classification. In: Proceedings of the 21rd annual international ACM SIGIR conference on Research and Development in Information Retrieval, ACM press (1998) 96–103

Using WWW-Distribution of Words in Detecting Peculiar Web Pages

Masayuki Hirose and Einoshin Suzuki

Electrical and Computer Engineering, Yokohama National University, Japan
mayusaki@slab.dnj.ynu.ac.jp, suzuki@ynu.ac.jp

Abstract. In this paper, we propose TFIGF, a method which detects peculiar web pages using distribution of words in WWW given a set of keywords. Our TFIGF detects a set of index words which represent a WWW page by estimating their importance in the WWW page and their rareness in WWW. Experiments using both English and Japanese WWW pages clearly show superiority of our approach over a traditional method which employs a limited number of WWW pages in the estimation.

1 Introduction

"Web mining" refers to discovery and extraction of knowledge from WWW with data mining techniques [7]. Web content mining [7], which represents a sub-category of Web mining, employs content of WWW pages as a resource, and can elucidate useful knowledge as well as detect interesting WWW pages. In web mining, we believe that detection of peculiar web pages can lead to valuable information. Even in a customary topic, a rare instance can lead to new business or new investment in order to make profit. Early detection of a unique technology or a service is expected to be valuable from technological or commercial viewpoint.

Conventional information retrieval techniques such as the TF-IDF method make use of word distribution in given document data for prediction. Such methods heavily rely on the precision of the word distribution in terms of predictive performance. We believe that using a word distribution on WWW through a search engine improves such performance. In this paper, we propose a method for detecting peculiar WWW pages with the word distribution on WWW. Examples found by our method include a wheel chair for a dog, a wheel chair made of bamboo, a gag strip for Hamlet, and an auto-parallel-parking device by Toyota[1].

2 Conventional Method

2.1 Problem Statement

A search engine accepts a set W of keywords as input and outputs the number of WWW pages each of which contains W and a ranked list of links to the WWW

---

[1] We believe that we need to analyze context in order to detect peculiar WWW pages for a more intellectual topic such as research themes.

E. Suzuki and S. Arikawa (Eds.): DS 2004, LNAI 3245, pp. 355–362, 2004.

pages. We represent the set of all WWW pages considered by a search engine and its subset of top a WWW pages by $D(W)$ and $D_a(W)$ respectively.

This paper tackles the problem of obtaining a set $D'(W_{\text{key}})$ of peculiar WWW pages from a set W_{key} of keywords given to a search engine. Here $D'(W_{\text{key}}) \subseteq D(W_{\text{key}})$ and whether a WWW page is peculiar is decided by the user. In this paper, the decision is made based on multiple-occurring words which seldom co-occur with the keyword but are relevant to the content of the WWW page.

2.2   Method Based on Collected WWW Pages

The problem in the previous section can be resolved by collecting the top a WWW pages $D_a(W_{\text{key}})$ and using the word distribution in $D_a(W_{\text{key}})$. We explain the TF-IDF method which is frequently used in information retrieval and introduce a method based on it for comparative purpose. It should be noted that this method employs only text in a WWW page thus neglects link information as well as multimedia data.

In the TF-IDF method, a WWW page $d_i \in D_a(W_{\text{key}})$ is modeled as a bag of words. Let n_{ij} be the number of occurrences of a word w_j in d_i, then a word frequency $f(d_i, w_j)$ represents the frequency of w_j in d_i.

$$f(d_i, w_j) = \frac{n_{ij}}{\sum_{j=1}^{m} n_{ij}} \qquad (1)$$

where $D_a(W_{\text{key}})$ contains m different kinds of words excluding stop words.

When w_j appears in many WWW pages in $D_a(W_{\text{key}})$, w_j can be considered as general and its $f(d_i, w_j)$ is typically large. The document frequency $g_a(w_j)$, which is defined as follows, intuitively represents generality of w_j in $D_a(W_{\text{key}})^2$.

$$g_a(w_j) = \frac{|D_a(W_{\text{key}} \cup \{w_j\})|}{|D_a(W_{\text{key}})|} \qquad (2)$$

where $|S|$ represents the cardinality of a set S. The TF-IDF value $h_a(d_i, w_j)$ represents, intuitively, importance of w_j in d_i. Several researchers use the inverse of $g_a(w_j)$ in its definition, but we here use a value normalized by its logarithm in order to reduce difference among words.

$$h_a(d_i, w_j) = f(d_i, w_j) \left[1 + \log\left(\frac{1}{g_a(w_j)}\right) \right] \qquad (3)$$

In this framework, d_i is modeled as follows.

$$d_i = (h_a(d_i, w_1), h_a(d_i, w_2), \cdots, h_a(d_i, w_m)). \qquad (4)$$

---

[2] We admit that we implicitly make several assumptions such as the search engine returns all documents which contain all keywords and $D_a(W_{\text{key}} \cup \{w_j\}) = D_a(W_{\text{key}}) \cup D_a(\{w_j\})$. A typical search engine employs a sophisticated algorithm thus is likely to violate these assumptions. We have, however, experienced no serious problems since our application typically employs at most a few words for W_{key}.

In information retrieval, the dissimilarity between a document d_i and a document d_j is typically represented by the angle of the corresponding vectors $(h_a(d_i, w_1) \; h_a(d_i, w_2), \cdots, h_a(d_i, w_m))$ and $(h_a(d_j, w_1), h_a(d_j, w_2), \cdots, h_a(d_j, w_m))$. We initially employed a method which clusters WWW pages using this dissimilarity measure and outputs WWW pages in small clusters as peculiar. Preliminary experiments, in which we used 3, 4, 5 as the number of clusters, showed that this method exhibits relatively low predictive performance.

Therefore, we considered to detect peculiar WWW pages by selecting a set of index words for characterizing each WWW page d_i and using rareness of these words in $D_a(W_{\text{key}})$. In this method, which we call TFIDF, the set of b words with the highest TF-IDF value in d_i is selected as index words[3]. If each w of these words has a relatively small $g_a(w)$, intuitively speaking if each w is rare in $D_a(W_{\text{key}})$, d_i can be regarded as peculiar. This method predicts d_i as peculiar if the sum of $g_a(w_j)$ weighted and normalized by its TF-IDF value $h_a(d_i, w_j)$ is smaller than a user-given threshold θ.

$$\frac{\sum_{j=1}^{b} g_a(w'_j) h_a(d_i, w'_j)}{\sum_{j=1}^{b} h_a(d_i, w'_j)} < \theta \Rightarrow d_i \text{ is peculiar} \tag{5}$$

Here w'_j represents a word that has the jth largest g_a.

2.3 Problems Specific to WWW

The word frequency in the collected set $D_a(W_{\text{key}})$ of WWW pages can be against the user's intuition. For instance, the set $D_{100}(\{\text{yokozuna}\})$ of the top 100 WWW pages contains 9 WWW pages each of which also contains "software". For a typical user, the term "software" is irrelevant to the term "yokozuna", which originally represents a grand champion of Japanese wrestling. Actually, with the search engine Google (http://www.google.com/), only 2,000 WWW pages contain both "yokozuna" and "software" although 40,000 and 18,900,000 WWW pages contain "yokozuna" and "software" respectively. In this case, the predictive performance of the method TFIDF in the previous section will be low if we settle $a = 100$. This problem comes from the fact that TF-IDF values are obtained from collected a WWW pages. Although increasing the value of a and making word frequencies more similar to those on the WWW is expected to reduce the problem, it is inefficient to collect a large number of WWW pages.

Another problem with TFIDF comes from its specification of stop words. If such stop words mismatch the user's intuition and/or tendencies on the WWW, the predictive performance will be low. For instance, terms such as "home" and "link" are not considered as stop words in a typical text classification, but occur frequently on WWW.

---

[3] Several web content mining methods such as [1] employ selection of index words based on TF-IDF value.

3   TFIGF: Proposed Method

In order to overcome the problems in the previous section, we have considered to employ a search engine for estimating word frequencies on WWW more accurately. A search engine would enable us to estimate such frequencies accurately without excessive costs for storing WWW pages. Our proposed method TFIGF employs a WWW frequency $g(w_j)$ instead of the document frequency $g_a(w_j)$ in Eq. (3) and Eq. (5)[4]. Since TFIDF and TFIGF estimate word frequencies from a documents and WWW, they employ g_a and g respectively.

We define the WWW frequency $g(w_j)$ for a word w_j contained in W_{key} as follows, where W' is a non-empty set of words.

$$g(w_j) = \frac{|D(W_{\text{key}} \cup \{w_j\})|}{\min(|D(W_{\text{key}})|, |D(\{w_j\})|)} \tag{6}$$

Intuitively, the WWW frequency $g(w_j)$ represents generality of a word w_j in the set $D(W_{\text{key}})$ of WWW pages. We expect that the WWW frequency $g(w_j)$ fits the user's intuition better than the document frequency $g_a(w_j)$, which is based on the top a WWW pages. Eq. (6) employs a minimum function in the denominator since the values for $|D(\{W_{\text{key}}\})|$ and $|D(\{w_j\})|$ can be substantially different and we want to reduce such effect. For example, the term "TCLA" (Toyota Land Cruiser Association) is relatively common in WWW pages with Toyota but is rare in WWW. If we substitute $|D(W_{\text{key}})|$ for $\min(|D(W_{\text{key}})|, |D(\{w_j\})|)$ in Eq. (6), $g(\text{TCLA})$ will be low and "TCLA" is mis-regarded as a rare word in WWW pages with Toyota. This will have negative effects since a WWW page which contains "TCLA" is likely to be judged as peculiar[5].

In order to reduce the second problem in section 2.3, we consider that a word which appears in many WWW pages rarely characterizes a WWW page. In our TFIGF, a word w is considered as a stop word when the number $|D(\{w\})|$ of WWW pages with w obtained with a search engine is above a predefined threshold value ξ.

$$|D(\{w\})| \geq \xi \Rightarrow w \text{ is a stop word} \tag{7}$$

4   Experimental Evaluation

4.1   Experimental Settings and Results

In the experiments, we use Google as the search engine since it is widely used. Five kinds of data have been collected by giving "wheel-chair + specification",

---

[4] Searching the document frequency of all non-stop words is not as slow as expected. In our experiments, the most time-consuming WWW page took less than 30 minutes in execution time. The page contains 762 kinds of words thus it took about 2.35 seconds for each kind of word in this case.

[5] On the other hand, replacing the denominator of Eq. (2) with $\min(|D_a(W_{\text{key}})|, |D_a(\{w_j\})|)$ makes $g_a(w_j) = 1$ for all w_j.

Table 1. Data sets used in the experiments, where # represents number.

# of pages	Japanese		English		
	wheel-chair + specifications	Hamlet	yokozuna	castle	toyota
all	86	94	86	94	70
peculiar	10	19	28	28	20

"Hamlet", "yokozuna", "castle", and "toyota" as keywords to the search engine, where a "+" represents that we performed an AND search[6]. In order to investigate the difference in languages, we gave the former two sets of keywords in Japanese and collected only WWW pages in Japanese while we used English in the latter three sets. Table 1 shows the data where we used $a = 100$ in order to collect WWW pages[7]. In the rest of this paper, we use these data sets in order to evaluate predictive performance of methods.

We performed morphological analysis for data in Japanese and split the text into words[8] while we used stemming for data in English. Stop words were identified based on the publicly available list of the SMART system [9] in TFIDF. In our TFIGF, we settled the value of the threshold $\xi = 10,000,000$ and $15,000,000$ for Japanese data and English data respectively.

We compared the two methods with $b = 10$, $\theta = 0.1$ based on their recalls and precisions[9]. In TFIDF, the number a of WWW pages is varied $a = 100, 200, \cdots, 700$. Since we discarded WWW pages without text content such as a WWW page which contains only links, the actual number of WWW pages employed by the method is a little smaller than a.

Figures 1 - 5 show the experimental results. From the figures, we see that our TFIGF outperforms TFIDF in precision in all cases and in recall in most of the cases. These empirical facts suggest that our TFIGF is more effective than TFIDF in detection of peculiar WWW pages. We attribute these results to our improved estimation of generality of a word. Using only a WWW pages can lead to a nonintuitive estimation of word distribution as we explained in Section 2.3.

The precisions in a Japanese problem (Figures 1 and 2) are lower than those in an English problem. Compared with a detection problem in English, a detection problem in Japanese has an additional problem of detecting word separation. In Japanese, several compound words are peculiar though their subwords are common, and our use of morphological analysis fails to recognize such peculiarity. For instance, in the problem of "wheel-chair + specification", both methods failed to recognize a WWW page with "auto-will-transmission-device" as peculiar.

---

[6] The keywords were chosen arbitrarily.

[7] The value 100 is chosen since Google can display links to at most 100 WWW pages on a screen as its result of search.

[8] Text in Japanese language is not split in words in its original form.

[9] Unfortunately performance of the methods depend on the values of the parameters. Since we settled their values after several working examples, the experimental results are subject to overfitting. Automatic determination of parameter values for a wide range of users is a challenging task.

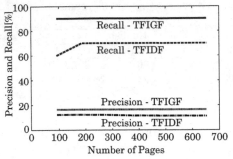

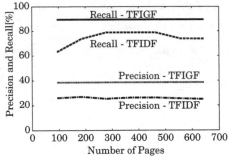

Fig. 1. Experimental results for the "wheel-chair + specifications" data set.

Fig. 2. Experimental results for the "Hamlet" data set.

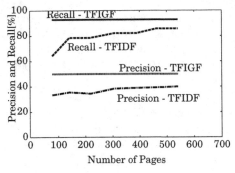

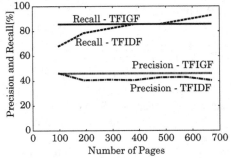

Fig. 3. Experimental results for the "yokozuna" data set.

Fig. 4. Experimental results for the "castle" data set.

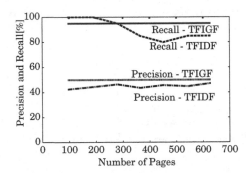

Fig. 5. Experimental results for the "toyota" data set.

We also performed experiments without using stop words removal in our TFIGF and obtained lower recall in all cases, lower precision twice, approximately the same precision twice, and higher precision once. The degradation in recall can be explained by the fact that remaining stop words can be mis-regarded as relevant and hinder detection of peculiar WWW pages since their document frequencies are relatively large. We therefore conclude that our stop word removal is effective at least in our experiments.

4.2 Detailed Analysis of the Experimental Results

In "castle", TFIDF performs poorly when a is small since it tends to recognize stop words excessively thus several WWW pages have less than ten index words. However, our TFIGF is outperformed by TFIDF in recall when a is large since the latter correctly classifies two WWW pages unlike the former. For instance, our TFIGF failed to predict a WWW page of the Lower House of the United States (http://www.house.gov/) as peculiar unlike TFIDF due to a problem related with geographical names. We consider that this WWW page is peculiar with a list of congresspersons and the names of their states. Both methods identify the name of the states as index words, but our TFIGF considers them as common. The reason is that the term "castle" appears frequently with the names of the states in lower-ranked WWW pages since many geographical names contain "castle" e.g. $D(\{castle + california\}) = 1.8$ million. Such usage is rare in higher-ranked WWW pages thus TFIDF correctly assumes that the names of states are rare in this problem.

The TFIDF is also superior to our TFIGF only in recall when a is small in toyota. The reason is that our method recognizes stop words excessively for a short WWW page, which comes to have less than ten index words. However, our stop word detection is typically superior to that of TFIDF, which is shown by our superior performance when a is large. For instance, our TFIGF succeeded in predicting a WWW page of a news article on a Toyota's robot (http://www.channelnewsasia.com/) as peculiar unlike TFIDF due our stop word detection method. The WWW page is considered as peculiar since this humanoid robot is much less known than Honda's ASIMO. Dtailed statistics show that TFIDF failed to recognize the peculiarity since it failed to detect stop words "news" and "market" inherent in WWW unlike our TFIGF. This analysis is justified since $D(\{market\}) = 89,200,000$ and $D(\{news\}) = 442,000,000$.

In both methods, most of the misclassifications are due to advertisements and headlines of other news articles. For instance, both methods failed to predict a memorial news article for a professional wrestler but with links to other articles (http://www.csulb.edu/~d49er/) as not peculiar. Both of the methods recognize words such as "headline" and "nylon", which are contained in links and irrelevant to the content of the article, as rare index words.

In summary, the predictive performance depends on recognition of index words as common or peculiar, which is influenced by the word distribution in the corresponding WWW pages as well as detection of stop words. Typically, increasing the number of WWW pages improves the predictive performance although polysemous words such as "castle" can result in the opposite results. Our TFIGF is particularly effective for WWW pages in English with relevant words although problems such as address, links, and advertisements remain. Our superiority in precision and often in recall comes from our improved estimation of generality of a word and our stop word detection with a search engine.

5   Conclusions

Discovery of peculiar instances has been addressed in outlier detection [5, 6], clustering [8], and classification [2–4]. While these methods detect peculiar instances from given data, our TFIGF relies on a large number of WWW pages via a search engine. This is made possible through our newly proposed WWW frequency $g(w)$ for a word w.

We consider that experimental results show effectiveness of a word-based approach which tries to estimate some sort of semantics from statistical information obtained with a search engine. We plan to support this endeavor by developing a visualization method for involving the users in the detection process.

References

1. D. Billsus and M. Pazzani: A Hybrid User Model for News Story Classification. *Proc. Seventh International Conference on User Modeling*, pp. 99–108, 1999.
2. N. V. Chawla, A. Lazarevic, L. O. Hall, and K. W. Bowyer: SMOTEBoost: Improving Prediction of the Minority Class in Boosting, *Principles of Data Mining and Knowledge Discovery, LNAI 2838 (PKDD)*, Springer-Verlag, pp. 107–119, 2003.
3. P. Domingos: MetaCost: A General Method for Making Classifiers Cost-Sensitive , *Proc. Fifth Intl. Conf. on Knowledge Discovery and Data Mining (KDD)*, pp. 155–164, 1999.
4. W. Fan, S. J. Stolfo, J. Zhang, and P. K. Chan: AdaCost: Misclassification Cost-sensitive Boosting, *Proc. Sixteenth Intl. Conf. on Machine Learning (ICML)*, pp. 97–105, 1999.
5. Y. Freund and R. E. Schapire: Experiments with a New Boosting Algorithm, *Proc. Thirteenth Int'l Conf. on Machine Learning (ICML)*, pp .148–156, 1996.
6. E. M. Knorr, R. T. Ng, and V. Tucakov: Distance-Based Outliers: Algorithms and Applications. *VLDB J.*, 8(3-4), pp. 237–253, 2000.
7. R. Kosala and H. Blockeel: Web Mining Research: A Survey, *ACM SIGKDD Exploration, Issue 2*, pp. 1–15, 2000.
8. M. Narahashi and E. Suzuki: Detecting Hostile Accesses through Incremental Subspace Clustering, *Proc. 2003 IEEE/WIC International Conference on Web Intelligence (WI)*, pp. 337–343, 2003.
9. G. Salton and M. J. McGill: *Introduction to Modern Information Retrieval*, McGraw-Hill, New York, 1983.

DHT Facilitated Web Service Discovery Incorporating Semantic Annotation*

Shoujian Yu, Jianwei Liu, and Jiajin Le

College of Information Science and Technology, Donghua University,
1882 West Yan'an Road, Shanghai, China, 200051
{Jackyysj,Liujw}@mail.dhu.edu.cn
Lejiajin@dhu.edu.cn

Abstract. Web services enable seamless application integration over the network regardless of programming language or operating system. A critical factor to the overall utility of Web services is an efficient discovery mechanism. In this paper, limitations of typical UDDI registry are analyzed. We propose a novel decentralized Web services registry infrastructure. We extract three levels metadata from Web services description. These metadata are annotated with service ontology, which acts as the bridge of P2P (peer to peer) and Web services. We have discussed multi-level Web services discovery under this infrastructure. Our experiment results testify that this infrastructure can effectively overcome bottlenecks of traditional UDDI registry. The evaluation shows that WordNet lexical dictionary facilitated and composite Web services discovery excel notably with large amounts of services.

1 Introduction

Web services are emerging as a powerful technology for organizations to integrate their applications within and across organizational boundaries. An important aspect is that the interactions should be done automatically by computers. Thus the critical factor is a scalable, flexible and robust discovery mechanism, which provides the desired functionality. At present, Web services are advertised in central registry. The initial focus of Universal Description Discovery and Integration (UDDI) specifications was proposed for this purpose [1]. It is a central registry. But with large number of private and semi-private registry implementations for electronic commerce, the challenge of dealing with hundreds of registries during service publication and discovery becomes critical. In this paper, we present a novel infrastructure for distributed Web services registry facilitated with DHTs (Distributed Hash Table). This paper is structured as follows. Section 2 briefly analyzes the limitations of typical UDDI registry. In section 3, the decentralized Web services registry infrastructure is proposed. We give detailed explanation for decentralized Web services publication and discovery. In section 4, scalability of our system and the semantic annotation method are evaluated. Section 5 concludes this paper and some future work are discussed.

---

* This work has been partially supported by The National High Technology Research and Development Program of China (863 Program) under contract 2002AA4Z3430.

E. Suzuki and S. Arikawa (Eds.): DS 2004, LNAI 3245, pp. 363–370, 2004.

2   Limitations of Typical UDDI Registry

UDDI is a specification for business registries from Ariba, IBM and Microsoft. The service providers publish their services in the registry and the requestors inquiry the registry to discover the location of desirable service and contact with it.

Web services discovery based on UDDI is in fact the traditional client/server model. With rapid development of electronic commerce applications, different organizations have specialized UDDI registries, which is called private registry. Also in e-marketplace applications, semi-private registry is used for theirs members to publish and discover Web services. With such increasing number of private or semi-private registries, typical centralized indexing scheme based on UDDI technology can't scale well. Physical distribution of the registries can quickly overwhelm this centralized configuration and can lead to serious performance bottlenecks.

Emerging P2P solutions particularly suit for the increasingly decentralized application. A P2P network is a distributed architecture where participants share a part of their own resources and play roles as resource providers as well as resource requestors. Recently a number of groups have proposed a new generation of P2P system like CAN [2], Pastry [3] and Chord [4], avoiding shortcomings of early P2P systems like Gnutella [5] and Napster [6]. While different implementation exists, these systems all support a DHT interface of *put (key, value)* and *get (key)*. In contrast to UDDI, P2P networks content is normally described and indexed locally to each peer and search queries are propagated across the network. No central index is required to span the network.

DHTs largely solve the problem of centralized UDDI registries. However, DHTs support only exact match, which is the same to UDDI based keywords search. Thus semantic Web services annotation is prerequisite for enhanced search capabilities. Semantically described services will enable better service discovery and easier interoperation among services. Web services must be described at both syntactic and semantic level. Syntactic information is concerned with the implementation aspects tailored towards the programmers' requirements. Semantic information is concerned with the conceptual aspects aiming at facilitating end users by shielding off the lower level technical details, as well as to facilitate developers to find services that best match their needs and to enable automatic service discovery [7]. Several approaches have been suggested for adding semantics to Web services [8, 9].

3   Decentralized Web Services Publication and Discovery

Fig. 1 shows a modular architecture of the DHT facilitated decentralized Web services infrastructure. We use service ontology to provide formal description of a domain. This is the basis for semantic Web service annotation. When a service provider or a service requestor wants to publish or discover Web services, he should first select suitable ontology. This infrastructure includes two main parts: DHTs facilitated decentralized Web services registry and upper Web services discovery components. Novel DHTs provide direct content routing and retrieval. D. Liben-nowell etc. points out that Chord's maintenance bandwidth to handle concurrent node arrivals and departures is near optimal [10]. Thus we choose Chord as our DHT protocol. Web services discovery involves three components. Direct Web services discovery is provided by DHT protocol itself. This search capability is limited. WordNet facilitated Web services discovery provides semantic level service matching. When single service can't fulfill service request, composite Web service discovery will try to compose several Web services.

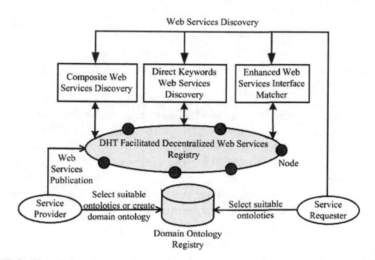

Fig. 1. DHT facilitated decentralized Web services registry. Detailed explanation of this infrastructure will be discussed in Sect. 3.3.

3.1 Service Ontology Facilitated Web Services Annotation

In a more dynamic environment, autonomous organizations must use the same vocabulary and definition of terms in order to understand each other. Ontological approach tries to address this problem. Ontology proposes a way of formal description of a domain to enable automated processes where understanding of content is needed. We use service ontology to specify a common language agreed by a community (see Fig. 2). The service ontology includes many *domains*. The *domain* defines basic concepts and terminologies that are used by all participants in that community. The domain concept is used for category of Web services published. It also uses *service class* to define the properties of services. The *service class* is used to describe functional properties of services. It is further defined by *attributes* and *operations*. For example, service class *Car* has attributes of *Brand Name* and operations of *Browse*. Each operation is specified by its name and signature (i.e. inputs and outputs). Commonly, the signature should be the combination of attributes. For flexible search capability, *service class*, *operation* and *attribute* can have a set of *synonyms*. For example, *Car* may have the synonym of *Automobile*.

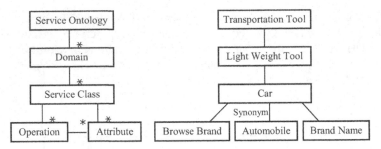

Fig. 2. Service ontology and an example: *Domain* is used for service category, *Service class* is used for service level annotation, *Operation* is used for operation level annotation and *Attribute* is used for signature level annotation.

In order to add semantics to Web services, service provider needs to give Web service description semantic annotation. WSDL (Web Services Description Language) is the current standard for Web services description [11]. We try to annotate Web services by annotating metadata of WSDL. The syntax of WSDL is defined in XML Schema. We extract three levels of metadata from this schema and give three-level annotations: service level, operation level and signature level. The following example code illustrates the Web services annotation method.

```
<types> <!-- signature level annotation -->
   <xsd:element name="brand" Attribute="Car:Brand Name"…
   <xsd:element name="price" Attribute="Car:Price"…
</types>   ……
<portType> <!-- operation level annotation -->
   <operation name="SearchBrand" Operation="Car:Browse
Brand">
   <operation name="SearchPrice" Operation="Car:Browse
Price">
   </operation>
</portType>   ……
<service name="Car Seller" Service class="Car"   ……
</service> <!-- service level annotation -->
```

3.2   Web Services Publication

A Web service consists of a set of operations, and each operation is associated with input and output parameters. According to DHT protocol and the multilevel service annotation in Sect. 3.1, *service class* should be the key of *put (key, value)*, and set of *operations* associated with *attributes* should be the *value* parameter. We assume that service provider N_n wants to publish his service: *car seller*. This service includes two operations: *SearchBrand* and *SearchPrice*. We assume this service is annotated as the example code in Sect. 3.1. Thus *Car* should be the key, *Browse Brand (Price)* and *Browse Price (Brand)* should be the value.

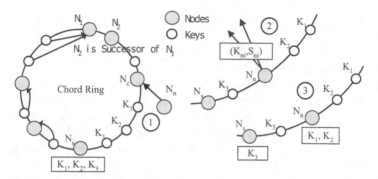

Fig. 3. Web services publication in decentralized registry. Nodes have the same name space with the keys. Chord uses hashing to map both keys and nodes onto the Chord ring. Each key is assigned to its successor node, which is the nearest node traveling the ring clockwise. All the service providers (node) and their service meta-description (key and value) are routed to the Chord ring according to DHT protocol.

There are three steps for N_n to publish his services (see Fig. 3). First N_n contacts any node N_c in the system. N_n gets a position in the Chord ring with hash algorithm. Chord finds N_n's successor N_s in the identifier ring. Second, the new node N_n, now part of the identifier ring, injects all its operations (k_{ni}, S_{ni}) into the system and the Chord protocol decides which nodes should receive the new catalog information. Third, N_n becomes part of the decentralized infrastructure and should share load of the registry by hosting parts of key-value sets already in the network. The Chord protocol will assign to N_n keys for which N_n should be the successor in the identifier ring. As Fig. 3 illustrates, k_1, k_2 and their corresponding operations should be reallocated from N_s to N_n when N_n joins.

3.3 Two Steps Web Services Discovery

For a service discovery, the requestor should first choose appropriate ontology in ontology registry to which his preferred service may refer. The service class should be determined. Based on *get (key)* interface of DHT protocol, the node that stores desirable Web services description is returned. We assume that service request is submitted on N_3 and N_2 returns with *get (key)*. This is the first step. Then the service requestor selects the interested operation and attribute as inquiry objectives. Also the requestor should provide some local information. For example, if a requestor wants to inquiry the price of some brand of car, then *Brand* should be provided as local parameter. All these information is the input for Web services discovery. On N_2, the inputted operation and attribute are matched against all the stored operation lists, i.e. signature matching. The enhanced Web services interface matcher component (see Fig. 1) carries out this procedure. The following section will discuss this. We assume that N_2 replies to N_3 with the node set $\{N_1\}$, on which the service matches the requestor with the biggest similarity. Finally N_3 can access the service in N_1. This is the final step.

WordNet Facilitated Interface Matching

During the service discovery procedure, step 2 intuitively executes exact string match for signature matching first. This is in fact inefficient because the service provider and service requestor may use approximate but not identical ontology information to express the same meaning. For example, *car* and *automobile* implies the same meaning. But exact match will fail because it can't utilize this semantic similarity. Word-Net is a comprehensive lexical reference system [12]. It catches three kinds of semantic relationship among concepts. We use it for improvement of exact signature matching. The following program code lists this algorithm.

```
double SemanticMatch (term1, term2) {
   maxScore = 1;
   if (term1 is identical to term2)
      score = maxScore;
   else if (term1 and term2 are synonymous
      score = 0.8* maxScore;
   else if (term1 and term2 have hierarchical relations)
      score = (0.6* maxScore) / N;
         // N is the number of hierarchical links;
   else score =0;
return score;}
```

Composite Web Services Discovery

It should be noted that a service request might not be provided by a single operation. Thus the matcher first looks for a direct match if a single operation meets the request. If this is not successful, then the composite Web services discovery component tries to look for composite services to fulfil the request.

On the other hand, the signature of operations in Web service tells us what type of parameter needs to be provided in order to execute it, as well as the types of output that will be returned. This gives us information on sequential compositions of services. For example, if a particular service has an operation *SearchPrice*, which takes as input a *Brand*, and another operation *SearchBrand* outputs values of *Brand*, then we know that the latter operation is a composition of the former operation. This process can be implemented automatically and recursively.

4   Implementation and Evaluation

We have implemented a TRP (Textile Resource Planning) system with Web service technology in our previous work. We extract 239 Web services, which include 600 operations from TRP system for experiments. The evaluation is based on the Chord protocol implementation found on the Chord project website. A virtual queue is used to simulate the nodes in Chord ring. WordNet 2.0 is embedded into the system through APIs for semantics interpretation of interface matching.

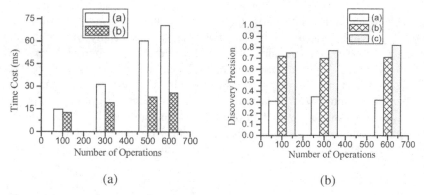

(a)   (b)

Fig. 4. Decentralized Web services registry infrastructure and semantic annotation method evaluation results.

We try to evaluate our decentralized Web service infrastructure from two aspects. We evaluate system scalability by evaluating the discovery time cost with increasing of number of operations. We change the number of operations from 100 to 600. We compare the results in two circumstances: (a) using typical UDDI registry and (b) using our method. From the result in Fig.4 (a), we can see that our method notable excels typical UDDI registry. Discovery time increases abruptly with number of operations increase in typical UDDI registry. Comparatively, out method keeps steady. This testifies that typical UDDI registries will determinately lead to bottlenecks.

We evaluate our semantic annotation method by evaluating discovery exactness with increasing of number of operations. We compute the ratio of the number of matches returned by our matching algorithm with the number of manually determined real matches, i.e. precision as our evaluation measure. We compare the result in three circumstances: (a) Web services are not semantic annotated, and parameter names are used for routing. (b) Web services are manually annotated. (c) Composite Web services discovery is used. From the result in Fig.4 (b), we can see that discovery precision is only 30% under circumstance (a). Semantic annotation method (b) will improve precision about 40%. With the increasing number of operations, composite Web service discovery can greatly increase discovery precision. This result makes out that our semantic method is effective.

5 Conclusions and Future Work

Emerging P2P technologies is appropriate and suitable to the increasingly decentralized nature of modern organizationss. A novel decentralized Web services registry infrastructure based on DHTs is proposed in this paper. We use service ontology as the bridge between DHTs and Web services. We extract three levels of metadata from WSDL: service level, operation level and signature level. These metadata are annotated with service ontology. Based on DHT, these metadata are directly routed to other service provider for publication. We have made improvements for DHT's exact content matching. We use WordNet lexical dictionary for flexible Web service discovery. The evaluation results show that DHT facilitated registry can reduce about 50% discovery time compared to typical UDDI registry, especially with large amount of Web services. The semantic annotation method can largely improve Web services discovery exactness. With large amount of Web services, composite service discovery can greatly increase discovery precision. While good results are obtained in our work, there is still much work to be done. The development of a toolkit is in progress. In this paper, we assume the domain ontology is well formed. If different organizations engineer ontologies for the same domain, then the interoperability problem is coming once again. Study of ontology matching will be incorporate in our work in the future.

References

1. UDDI. UDDI white papers. http://www.uddi.org/whitepapers.html
2. S. Ratnasamy, P. Francis, M. Handley, R. Karp, S. Schenker: A Scalable Content-Addressable Network. Proceedings of ACM SIGCOMM2001, San Diego CA USA (2001)
3. A. Rowstron, P. Druschel: Pastry: Scalable, distributed object location and routing for large-scale peer-to-peer systems. IFIP/ACM International Conference on Distributed Systems Platforms. Heidelberg Germany (2001) 329–350
4. I. Stoica, R. Morris, D. Karger, M.F. Kaashoek, H. Balakrishnan: Chord: A Scalable Peer-to-Peer Lookup Protocol for Internet Applications. IEEE/ACM Transactions on Networking. Vol. 11, 1 (2003) 17–32
5. Gnutella Resources. http://gnutella.wego.com/
6. Napster. http://www.napster.com
7. M. Paolucci, K. Sycara, T. Nishimura, N. Srinivasan: Using DAML-S for P2P Discovery. Proceedings of the First International Conference on Web Services (ICWS'03), Las Vegas USA (2003) 203–207
8. DAML-S Coalition: DAML-S: Web Service Description for the Semantic Web. In: Ian Horrocks, James Hendler (eds.): The Semantic Web - ISWC 2002. Lecture Notes in Computer Science, Vol.2342. Springer-Verlag, Berlin Heidelberg New York (2002) 348–363
9. Rama Akkiraju, Richard Goodwin, Prashant Doshi, Sascha Roeder: A Method for Semantically Enhancing the Service Discovery Capabilities of UDDI. Proceedings of the Workshop on Information Integration on the Web, IIWeb-03. Acapulco Mexico (2003)
10. D. Liben-nowell, H. Balakrishnan, D. Karger: Observations on the Dynamic Evolution of Peer-to-Peer Networks. Proceedings of the First International Workshop on Peer-to-Peer Systems (IPTPS'02), Cambridge MA (2002) 22–33
11. E. Christensen, F. Curbera, G. Meredith, and S. Weerawarana: Web Services Description Language (WSDL) Version 2.0 Part 1: Core Language. http://www.w3.org/TR/2004/WD-wsdl20-20040326/ (2004)
12. WordNet 2.0. http://www.cogsci.princeton.edu/~wn/

Discovering Relationships Among Catalogs

Ryutaro Ichise[1,2], Masahiro Hamasaki[2], and Hideaki Takeda[1,2]

[1] National Institute of Informatics, Tokyo 101-8430, Japan
[2] The Graduate University for Advanced Studies, Tokyo 101-8430, Japan
{ichise,takeda}@nii.ac.jp, hamasaki@grad.nii.ac.jp

Abstract. When we have a large amount of information, we usually use categories with a hierarchy, in which all information is assigned. The Yahoo! Internet directory is one such example. This paper proposes a new method of integrating two catalogs with hierarchical categories. The proposed method uses not only the contents of information but also the structures of both hierarchical categories. In order to evaluate the proposed method, we conducted experiments using two actual Internet directories, Yahoo! and Google. The results show improved performance compared with the previous approaches.

1 Introduction

The progress of information technologies has enabled us to access a large amount of information. This naturally demands a method of managing such information. One popular method of information management systems is to utilize a concept hierarchy and categorize all information into the concept hierarchy. Examples include catalogs of publications, file directories, Internet directories and shopping catalogs. A catalog of publications uses one standardized concept hierarchy for categorization. However, most concept hierarchies for information management are hard to standardize because each concept hierarchy has its own purpose and user. This situation produces difficulties when we use multiple information sources with different concept hierarchies. Hence, technologies for integrating multiple catalogs with different concept hierarchies are necessary for seamless access among them. Similarity-based integration (SBI) [7] is an effective method for solving this problem. The main idea of SBI is to utilize only the similarity of categorizations across concept hierarchies. Namely, SBI does not analyze the contents of information assigned to the concept hierarchies. In this paper, we propose an extension of SBI which uses the contents in information instances. The basic idea of our approach is to combine SBI and Naive Bayes [12].

This paper is organized as follows. Section 2 characterizes the catalog integration problem, which is the subject of this paper. Section 3 describes related work. Section 4 presents new catalog integration algorithms and mechanisms. Section 5 applies the algorithms to real world catalogs, which are Internet directories, to demonstrate our algorithm's performance, and discusses our experimental results and methods. Finally, in Section 6 we present our conclusions and future work.

E. Suzuki and S. Arikawa (Eds.): DS 2004, LNAI 3245, pp. 371–379, 2004.
© Springer-Verlag Berlin Heidelberg 2004

2   Catalog Integration Problem

In order to state our problem, we introduce a model for the catalogs we intend to integrate. We assume there are two catalogs: a *source* catalog and a *target* catalog. The information instances in the source catalog are expected to be assigned to categories in the target catalog. This produces a *virtually* integrated catalog in which the information instances in the source catalog are expected to be members of both the source and target catalogs. This integrated catalog inherits the categorization hierarchy from the target catalog.

Our catalog model for the source and target is as follows:

- The source catalog S_C contains a set of categories $C_{s1}, C_{s2}, \ldots, C_{sn}$ that are organized into an "is-a" hierarchy. Each category can also contain information instances.
- The target catalog T_C contains a set of categories $C_{t1}, C_{t2}, \ldots, C_{tm}$ that are organized into an "is-a" hierarchy. Each category can also contain information instances.

We will assume that all information instances in both catalogs are supposed to have some attributes for each. The problem addressed in this paper is finding an appropriate category C_t in the target catalog T_C for each information instance I_{si} in the source catalog S_C. What we need to do is determine an appropriate category in T_C for an information instance which appears in S_C but *not* in T_C, because mapping is not necessary if the information instance is included in both the source and the target catalogs.

3   Related Work

One popular approach to this kind of problem is to apply standard machine learning methods. This requires a flattened class space which has one class for every leaf node. Naive Bayes (NB) [12] is an established method used for this type of instance classification framework. However, this classification scheme ignores the hierarchical structure of classes and, moreover, cannot use the categorization information in the source catalog. Enhanced Naive Bayes (E-NB) [1] is a method which does use this information. Although E-NB has a better performance than NB, it does not achieve the performance of SBI [7]. SBI will be discussed in the next section. GLUE [4] is another type of system employing NB. GLUE combines NB and a constraint optimization technique. Unlike normal NB, the document classification systems in [8, 10, 16] classify documents into hierarchical categories, and these systems use words in the documents for classification rules. These systems can be applicable to the catalog integration problem. However, these systems cannot use the categorization information in the source catalog.

Another type of approach is ontology merging/alignment systems. These systems combine two ontologies, which are represented in a hierarchal categorization. Chimaera [11] and PROMPT [13] are examples of such systems and assist in the combination of different ontologies. However, such systems require human interaction for merging or alignment. FCA-Merge [15] is another type of ontology merging method. It uses the attributes of concepts to merge different ontologies. As a result, it creates a new concept without regarding the original concepts in both ontologies.

From the viewpoint of catalog integration, an approach besides E-NB is to construct the abstract-level structure of two hierarchies [14]. This approach does not direct the transformation of source and target information, but transforms the information via the abstract-level structure. It is relatively easy to transfer information through many hierarchical structures, but it is hard to create a common structure for these hierarchies.

4   A Multi-strategy Approach

In order to solve the problem stated in Section 2, we proposed the SBI system [7]. SBI used only the similarity of the categorization of instances. In other words, SBI did not use the contents of the instances. In this paper, we propose a method which combines the SBI approach and the NB approach. This approach naturally implies that the new approach is able to use more information than the previous approaches; thus, we can expect to obtain a better performance compared with previous methods. In this section, we briefly explain the SBI and NB methods, and then we show the new combination approach, called SBI-NB.

4.1   Similarity-Based Integration

SBI focuses on the similarity of the way of categorization, not on the similarity of instances. Then, how do we measure the similarity of categorization? We utilize the shared instances in both the source and target catalogs as our measurement standard. If many instances in category C_{si} also appear in category C_{tj} at the same time, we consider these two categories to be similar, because the ways of categorization in C_{si} and C_{tj} are supposed to be similar, i.e., if another instance I is in C_{si}, it is likely that I will also be included in C_{tj}.

SBI adopts a statistical method to determine the degree of similarity between two categorization criteria. The κ-statistic method [5] is an established method for evaluating the similarity between two criteria. Suppose there are two categorization criteria, C_{si} in S_C and C_{tj} in T_C. We can determine whether or not a particular instance belongs to a particular category. Consequently, instances are divided into four classes, as shown in Table 1. The symbols N_{11}, N_{12}, N_{21} and N_{22} denote the number of instances for these classes. For example, N_{11} denotes the number of instances which belong to both C_{si} and C_{tj}. We may logically assume that if categories C_{si} and C_{tj} have the same criterion of categorization, then N_{12} and N_{21} are nearly zero, and if the two categories have a different criterion of categorization, then N_{11} and N_{22} are nearly zero. The κ-statistic method uses this principle to determine the similarity of the categorization criteria. If you are interested in more details of the SBI mechanism, please refer to the work of [7].

Table 1. Classification of instances by two categories.

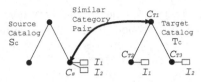

Fig. 1. An example of a learned rule.

		Category C_{tj}	
		Belongs	Not belongs
Category	Belongs	N_{11}	N_{12}
C_{si}	Not belongs	N_{21}	N_{22}

4.2 SBI-NB

A problem of SBI is that it is hard to learn a mapping rule when the destination category is in a lower category in the target concept hierarchy. In other words, the learned rules are likely to assign relatively general categories in the target catalog. Let us consider an example of this problem. Assume that a category in the source directory has two instances, and these instances are categorized in different categories in the target catalogs. We also assume that the two categories in the target catalog have the same parent category. This situation is shown in Figure 1. In this case, the SBI system could induce a mapping rule between C_s and C_{T1}. According to the problem definition, this rule is reasonable since we assume that the hierarchy relationship is "is-a" only. However, the best rules for these instances are rules which categorize them directly into category C_{T2} and C_{T3}. In order to induce this type of rules, we propose to combine a contents-based classification method after we apply the SBI algorithm. Since NB is very popular and easy to use, we adopt NB as the contents-based classification method.

The NB method is used to create classifiers from instances and their categories. The basic concept of the learned classifier is that the classifier assigns the category of maximum probability for the instance we want to classify. In the context of the problem defined in Section 2, the classifier is constructed using categories $C_{t1}, C_{t2}, \ldots, C_{tm}$ as well as the attribute values in instances $I_{t1}, I_{t2}, \ldots, I_{tv}$ in target catalog T_C. The classifier is then applied to instances $I_{s1}, I_{s2}, \ldots, I_{su}$ in source catalog S_C and assigns a category in T_C for each instance. On the other hand, in order to apply the NB algorithm for hierarchical classification, we utilize the simple method of the *Pachinko Machine* NB [10]. The Pachinko Machine classifies instances at internal nodes of the tree, and greedily selects sub-branches until it reaches a leaf [10]. In the Pachinko Machine NB method, when the system selects categories in the internal node, NB is used as the classification method. In addition, our system makes a virtual category in an internal node, and treats it in the same manner as a normal category. The virtual category has instances assigned to the internal node. This is because of the capability of classification for the internal node. This method is applied after the rule induced by SBI decides the starting category for the Pachinko Machine NB. To implement our system, we utilize the NB system called Rainbow [9] and SBI.

Table 2. Statistics on the experimental data.

	Yahoo!		Google		Shared
	Categories	Instances	Categories	Instances	instances
Autos	885	5134	874	9702	544
Movies	5297	19192	7947	36288	1480
Outdoors	2590	7960	1221	17065	362
Photography	578	5548	278	5443	305
Software	513	3268	2339	41883	353

5   Experiment

5.1   Experimental Settings

In order to evaluate the proposed algorithm, we conducted experiments using real Internet directories collected from Yahoo! [17] and Google [6][1]. The locations in Yahoo! and Google are as follows:

- Yahoo! : Recreation / Automotive & Google : Recreation / Autos
- Yahoo! : Entertainment / Movies_and_Film & Google : Arts / Movies
- Yahoo! : Recreation / Outdoors & Google : Recreation / Outdoors
- Yahoo! : Arts / Visual_Arts / Photography & Google : Arts /Photography
- Yahoo! : Computers_and_Internet / Software & Google : Computers / Software

Table 2 shows the numbers of categories, the instances in each Internet directory and the instances included in both Internet directories. From Table 2, we can see that each Internet directory tends to have its own bias in both collecting and categorizing pages. This fact proves the necessity for catalog integration.

We conducted ten-fold cross validations for the shared instances. Ten experiments were conducted for each data set, and the average accuracy is shown in the results. We compared the SBI-NB system with the SBI system and the NB system with flattened classes. Keyword extraction from documents and keyword indexing are necessary to use NB. In our experiment, we obtained web pages from the Internet and indexed all of them by using Rainbow. The same indexing method was used for NB and SBI-NB. The accuracy is measured for each depth of the Internet directories. If the correct categories are deeper than the depth which is used in the experiment, the bottommost categories are considered as the answers instead of the actual answer. In our experimental settings, other categories, such as a parent of the actual answer, can also be a correct category in semantic meaning. However, in this experiment, we utilized strict answer criteria[2]. The number of categories in the Internet directories is shown in Table 3. As one can see from this table, categorization for a lower category is more difficult than it is for upper categories. The significance level for the κ-statistic was set at 5%.

---

[1] Since the data in Google is constructed by the data in dmoz [3], we collected data through dmoz.

[2] The criteria of accuracy in this paper is more strict than that in the experiment of [7]. In the previous experiment, we chose other criteria to compare the result with those of another system. Therefore, the accuracy in this paper looks lower than that in [7] because of different criteria.

Table 3. Number of classes at each depth.

Depth	Autos		Movies		Outdoors		Photography		Software	
	Yahoo!	Google	Yahoo!	Google	Yahoo!	Google	Yahoo!	Google	Yahoo!	Google
1	37	15	38	31	46	27	31	10	32	89
2	240	227	214	202	224	173	83	39	125	494
3	475	690	730	5345	541	441	328	162	258	1154
4	765	833	3659	7239	1286	748	578	267	403	1817
5	867	852	4979	7789	2120	1112	578	277	489	2171
6	883	874	5277	7947	2391	1221	578	278	507	2316
7	885	874	5293	7947	2590	1221	578	278	513	2337
8	885	874	5296	7947	2590	1221	578	278	513	2339
9	885	874	5297	7947	2590	1221	578	278	513	2339
10	885	874	5297	7947	2590	1221	578	278	513	2339

5.2   Experimental Results

The experimental results are shown in Figure 2. The vertical axes show the accuracy and horizontal axes show the depth of the concept hierarchies. The experimental domains of the graphs are Autos, Movies, Outdoors, Photography and Software, from top to bottom. The left side of Figure 2 shows the results obtained using Google as the source catalog and Yahoo! as the target catalog, and the right side of Figure 2 shows the results obtained using Yahoo! as the source catalog and Google as the target catalog. For comparison, these graphs also include the results of SBI and NB. SBI-NB denotes the results of the method proposed in this paper. Since it is impossible to calculate the accuracy in the Movie domain on Yahoo! as a target catalog by using NB because of the large number of categories, the accuracy is not shown in Figure 2.

The proposed SBI-NB algorithm has high accuracy compared with NB for all experimental domains. In particular, our algorithm finds a solution in the Movie domain for Yahoo! whereas NB cannot find a solution. From this we can conclude that our algorithm is better than the NB method. On the other hand, the proposed algorithm is better than SBI for all domains except Automotive and Outdoors for Yahoo!. In addition, in these two domains, the difference of accuracy between our algorithm and SBI is very small. Hence, we can conclude that our algorithm is better than SBI. From the comparison of the results of both NB and SBI with SBI-NB, we can expect high accuracy when NB produces a good result. For example, in the Photography domain, the proposed algorithm performs much better in accuracy than the original SBI. One reason for this is that the NB works well. In other words, the contents-based classification is suited for this domain. Our proposed algorithm effectively combines the contents-based method with the category similarity-based method. On the other hand, when the NB method has poor accuracy, our method has at least the same performance of SBI and does not have any side effect from NB. For example, in the Autos domain, the performance is similar for SBI and SBI-NB, regardless of the poor performance of NB. Since the SBI-NB method proposed in this paper has a high performance in many domains compared with NB and SBI, we can conclude that our approach is a good method for integrating the approach of document classification based on the attribute of an instance and the approach of the similarity-based integration method.

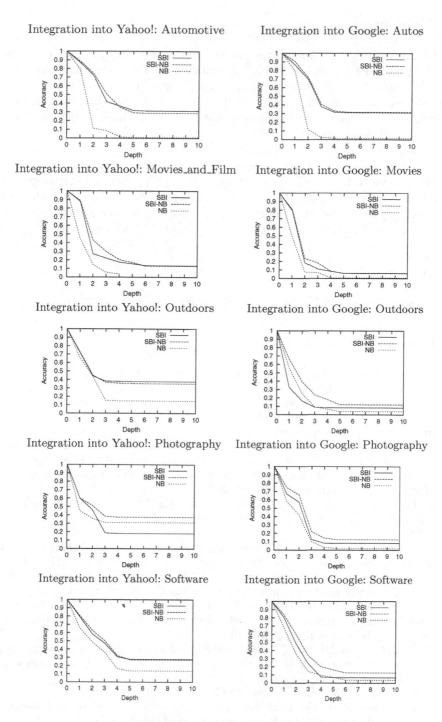

Fig. 2. Experimental Results.

6 Conclusion

In this paper, a new technique was proposed for integrating multiple catalogs. The proposed method uses not only the similarity of the categorization of catalogs but also the contents of information instances. The performance of the proposed method was tested using actual Internet directories, and the results of these tests show that the performance of the proposed method is more accurate for most of the experiments.

Although the present results are encouraging, much has yet to be done. For this research, we applied the Pachinko NB method as the contents-based classification method. However, other methods such as SVM [2] and shrinkage [10] are adoptable for our system because of the independency gained from the SBI method. We plan to test such combinations. Other future work includes expanding the proposed method so that it can apply to more than three catalogs.

Acknowledgment

We would like to thank Dr. McCallum for providing the Naive Bayes program.

References

1. R. Agrawal and R. Srikant. On integrating catalogs. In *Proc. of the Tenth Int. WWW Conf.*, pp. 603–612, 2001.
2. N. Cristianini and J. Shawe-Taylor. *An Introduction to Support Vector Machines.* Cambridge University Press, 2000.
3. dmoz. http://dmoz.org/, 2003.
4. A. Doan, J. Madhavan, P. Domingos, and A. Halevy. Learning to map between ontologies on the semantic web. In *Proc. of the 11th Int. WWW Conf.*, 2002.
5. J. Fleiss. *Statistical Methods for Rates and Proportions.* John Wiley & Sons, 1973.
6. Google. http://directory.google.com/, 2003.
7. R. Ichise, H. Takeda and S. Honiden. Integrating multiple internet directories by instance-based learning. In *Proc. of the 18th Int. Joint Conf. on AI*, pp. 22-28, 2003.
8. D. Koller and M. Sahami. Hierarchically classifying documents using very few words. In *Proc. of the 14th Int. Conf. on Machine Learning*, pp. 170–178, 1997.
9. A. K. McCallum. Bow: A toolkit for statistical language modeling, text retrieval, classification and clustering, http://www.cs.cmu.edu/ mccallum/bow/, 1996
10. A. K. McCallum, R. Rosenfeld, T. M. Mitchell and A. Y. Ng. Improving text classification by shrinkage in a hierarchy of classes. In *Proc. of the 15th Int. Conf. on Machine Learning*, pp. 359–367, 1998.
11. D. L. McGuinness, R. Fikes, J. Rice, and S. Wilder. An environment for merging and testing large ontologies. In *Proc. of the Conf. on Principles of Knowledge Representation and Reasoning*, pp. 483–493, 2000.
12. T. M. Mitchell. *Machine Learning.* McGraw-Hill, 1997.
13. N. F. Noy and M. A. Musen. Prompt: Algorithm and tool for automated ontology merging and alignment. In *Proc. of the 17th National Conf. on AI*, pp. 450-455, 2000.

14. B. Omelayenko and D. Fensel. An analysis of B2B catalogue integration problems. In *Proc. of the Int. Conf. on Enterprise Information Systems*, pp. 945–952, 2001.
15. G. Stumme and A. Madche. FCA-Merge: Bottom-up merging of ontologies. In *Proc. of the 17th Int. Joint Conf. on AI*, pp. 225–230, 2001.
16. A. Sun and E. Lim. Hierarchical Text Classification and Evaluation. In *Proc. of IEEE Int. Conf. on Data Mining*, pp. 521–528, 2001.
17. Yahoo! http://www.yahoo.com/, 2003.

Reasoning-Based Knowledge Extraction for Text Classification

Chiara Cumbo[2], Salvatore Iiritano[1], and Pasquale Rullo[1,2]

[1] Exeura s.r.l.
iiritano@exeura.it
[2] Dipartimento di Matematica, Università della Calabria, 87030 Rende (CS), Italy
{cumbo,rullo}@mat.unical.it

Abstract. We describe a reasoning-based approach to text classification which synergically combines: (1) ontologies for the formal representation of the domain knowledge; (2) pre-processing technologies for a symbolic representation of texts and (3) logic as the categorization rule language.

1 Introduction

Because of the huge amount of documents available on both the Web and the intranets, text classification is gaining more and more interest. As a consequence, a number of statistical and machine learning approaches have been proposed in the last few years [1–5].

In this paper we describe a novel reasoning-based approach to text classification which synergically combines: (1) ontologies for the formal representation of the domain knowledge; (2) pre-processing technologies for a symbolic representation of texts and (3) logic as the categorization rule language. Logic, indeed, provides a natural and powerful way to describe features of document contents that may relate to concepts. To this end, we equip each concept of the domain ontology with a Datalog^{TC}(Datalog for Text Classification) program, a suitable extension of Datalog [6] by aggregate and built-in predicates [7]. Based on the proposed approach, we have implemented a prototypical system for text classification, called *OLEX* [8].

In the next sections we shall concentrate on the following aspects of *OLEX*: ontology management, pre-processing and classification.

2 Ontology Management

Ontologies in our system provide the domain knowledge needed for a high-precision classification. The ontology language of *OLEX* supports the specification of the following basic constructs: Concepts, Attributes, Properties (attribute values), Taxonomic (is-a) and Non-Taxonomic binary associations, Concept Instances, Links (association instances), Constraints (i.e., association cardinalities, exists link, for all links), Synonyms.

E. Suzuki and S. Arikawa (Eds.): DS 2004, LNAI 3245, pp. 380–387, 2004.

Example 1. EEL (Electronic Exeura Library) is an ontology developed at Exeura with the purpose of classifying documents concerning the different areas of Knowledge Management. It consists of 16 concepts (plus a number of instances), among which KRR (Knowledge Representation and Reasoning), KD (Knowledge Discovery), TM (Text Mining). A fragment of EEL is shown in Figure 1. As we can see, the concept KRR (to be intended as "techniques for KRR") has the sub-concept "Ontology" (ontologies represent a knowledge representation technique) and is linked to both "Language" and "Formalism" (KRR techniques rely on both languages and formalisms), each other related by the relation "Supported_by" (a language is supported by a formalism). □

An ontology O is internally represented as a set of facts F_O (the representation of O) of the following type (the list is not complete):

- *concept(conc_Name)*, e.g., *concept(Ontology)*;
- *is-a(conc_Name1, conc_Name2)*, e.g., *is-a(Ontology, KRR)*
- *relation(rel_Name,conc_Name1,conc_Name2)*,
  e.g., *relation(Supported_by,Language,Formalism)*
- *cardinality(rel_Name, cond1, cond2)* e.g., *cardinality(Supported_by, "⩾= 1", "⩾= 0")*, expressing that each language is supported by at least one formalism and that each formalism may support more languages
- *instance_of(conc_Name, inst_Name)*, e.g., *instance_of(Language,Datalog)*
- *synonym(conc_Name1,conc_Name2)*.

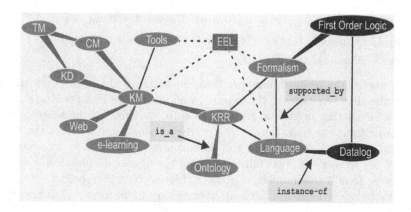

Fig. 1. The EEL Ontology.

3   Pre-processing

The aim of the pre-processing is to obtain a machine-readable representation of textual documents [9]. To this end, the following tasks are performed: (1) Pre-Analysis, based on three main activities: Document Normalization, Structural Analysis and Tokenization; (2) Linguistic Analysis, based on the following steps:

Lexical Analysis, which determines the Part of Speech (PoS) of each token, Reduction, for the elimination of the stop words, and Frequency Analysis.

The output of the pre-processing phase for a document D is a set of facts F_D (the representation of D) of the following type: (1) PoS facts of the form $p(idt, token)$, where p is either one of *noun, properNoun, verb, adjective*, and *idt* is the position of *token* within the text; (2) formatting facts like *title(first-idt,last-idt)* and *paragraph(idp,first-idt,last-idt)*, where *idp* is the paragraph identifier; (3) frequency facts, namely, $frequency(token, number)$ and $numberOfTokens$ $(number)$ representing the frequency of *token* and the total number of tokens within a document, respectively. Note that F_D is actually a compressed representation of D, as only the relevant information of D are represented.

4   Classification

4.1   The Logic Language Datalog^{TC}

The basic idea is that of using logic programs to embed in our system the knowledge needed to recognize concepts within texts. To this end, we use the logic language Datalog^{TC} [7], an extension of stratified Datalog [6] by constructs suitable for text processing, namely, *aggregate functions* and *external predicates* (see [7]). Datalog^{TC}is supported by the logic system DLV [10]. The following example provides an informal presentation of Datalog^{TC}.

Example 2. The following (Datalog^{TC}) rules are aimed at discovering the concept "text mining" of the EEL ontology within a document:

$$r_1 : p_1 : -noun(I, text), verb(J, mining), J = I + 1, title(X, Y), J \leq Y$$
$$r_2 : p_2(I) : -noun(I, text), verb(J, mining), J = I + 1, same\_par(K, I, J)$$
$$r_3 : p_3 : -\#count\{I : p_2(I)\} > t.$$

In rule r_1 above, the propositional predicate p_1 is true if the term "text mining" occurs in the title. We observe that $noun()$ and $verb()$ are PoS predicates, while $title()$ is a formatting predicate. Rule r_2, in turn, "verifies" if the searched term occurs throughout the text. Herein, the predicate *same_par* requires the tokens "text" and "mining" to occur in the same paragraph. Rule r_3 "counts" the number of occurrences of "text mining", using the aggregate function $\#count()$, and states p_3 true if this number is greater than a given threshold t.

Rule r_4 below is used to match an expression of the form "... discover(ing) knowledge within text(s)...".

$$r_4 : \ p_4(K) \ :- \ verb(I, X), same\_stem(X, discover), noun(J, knowledge),$$
$$J = I + 1, noun(K, Y), same\_stem(Y, text), K > J,$$
$$same\_par(N, I, J), same\_par(N, J, K)$$

where $same\_stem()$ is defined in terms of the external predicate $\#stem()$[1]:

$$r_5 : \ same\_stem(X, Y) :- \ \#stem(X, Z), \#stem(Y, Z)$$

---

[1] This predicate is implemented in C++ by using the Porter algorithm [11].

Finally, the following rules are aimed at capturing the concept "Ontology" in a document talking of ontological languages (such as OWL, RDF, etc., that are instances of KRR Languages in the EEL ontology):

$$r_6 : p_6(I, F) :\text{-} \; instance\_of(Language, I),$$
$$relation(onto\_lang, I, ontology), frequency(I, F)$$
$$r_7 : p_7(F) :\text{-} \; F = \#sum\{F, I : p_6(I, F)\}$$

Here, r_6 provides the number of occurrences of each ontological language mentioned in the document (i.e, each instance of "Language" linked to "Ontology" through the association "Onto-Lang"), while r_7 gives the total number of such occurrences (through the aggregate function $\#sum$).   □

4.2   Categorization Programs and Their Evaluation

In order to classify documents w.r.t. an ontology O, we equip each concept C of O with a *categorization program* used to provide evidence that D is relevant for C. The formal definition of categorization program stems from the following basic observations: (1) there are documents that are *easy* to classify, i.e., for which few simple rules (such as $r_1 - r_3$ in example 2) are enough; for instance, if the term "text mining" is contained in the title or occurs frequently throughout the text, we can confidently classify this text as relevant for the concept "text mining", and (2) a deeper semantic analysis is needed only for documents that are *difficult* to classify, i.e., those for which the recognition of concepts requires more complex rules (like $r_4 - r_7$ in the previous example).

Having this in mind, we define the categorization program P_C of a concept C as a totally ordered set of components $(c_1, .., c_n, < *)$, where c_i, $1 \leq i \leq n$, is a Datalog^{TC}program and $< *$ is such that $c_i < *c_j$, for any i, j with $i < j$. Informally, each component groups rules performing some specific retrieval task, such as word-based search, term matching, etc., of increasing semantic complexity – that is, each component is capable to recognize texts that are possibly inaccessible to the "previous" ones. We call c_1 the *default component* of P_C, as it is automatically generated by the system.

Example 3. We next show a simplified version of the default component of a generic concept with name C:

$r_0 : gen\_syn(I, C) : -word(I, X), same\_stem(X, Z), synonym(Z, C), concept(C).$
$r_1 : occur(I, C) : -gen\_syn(I, C).$
$r_2 : occur(I, C) : -gen\_syn(I, W), instance\_of(C, W).$
$r_3 : in\_text(C) : -\#count\{I, C : occur(I, C)\} \geq T, concept(C), threshold(T).$
$r_4 : in\_title(C) : -occur(I, C), title(J, K), J \leq I, I \leq K.$
$r_5 : in\_sect(C) : -occur(I, C), section(J, K), I \geq J, I \leq K.$
$succ(C) : -in\_text(C). \quad succ(C) : -in\_title(C). \quad succ(C) : -in\_sect(C).$

Rule r_0 above generalizes the notion of synonymity making it independent of word suffixes. The predicate $word()$ is suitably defined as the union of all PoS predicates (*noun*, *verb*, etc.). Rules r_1 and r_2 state that the concept C occurs at position I if either a generalized synonym of C or a generalized synonym of any of its instances occurs at position I. Rule r_3 states $in\_text(C)$ true if the number of times C appears in the text is greater than a given threshold; rules r_4 and r_5 verify whether the concept C appears in the title or in some section title, respectively. The predicate $succ(C)$ is true if either one of r_3, r_4 or r_5 succeeds. The above program can be easily generalized to the case of multi-words concept/synonym names.   □

Now, given a document D, the evaluation of P_C (w.r.t. D) starts from c_1 and, as soon as a component c_i, $1 \leq i \leq n$, is "satisfied" (by D), the process stops successfully – i.e., D is recognized to be relevant for C; if no such a component is found, the classification fails (note that, if D is not relevant for C, then all components of P_C get evaluated).

The algorithm for the evaluation of P_C w.r.t. D is sketched in Figure 2. As we can see, the evaluation of a single component is carried out (in a bottom-up fashion) by the DLV system (line 4). As regards the computational complexity of this task, it is well known that the data complexity of the evaluation of a stratified Datalog program is polynomial in the size of the extensional database (i.e., the set of the input facts). Thus, we can state that *Algorithm 1* is polynomial in the size of $F_O \cup F_D$, where F_O and F_D are the representations of the ontology O and the document D, respectively.

Fact 1. Under data complexity, *Algorithm 1* runs in polynomial time.   □

Algorithm 1

Input: the categorization program $P_C = (c_1, .., c_n, < *)$; the representations F_O and F_D of the ontology O and the document D, respectively
var M_i, $i \geq 1$, is the model generated by DLV for the component c_i; $success\_c_i$ is a propositional predicate which is true when the evaluation of c_i succeeds.
Output: *true* if some component c_i succeeds (i.e., the document D is relevant for the concept C), *false* otherwise.

bool Program_Eval(setOfFacts F_D, F_O)
begin
1. $i = 1$; $success = false$; $M_0 = F_O \cup F_D$
2.   **while** (*not success and* $i \leq n$) **do**
3. $P_i = c_i \cup M_{i-1}$;
4. $M_i = DLV(P_i)$;
5. $success = (success\_c_i \in M_i)$;
6. $i = i + 1$;
7.   **end_while**
8.   **return** *success*
end

Fig. 2. The Categorization Program Evaluation Algorithm.

4.3   Ontology-Driven Classification Strategy

Let χ be a corpus of documents to be classified w.r.t. an ontology O. As we have seen in the previous section, each concept C of O is equipped with a suitable categorization program P_C whose evaluation determines, for each $D \in \chi$, if D is relevant for C or not. Now, in order to classify χ w.r.t. O, the categorization program of *each* concept of O should be evaluated, according to the algorithm of Figure 2, against *each* document $D \in \chi$. Unfortunately, such an approach could result in a rather heavy computation. To overcome this drawback, we adopt an *ontology-driven* classification heuristics which, by exploiting the presence of taxonomic hierarchies, allows us to drastically reduce the "concept space". This technique is based on the following inheritance principle: if a document D "talks" about a concept C which specializes some other concept B, then D (implicitly) "talks" of B too. For an instance, a document talking of "text mining" talks of "knowledge discovery" too, the former being a specialization of the latter. Thus, a document which is relevant for a concept is (implicitly) relevant for each of its ancestors within an is-a taxonomy (this is much alike the folder-subfolder philosophy of file systems). This inheritance principle suggests us a classification strategy where concepts within a sub-class hierarchy are processed in a bottom-up fashion. As soon as $D \in \chi$ is found to be relevant for a concept C in the hierarchy H, it is no more processed w.r.t. any of the ancestors of C in H.

A sketch of the algorithm is shown in Figure 3. Here, lines 1-3 invoke the recursive function $Classify$, once for each node u of the ontology graph $G(N, E', E'')$ (i.e., for each concept of the ontology O - see Figure 3 for the definition of

---

Algorithm 2

Input: The set χ of documents (corpus) to be classified w.r.t. O. The graph $G(N, E', E'')$ of the ontology O, where (1) N is the set of nodes, one for each concept of O, (2) E' is the set of directed edges representing the i-sa relationships; in particular, $(u, v) \in E'$ if v *is-a* u holds in O, and (3) E'' is the set of arcs representing any other kind of relationship in O.
var: χ_v represents the subset of χ relevant for concept v; χ^u represents the subset of χ to be processed w.r.t. u
Output: the classification of χ w.r.t. O, i.e., χ_u, for each $u \in N$.

```
1.     for each node u ∈ N
2.        if not processed[u] then
3.           χu = Classify(χ,u);
```

setOfDocs **Classify**(setOfDocs χ, node u)
begin
```
4.     χ^u = χ;
5.     for each node v ∈ N s.t. (u, v) ∈ E'
6.        if not processed[v] then χv = Classify(χ, v);
7.        χ^u = χ^u − χv;
8.     end_for
9.     processed[u] = true;
10.    return Classifier(χ^u, u)
```
end

Fig. 3. The Ontology-driven Classification Algorithm .

the graph G). This function (lines 4-10) essentially performs a depth-first search of G on the edges of E' (i.e., on the i-sa taxonomies of O), starting from any given node u. The set of documents χ^u that are to be processed w.r.t. u is computed as follows: χ^u is initially set to χ, the corpus of the input documents, and then, for each node $v \in N$ such that $(u, v) \in E'$ (i.e., s.t. v i-sa u holds in O), the set $\chi_v \subseteq \chi$ of the documents relevant for v is recursively evaluated; now, since each document in χ_v must'nt be evaluated w.r.t. u (according to our inheritance principle), χ_v is subtracted to χ^u (line 7). The set of documents χ_u relevant for node u is computed by the classification engine function $Classifier$ (line 10) which evaluates each document in χ^u by using $Algorithm\ 1$ of Figure 2.

Concerning the complexity of $Algorithm\ 2$, we just need to observe that the function $Classifier$ is invoked once for each node of G; thus, by Fact 1, the following holds:

Fact 2. Under data complexity, $Algorithm\ 2$ runs in polynomial time.   □

4.4   Shallow and Deep Classification Strategies

So far we have seen that, in order to classify a corpus χ w.r.t. an ontology O, each document $D \in \chi$ is to be processed, using $Algorithm\ 1$, w.r.t. a subset of the concepts of O determined by the ontology-driven evaluation strategy. However, as already noticed, the evaluation of a categorization program by $Algorithm\ 1$ succeeds to be efficient only for those concepts for which D is relevant (particularly those that are "easy" to discover), while it fails in all other cases (which means, generally, for a great many documents). Indeed, if D is not relevant for a concept C, then all components of P_C get evaluated. Thus, in order to make classification faster (possibly at the price of a lower precision), we may think of a sort of *shallow* evaluation, as opposed to the *deep* evaluation so far considered. This strategy works as follows: (1) based on the ontology-driven strategy, the documents in χ are first evaluated w.r.t. the concepts of O using only the respective default components (thus, all documents that are "easy" get classified), and (2) the remaining documents of χ, i.e., the "difficult" ones, are classified using the deep evaluation strategy.

We notice that both shallow and deep evaluation strategies are orthogonal to the ontology-driven classification strategy.

5   Conclusions and Future Work

Some preliminary tests of the current implementation of $OLEX$ have been carried out on a 2.2GHz Linux PC, using a corpus of 71 scientific papers that has been classified w.r.t. the EEL ontology of example 1. To this end, we have used only the default components automatically generated by the system (i.e., we have added no manual knowledge, so completely delegating the system in the task of classification). Concerning the quality of classification, we have got (macro) averages for precision and recall of 0,79 and 0,87, respecively. Cpu time for the classification of one document (w.r.t. the whole ontology) ranges from

0.04 seconds (for a document of 800 words) to 3.62 seconds (for a document of 58.500 words). The time needed to classify the entire corpus is of 30 seconds, with an average time of 0.42 seconds/document.

Our future work concentrates on several topics: (1) Definition of suitable linguistic interfaces for the ontology objects; (2) extension of Datalog^{TC} by new built-in predicates, for instance, for the integration of Word-Net; (3) strengthening of default components by a deeper exploitation of both domain and linguistic knowledge, and (4) thorough experimentation based on standard testbeds.

References

1. Cohen, W.W., Singer, Y.: Context-sensitive learning methods for text categorization. ACM Transactions on Information Systems **17** (1999) 141–173
2. Yang, Y., Liu, X.: A re-examination of text categorization methods. In Hearst, M.A., Gey, F., Tong, R., eds.: Proceedings of SIGIR-99, 22nd ACM International Conference on Research and Development in Information Retrieval, Berkeley, US, ACM Press, New York, US (1999) 42–49
3. Wiener, E.D., Pedersen, J.O., Weigend, A.S.: A neural network approach to topic spotting. In: Proceedings of SDAIR-95, 4th Annual Symposium on Document Analysis and Information Retrieval, Las Vegas, US (1995) 317–332
4. Lewis, D.D., Ringuette, M.: A comparison of two learning algorithms for text categorization. In: Proceedings of SDAIR-94, 3rd Annual Symposium on Document Analysis and Information Retrieval, Las Vegas, US (1994) 81–93
5. Weiss, S.M., Apté, C., Damerau, F.J., Johnson, D.E., Oles, F.J., Goetz, T., Hampp, T.: Maximizing text-mining performance. IEEE Intelligent Systems **14** (1999) 63–69
6. Ullman: Principles of Database and Knowledge-Base Systems. Computer Science Press, Rockville (Md.) (1988)
7. Dell'Armi, T., Faber, W., Ielpa, G., Leone, N., Pfeifer, G.: Aggregate Functions in Disjunctive Logic Programming: Semantics, Complexity, and Implementation in DLV. In: Proc. IJCAI 2003, Acapulco, Mexico, Morgan Kaufmann Publishers (2003)
8. Cumbo, C., Iiritano, S., Rullo, P.: Olex – a reasoning-based text classifier. In Alferes, J., Leite, J., eds.: Logics in Artificial Intelligence, Ninth European Conference, JELIA'04. (2004) Forthcoming.
9. Yiming, Y.: A comparative study on feature selection in text categorization. In: International Conference on Machine Learning, ICML (1997) 412–420
10. Faber, W., Pfeifer, G.: DLV homepage (since 1996) http://www.dlvsystem.com/.
11. Porter, M.: An algorithm for suffix stripping. Program **3** (1980) 130–137

A Useful System Prototype for Intrusion Detection – Architecture and Experiments

Ye Du, Huiqiang Wang, and Yonggang Pang

College of Computer Science and Technology, Harbin Engineering University,
150001 Harbin, China
{duye,hqwang,ygpang}@hrbeu.edu.cn

Abstract. With the ever increasing sophistication of attacking techniques, intrusion detection has been a very active field of research. We are designing and implementing a prototype intrusion detection system (IADIDF) that is composed of distributed agents. This paper describes the function of entities, defines the communication and alert mechanism. There are three main agents include Detectors, Managers and Communicators. Each agent operates cooperatively yet independently of the others, providing for efficiency alerts and distribution of resources. Communication mechanism is composed of three layers, which are Transport, Hosts and MessageContent layers. All entities of the prototype are developed in C program under Linux platform. Then, we analyze system performance, advantages of the prototype, and come to a conclusion that the operating of agents will not impact system heavily.

1 Introduction

Intrusion detection has been an active topic for about two decades, starting in 1980 with the publication of John Anderson's "Computer Security Threat Monitoring and Surveillance" [1]. In that paper, an intrusion attempt is the potential possibility of a deliberate unauthorized attempt to access information, manipulate information, or render a system unreliable. These are three main goals of computer security — CIA (Confidentiality, Integrity, Availability). The system designed to detect these types of activities is Intrusion Detection System (IDS).

We measure the quality of an IDS by its effectiveness, adaptability and Flexibility. There are a number of problems with existing IDSs [2], including a single point of failure, limited scalability and difficult reconfiguration.

In order to remedy the insufficient, some IDSs have been designed to do distributed collection and analysis of information. For instance, some systems are agents-based and employ hierarchical structures with data collected and analyzed in local. Nevertheless, there are still highest-level entities in these architectures, which are the bottlenecks of systems and lead to the matter of a single point of failure rise inevitably. For the purpose of solving above-mentioned problems, a distributed IDS framework model based on independent agents – IADIDF was proposed.

E. Suzuki and S. Arikawa (Eds.): DS 2004, LNAI 3245, pp. 388–395, 2004.

2   An Independent Agents-Based Distributed Intrusion Detection Framework

In the system, an application is composed of a series of interconnected entities. These entities are "agents" that can make a response to behavior (activity). By adding categorized agents to corresponding entity models, users may append new function through distributed application, while needn't to change other main parts. Agents for an application may be distributed on different network nodes. Thus, the task of this application may be operated distributedly. By this way, network bottleneck of data transmission problem can be solved, and the real-time character and dependability are strengthened. In the article, we define independent agent as software agent, for it can carry out certain security control function on host.

2.1   System Architecture

The structure of IADIDF is shown in Figure 1. In IADIDF, every entity of the same host is in the organization of hierarchical structure. Of all these entities, manager is in the supreme level, and detector is in lower level. The cooperating entities among different machines are in the equity position, and there is no control center among these entities. That is, intra-host entities employ hierarchical structure, while inter-host entities are egality. The function of entities was first described in my paper [5], but it was not explicated clearly. And then, the detailed description is given in this paper.

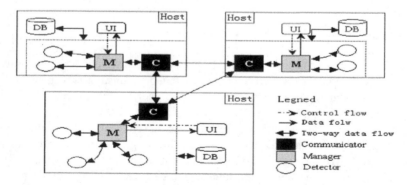

Fig. 1. The architecture of IADIDF.

2.2   Detector

Detector is the basic detection unit in this framework. There may be as many as possible detectors in one host, with their responsibilities for monitoring the operating status of host, and reporting abnormal or interesting behaviors to higher-level entity.

Detector is a component that operates independently with self-governed data sources and detection modes, and it can be written in any programming language. It is able to fulfill the task of detection alone, or many detectors cooperate one another to take actions. Detector does not produce any alarm. Generally, if the manager receives one or more reports from detectors, it will send an alarm to users.

The architecture does not specify any limitations for the functionality of a detector, so it is a part that has more flexibility in the model. It can monitor a certain special incident of host (for example, counting the number of tcp connections per unit time), or be complicated software too (for example, a lightweight intrusion detection system Snort). So long as a procedure can read information and write these to manager in the appropriate format, it can run as a detector of the system.

2.3 Manager

Inside one host, manager is the highest-level entity. It controls the operation of every detector, which includes starting and stopping a detector; checking the working state of a detector and reporting to system administrator; gathering, analyzing and simplifying the reports generated by subordinate entity detector, and producing an alarm. Furthermore, it plays the important role of transmitting messages between detectors that are in the same host. When a detector getting initialized, it will register itself to its upper entity manager. In this way, manager will hold all IDs of the detectors it controlled. When a detector needs intra-host communication, it merely gives the ID of the aim detector, then manager will redirect the standard input and output of source detector to aim detector. By this mechanism, intra-host communication and cooperation are realized.

2.4 Communicator

If a communicator and a manager are inter-host entities on different hosts, they are in equal position, while not in layer structure on logic. Communicator is responsible for setting up and maintaining communication channels. There is only one communicator on each host, which acts as the bridge of communications between cooperating hosts. When a detector has data to send, it appoints the aim detector (including identification of aim host and that detector), converts data into message format according to the definition of MESSAGE class, and then forwards it to upper entity manager. After receiving this message, manager shall judge if the destination of this message is in local machine or not. If not, manager passes it to communicator. Communicator firstly checks if there is a communication channel (TCP connection) between source and goal host. If there is one, messages are sent directly, otherwise establishing connection after consulting. When the communicator on remote host accepts messages, it will pass it in reverse way. As the connection channel is established, it keeps open, thus, the following messages can be transmitted simply.

Communicator is the intermediary part and offers route service in the course of messages conveying. In the system, two kinds of data package exist: broadcast package and directional package. Broadcast package is the data package that is sent to all hosts in network besides its own host, while directional package is sent to a certain specific host. Communicator must be capable of discerning and receiving these two kinds packages. Broadcasting of messages would be the easiest. It is because that a TCP/IP network would require a point-to-point protocol, a host name file needs to be maintained by each host.

The inter-host and intra-host communication services of communicators and managers have similar place, but the key difference is that, managers just redirect standard input and output during messages transmitting, while communicators need to choose and update route information, establish and maintain connection channels etc. This is the reason that we select an independent agent for the working of inter-host communication.

2.5 Other Entities

The components of system include user interface (UI) and database (DB) also. Administrator may issue orders in real-time and intercommunicate with system through UI. Database is the entity, which is used to store activities of users and data forwarded by managers, for the sake of being a knowledge base for entity to study and incident matching.

2.6 Agent Communication

Agents in this framework communicate and collaborate with peers, superiors and subordinates using an agent communication language and protocol called Knowledge Query and Manipulation Language (KQML). The KQML is an initiative of the DARPA Knowledge sharing Initiative External Interfaces Working Group and is designed specially for supporting communications between software agents [10]. It supports a variety of message types called performatives. In the original language specification, there were 37 different performative types. This virtual knowledge contains information about a given agent's beliefs and goals [11].

The communication mechanism is composed of three layers: Transport, Hosts and MessageContent, which is shown in Figure 2. Transport layer provides the most basic communication services. Commercial standards offer a number of mechanisms for basic communication [4], such as TCP, SNMP, CORBA. In our system, Unix pipe is selected for communication within the same host, and TCP connections for communication over the network, which is used in the communication mechanism in AAFID [2]. Hosts layer is constitutive of two parts, which identify the hosts that send or receive messages respectively. MessageContent layer is composed of three fields, that is <MessageContent>::=<MType><MData><MTimestamp>. This generic structure can be used by entities to represent any type of information. The MType field includes all predefined performatives of KQML, and determines the format of MData field.

MessageContent	MType
	MData
	MTimestamp
Hosts	Receiver
	Sender
Transport	TCP
	Pipe

Fig. 2. Three layers of communication mechanism.

2.7 Alert Mechanism

Detectors are responsible for monitoring specific activities and events, while not raising any alarm. This assignment will be done by monitors based on the *State* value reported from detectors. Two main factors are used to describe *State* value: *danger* and *transferability* [9]. The *danger* is defined as the potential damage that can occur if the attack at a particular stage is allowed to continue, with values from 1 to 4. The transferability is defined as the applicability of an attack in other nodes in a network, with values form 1 to 3. The detailed description was omitted here, which can be found in my paper [5].

Then, we get *State* value by the following formula:

$$State = danger * transferability .\qquad(1)$$

With the values given above, *State* will range from 1 to 12. Then, the alert level can be evaluated:

√ Normal: *danger*<2;   √ Partial alert: *danger*=2;

√ Alert: *danger*=3;   √ Full alert: *danger*=4;

The higher alert levels should produce alerts. According to the *transferability*, we deploy the scope of alarms (1: local machine; 2: multicasting; 3: broadcasting).

3   Representative Agents Realization

Three agents, which used the methods of misuse detection (3.1, 3.3) and anomaly detection (3.2) individually, were realized in this paper.

3.1   File Access Privilege Detector

Our system was deployed under Linux platform. As a multi-user operating system, access rights are the first line to protect file security. However, rsh and rlogin commands can be used to login without a password using the '.rhosts' authority file for trusted hosts and users. Thus, the access rights of some critical files are what we will pay close attention to, such as the rhosts file we mentioned above.

Regards checking the SUID status as the example. SUID permission allows unprivileged users the ability to accomplish certain, privileged tasks, and SUID enabled programs will execute with the access rights of the owner and not the executing user. If the owner has the privilege of root, the executer will be a supervisor to take any actions. We monitor it according to the follows: (1)No illegal settings, *Danger* equals 0; (2)One file set, *Danger* equals 1; (3)Multi files set, *Danger* equals 2; (4)If one of above files has root privilege, *Danger* equals 4. The detectors merely run in some special host, so we set *transferability* equal to 1, which means only alert in local machine.

3.2 Login Detector

User login information are usually the data source for statistic-based analyse. Firstly, the threshold of normal behavior is set in advance, then if the actual behavior approaches or exceeds that value, alerts will be raised. A simple example is doorknob attack, the intruder generally tries a few common account and password combinations on each of a number of computers. These simple attacks can be remarkably successful. For instance, 2 times inaccurate loggings are taken for normal, while 30 times may be intrusive. Consider a situation that the intruder distributes the attacks into 15 hosts and each host performs 2 attempts, no abnormal actions are found. Obviously, attacking is under way.

During the course of detecting, if the detector finds many login attempts to host, and the soure addresses are same, but the suspicion degree (logging times) is lower than threshold, we assume that may be a instance of doorknob attack. We set *danger* value to 2, since it is not yet sure to be an attack. Simultaneously, we record login data (user name, frequency etc.), and register an access list. As both Unix/Linux and Windows operating systems are analogous, and all machines in the network are likely to be affectd, detector sets *transferability* equal to 3. That means, the suspicious user's logging information are sent to all the other hosts. After certain host receives this information, local detector will check if it has a record of that user and plus these two parts together. In case the suspicion degree is bigger than threshold, that user are considered as an intruder and *danger* is set to 3. Detector boardcasts this intrusive message to all the other hosts in the network, and asks for attention.

3.3 Network Data Detector

Network intrusion detection uses network traffic as data source, and employs techniques like 'packet-sniffing' to pull data from TCP/IP or other protocol packets traveling along the network. In our framework, we adopt libpcap (lib for packet capture), a system-independent interface for user-level packet capture, as our tool to monitor traffic. Detectors are realized to surveil below attacks, including TCP/UDP port scan, SYN flooding, Ping flood, Teardrop, Land, and Smurf.

Give an example of detecting SYN flooding attack, the course is described with the following steps: (1) Insert new receiving packet into destination address list based on the destination address (*daddr*) of that packet. The list records all the packets with the destination addr. (2) Take out the fame header and suppose we get a TCP entity. If *RST* bit is set to 1, then we drop it; else if *SYN*=1, then we judge whether Land attack happens; else go to next setp; (3) Insert that segment into *TCPList* and sort by source port and destination port. (4) If *SYN* equals 1, then insert it into *SYNpktList*; else if

FIN=1 or *ACK*=1, then clear that node form List. (5) In time interval *T*, searches *SYNpktList*. If the nodes number is bigger than the threshold we set before for allowable waiting TCP number, we believe SYN flooding attack happens. Since this type of attack is concerned with OS and protocol, *Transferability* is set to 2, and multicasts that intrusive message.

4   Evaluation

In the system, detector is an independent entity, which can analyze data and detect intrusions individually. If a manager or a communicator is out of action, only detectors that belong to the same host are affectd, for example data collecting and message transferring are blocked, while other detectors still run well without any trouble. In addition, if there is a highest-level entity in a system, the problem of bottleneck will occur inevitably, which is opposite to the principle of distributed detection. Fortunately, no such entity exists in the prototype. Manager provides a user interface, and is distributed. It can record the detection information and system status of corresponding host, and ask for the statuses of other mangers through the services offered by communications. Therefore, even if a manager is destroyed, it will not influence the operating of whole system. From above mentioned, we can conclude that the prototype can avoid the problem of a single point of failure effectively.

In experiments, we use Red Hat Linux 7.2 as the operating system, and hardware devices are CPU: PIII800, Memory: 128M, Harddisk: 20G, and 10M D-link network card. Program language is C, and Libpcap is used to monitor network traffic. In order to receive data continuously, we firstly capture an amount of packets as source data, and then deal with these. Rate of utilization before and after agents running is shown in Table 1. Results indicate that the operating of agents will not impact system performance heavily.

Table 1. CPU and Memory utilization rate.

	CPU	Memory
Before running	33.5%	76.9%
After running	36.2%	79.4%

5   Conclusion

We propose architecture for intrusion detection called IADIDF, which is based on independent agents and employs distributed structure. The functionality of entities is designed, also communication and alert mechanism is defined. This model is an open system with fine extensibility. The system must scale as the network grows. Since the agents run independently of each other, they can be added without adversely affecting the performance of the other agents. The IADIDF performs distributed data collection and analysis. Cooperative agents in system collaborate each other equally. We adopt the mode that does not have control center, which avoid the matter of a single point of failure. The architecture does not specify any limitations for the functionality of detector, so the power and scale of it can be great or trivial, and any intrusion tector, so the power and scale of it can be great or trivial, and any intrusion detection technique may be applied. Agents are independent generally, however to the complicated attack, they can cooperate one another, and the cooperation is realized by communication. When running, agents are daemons and cost low system resources and network bandwidth. Although the prototype is realized under Linux platform, it is easy to migrate into other platforms for independence of system environment.

References

1. J. P. Anderson: Computer Security Threat Monitoring and Surveillance. James P. Anderson Co., Fort Washington (1980)
2. Jai S. Balasubramaniyan, Jose O. Garcia-Fernandez, David Isacoff, Eugene Spafford, Diego Zamboni: An Architecture for Intrusion Detection Using Autonomous Agents. CERIAS Technical Report, Purdue University (1998)
3. Wenke Lee: A Data Mining Framework for Constructing Features and Models for Intrusion Detection Systems. Computer science department, Columbia University (1999)
4. Ming-Yuh Huang, Robert J. Jasper, Thomas M. Wicks: A Large Scale Distributed Intrusion Detection Framework based on Attack Strategy Analysis. Computer Networks, Vol. 31. Elsevier Science (1999) 2465-2475
5. Ye Du, Huiqiang Wang, Yonggang Pang: A Framework for Intrusion Detection. In: Proceedings of 3rd International Conference on Grid and Cooperative Computing, Wuhan, China (2004)
6. Theuns Verwoerd, Ray Hunt: Intrusion Detection Techniques and Approaches. Computer Communications, Vol. 25 (2002) 1356-1365
7. T.H. Ptacek, T.N. Newsham: Insertion, Evasion, and Denial of Service: Eluding Network Intrusion Detection. Secure Networks Inc. (1998)
8. Müller H J: Negotiation principles. In: O'Hare, Jennings N. R. (eds.): Foundation of Distributed Artificial Intelligence 6th Generation Computer Technology Series. John Wiley & Sons Inc., New York (1996) 211-229
9. Joseph Barrus, Neil C. Rowe: A Distributed Autonomous-Agent Network-Intrusion Detection and Response System.
http://www.cs.nps.navy.mil/people/faculty/rowe/barr- uspap.html
10. Y. Labrou, T. Finin: A Proposal for A New KQML Specification. Technical TR CS-97-03, University of Maryland (1997)
11. T. Finin, Y. labrou, J. Mayfield: KQML As An Agent Communication Language. In: J. M. Bradshaw (eds.): Software Agents. Menlo Park, AAAI Press (1997) 291-316

Discovery of Hidden Similarity on Collaborative Filtering to Overcome Sparsity Problem

Sanghack Lee, Jihoon Yang, and Sung-Yong Park

Department of Computer Science, Sogang University
1 Shinsoo-Dong, Mapo-Ku, Seoul 121-742, Korea
shlee@mllab.sogang.ac.kr
{jhyang,parksy}@ccs.sogang.ac.kr

Abstract. This paper presents a method for overcoming sparsity problem of collaborative filtering system. The proposed method is based on an intuition on a network of human (such as a friendship network with friends of a friend). This method increases the density of similarity matrix and the coverage of predictions. We use sparse training data to test the sparsity of real-world situation. Consequently, experimental results show that this method increases coverage and f-measure especially for sparse training data.

1 Introduction

To get high quality information in some domains, we have to experience the information directly or indirectly. But we can't experience all the information directly. Word-of-mouth information are often helpful when we choose a decision without direct experience.

A recommender system discovers and renders items (e.g. movies, books, etc.) which suits our preferences. Many recommender systems have been developed with collaborative filtering techniques. The terms collaborative filtering and recommender systems are used interchangeably. Content based filtering is one of information filtering techniques which uses context of items. On the other hand, collaborative filtering doesn't make use of context of items which makes it possible to recommend serendipitous items [1, 2]. Collaborative filtering is based on opinions of similar users who buys or rates same items.

However, there are two main limits in collaborative filtering: *scalability* and *sparsity* problems.

- *Scalability*: In e-commerce environment, there might be millions of customers and catalog items. Traditional collaborative filtering algorithms thus require expensive time and space complexity [3].
- *Sparsity*: With millions of items, we can't experience even 1% of them. For instance, Amazon.com has several millions of catalog items and 1% of them are more than 10,000 items which people can hardly access [3].

We will focus on the sparsity problem and suggest an approach to overcoming it. Previous works for overcoming the sparsity problem include combining content

E. Suzuki and S. Arikawa (Eds.): DS 2004, LNAI 3245, pp. 396–402, 2004.
© Springer-Verlag Berlin Heidelberg 2004

based filtering and collaborative filtering to link contextual information among items [4, 5] and clustering items or users to reduce dimensionality [6]. (You can find more detailed descriptions in Section 4.) In this paper, we will explain how to increase coverage based on a network of similar users and the transitivity of similarity.

2 Method

A main idea of this paper is connecting or relating two users who don't have any co-rated items. We keep meeting people as friends of a friend in everyday life. Friends of a friend construct a network of human with friendship. Changing a friendship network into a similarity network, users that are connected to similar set of neighbors (middlemen) construct a network of users with similarity. Using this scheme any two users with common middlemen can be connected in the network.

As shown in Fig. 1, discovering hidden similarity (hereinafter referred to DHS) computes a similarity matrix among users based on their rates on items, and modifies the matrix incorporating the discovered hidden similarity.

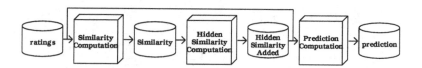

Fig. 1. A simple architecture of DHS based collaborative filtering system.

2.1 Similarity Computation

The Pearson correlation and the cosine based similarity computation is mostly used in traditional collaborative filtering recommendation systems [7]. In this paper, we will use these two basic similarity computation methods:

– *Pearson Correlation*

$$w_{a,i} = \frac{\sum_{j=1}^{N} (v_{a,j} - \bar{v}_a)(v_{i,j} - \bar{v}_i)}{\sqrt{\sum_{j=1}^{N} (v_{a,j} - \bar{v}_a)^2} \sqrt{\sum_{j=1}^{N} (v_{i,j} - \bar{v}_i)^2}} \tag{1}$$

– *Cosine Based*

$$w_{a,i} = \frac{\sum_{j=1}^{N} v_{a,j} v_{i,j}}{\sqrt{\sum_{j=1}^{N} v_{a,j}^2} \sqrt{\sum_{j=1}^{N} v_{i,j}^2}} \tag{2}$$

where, $w_{a,i}$ is a similarity between two users a and i, N is the number of users in the system, $v_{i,j}$ is rating for item j by user i, and $\bar{v}_i$ is the average rating of user i. (See [7] for detailed descriptions.)

2.2   Hidden Similarity Computation

The discovery hidden similarity (DHS) is computed as follows:

$$w_{a,i} = \frac{\sum_k |w_{a,k}| \frac{\sum_k w_{a,k} w_{k,i}}{\sum_k |w_{a,k}|} + \sum_k |w_{k,i}| \frac{\sum_k w_{a,k} w_{k,i}}{\sum_k |w_{k,i}|}}{\sum_k |w_{a,k}| + \sum_k |w_{k,i}|}$$

$$= \frac{2 \sum_k w_{a,k} w_{k,i}}{\sum_k |w_{a,k}| + \sum_k |w_{k,i}|} \tag{3}$$

where, a and i are users without co-rated items, k is a middleman who have co-rated items for both a and i, and $w_{a,i}$ is the weighted sum of two similarity predictions. The numerator consists of two terms: one is the weighted sum of $w_{i,k}$ and the other is the weighted sum of $w_{a,k}$. The similarity predictions among users in DHS are similar to the prediction computation in Section 2.3.

2.3   Prediction Computation

– *Prediction for Pearson Correlation*

$$p_{a,j} = \bar{v}_a + \frac{\sum_{i=1}^{N} w_{a,i}(v_{i,j} - \bar{v}_i)}{\sum_{i=1}^{N} |w_{a,i}|} \tag{4}$$

– *Prediction for Cosine Based*

$$p_{a,j} = \frac{\sum_{i=1}^{N} w_{a,i} v_{i,j}}{\sum_{i=1}^{N} w_{a,i}} \tag{5}$$

where $p_{a,j}$ is the prediction for an item j of a user a. Predicted items are recommended when predicted ratings of the items are over a threshold. In our experiments, the recommendation threshold is set to 3.5 because a range of the ratings is from 1 to 5.

The hidden similarity computation in Section 2.2 has the same purpose as that for the similarity computation. Both computations yield a similarity matrix among users. By computing the hidden similarity, the similarity matrix becomes less sparse and the system overcomes the sparsity of the user similarities and the predicted ratings. Note that the equation for the hidden similarity computation (3) is similar to the equations for the prediction computation ((4) and (5)). Prediction computation predicts a rating as the weighted sum of ratings based on the similarity, and the hidden similarity computation predicts a hidden similarity as the weighted sum of similarities based on similarity. The difference between two computations is that the former is directional while the latter is not. Fig. 2 depicts these three computations.

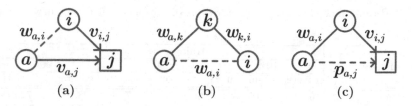

Fig. 2. Comparison among (a) similarity computation, (b) hidden similarity computation, and (c) prediction computation. Circles and rectangulars are users and items, respectively. Solid lines and dotted lines are known and unknown values, respectively.

3 Experiments

3.1 Dataset

In our experiments, we used MovieLens[1] data which contains 100,000 ratings from 943 users and 1682 movies. The sparsity of the data is 93.7%, and the number of average rated movies per user is about 106.

We then randomly divided the ratings into two parts, training and test data. We performed experiments from 2-fold to 10-fold cross validations, and additional cross validations interchanging the training and test data in order to evaluate our approach in a real-world situation where training data is very sparse.

Several different sizes of middlemen (k) in Equation 3 was tested with 5, 10, and without any restrictions.

3.2 Evaluation Metrics

There are several metrics to evaluate a recommender system. The recommendation quality can be measured by *precision* and *mean absolute error* (MAE). *Precision* is the number of correctly recommended movies over the number of all recommended movies. *MAE* is the average difference between the real ratings and predictions. *Recall* and *coverage* are used for evaluating quantity of recommendation. *Recall* is the number of correctly recommended movies over the number of all liked movies. *Coverage* is the number of predicted movies over the number of all movies in the test data. In addition, *f-measure* is the weighted combination of *precision* and *recall* as $f_1 = \frac{2 precision \times recall}{precision + recall}$ which evaluates quality and quantity of the recommendation. We will use *coverage* and *f-measure* to evaluate both quality and quantity of our recommender system.

3.3 Results

Table 1 and Fig. 3 show the effect of DHS on coverage and f-measure. Experiments with a restriction on the size of middlemen makes a slight difference in

---

[1] http://movielens.umn.edu

Table 1. Performance of DHS-based algorithms on two methods of similarity computation.

method		training percentage					
		10%		50%		90%	
		coverage	f-measure	coverage	f-measure	coverage	f-measure
Pearson	pure	68.20%	63.00%	96.79%	70.70%	98.16%	72.87%
correlation	DHS	**93.38%**	63.76%	**99.65%**	**72.05%**	**99.86%**	**73.93%**
based	10-DHS	**93.38%**	63.78%	**99.65%**	72.03%	**99.86%**	**73.93%**
	5-DHS	**93.38%**	**63.83%**	**99.65%**	72.03%	**99.86%**	**73.93%**
	pure	70.55%	65.88%	96.79%	69.22%	98.16%	69.65%
cosine	DHS	**96.66%**	**68.62%**	**99.65%**	**70.21%**	**99.86%**	**70.50%**
based	10-DHS	**96.66%**	**68.62%**	**99.65%**	70.20%	**99.86%**	**70.50%**
	5-DHS	**96.66%**	68.60%	**99.65%**	70.20%	**99.86%**	70.49%

f-measure. Without restriction, the average size was 9.42, 211.58, and 286.57 for 10%, 50%, and 90%, respectively.

As we can see, the coverage increased in DHS-based methods. In particular, the coverage increased significantly when the training data is sparse. The coverage strongly depends on the density of similarity matrix (i.e. Hidden Similarity Added in Fig. 1) because predictions per user is a union of set of movies rated by neighbors.

Surprisingly, f-measure was increased slightly against our expectation. We found that the predictions via the hidden similarity added would not be as good as via pure similarity at first. Because the hidden similarity is predicted based on the pure similarity and then the quality of the hidden may be worse than the pure. But actually, the reliability of user similarity by the Pearson correlation or the cosine based is determined by the number of co-rated items [8]. In that manner, the initially derived similarity (i.e. pure similarity) has low reliability because of sparse ratings, and adding combined similarities (i.e. hidden similarity added) yields higher reliability. Also the increase of the number

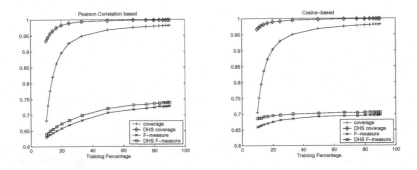

Fig. 3. Performance graphs with training percentage from 10 to 90.

of available neighbors for the prediction causes better results [1]. Consequently, the DHS may have higher reliability and quality than the similarity based on sparse ratings.

4   Related Work

Collaborative filtering has been studied and varied in many ways. Its diverse methods can be categorized by several criteria [9]. One of the criteria is hybrid algorithms versus plain algorithms.

Plain algorithms are based on the user-item matrix, the user correlation, and their revised information. For example, item based collaborative filtering which alternates the role of users and items [10], LSI/SVD as a dimensionality reduction technique [11], and clustering algorithms for large dimensional sparse vectors without contextual information [12] have been developed.

Hybrid algorithms often use demographic or content data to achieve high quality recommendation. For instance, collaborative filtering combined with content based methods [5, 13], a content-boosted method to create a pseudo user-ratings vector for overcoming sparsity [4], collaborative filtering via content and demographic data based approach [14], and clustering items to improve the quality [6] have been designed.

Our method is a plain algorithms based on the DHS that renders denser user similarity.

5   Summary and Future Work

In this paper, we have presented the DHS algorithm to overcome the sparsity problem by connecting more users. We have found that similarities through not only movie ratings but also user similarities work well. Major contribution is that hidden similarity computation have showed how to capture a transitive similarity relationship with simple and neat equations [15]. Even though our proposed method is based on the intuition of the human network, it increases coverage of traditional collaborative filtering systems without losing precision and recall. Future works will include extension of DHS:

- *System integration:* We suggest integrating systems with number of co-users in the same domain. Systems with content based filtering and with collaborative filtering also can be integrated by combining their similarity matrices into a single similarity matrix.
- *Repeated DHS:* DHS discovers the similarities between two users when the length of a shortest path between two users is 2. This fact implies two users of d-length shortest path can compute the similarity by repeating DHS at most $\lceil \log_2 d \rceil$ times because the length of shortest path is halved in every execution of DHS. In this manner, we can discover similarities between any connected users.

References

1. Herlocker, J.L., Konstan, J.A., Borchers, A., Riedl, J.: An algorithmic framework for performing collaborative filtering. In: SIGIR '99: Proceedings of the 22nd Annual International ACM SIGIR Conference on Research and Development in Information Retrieval, Berkeley, CA, ACM Press (1999) 230–237
2. Polcicova, G., Slovak, R., Navrat, P.: Combining content-based and collaborative filtering. In: ADBIS-DASFAA Symposium. (2000) 118–127
3. Linden, G., Smith, B., York, J.: Amazon.com recommendations: Item-to-item collaborative filtering. IEEE Internet Computing 7 (2003) 76–80
4. Melville, P., Mooney, R.J., Nagarajan, R.: Content-boosted collaborative filtering for improved recommendations. In: Proceedings of the Eighteenth National Conference on Artificial Intelligence, Edmonton, Canada (2002) 187–192
5. Maneeroj, S., Kanai, H., Hakozaki, K.: Combining dynamic agents and collaborative filtering without sparsity rating problem for better recommendation quality. In: Proceedings of the Second DELOS Network of Excellence Workshop on Personalisation and Recommender Systems in Digital Libraries. (2001)
6. O'Connor, M., Herlocker, J.: Clustering items for collaborative filtering. In: Recommender Systems Workshop at 1999 Conference on Research and Development in Information Retrieval, Berkeley, CA (1999)
7. Heckerman, D., Kadie, C., Breese, J.S.: Empirical analysis of predictive algorithms for collaborative filtering. In: Uncertainty in Artificial Intelligence. Proceedings of the Fourteenth Conference. (1998) 43–52
8. Ariyoshi, Y.: Improvement of combination information filtering method based on reliabilities. IPSJ SIGNotes Fundamental Infology (1998)
9. Vozalis, E., Margaritis, K.G.: Analysis of recommender systems' algorithms. In: Proceedings of the Sixth Hellenic-European Conference on Computer Mathematics and its Applications - HERCMA 2003. (2003)
10. Sarwar, B.M., Karypis, G., Konstan, J.A., Riedl, J.: Item-based collaborative filtering recommendation algorithms. In: World Wide Web. (2001) 285–295
11. Sarwar, B., Karypis, G., Konstan, J., Riedl, J.: Application of dimensionality reduction in recommender systems – a case study. In: ACM WebKDD 2000 Web Mining for E-Commerce Workshop. (2000)
12. Kohrs, A., Merialdo, B.: Clustering for collaborative filtering applications (1999)
13. Good, N., Ben Schafer, J., Konstan, J.A., Borchers, A., Sarwar, B., Herlocker, J., Riedl, J.: Combining collaborative filtering with personal agents for better recommendations. In: Proceedings of the 6th National Conference on Artificial Intelligence (AAAI-99); Proceedings of the 11th Conference on Innovative Applications of Artificial Intelligence. (1999) 439–446
14. Pazzani, M.J.: A framework for collaborative, content-based and demographic filtering. Artificial Intelligence Review 13 (1999) 393–408
15. Billsus, D., Pazzani, M.J.: Learning collaborative information filters. In: Proceedings of the Fifteenth International Conference on Machine Learning (ICML-98), Madison, WI, Morgan Kaufmann (1998) 46–54

Seamlessly Supporting Combined Knowledge Discovery and Query Answering: A Case Study

Marcelo A.T. Aragão and Alvaro A.A. Fernandes

Department of Computer Science
University of Manchester
{m.aragao,a.fernandes}@cs.man.ac.uk

Abstract. Inductive and deductive inference are both essential ingredients of scientific activity. This paper takes a database-centred view some of the crucial issues arising in any attempt to combine these two modes of inference. It explores how a recently-proposed class of database systems (that support the execution of composite tasks, each of whose steps may involve knowledge discovery, as an inductive process, and or query answering, as a deductive one) might deliver significant benefits in the context of a case study where the specific characteristics of such systems can be more vividly perceived as being relevant and nontrivial.

Introduction

One of the central obstacles to improving the level of automated support for scientific discovery in data-rich and knowledge-poor areas such as genomics is that only in the last few years has there been a concerted attempt to reconcile inductive and deductive inferential processes. As a result of this, the rate of growth of data stocks (e.g., in genomics, nucleotide sequences) has outstripped by far the rate of growth in knowledge stocks (e.g., in genomics, knowledge of protein function). This paper uses a case study that, although not set in a scientific domain, crisply illustrates the significant difference (in comparison with the decoupled tools currently available) made by a recently proposed class of logic-based database systems that seamlessly combines arbitrarily complex inductive (in the form of knowledge discovery) and deductive (in the form of query answering) steps. These systems are referred to as **combined inference database** (CID) **system(s)** (CID(S)) and are founded on *parametric Datalog* (or p-Datalog, for short) [12]. The detailed technical development of CIDSs was reported in [5, 4, 6] to which the reader is referred.

A Case Study

In open economies, banks hold foreign currency stocks resulting from their customers' trade on imports and exports. This gives rise to an interbank market where currency surpluses and deficits are traded. This market acts to balance currency supply and demand and provides banks with profit opportunities when

E. Suzuki and S. Arikawa (Eds.): DS 2004, LNAI 3245, pp. 403–411, 2004.

prices fluctuate and is regulated by each country's *bank supervising authority* (BSA). BSAs set limits and margins which banks must abide by so that risks are managed, as stipulated in international agreements. However, in their attempt to maximize profits, banks often resort to speculative practices that can lead to severe losses if the market shifts suddenly. Since, in the limit, such practices may undermine the economic stability of entire countries and even propagate to linked economies, BSAs must elicit the knowledge that will justify intervention, coercion and, ultimately, revision of regulations.

BSAs have a team of inspectors to analyse the data on banks and on their transactions throughout the day. Their goal is to shortlist those banks, if any, whose pratices merit closer scrutiny. However, these practices are, by definition, hard to spot, because they often exploit loopholes in the regulations and typically arise by composition of several, individually-legitimate transactions. They tend, therefore, to lie hidden in the huge volume of interbank trade. Worse still, banks roughly know (or are adept in predicting) where inspection is likely to focus. Provided there is the intention to do so, they are often able to cover signals that could be more easily identified.

From the viewpoint of the inspectors, the target is clearly an elusive, and moving, one: they must learn quickly, and then learn again and again, which practices are being deployed that may merit intervention. Thus, in trying to keep abreast with constantly changing practices, BSA inspectors would benefit from automated, flexible, expressive support for composing knowledge discovery with query answering steps. The ability to do so could underpin an on-demand, exploratory mode of interaction with the data that best serves the aim of eliciting new (or refreshing stale) knowledge about potentially risky practices.

The problem faced by BSAs is that eliciting this knowledge in this particular way requires combining knowledge discovery and query answering steps seamlessly (i.e., without incurring costs and delays that would defeat the purpose of swift response) and reliably (i.e., in a principled, justifiable manner as behoves the BSA's crucial role in promoting economic stability). Unfortunately, if one takes a representative range of knowledge discovery techniques, one finds that state-of-the-art support for combining knowledge discovery and query answering steps is neither seamless (insofar as existing tools are independent, totally decoupled, from one another and hence offer no compositionality unless glueware is specifically developed for that purpose) nor reliable (insofar as there is no principled way to automate the piping of the outputs of one tool into another that considers the validity assessments of the original and intermediate outcomes) across that range.

Alternative Task Models

Table 1 contrasts two task models. To the left, one based on the best support that can be offered by decoupled tools (granting that specific products may offer specifically integrated support for specific combinations of knowledge discovery and query answering steps, e.g., a classification model can be constructed and

Table 1. Task Models for BSA Inspectors: Contrasting Quality of Support.

Decoupled Tools	CIDS
1. Inspectors have at their disposal DBMSs and data mining workbenches. 2. Throughout the day, they make use prepared interfaces to the tools in (1) that they have commissioned the IT staff to implement so that they can retrieve, clean and transform the data they wish to analyse. 3. They may, e.g., execute prepared queries over the DBMSs. For example, they may seek for specific indicators, e.g., banks effecting unusually high-valued transactions, or operating too close to their legal limit. 4. They may also execute mining scripts that, again, the IT staff has prepared to help them search for promising models of the data in (2). 5. Steps (3) and (4) above may lead an inspector to request the IT staff to integrate a promising model so that it can be invoked from the queries in (3) and to expect this model to be deployed with minimal delay. 6. Inspectors constantly make notes about their explorations, since they will need to explain the knowledge they have elicited and provide the data that supports the former. Such explanations will perforce rely on the various tasks on various tools in (1) that will need to be structured, somehow, so as to aid understanding. 7. At close of play, the inspectors draw a shortlist of any targets for inspection they have succeeded in identifying.	1. Inspectors have at their disposal a CIDS that is configured with the necessary inference rules and populated with the same data as in (2) to the left and, in addition, domain knowledge (e.g., regulations, best practices) in the form of logical rules. 2. Throughout the day, inspectors rely only on this CIDS. Because combined inference tasks are declarative, expressive and capable of being composed with closure, and because the interaction model supports on-demand explorations, each inspector can him or herself devise and evaluate CID tasks without resorting to IT staff. 3. They may, e.g., interactively explore the data either making no assumptions about it or starting from accumulated background knowledge. They can compose querying and discovering steps under the assurance that closure is guaranteed by uniform representations, and compositionality, by a principled treatment of validity assessments. 4. The declarative nature of CID tasks and the uniform clausal representation mean that a simple log of tasks executed and results obtained comes close to being self-explanatory and may stand in lieu of ad-hoc notes. 5. At close of play, the inspectors draw a shortlist of any targets for inspection by simply consolidating their explorations into one combined CID task that, upon execution, derives that shortlist.

then invoked in a prediction join in Microsoft SQL Server). To the right, the alternative offered by CIDS, which can seamlessly and reliably combine knowledge discovery and query answering steps.

Table 1 both describes and comments on the alternative task models, thus hinting at usability benefits in deploying CIDS when compared with decoupled tools for query answering and knowledge discovery. The remainder of this section discusses in detail one instance of what inspector might use a CIDS to explore and how they can go about it. The problem is real and is known in the relevant literature as *the tandem strategy* since it involves co-ordinated action between different banks.

Characterizing Cases of Tandem Strategy

The task for a BSA inspector is to shortlist banks for inspection. The case study describes how CIDs would help identify candidates for inspection. Insofar as a goal of the paper is to argue for the usability benefits of CIDS, the description below is cast in the form of an inspector's possible train of thought. The inspector starts from the belief that transactions between medium-sized banks with similar business profile are often similar. Violation of this assumption raises the suspicion of collusion between the banks involved. The inspector is after sophisticated speculative practice in which (often volatile) close partnerships are formed between several banks.

The task model is the one to the right in Table 1. The CIDS, in this case, stores data on banks and their assets, their buying and selling limits, their exposure to risk and their position (i.e., their balance) for a given time frame in days. It also stores data on transactions (e.g., the banks involved, the currency in question, the value, the contracted date, the time frame for clearance). Transactions that only differ in their id are grouped into a entry in a transaction group relation and annotated with a count of the number of tuples that share those attribute values. The extent of such relations would contain p-Datalog clauses such as:

bank(b1, commercial, gru, international, 2000M)↤1.0⊢
risk_exposure(b1, 30, 70M)↤0.9⊢
balance(b1, 30, 40M, bought)↤1.0⊢
transaction(b1, b3, usd, low, 20031107, 30, 0.41M)↤1.0⊢
limits(Bank, 1M, 10M)↤1.0⊢bank(Bank, _, _, _, Assets), Assets> 1000M

Note that numeric literals are simplified and note the validity assessments [5, 4, 6] that annotate each p-Datalog clause. In particular, the validity assessment of 0.9 in the second fact above asserts that the computed risk exposure measure is thought to be less than certain. One of the most important contributions of CIDS is to provide a foundation for the principled treatment of validity assessments.

Note also that while all other relations are extensionally defined in the EDB, limits/3 is a rule-defined view in the IDB. Thus, CIDs can use rules to capture simple norms and regulations, such as exemplified above. As an another example, the inspector can explicitly represent the operating assumption with the following clause:

supervision_target(Bank)←
    mid_sized(Bank),
    mid_sized(Partner_Bank),
    similar_banks(Bank, Partner_Bank),
    transaction_group(Transaction1, Bank, Partner_Bank, 20030510, _, _, _, _, Count1),
    transaction_group(Transaction2, Bank, Partner_Bank, 20030510, _, _, _, _, Count2),
    dissimilar_transactions(Transaction1, Transaction2).

The extent of this rule contains mid-sized banks that have engaged in dissimilar transaction with similar partners. The inspector might then formalize the intuition about what is a mid-sized bank as follows:

mid_sized(Bank)←1.0⊢bank(Bank,_,_,_,_,Assets),Assets > 10M,Assets < 100M

So far, all the specified information can be obtained via query answering, i.e., deductively. However, the concepts of *similar_banks* and *dissimilar_transactions* appeal, in this case, to knowledge that is sufficiently complex and volatile to merit being induced on the fly, instead of axiomatically asserted once and for all. Thus, the inspector plans for these relations to be defined in terms of inducible ones. These inducible relations can be taken to be those resulting from a clustering step, in the sense that banks that are placed in the same cluster are taken thereby to be similar, and likewise for transactions. This means that, provided the inspector is able to induce clusters on banks and on transactions (i.e., bank_cluster/2 and transaction_cluster/2 resp., below), s/he can capture bank similarity and transaction dissimilarity in the obvious way, as shown to the left and right below, respectively:

similar_banks(Bank1, Bank2)←1.0⊢   dissimilar_transactions(Transct1, Transct2)←1.0⊢
    bank_cluster(Bank1, Cluster1),   transaction_cluster(Transct1, Cluster1),
    bank_cluster(Bank2, Cluster2),   transaction_cluster(Transct2, Cluster2),
    Cluster1 = Cluster2.   Cluster1 ≠ Cluster2.

Thus, the computational implementation of this investigative approach can be seen to depend on deductive and inductive steps. Indeed, the case above is an instantiation of a hypothetico-deductive cycle not unlike those that empirical science often relies upon, with the induction of the clusters being followed by their use in deducing similarity relations over a particular body of evidence.

Specifying CID Tasks

CIDS, in this case study, have a significant, positive usability impact because they support the seamless combination of inductive, e.g., clustering, steps and classical querying, i.e., deductive, steps. Once the inspector has come up with a problem-solving strategy, specifying it as a CID task will lead to an executable request to identify banks that are supervision targets because of suspicions that they are engaging in a tandem strategy. This task can be formulated and submitted for a CIDS (e.g., the proof-of-concept prototype developed by the authors and described in [5]) to execute.

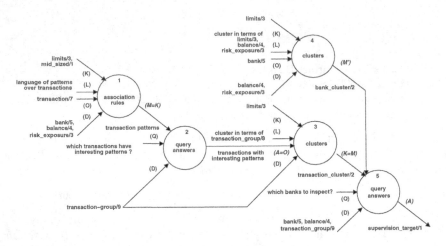

Fig. 1. Shortlisting Banks Suspected of Engaging in Tandem Strategies.

A refinement would be to try and filter the transaction according to some criterion. One option is to refine the hypothesis definition directly, but this may be too blunt in cases, such as the one in hand, where the knowledge is volatile. So, instead, the inspector decides to only cluster transactions that are, in a sense to be defined, unusual. One way to characterize what is a unusual transaction is to use association mining to find interesting co-occurrence patterns. Then, the inspector would only need to cluster transactions that instantiate such patterns.

The inspector first tries to induce a set of association rules to check whether they define an appropriate selection criterion. Satisfied with that, the inspector can then proceed to refine the task by inserting an association analysis step before clustering the transactions thus filtering out those that less likely to be unusual. This task is a complete specification of the inductive-deductive argument for shortlisting a bank on the grounds that it is suspected of engaging in tandem strategies. Figure 1 depicts the complete CID task in graphical form. Nodes denote inferential rules (and hence, steps in CID tasks). Directed edges denote inputs and outputs. Edges are labelled with the p-Datalog clause collections they denote. When the output of one step is used as input of another the the two edges are unified, as indicated by their labels. This is possible because of closure and leads to compositionality, since such graphs can be arbitrarily complex. Every occurrence of edge unification is manifested as variable sharing between nested comprehensions in the textual form of a CID task. Note that a bracketed letter indicates which kind of p-Datalog clause collections instantiate the corresponding edge. Thus, K, Q, L, and M are p-Datalog rule collections (with Q a singleton) and stand for background knowledge, query, language bias and model, respectively. D, A, and O are p-Datalog fact collections and stand for data, answers and observations (i.e., examples, or training instances), respectively. For details, see [5, 4, 6].

Evaluating CID Tasks

As explained above, a CID task such as depicted in graphical form in 1 can be expressed as a monoid comprehension. Two of the CID inference rules involved (viz., *query_answer* to a monoid calculus expression that nests the definition of the CID inference rules presented in [5, 4, 6]. The translation of this expression into an equivalent expression in the monoid algebra is performed exactly as described in [8]. Then, the resulting logical plan is optimized into an execution plan consisting of operators in a physical algebra and then evaluated using an iterator-based paradigm [9]. The execution of a physical plan retrieves stored p-Datalog clauses (including the lazy generation of p-Datalog clauses specified in language biases); propagates clauses upwards with the validity assessments arising from intermediate operations; and finally, outputs the set of p-Datalog clauses produced by the root operator as the final answer. A prototype implementation of a CID engine was built, and is described in [5] was build, that supports exactly the evaluation strategy for CID tasks. The execution of the above task in the prototype implementation has proved the CID task in Figure 1 effective in detecting tandem strategies in data sets known to contain concrete historical examples of such phenomena. For instance, a p-Datalog fact such as `supervision_target(b2)`$\leftarrow$`0.7`$\vdash$ might be part of the task outcome. Note that the validity assigned to such p-Datalog facts embodies the principled treatment of the validity asserted of the axiomatic data, and computed for the induced association and clustering models.

Related Work

Some extensions of SQL, and even new SQL-like languages, with constructs that denote mining algorithms have been proposed [10, 11, 15]. Since these approaches involve hard-wired knowledge discovery functionalities, they deliver modest results in terms of closure, compositionality and seamlessness. Other researchers [7, 13, 14] have proposed integrating knowledge discovery capabilities into database systems. , including: inductive relations [7], inductive databases [14], inductive queries [13]. These proposals have in common the lack of a foundation that reconciles query answering and knowledge discovery of the kind that CIDS provide. The use of inductive inference in deductive databases to derive information in intensional form was first suggested in [7]. However, to the best of our knowledge, the technical development of this idea was never pursued further. In another approach, [14] hints of the idea of combining data and knowledge stocks into a class of information management tools called *inductive databases*. Again, the published details make it difficult to estimate the nature of the task of, and the effort involved in, formalizing and implementing this class of systems. Finally, the proposal in [13] instantiates some ideas in [14]. It uses version spaces to induce strings over boolean constraints that capture simple forms of knowledge. In contrast, CIDS use a fully-fledged database calculus and algebra, and operate on a more expressive knowledge representation, viz., Datalog clauses. It is difficult to be more specific as to the relevant contrasts between these proposals and CIDS as the published detail is not plentiful and implementations are not publicly available. In general, the lack of a reconciliation of the inductive and deductive methods into a unified foundation has led all of [7, 14, 13] to lack both flexibility and a clear route for incorporation into mainstream database technology.

The capabilities that CIDS make available are not matched by commercial database and data-mining systems. Modulo user-interfacing facilities, the state-of-the-art for commercial database systems is to equate data mining algorithms to user-defined functions invokable in SQL queries. IBM Intelligent Miner allows classification and clustering models coded in XML to be deployed and applied to row sets. Microsoft SQL Server with OLE DB DM provides a comparable mechanism, viz., *prediction join*, to build and apply classification models. In SPSS Clementine, one can specify compositions of data preparation, mining, and visualization steps, but no more than one mining step, moreover, the evaluation of each step is decoupled, rather than seamless as in CIDS. This is because different kinds of tasks are supported by different components and the corresponding switched in execution context give rise to impedance. In all the above, compositionality and closure are very restricted and evaluation is far from seamless across the two inference modes involved. Coupling is often so loose that significant engineering effort is required for interoperation to be possible.

Conclusions

The main contribution of this paper is detailed argument, based on a realistic case study, for the significant value added by the use of CIDS in lieu of the decoupled tools that, bundled in suites and workbenches or not, represent the state-of-the-art in combining query answering and knowledge discovery steps. The paper has shown that CIDS exhibit a collection of properties that make them a good candidate for supporting the combination of inductive and deductive inference that is an essential ingredients of scientific activity. Crucially, CIDS have a database-centred operational semantics, i.e., they rely on optimizing execution plans for scalability. They have better chances, therefore, of avoiding the pitfalls inherent in approaches based on heuristic search. This principled, well-founded reconciliation of inductive and deductive inferential processes contributes to the removal of one of the central obstacles to improving the level of automated support for scientific discovery in data-rich and knowledge-poor areas such as genomics. As a result of this step forwards, the way is open for the engineering of industrial-strength CIDS whose impact may be to attenuate the gap between very steep growth rates for data stocks and very flat ones for knowledge stocks. It is thus a concrete hope that CIDS may become important tools in the broad are of e-science. In particular, the way is wide open for the deployment of CIDS on computational grids as they share the same semantics and the same architectural patterns as high-level Grid middleware such as [1], which the authors have helped design and implement.

Acknowledgement

Marcelo A.T. Aragão gratefully acknowledges the support from the Department of Computer Science and the University of Manchester and from the Central Bank of Brazil.

References

1. M. Alpdemir, A. Mukherjee, A. Gounaris, N. Paton, P. Watson, A. Fernandes, and D. Fitzgerald. OGSA-DQP: Service-based distributed querying on the grid. In *Proc. EDBT 2004*, volume 2992 of *LNCS*, pp. 858–861.
2. M. A. T. Aragão and A. A. A. Fernandes. Characterizing web service substitutivity with combined deductive and inductive engines. In *Proc. ADVIS'02*, volume 2457 of *LNCS*, pp. 244–254.
3. M. A. T. Aragão and A. A. A. Fernandes. Inductive-deductive databases for knowledge management. In *Proc. ECAI KM&OM'02*, pp. 11–19.
4. M. A. T. Aragão and A. A. A. Fernandes. A case study on seamless support for combined knowledge discovery and query answering. Longer version of this. Available from http://www.cs.man.ac.uk/~alvaro/, 2003.
5. M. A. T. Aragão and A. A. A. Fernandes. Combining query answering and knowledge discovery. Technical report, University of Machester, 2003. Available from http://www.cs.man.ac.uk/~alvaro/.
6. M. A. T. Aragão and A. A. A. Fernandes. Logic-based integration of query answering and knowledge discovery. In *Proc. 6th Flexible Query Answering Systems (FQAS'2004)*, LNCS, pp. 68–83.
7. F. Bergadano. Inductive database relations. *IEEE TKDE*, 5(6):969–971, 1993.
8. L. Fegaras and D. Maier. Optimizing object queries using an effective calculus. *ACM TODS*, 25(4):457–516, 2000.
9. G. Graefe. Query evaluation techniques for large databases. *ACM Computing Surveys*, 25(2):73–170, June 1993.
10. J. Han, Y. Fu, W. Wang, J. Chiang, O. R. Zaïane, and K. Koperski. DBMiner: interactive mining of multiple-level knowledge in relational databases. In *Proc. SIGMOD'96*, pp. 550–550s.
11. T. Imielinski and A. Virmani. MSQL: A query language for database mining. *DMKD*, 3(4):373–408, Apr. 1999.
12. L. V. S. Lakshmanan and N. Shiri-Varnaamkhaasti. A parametric approach to deductive databases with uncertainty. *IEEE TKDE*, 13(4):554–570, 2001.
13. S. D. Lee and L. de Raedt. An algebra for inductive query evaluation. In *Proc. ICDM'03*, 2003.
14. H. Mannila. Inductive databases and condensed representations for data mining. In *Proc. ILP'97*, volume 13, pp. 21–30.
15. R. Meo, G. Psaila, and S. Ceri. A new SQL-like operator for mining association rules. In *Proc. VLDB'96*, pp. 122–133.

A Structuralist Approach
Towards Computational Scientific Discovery

Charles-David Wajnberg[1,3], Vincent Corruble[1],
Jean-Gabriel Ganascia[1], and C. Ulises Moulines[2,3]

[1] LIP6 – Université P. et M. Curie, 8 rue du Capitaine Scott, 75015 Paris, France
{wajnberg,corruble,ganascia}@ia.lip6.fr
[2] L.-M.-Universität München, Ludwigstr. 31, 80539 München, Deutschland
moulines@lrz.uni-muenchen.de
[3] Ecole Normale Supérieure*, 45 rue d'Ulm, 75230 Paris Cedex 05, France

Abstract. This paper introduces a new collaborative work between AI and philosophy of science, and an original system for machine discovery. Our present goal is to apply the precepts of a major philosophical methodology, namely *structuralism*, in order to build *theory-nets*. The proposed framework handles many kinds of operators, including some that lead to the creation of concepts, and we illustrate how it can create theories with physical concepts in an example from collision mechanics.

1 Introduction

1.1 Computational Scientific Discovery and Philosophy of Science

Newell and Simon introduced in the 60's the relevance of problem solving to study scientific discovery and creativity [1]. Both subjects are connected, as soon as one wants the computer to provide relevant and original solutions.

Discovery systems often start with initial knowledge, namely examples or basic theories, on which they try to apply operators (in a very general sense) in order to build a new "theory". This contribution can be evaluated in terms of coverage of the input examples as well as in terms of aesthetics (Ockham's razor, for example). Moreover, the way which leads to this final state can be considered as a plausible model of scientific reasoning, or at least of scientific inference.

Starting from Simon, who was a student of Carnap, the study of computational scientific discovery has been a topic of interest to both AI and philosophy of science. Philosophers and historians could provide consistent representations of science, historical case studies and even general methodologies. In return, AI would provide simulation capabilities for such philosophical scientific theories. Hence, there is a fruitful tradition of researchers working at the crossroad of both areas. We can cite Zytkow [2], Glymour [3] or Thagard [4]. Most of them have transfered theories from philosophy of science to AI and *vice versa*. Even computer science representations have been sometimes considered as scientific meta-theories themselves.

* This work was made possible thanks to the *Fondation de l'Ecole Normale Supérieure*, within the framework of the *Blaise Pascal Research Fellowships*.

E. Suzuki and S. Arikawa (Eds.): DS 2004, LNAI 3245, pp. 412–419, 2004.
© Springer-Verlag Berlin Heidelberg 2004

The present paper is in keeping with this general collaboration between problem solving and philosophical theories. In order to discuss important evolutions of philosophy of science, we aim to design a system which is compatible with a major stream of modern philosophy of science, namely *structuralism*. We will introduce some concepts of such theoretical framework and then show how a computer program can build structured theories starting from descriptions of scientific experiments.

1.2   Main Ideas of the Structuralism

Structuralism was born with Joe Sneed's and Wolfgang Stegmüller's work in the beginning of the 70's [5, 6]. It belongs to the *non-statement view* major stream of contemporary philosophy of science, which claims that the inner structure of a theory is more important than the statements, namely the laws, of the theory. That means also that the method to discover the laws is not of primary importance as long as we can organize them in a coherent structure. Such a theoretical structure is called a *theory-net*, which is a semi-lattice structure organizing laws of the theory (cf. §2.5).

The Fig. 5. shows a simple theory-net for classical collision mechanics. At the top of the net stands the main law, namely the general momentum conservation law. This law can be easily specialized into the two-particle momentum conservation law which covers only two-particle experiments. But the main law also generalizes (not in a logic or formal sense) two special cases of collision mechanics, namely inelastic collision (the colliding objects stick together after the collision) and elastic (kinetic energy conservation). It means that any experiment which fulfills special inelastic or elastic collision laws, fulfills also the momentum conservation law.

Such a theory-net is associated with a set of experiments, which are explicitly selected to be relevant with respect to the theory, and correspond more or less to the description of some real-world situations. This domain of *intended applications* represents the phenomena a theory deals with, in a descriptive way.

The structuralist model of the real world is merely a set-theoretical structure in which all components of the theory are described. For example, a model for collision mechanics would be a sequence $\langle \mathbb{P}, \mathbb{T}, \mathbb{R}, m, \boldsymbol{v} \rangle$, where

- $\mathbb{P}$ is a finite set of particles (whatever they are, billard balls or atoms),
- $\mathbb{T}$ is a couple of discrete times $\{t_1, t_2\}$ (before and after the collision),
- $\mathbb{R}$ is an auxiliary domain to represent mass and space as physical measures,
- $m : \mathbb{P} \to \mathbb{R}^+$ represents the mass of each particle,
- $\boldsymbol{v} : \mathbb{P} \times \mathbb{T} \to \mathbb{R}^3$ represents the velocity of each particle at a given time.

An intended application could be described as a *substructure* of this class of models, for example $\langle \mathbb{P} = \{p_1, p_2\}, \boldsymbol{v}(p_1, t_1) = (0, -2, 3) \rangle$, and so on.

We cannot go into deeper details in the present paper, but interested readers should consult [5] which is a key reference for the structuralist program. Our goal is to automatically build theory-nets, by discovering laws and then organizing them. We discuss in the following sections the content of such a program.

2   A Structuralist-Compatible Problem Solving Discovery System

2.1   The Algorithm Skeleton

In order to build a theory-net, the system has first to discover many laws before organizing them. This loop-process has no explicit final goal state, exactly as the scientific process, and produces laws until the user stops it. The Fig. 1. is a basic representation of the system, where the large arrow represents the successive reformulations of the theory.

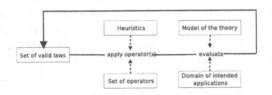

Fig. 1. The problem solving process in the case of structuralist-compatible computational scientific discovery.

The laws are represented by runnable programs, namely functional programs, and intended applications, as well as models of the theory, are viewed as environments for program execution, namely a set of definitions (cf. §1.2). Formally, a law is a predicate. If it equals true under a given execution environment, then the associated application is deemed covered. There are only two criteria for a resulting law to be valid: first, the associated functional program must be runnable, and then it must cover at least one application.

The operators are transformation rules over these programs.

2.2   Rewritting Operators

Our system can deal with many kinds of operators, since the only oracle of the system is the domain of intended applications. This is rather close to the genetic programming approach, in which the construction of sucessive programs is only based on their validity with respect to the concerned examples. On the other hand, genetic programming suffers of low efficiency [7] and no *explicit inference*. Here, we use a problem solving approach which allows as well the handling of different operators as the clear representation of scientific inference.

We use more flexible operators (non-arithmetic, random, approximation, concept handling, etc.) to rewrite laws of the theory. The evolution process is also quite close to the synthesis of functional programs [8]. Of course, any kind of operator can be created, up to the needs of science, but with care to the complexity of the corresponding search space.

As one can notice, some operators are truth preserving (3,5), some others are not (2,4). The first operator (commutativity) depends on the nature of the

Table 1. Examples of operators.

	Reasoning	Operator	Example	
			original	rewritten
1	Commutativity	$x \star y \to y \star x$	$P = m \cdot V$	$P = V \cdot m$
2	1-to-n generalization	$x_1 + x_2 \to \sum_X x_i$	$F_1 + F_2 = m \cdot A$	$\sum_F F_i = m \cdot A$
3	Equality to distance	$x = y \to \lvert x - y \rvert \le 0$	$u = r \cdot i$	$\lvert u - r \cdot i \rvert \le 0$
4	Approximation	$x \to eval(x + 0.1)$	$\lvert x - y \rvert \le 2.0$	$\lvert x - y \rvert \le 2.1$
5	Concept creation	$x \to concept(x)$	$m(p_1) \cdot v(p_1, t_1) = (0, -2, 3)$	$f_1(p_1, t_1) = (0, -2, 3)$ with $f_1(p, t) = m(p) \cdot v(p, t)$

mathematical operator $\star$: if "$\star$" $=$ "$+$", the commutativity operator will be purely deductive, but if "$\star$" $=$ "$/$", the commutativity will be false. The fourth operator (approximation) is quite interesting because it uses some computation outside of the program representing the law, namely there is a *meta-level* in the program execution. For example, a part of a law can be replaced by its value in a given environment: "$u = r \cdot i$" can be formally replaced by "$u = 4.2$" where "$r = 2$" and "$i = 2.1$". This allows some numerical induction and law approximation.

Concept Creation: Finally, the fifth operator is very special: it creates a concept starting from a part of the program. It is *not* a deterministic operator since, in our example, it could achieve as well $f_1(p, t) = m(p) \cdot v(p, t)$ as $f_1(p) = m(p) \cdot v(p, t_1)$ or $f_1(t) = m(p_1) \cdot v(p_1, t)$. Such operators make the search very complex, so the system needs efficient heuristics (cf. §2.4). Simply speaking, the *create* operator adds a new function to current definitions, and replaces the selected occurence by one instance of this function. This means that, in our problem solving framework, we are able to create some concepts as functions of other existing concepts/functions. basic example.

2.3   A Simple Example

We show here a short application of such operators in the case of classical collision mechanics. Moreover, we use the demonstration to give an outline of concept creation in such a system. The initial theory is rather simple and contains both special laws of momentum conservation for two and three particles (Fig. 2). It means that our program will have to generalize these laws. According to the nature of the operators, we could do exactly the opposite: start with a general law and then specialize it to these two special cases. Of course, in this case, generalization is more interesting.

$$(1)\ m(p_1) \cdot v(p_1, t_1) + m(p_2) \cdot v(p_2, t_1)$$
$$= m(p_1) \cdot v(p_1, t_2) + m(p_2) \cdot v(p_2, t_2)$$

$$(2)\ m(p_1) \cdot v(p_1, t_1) + m(p_2) \cdot v(p_2, t_1) + m(p_3) \cdot v(p_3, t_1)$$
$$= m(p_1) \cdot v(p_1, t_2) + m(p_2) \cdot v(p_2, t_2) + m(p_3) \cdot v(p_3, t_2)$$

Fig. 2. Initial local laws for momentum conservation.

Table 2. The discovery of conservation momentum general law, with the creation of the momentum concept (f_3).

It.	Rule		Result
1	$m(p_3) \cdot v(p_3, t_1)$ $\longrightarrow f_1(t_1)$	(2)	$m(p_1) \cdot v(p_1, t_1) + m(p_2) \cdot v(p_2, t_1) + f_1(t_1)$ $= m(p_1) \cdot v(p_1, t_2) + m(p_2) \cdot v(p_2, t_2) + f1(t_2)$
2	$m(p_1) \cdot v(p_1, t_1)$ $\longrightarrow f_2(p_1)$	(1) (2)	$f_2(p_1) + f_2(p_2) = m(p_1) \cdot v(p_1, t_2) + m(p_2) \cdot v(p_2, t_2)$ $f_2(p_1) + f_2(p_2) + f_1(t_1)$ $= m(p_1) \cdot v(p_1, t_2) + m(p_2) \cdot v(p_2, t_2) + f_1(t_2)$
3	$m(p_1) \cdot v(p_1, t_1)$ $\longrightarrow f_3(p_1, t_1)$	(1) (2)	$f_2(p_1) + f_2(p_2) = f_3(p_1, t_2) + f_3(p_2, t_2)$ $f_2(p_1) + f_2(p_2) + f_1(t_1)$ $= f_3(p_1, t_2) + f_3(p_2, t_2) + f_1(t_2)$
4	$f_1(t_1) \longrightarrow$ $m(p_3) \cdot v(p_3, t_1)$	(2)	$f_2(p_1) + f_2(p_2) + m(p_3) \cdot v(p_3, t_1)$ $= f_3(p_1, t_2) + f_3(p_2, t_2) + m(p_3) \cdot v(p_3, t_2)$
n-1		(1) (2)	$f_3(p_1, t_1) + f_3(p_2, t_1) = f_3(p_1, t_2) + f_3(p_2, t_2)$ $f_3(p_1, t_1) + f_3(p_2, t_1) + f_3(p_3, t_1)$ $= f_3(p_1, t_2) + f_3(p_2, t_2) + f_3(p_3, t_2)$
n	$f_3(p_1, t_1) + f_3(p_2, t_1)$ $\longrightarrow \sum_P f_3(p, t_1)$	(1)	$\sum_P f_3(p, t_1) = \sum_P f_3(p, t_2)$

Then, we provide the same operators as mentionned in Table 1, but only *1-to-n generalization* and *concept creation* operators are useful here, since it is a typical 1-to-n generalization over differents situations. The log of the system is shown in Table 2.

The system applies some rules on equations (1) and (2) so that they are subsumed under a more general one, which covers the whole set of intended applications. After a few tens of iterations, the system obtains a general law, namely the conservation momentum law, which covers all our collision mechanics applications. Since the *create* operator is not deterministic, we can see how the computer tries to find the concept of momentum, without which it cannot formulate the general law. Indeed, creating the f_3 concept (namely the momentum), is the only way it has to use the *1-to-n* operator.

2.4   Search Heuristics

The search space is obviously wide and many different laws could be obtained, including some uninteresting ones. Of course, the validity of a program (in the sense of coverage and execution) prunes already much of the search space. Nevertheless, we need another efficient bias.

The first we chose, for mostly intuitive and pragmatic reasons, is to keep a certain consistency in the set of laws, in order to progressively explore the space of possible laws instead of doing jumps (which is the case of genetic programming). This consistency of the theory is provided by a syntactical similarity measure for each pair of laws, based on a classical edit distance [9]. Hence, the neighborhood of an initial law (*i.e.* the laws generated by applying any operator on the initial law) is limited by this purely syntactical criterion.

At each iteration, a law λ_L is picked out of the theory and another λ_R is selected such that $\lambda_R = \arg\max_{\lambda \in \Lambda} similarity(\lambda_L, \lambda)$, where Λ is the set of valid laws in the theory. Then, two new sets of laws are created, namely

$$\Lambda_L = \bigcup_{\omega \in \Omega} \omega(\lambda_L) \quad \text{and} \quad \Lambda_R = \bigcup_{\omega \in \Omega} \omega(\lambda_R)$$

$$\Lambda_1 = \omega_{com}(m(p_1) \cdot v(p_1, t_1) + m(p_2) \cdot v(p_2, t_1) \ldots) = \left\{ \begin{array}{l} v(p_1, t_1) \cdot m(p_1) + m(p_2) \cdot v(p_2, t_1) \ldots \\ m(p_1) \cdot v(p_1, t_1) + v(p_2, t_1) \cdot m(p_2) \ldots \\ m(p_2) \cdot v(p_2, t_1) + m(p_1) \cdot v(p_1, t_1) \ldots \\ \vdots \end{array} \right\}$$

Fig. 3. An example of a new set of laws created by the commutativity operator ω_{com} applied on the law (1) (cf. Fig. 2).

where Ω is the set of applicable operators. It is clear that an operator can be applied on different terms of a given equation (cf. Fig. 3.).

Then, the best law is selected in order to bring the laws still closer:

$$or \left\{ \begin{array}{l} \lambda_L \leftarrow \arg\max_{\lambda \in \Lambda_L} \ similarity(\lambda, \lambda_R) \\ \lambda_R \leftarrow \arg\max_{\lambda \in \Lambda_R} \ similarity(\lambda, \lambda_L) \end{array} \right.$$

We add also a tabu meta-heuristic which ensures that the same operator is never re-applied on the same two laws – except if the result of the operator is not deterministic (cf. §2.2.).

2.5 Building Theory-Nets

The process described above generates several laws (potentially one at each iteration), and they need to be organized in order to build a relevant theory. Structuralism does only provide a formal definition of a theory-net, but no algorithm to do so. From a machine learning point of view, such structure could be merely built as a coverage lattice over the set of discovered laws. It means that the laws which cover most examples will stand at the top of the net, and very special ones will move to the bottom.

Unfortunately, the coverage criterion is not enough. For many examples, we can discover laws which are compatible with many applications but have no scientific relevance. For example, the law $m(p_1) = m(p_1)$ will match all intended applications, since it is valid for any function m and any particle p_1 (if the set of particles contains a p_1 particle). But the general law of momentum conservation, which has the same coverage, is obviously more scientifically relevant.

One the Fig. 4., the three laws on the top of the net cover all applications, the middle one (d) covers only inelastic collisions, and the most special (e) covers only i_1, in which the mass of the particle p_1 is 4 (in our example).

Although this structure is already more interesting than isolated statements, there are several things to change in order to build a scientifically relevant theory-net. We can distinguish many problems:

- Some laws can be syntactically different but physically identical (e.g. (b) and (c));
- some others can be tautologies and have a full-coverage. They are nevertheless absolutely not relevant (e.g. (a));
- finally, some laws can be uninteresting since they are totally redundant with the description of intended applications (e.g. (e))

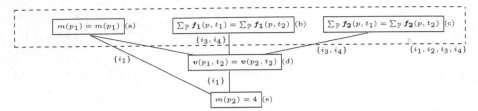

Fig. 4. A part of a theory-net for classical collision mechanics, with a tautology and equivalent laws. This is only built with a coverage criterion.

- $\boldsymbol{f_1}(p,t) = m(p) \cdot \boldsymbol{v}(p,t)$ and $\boldsymbol{f_2}(p,t) = \boldsymbol{v}(p,t) \cdot m(p)$.
- $\{i_1...i_4\}$ are four different experiments with two particles (i_1, i_3) or three particles (i_2, i_4), with elastic (i_3, i_4) or inelastic collision (i_1, i_2).

The structuralist approach provides here a simple method to avoid at least the first two situations, by using the initial definition of the class of models (cf. §1.2). Without going into further detail, a law is relevant if and only if it does not fulfill all the possible models of the class, namely for any function $\boldsymbol{p}, m$, any set of particles, and so on.

After resolving such ambiguous cases, we obtain a "nicely prunned" theory-net, as a structured representation of a theory (Fig. 5).

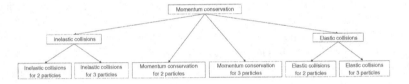

Fig. 5. The obtained theory-net for our model of classical collision mechanics.

3   Conclusion and Ongoing Work

This paper has shown how modern ideas from philosophy of science can help us to design scientific discovery systems in the framework of problem solving. We designed an original system, which is neither a numerical inductive one, nor a pure symbolic one and which is compatible with philophical assumptions as well as scientific requirements. The strucuralist view of theories, as a *non-statement* view, is relevant with respect to the computational problem-solving approach.

Moreover, we show through a basic example how such a system can create physical concepts in some simple (but non-trivial) cases, by means of various operators. The rule of inductive, approximative or chancy operators should be increased in order to plausibly and efficiently model the scientific reasoning in discovery systems.

We currently work on a larger application, which aim is to apply our system to Gibbsian phenomenological thermodynamics. Of course, the ultimate goal of such a work would be to improve modern theories and scientific activities by means of computer assistance.

References

1. Newell, A., Simon, H.A.: Gps: A program that simulates human thought. In Billings, H., ed.: Lernende Automaten. R. Oldenbourg, Munchen (1961) 109–124
2. Zytkow, J.M.: Deriving laws by analysis of processes and equations. In Langley, P., J., S., eds.: Computational Models of Scientific Discovery and Theory Formation. Morgan Kaufmann, San Mateo (1990)
3. Spirtes, P., Glymour, C., Scheines, R.: Causation, Prediction, and Search. 2nd edn. MIT Press (2001)
4. Thagard, P.R.: Computational Philosophy of Science. MIT Press (1993)
5. Balzer, W., Moulines, C.U., Sneed, J.D.: An architectonic for science – The structuralist program. D.Reidel Publishing Company (1987)
6. Moulines, C.U.: Structuralism as a program for modelling theoretical science. SYNTHESE **130** (2002) 1–11
7. Michalski, R.S.: Learning and evolution: An introduction to non-darwinian evolutionary computation. Twelfth International Symposium on Methodologies for Intelligent Systems (2000) Invited paper.
8. Schmid, U.: Inductive Synthesis of Functional Programs. Springer-Verlag (2003)
9. Gusfield, D.: Algorithms on Strings, Trees and Sequences. Cambridge University Press (1997)

Extracting Modal Implications and Equivalences from Cognitive Minds

Radoslaw Piotr Katarzyniak

Wroclaw University of Technology, Wybrzeze Wyspianskiego 27, 50-350 Wroclaw, Poland
radoslaw.katarzyniak@pwr.wroc.pl

Abstract. An original approach to modelling the correspondence between internal cognitive state and external linguistic representation is proposed for the case of modal implications and modal equivalences. The so called epistemic satisfaction relations are given that describe the correspondence between internal states and external representations in technical terms. The phenomenon of language grounding, language extraction and linguistic representation discovery from artificial cognitive mind are treated as related.

1 Introduction

According to cognitive linguistics external representations of internal knowledge need to be grounded in stored empirical experiences of agents in order to preserve the semantic nature of linguistic communication [2] [14] [15]. This natural requirement for language grounding in living cognitive systems should also be assumed for artificial cognitive systems if their language behavior is expected to be rational and human understandable [1] [3] [11] [12]. In some previous works certain cases of modal formulas have already been considered in the context of semantic communication and artificial cognitive systems [5] [6] [7] [8] [9] [10] [11] . For each class of considered modal formulas the so called epistemic satisfaction relation has been introduced to describe states of cognition corresponding to them. The only case that has not been covered by the epistemic satisfaction relations has been the modal implications and modal equivalences.

Definitions for grounding are important for practical reasons. Namely, they can give precise descriptions of cognition states for which particular linguistic representations are appropriate. The practical importance of epistemic satisfaction relations follows from the underlying commonsense assumption that in a particular cognition state a certain logic formula should be discovered by cognitive processes of language extraction if and only if this formula is well grounded in all data stored internally in the agent (cognitive system). This fact has already been used in order to model the choice of modal responses produced by cognitive agents [7].

Each epistemic satisfaction relation between a certain external linguistic representation and internally represented content is usually quite complex. In particular, for each case of modal formulas this relation is defined by a list of complex requirements for certain distribution and certain content of internally stored empirical data. In case of artificial cognitive systems that are required to be rational and intentional [1] [12] this complexity results from additional commonsense necessity of preserving strict

E. Suzuki and S. Arikawa (Eds.): DS 2004, LNAI 3245, pp. 420–428, 2004.

correspondence between human and artificial reactions. To fulfill this requirement it is necessary to organize internal cognitive processes of an artificial system in a way that makes it possible to conceptualize them similarly to the natural cognitive processes.

2 Cognitive Agent

2.1 General Assumptions

The language grounding is a purely cognitive process. Therefore it is assumed in the given approach that artificial cognitive agents are internally organized in a way suggested by cognitive models for human mind e.g. [13] and the related theory of mental models [3]. The accepted assumptions can be summarized as follows: (1) Each interpreted formula of a semantic language of communication is always correlated (at the conscious level of knowledge processing) with its related mental model. (2) Each mental model an abstraction extracted from previous empirical experience of the agent. This abstraction is extracted from knowledge base by dedicated cognitive processes. Each mental model exist of and only if there exists related empirical material stored in internal knowledge bases of the agent. (3) The internal mental space of each cognitive agent is divided into two autonomous cognitive subspaces. The smaller cognitive subspace represents the so called working memory in which these mental activities take place which are experienced by the agent as conscious ones. The other cognitive subspace represents mental processes and semantic content stored internally in the cognitive agent and experienced by this agent as preconscious ones. Such two level organization of cognition is well known from the majority of models for natural human cognition e.g. [13]. (4) According to the cognitive linguistics both cognitive subspaces are involved in producing external language behavior. In particular, it is accepted that the content stored at the preconscious level of knowledge representation influences the agents' language behaviour, too, and participates in creating the meaning of language [11]. In this particular sense all interpreted formulas are grounded in the empirical material located in the working memory and in the preconscious cognitive subspace. The meaning of formulas results from the overall knowledge base.

The original epistemic satisfaction relations given below define the distribution of knowledge in the working memory and the preconscious cognitive subspace that represents this cognitive state of the agent that is the actual realization of the epistemic meaning of corresponding formula. Any complete definition of such a grounding state requires a dedicated system of concepts. Such a system is defined below.

2.2 External World

Similarly to [4] [6] it is assumed that the external world W in which cognitive systems are located is a dynamic environment built of atomic objects. Particular states of the world are related to time points $T=\{t_0,t_1,t_2,..\}$ for which a strong linear temporal order $\leq^{TM}$ is given. At each time point t the state of the world is described by the so called t-related world profile:

Definition 1. *The states of the real world W are represented by the following relational systems WorldProfile(t)=<O,P$_1$(t),...,P$_K$(t)>.*

The interpretation of *WorldProfile(t)* is given as follows: The set $O=\{o_1,o_2,...,o_M\}$ consists of all atomic objects of the world *W*. The symbols in the set $\Delta=\{P_1, P_2, ..., P_K\}$ are unique names of properties that can be attributed to the objects from *O*. Namely, each object $o \in O$ may or may not exhibit a particular property $P \in \Delta$. For $i=1,...,K$ and $t \in T$ the symbol $P_i(t)$ denotes a unary relation $P_i(t) \subseteq O$. For $i=1,...,K$ and $o \in O$ the condition $o \in P_i(t)$ holds if and only if the object *o* exhibits the property P_i at the time point *t*. If the condition $o \notin P$ holds, than the object *o* is understood as not exhibiting the property P_i at the time point *t*.

2.3   Internal Empirical Knowledge

It is also assumed that the sensors and internal system of concepts make it possible for the cognitive system to observe and represent states of properties $P \in \Delta$ in atom objects $o \in O$ at particular time points $t \in T$. The basic knowledge of the cognitive system consists of all empirical data that have ever been collected by cognitive processes in encapsulated databases. Let the basic piece of empirical knowledge be given as the result of a separate perception of the world at a particular time point *t* and stored as the so-called base profile (see [4] [6]):

Definition 2. *The internal representation of observation of the world W realized by the agent at the moment t is represented by the following relational system:*
$$BP(t)= <O,P^+{}_1(t),P^-{}_1(t),...,P^+{}_K(t),P^-{}_K(t)>$$

The interpretation of its elements is given as follows: $O=\{o_1,o_2,...,o_M\}$ consists of all representations of atom objects o_i, when o_i used in the context of base profiles and denotes a unique internal cognitive representation of the atomic object of the external world. For each $j=1,2,...,K$, both $P^+{}_j(t) \subseteq O$ and $P^-{}_j(t) \subseteq O$ hold. For each $o \in O$ the relation $o \in P^+{}_j(t)$ holds if and only if the agent's point of view is that the object *o* exhibited the atomic property P_j and this fact was empirically verified at the time point *t* by the agent itself. For each $o \in O$ the relation $o \in P^-{}_j(t)$ holds if and only if the agent's point of view is that the object *o* did not exhibit the atomic property P_j and this fact was empirically verified at the time point *t* by the agent itself. For each $j=1,2,...,K$, the condition $P^+{}_j(t) \cap P^-{}_j(t)=\varnothing$ holds. Obviously, it reflects in a formal way that it is not possible for the cognitive agent to assign and not to assign the same property P_j to the same object *o* at the same time point *t*. It is quite a natural feature of cognitive minds.

The so called incompetence area can also be defined in relation to the idea base profile:

Definition 3. *The area of the cognitive agent's incompetence related to a property P$_i$ in t-related agent-encapsulated base profile is given as the following set of objects:*

$$P^{\pm}_i(t)=O \mathbin{/} (P^+_i(t) \cup P^-_i(t))$$

The set of all stored perceptions constitutes the t-related knowledge state:

Definition 4. *At each time point $t \in T$ the state of basic empirical knowledge is given as the following temporal collection of encapsulated base profiles:*
$$KS(t) = \{BP(t_n): t_n \in T \text{ and } t_n \leq^{TM} t\}$$

At each state of cognitive processing the knowledge content is distributed between the working memory and the preconscious module. Therefore in order to completely describe the state of knowledge processing additional classification of base profiles needs to be introduced. The related definition is:

Definition 5. *At each time point $t \in T$ the state of cognition is given by a binary partition $CS(t)$ of $KS(t)$ provided that $CS(t)=\{WM(t),PM(t)\}$, $WM(t) \cup PM(t)=KS(t)$ and $WM(t) \cap PM(t)=\varnothing$.*

Obviously, $WM(t)$ denotes the set of base profiles located in the working memory and $PM(t)$ the set of remaining base profiles located in preconscious module.

3   The Language of Modal Implications and Equivalences

The semantic language L_C considered in this paper consists of a few types of non-modal and modal formulas. Each formula of L_C is assumed to be intentionally interpreted [1] and corresponds to a particular natural language statement. The formal description of L_C is given as follows: Let the set $O=\{o_1, o_2, ..., o_M\}$ consist of all individual constants. Let the symbols $p_1, p_2, ..., p_K$ be unary predicate symbols interpreted as unique names of properties $P_1, P_2, ..., P_K$, respectively. Let the symbols $\Rightarrow$ and $\Leftrightarrow$ be the logic connectives of implication and equivalence, respectively. Let the symbols *Pos, Bel, Know* be modal operators called the modal operator of possibility, belief and knowledge, respectively. Each symbol $p_i(o_m)$ represents the statement *Object o_m exhibits the property P_i* and is called a non-modal atom formula.

The non-modal atom formulas can be used to create the so-called non-modal binary δ-compound formulas, where δ states for a binary logic connective $\Rightarrow$ or $\Leftrightarrow$. Possible non-modal binary δ-compound formulas are $p_i(o_m) \Rightarrow p_j(o_m)$ and $p_i(o_m) \Leftrightarrow p_j(o_m)$ and their commonsense interpretation is *If object o_m exhibits the property P_i, then object o_m exhibits the property P_j.* and *Object o_m exhibits the property P_i if and only if object o_m exhibits the property P_j,* respectively.

The complete definition for L_C is as follows:

Definition 6. *The semantic language of communication L_C is given as follows:*
  a)   Each non modal binary δ-compound formula is an element of L_C, $\delta \in \{\Rightarrow, \Leftrightarrow\}$.
  b)   For each non modal binary δ-compound formula φ, the modal extensions $Pos(\varphi)$, $Bel(\varphi)$, and $Know(\varphi)$ are formulas of L_C.

The intentional interpretation of the above modal formulas $Pos(\varphi)$, $Bel(\varphi)$, $Know(\varphi)$ are *It is possible that φ.*, *I believe that φ.* and *I know that φ.*, respectively.

4   The Epistemic Satisfaction of Modal Formulas

4.1   Modal Implications

In order to define cognitive requirements for modal implications to be grounded in internal cognitive states one needs to refer to basic assumptions of the theory of mental models [3]. Let the following symbols be introduced: m^I_1 denoting a mental model for the conjunction $p(o) \wedge q(o)$, m^I_2 denoting a mental model for the conjunction $p(o) \wedge \neg q(o)$, m^I_3 denoting a mental model for the conjunction $\neg p(o) \wedge q(o)$, m^I_4 denoting a mental model for the conjunction $\neg p(o) \wedge \neg q(o)$, m^I_5 denoting a mental model for the implication $p(o) \Rightarrow q(o)$. The semantic relation between the above conjunctions and implication can easily be described by natural relations between their mental models. Namely, the mental model m^I_5 for the implication $p(o) \Rightarrow q(o)$ can be given as the set of mental models $\{m^I_1, m^I_3, m^I_4\}$. Each mental model m^I_i, i=1,2,3,4, is induced by certain collections of empirical data. Let the following definition for this empirical material be given:

Definition 7. *Let the sets $C^i(t)$ (further denoted by C^i) from which the mental models m^I_i, i=1,2,3,4, are extracted be given as follows:*

$$C^1(t) = \{BP(t_n): t_n \leq^{TM} t \text{ and } BP(t_n) \in KS(t) \text{ and } o \in P^+(t_n) \text{ and } o \in Q^+(t_n) \text{ hold}\}$$
$$C^2(t) = \{BP(t_n): t_n \leq^{TM} t \text{ and } BP(t_n) \in KS(t) \text{ and } o \in P^+(t_n) \text{ and } o \in Q^-(t_n) \text{ hold}\}$$
$$C^3(t) = \{BP(t_n): t_n \leq^{TM} t \text{ and } BP(t_n) \in KS(t) \text{ and } o \in P^-(t_n) \text{ and } o \in Q^+(t_n) \text{ hold}\}$$
$$C^4(t) = \{BP(t_n): t_n \leq^{TM} t \text{ and } BP(t_n) \in KS(t_k) \text{ and } o \in P^-(t_n) \text{ and } o \in Q^-(t_n) \text{ hold}\}$$

Unfortunately, the assumption that the mental model m^I_5 can be given by the set of simpler mental models $\{m^I_1, m^I_3, m^I_4\}$ needs to be enriched with additional requirements referred to particular models m^I_1, m^I_3, m^I_4. For instance, it seems rational to assume that the cognitive agent is equipped with actual mental model of implication if and only if its mental model for $p(o) \wedge q(o)$ is not empty. Moreover, if modal extensions of the implication $p(o) \Rightarrow q(o)$ are considered it becomes necessary to measure the relative strength of particular mental models m^I_1, m^I_3, m^I_4 in m^I_5. To capture the latter phenomenon the following definition for the relative grounding value can be introduced:

Definition 8. *Let $G^1 = card(C^1)$, $G^3 = card(C^3)$, $G^4 = card(C^4)$. The relative grounding value $\lambda(p(o) \Rightarrow q(o))$ is given as* $\lambda(p(o) \Rightarrow q(o)) = \dfrac{G^1}{G^1 + G^3 + G^4}$.

This value describes the subjectively experienced strength of the most important grounding material C^1 in the whole mental model of $p(o) \Rightarrow q(o)$. It is quite clear that the above relative grounding value cannot be determined in a direct way due to the fact that involved parts of mental model are additionally distributed among the working memory and the preconscious cognitive subspace. In consequence, in practical settings the above equation for $\lambda(p(o)\Rightarrow q(o))$ need to be reformulated in order to make this distribution apparent and preserve the nature of cognitive minds:

Definition 9. *Let the symbols $WC^i=WM(t)\cap C^i$, $PC^i=PM(t)\cap C^i$ be introduced for each $i=1,2,3,4$, where for $i=1,2,3,4$ the equations $WC^i\cap PC^i=\varnothing$ and $WC^i\cup PC^i=C^i$ hold. The aspects of the state of cognition $CS(t)$ related to the implication are given by the following partition of empirical knowledge: $CS(t) = \{WC^1(t), PC^1(t), WC^2(t), PC^2(t), WC^3(t), PC^3(t), WC^4(t), PC^4(t)\}$.*

The above set of simple concepts makes it possible to capture in a formal way the structure of cognitive states in which implication $p(o)\Rightarrow q(o)$ and its extensions are well grounded:

Definition 10. *(grounding implication $p(o)\Rightarrow q(o)$) Let the following be given: a time point $t\in T$, a state of cognition $CS(t) = \{WC^1(t), PC^1(t), WC^2(t), PC^2(t), WC^3(t), PC^3(t), WC^4(t), PC^4(t)\}$ and a system of modality thresholds $0<\lambda_{minPossibility}< \lambda_{maxPosssibility}<\lambda_{minBelief}<\lambda_{maxBelief}<1$. It is assumed for each pair of different properties P, $Q\in\{P_1,...,P_K\}$ and each object $o\in O$ that:*

1. *The epistemic satisfaction relation $CS(t)|=_G Pos(p(o)\Rightarrow q(o))$ holds if and only if $C^3=\varnothing$, $WC^1\neq\varnothing$, $PC^1\neq\varnothing$ and $\lambda_{minPos}\leq\lambda(p(o)\Rightarrow q(o))\leq\lambda_{maxPos}$.*
2. *The epistemic satisfaction relation $CS(t)|=_G Bel(p(o)\Rightarrow q(o))$ holds if and only if $C^3=\varnothing$, $WC^1\neq\varnothing$, $PC^1\neq\varnothing$ and $\lambda_{minBel}\leq\lambda(p(o)\Rightarrow q(o))\leq\lambda_{maxBel}$.*
3. *The epistemic satisfaction relations $CS(t)|=_G Know(p(o)\Rightarrow q(o))$ and $CS(t)|=_G p(o)\Rightarrow q(o)$ hold if and only if $C^3=\varnothing$, $WC^1\neq\varnothing$, $PC^1=\varnothing$.*

The rationale underlying the above requirements are as follows: Each of the above implications are acceptable as relevant and well grounded in the previous empirical experiences of the agent if and only if this agent has experienced at least once the actual correlation between the existence of P in o followed by the existence of Q in o. Therefore $WC^1\neq\varnothing$ and $PC^1\neq\varnothing$ are required. Each of the above implications are acceptable as relevant and well grounded in the previous empirical experiences of the agent if and only if this agent has never experienced the situation in which o exhibited P and did not exhibit Q. Therefore in each case $C^3=\varnothing$ is required. In point 1 and 2 it is also required that $PC^1\neq\varnothing$ holds. This piece of relevant data located in preconscious cognitive subspace forces the agent to explore the preconscious areas of its knowledge and in this sense brings about the modality of cognitive state. If $PC^1=\varnothing$ then it can be said that all relevant material has been extracted from preconscious cognitive subspace and there exists no empirical material that forces the agent to explore preconscious levels of knowledge representation. Therefore the agent is certain as regards to its empirically induced view on grounding implication. No belief and possibility operator is necessary for they are irrelevant to the situation.

4.2 Modal Equivalences

Similar approach to defining the rules for grounding can be proposed for modal equivalences. One can refer to the following set of mental models: $m^E_1 = m^I_1$, $m^E_2 = m^I_2$, $m^E_3 = m^I_3$, $m^E_4 = m^I_4$, m^E_5 denotes a mental model for the equivalence $p(o) \Leftrightarrow q(o)$. Again, the semantic relation between the mental models for conjunctions and equivalence can be given. The mental model m^E_5 for the equivalence $p(o) \Leftrightarrow q(o)$ is the set of mental models $\{m^E_1, m^E_4\}$. The relative grounding value can be defined:

Definition 11. *Let $G^1 = card(C^1)$, and $G^4 = card(C^4)$. The relative grounding value $\lambda(p(o) \Leftrightarrow q(o))$ is given as $\lambda(p(o) \Rightarrow q(o)) = \dfrac{G^1}{G^1 + G^4}$.*

Additional requirements for modal equivalence grounding are similar:

Definition 12. *(grounding equivalence $p(o) \Leftrightarrow q(o)$) Let the following be given: a time point $t \in T$, a state of cognition $CS(t) = \{WC^1(t), PC^1(t), WC^2(t), PC^2(t), WC^3(t), PC^3(t), WC^4(t), PC^4(t)\}$ and a system of modality thresholds $0 < \lambda_{minPossibility} < \lambda_{maxPosssibility} < \lambda_{minBelief} < \lambda_{maxBelief} < 1$. It is assumed for each pair of different properties P, $Q \in \{P_1, ..., P_K\}$ and each object $o \in O$ that:*

1. *The epistemic satisfaction relation $CS(t)|=_G Pos(p(o) \Leftrightarrow q(o))$ holds if and only if $C^2 = \varnothing$, $C^3 = \varnothing$, $WC^1 \neq \varnothing$, $PC^1 \neq \varnothing$ and $\lambda_{minPos} \leq \lambda(p(o) \Leftrightarrow q(o)) \leq \lambda_{maxPos}$.*
2. *The epistemic satisfaction relation $CS(t)|=_G Bel(p(o) \Leftrightarrow q(o))$ holds if and only if $C^2 = \varnothing$, $C^3 = \varnothing$, $WC^1 \neq \varnothing$, $PC^1 \neq \varnothing$ and $\lambda_{minBel} \leq \lambda(p(o) \Leftrightarrow q(o)) \leq \lambda_{maxBel}$.*
3. *The epistemic satisfaction relations $CS(t)|=_G Know(p(o) \Leftrightarrow q(o))$ and $CS(t)|=_G p(o) \Leftrightarrow q(o)$ hold if and only if $C^2 = \varnothing$, $C^3 = \varnothing$, $WC^1 \neq \varnothing$, $PC^1 = \varnothing$.*

The rationale for the first point of this definition are given as follows: At first, the definition represents the requirement stating that to construct the meaning of equivalence the cognitive agent needs to have at least one relevant and positive perception stored in its empirical database. If such a perception did not exist the agent would not have any empirical material by which the meaning of the equivalence $p(o) \Leftrightarrow q(o)$ could be induced (shaped, generated). At second, it represents the requirement that the cognitive agent needs to have no relevant negative perceptions stored in the empirical database because it would make the equivalence grounding impossible. At third, it represents the requirement that at the time point of equivalence validation and grounding that an uncovered set of relevant perceptions representing any empirical material stored at the unconscious layer needs to exist. The existence of such a set of unprocessed perceptions constitutes the integral dimension of cognitive state referred by the formula $Pos(p(o) \Leftrightarrow q(o))$. The meaning of possibility is partly built of the presence of empirical material unknown at the conscious layer. At fourth, it represents the requirement that the strength of uncovered material needs to be high enough to make the use of possibility operator well based. If this strength were to low, the use of possibility operator were not acceptable.

Similar requirements can be formulated for the case of belief, provided that the acceptable level of relative value is changed. The case of knowledge operator is similarly clarified.

5 Final Remarks

An original approach to modelling the correspondence between internal cognitive state and external linguistic representation has been proposed for the case of modal implications and modal equivalences. This approach has been strongly based on basic assumptions accepted in cognitive linguistics in which two level organization of mind is required.

The so called epistemic satisfaction relations have been given that describe the correspondence between internal state and external representations in technical terms and make the rational conceptualization of artificial language behaviour possible.

The phenomenon of language grounding, language extraction and linguistic representation discovery from artificial cognitive mind are treated as strictly related aspects of the same cognitive process. The proposed model defines a certain class of technical implementations preserving the rationality and intentionality of artificial agents.

The proposed application of the given definitions for grounding results from the fact that they describe in technical terms processes for discovering appropriate and acceptable linguistic representations.

References

1. Denett, D.C.: Kinds of minds. Toward an understanding of consciousness., Orion Publishing Group Ltd. (1996)
2. Harnad S.: The symbol grounding problem, Physica D, 42, (1990) 335-346
3. Johnson-Laird, P.N.: Mental models. Towards a cognitive science of language, inference, and consciousness. Cambridge University Press, Cambridge (1983)
4. Katarzyniak, R., Nguyen, T.N.: Reconciling inconsistent profiles of agent's knowledge states in distributed multiagent systems using consensus methods., Systems Science, Vol. 26 No.4 (2000) 93-119
5. Katarzyniak, R., Owczarek, R.: Applying a cognitive ranking of stored perceptions to approximate procedure for uttering a class of belief formulas. In: Proc. 6th Hellenic European Conference on Computer Mathematics and its Applications'2003, Athens, Greece, Vol.1 (2003) 184-189.
6. Katarzyniak, R., Pieczynska-Kuchtiak, A.: A consensus based algorithm for grounding belief formulas in internally stored perceptions. International Journal of Neural Network World, Vol. 5. (2002) 461-472
7. Katarzyniak, R., Pieczyńska-Kuchtiak, A.: Grounding and extracting modal responses in cognitive agents: AND query and states of incomplete knowledge, International Journal of Applied Mathematics and Computer Science, Vol. 14 No. 2 (2004) 249-263
8. Katarzyniak, R., Pieczyńska-Kuchtiak, A.: Grounding languages in cognitive agents and robots", Systems Science, vol. 29, no 2 (2004)
9. Katarzyniak, R.: A model for extracting a class of knowledge operators, logic disjunctions and logic alternatives from BDI agents. Intelligent Information Systems X'2001, Advances in Soft Computing, Physica-Verlag, (2001) 257-269

10. Katarzyniak, R.: Grounding atom formulas and simple modalities in communicative agents. In: Proc. 21st IASTED Int. Conf. Applied Informatics (2003) 388-392
11. Lakoff, G. , Johnson, M.: Philosophy in the Flesh: the embodied mind and its challenge to Western thought, New York, Basic Books (1999)
12. Newell, A.: The unified theories of cognition., Harv. Univ. Press, Cambridge (1990)
13. Paivio, A.: Mental representations: A dual coding approach., Oxford University Press, New York (1986)
14. Vogt, P.: The Physical Grounding Symbol Problem, *Cognitive Systems Research*, Vol. 3, (2002) 429-457
15. Vogt, P.: Anchoring of semiotic symbols, Robotics and autonomous systems, Vol. 43, (2003) 109-120

Author Index